T0122372

Springer Optimization and Its Applications

VOLUME 127

Aims and Scope
Optimization has been expanding in all directions at an astonishing rate during the last few decades. New algorithmic and theoretical techniques have been developed, the diffusion into other disciplines has proceeded at a rapid pace, and our knowledge of all aspects of the field has grown even more profound. At the same time, one of the most striking trends in optimization is the constantly increasing emphasis on the interdisciplinary nature of the field. Optimization has been a basic tool in all areas of applied mathematics, engineering, medicine, economics and other sciences.

The series *Springer Optimization and Its Applications* publishes undergraduate and graduate textbooks, monographs and state-of-the-art expository works that focus on algorithms for solving optimization problems and also study applications involving such problems. Some of the topics covered include nonlinear optimization (convex and nonconvex), network flow problems, stochastic optimization, optimal control, discrete optimization, multi-objective programming, description of software packages, approximation techniques and heuristic approaches.

More information about this series at http://www.springer.com/series/7393

Nikolaos Ploskas • Nikolaos Samaras

Linear Programming Using MATLAB®

Springer

Nikolaos Ploskas
University of Macedonia
Department of Applied Informatics
Thessaloniki, Greece

Nikolaos Samaras
University of Macedonia
Department of Applied Informatics
Thessaloniki, Greece

ISSN 1931-6828   ISSN 1931-6836   (electronic)
Springer Optimization and Its Applications
ISBN 978-3-319-88131-7   ISBN 978-3-319-65919-0   (eBook)
DOI 10.1007/978-3-319-65919-0

Mathematics Subject Classification: 90C05, 90C06, 90C49, 90C51, 49M15, 49M29, 65K05

Printed on acid-free paper

This Springer imprint is published by Springer Nature
The registered company is Springer International Publishing AG
The registered company address is: Gewerbestrasse 11, 6330 Cham, Switzerland

To my family
– Nikolaos Ploskas

To my parents, to my wife
  Lydia and to my son Stathis
– Nikolaos Samaras

Preface

Linear Programming (LP) is a significant area in the field of operations research. The simplex algorithm is one of the top ten algorithms with the greatest influence in the twentieth century and the most widely used method for solving linear programming problems (LPs). Since the introduction of the simplex algorithm in 1947, LP has been widely used in many practical problems. However, the size of practical LPs grew up significantly. Consequently, the simplex algorithm began to encounter computational issues in the solution of large LPs. A variety of methods have been proposed to strengthen the computational performance of simplex algorithm. Furthermore, new algorithms have been proposed to solve LPs, like the dual simplex algorithm, interior point methods, and exterior point simplex algorithms.

The main feature of this book is the presentation of a variety of LP algorithms and methods and especially the revised simplex method and its components. The computational performance of simplex algorithm on practical problems is usually far better than the theoretical worst case. This book includes the thorough theoretical and computational presentation of four LP algorithms:

- the revised primal simplex algorithm,
- the revised dual simplex algorithm,
- the exterior point simplex algorithm, and
- Mehrotra's interior point method.

Furthermore, this book presents:

- 11 presolve techniques,
- 11 scaling techniques,
- 6 pivoting rules, and
- 4 basis inverse and update methods.

The novelty of this book is that the presentation of each LP algorithm or method is focused on three aspects:

- Initially, the theoretical background is presented for each algorithm or method including its mathematical formulation.
- Secondly, a thorough numerical example is presented for each algorithm or method.
- Finally, a MATLAB code is given to fully cover the presentation of each algorithm or method. The MATLAB implementations that are presented in this book are sophisticated and allow the solution of large-scale benchmark LPs.

This book is addressed to students, scientists, and mathematical programmers. Students will learn various aspects about LP algorithms and especially the revised simplex algorithm through illustrative examples, while they can solve the examples using the MATLAB codes given in this book. This book covers thoroughly a course on Linear Programming whether MATLAB is used or not. Scientists and mathematical programmers will have a book in their library that presents many different components of simplex-type methods, like presolve techniques, scaling techniques, pivoting rules, basis update methods, and sensitivity analysis. Moreover, the presentation of each component or algorithm is followed by a computational study on benchmark problems to present the computational behavior of the presented methods or algorithms.

Even though there is an abundance of books on LP, none of them comprehensively contains both the theoretical and practical backgrounds of each method. Our dual goal is to fill this gap and provide a book to students and researchers that can be used to understand many widely used LP methods and algorithms and find codes to implement their own algorithms. All codes presented in this book are sophisticated implementations of LP algorithms and methods, aiming toward a balance between speed and ease of use. Experienced programmers can further optimize the presented implementations to achieve better execution times. Moreover, we preferred not to present the simplex algorithm in tableau format, because it is not efficient in practice.

Our intention is to present many different LP methods and algorithms and especially present different computational techniques for the revised simplex algorithm. On the other hand, we decided not to include some other significant techniques, like LU basis update techniques and crash procedures for finding an initial basis. However, the inclusion of a part in this book presenting more advanced techniques is to be reconsidered in future editions.

We have selected to use MATLAB in order to implement the codes presented in this book for several reasons. First of all, MATLAB is a matrix language intended primarily for numerical computing. MATLAB is especially designed for matrix computations like solving systems of linear equations or factoring matrices. Secondly, MATLAB gives us the ability to create concise codes. Hence, readers will focus on the implementation of the different steps of the algorithms and not on how linear algebra operations (e.g., matrix operations, decompositions, etc.) are implemented. Finally, MATLAB provides sophisticated LP algorithms, and users can use them to solve their LPs. Readers that are not familiar with programming can also use MATLAB's graphical user interface for solving LPs.

We are well aware that errors, ambiguous explanations and misprints are still part of this book. Please, let us know about any errors you found. We will be thankful to receive your email at ploskasn@gmail.com and/or samaras@uom.gr.

We are thankful to MathWorks for providing us an academic license for MATLAB through their MathWorks Book Program. We also thank Charalampos Triantafyllidis, Themistoklis Glavelis, and the students of the Department of Applied Informatics, University of Macedonia, who have given us feedback on the manuscript. Also, we are thankful to many colleagues who have been part of our exciting involvement with the scientific field of Operations Research. This book is devoted to the loving memory of Professor Konstantinos Paparrizos. We feel truly blessed to have received his mentorship and friendship. Finally, we thank our families for their love and support over many years.

Thessaloniki, Greece   Nikolaos Ploskas
Thessaloniki, Greece   Nikolaos Samaras
November 2017

Contents

The original version of this book was revised. An erratum to this book can be found at
https://doi.org/10.1007/978-3-319-65919-0_13

Chapter 1
Introduction

Abstract The Linear Programming (LP) problem is perhaps the most important and well-studied optimization problem. Numerous real world problems can be formulated as Linear Programming problems (LPs). LP algorithms have been used in many fields ranging from airline scheduling to logistics, transportation, decision making, and data mining. This chapter introduces some key features of LP and presents a brief history of LP algorithms. Finally, the reasons of the novelty of this book and its organization are also presented.

1.1 Chapter Objectives

- Justify the significance of linear programming algorithms.
- Provide the mathematical formulation of the linear programming problem.
- Introduce different linear programming algorithms.
- Present the reasons of the novelty of this book.
- Present the organization of this book.

1.2 Linear Programming

The Linear Programming (LP) problem is perhaps the most important and well-studied optimization problem. Numerous real world problems can be formulated as Linear Programming problems (LPs). LP is the process of minimizing or maximizing a linear objective function subject to a number of linear equality and/or inequality constraints.

Here is a simple example. *A store produces and sells pants and jackets. The price of a pair of pants is at \$60, while the price of a jacket is at \$50. Every pair of pants needs $1m^2$ of cotton and $4m^2$ of polyester, while every jacket needs $2m^2$ of cotton and $2m^2$ of polyester. The store can buy from its supplier $1,000m^2$ of cotton and*

The original version of this chapter was revised. An erratum to this chapter can be found at https://doi.org/10.1007/978-3-319-65919-0_13

© Springer International Publishing AG 2017   1
N. Ploskas, N. Samaras, *Linear Programming Using MATLAB®*,
Springer Optimization and Its Applications 127, DOI 10.1007/978-3-319-65919-0_1

$1,600m^2$ *of polyester every month. What is the number of pants and jackets that the* *store should produce and sell each month in order to maximize its revenue?*

In this problem, there are two unknown variables, the number of pants and the number of jackets that the store should produce and sell each month. These two variables are called decision variables. Let x be the number of pants and y the number of jackets. The main goal of the store is to maximize its revenue. So, the objective function is: max $60x + 50y$.

The store would produce and sell infinite amount of pants and jackets if it would be possible. However, there are some technological constraints in the number of pants and jackets that it can produce. First of all, it has only $1,000m^2$ of cotton and $1,600m^2$ of polyester available every month and every pair of pants needs $1m^2$ of cotton and $4m^2$ of polyester, while every jacket needs $2m^2$ of cotton and $2m^2$ of polyester. Hence, we can derive two technological constraints, one for the available cotton and one for the available polyester:

$$x + 2y \leq 1,000 \text{ (the technological constraint for the available cotton)}$$

$$4x + 2y \leq 1,600 \text{ (the technological constraint for the available polyester)}$$

The aforementioned technological constraints are the explicit constraints. The explicit constraints are those that are explicitly given in the problem statement. This problem has other constraints called implicit constraints. These are constraints that are not explicitly given in the problem statement but are present nonetheless. These constraints are typically associated with natural restrictions on the decision variables. In this problem, it is clear that one cannot have negative values for the amount of pants and jackets that are produced. That is, both x and y must be nonnegative constraints and integer.

The entire model can be formally stated as:

$$\begin{aligned} \max z = {}& 60x + 50y \\ \text{s.t.} \quad & x + 2y \leq 1,000 \\ & 4x + 2y \leq 1,600 \\ & x \geq 0, y \geq 0, \{x, y\} \in \mathbb{Z} \end{aligned}$$

Both the objective function and the constraints are linear. Hence, this is a linear programming problem. Note that x and y are integer variables. This book is focused on algorithms that solve continuous LPs, i.e., x and y will be treated as continuous variables.

This model is particularly easy to solve. Chapter 2 describes how to solve LPs graphically. Without giving many details on how to solve an LP problem graphically (for a detailed analysis see Chapter 2), let's graph the constraints, find the half-planes that represent them, and identify the feasible region (Figure 1.1).

We have identified the feasible region (shaded region in Figure 1.1) which satisfies all the constraints simultaneously. It is easy to see in general that the objective function, being linear, always takes on its maximum (or minimum) value at a corner point of the constraint set, provided the constrained set is bounded. Hence,

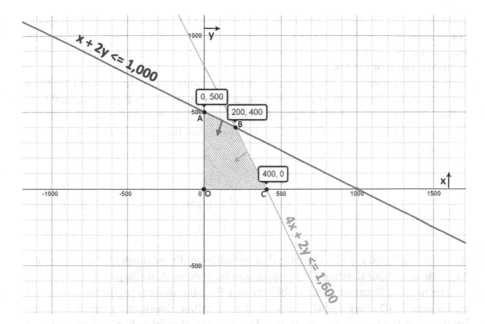

Fig. 1.1 Feasible region of the LP problem

we can identify the maximum value if we evaluate the objective function at each corner point (points $O, A, B,$ and C) and find the maximum value. The optimal point is $B(200, 400)$. That means that the store should produce and sell 200 pairs of pants and 400 jackets. Its revenue will be $60 \times 200 + 50 \times 400 = \$32,000$.

It is relatively easy to solve LPs with two or even three variables following this procedure, but we cannot use the graphical procedure to solve LPs with more than three variables. Chapters 8, 9, 10, and 11 present algorithms to solve large LPs with thousands of variables and constraints.

Considering now a general form to model LPs, an LP problem can be formulated in its standard form as shown below:

$$\min \quad c^T x$$
$$\text{s.t.} \quad Ax = b \qquad\qquad (LP.1)$$
$$x \geq 0$$

where $A \in \mathbb{R}^{m \times n}, (c, x) \in \mathbb{R}^n, b \in \mathbb{R}^m,$ and T denotes transposition. We assume that A has full rank, $rank(A) = m, m < n$. Consequently, the linear system $Ax = b$ is consistent.

The word *min* shows that the problem is to minimize the objective function and *s.t.* is the shorthand for *subject to* that is followed by the constraints. x is the vector of the decision variables (size $n \times 1$). These variables are the unknown variables of the problem. When solving an LP problem, we want to find the values of the decision variables while minimizing the value of the objective function and satisfying all the constraints. A is the matrix of coefficients of the constraints (size $m \times n$), c is the vector of coefficients of the objective function (size $n \times 1$), and b is the vector of the right-hand side of the constraints (size $m \times 1$).

It turns out that LPs come in pairs. We can automatically derive two LPs from the same data. The relation of these problems is close. The LP problem (LP.1) is called the primal problem, while the LP problem (DP.1) is called the dual problem. If the primal LP problem refers to the employer, then the dual LP problem refers to the employees.

$$\max \qquad b^T w$$
$$\text{s.t.} \qquad A^T w + s = c \qquad\qquad (DP.1)$$
$$s \geq 0$$

where $w \in \mathbb{R}^m$ and $s \in \mathbb{R}^n$. w are the dual variables and s are the dual slack variables. More details about the dual LP problem will be presented in Chapter 2.

The most well-known method for solving LPs is the simplex algorithm developed by George B. Dantzig [3, 4]. It is one of the top ten algorithms with the greatest influence in the twentieth century [5]. The simplex algorithm begins with a primal feasible basis and uses pricing operations until an optimum solution is computed. It searches for an optimal solution by moving from one adjacent feasible solution to another, along the edges of the feasible region. It also guarantees monotonicity of the objective value. It has been proved that the expected number of iterations in the solution of an LP problem is polynomial [2]. Moreover, the worst case complexity has exponential behavior [15].

Since Dantzig's initial contribution, researchers have made many efforts in order to enhance the performance of the simplex algorithm. At the same time when Dantzig presented the simplex algorithm, Lemke [16] proposed the dual simplex algorithm and other researchers proposed the interior point algorithms that traverse across the interior of the feasible region [8, 12, 25]. The dual simplex algorithm was not considered as an alternative to the primal simplex algorithm for nearly 40 years. In the early 90s, the contributions of Forrest and Goldfarb [6] and Fourer [7] motivated many researchers and optimization software developers to develop new efficient implementations of the dual simplex algorithm. On the other hand, the interior point methods did not compete with the simplex algorithm in practice due to the expensive execution time per iteration and the possibility of numerical instability. The first polynomial algorithm for LP was proposed by Khachiyan [14]. The development of the ellipsoid method had a great impact on the theory of LP, but the proposed algorithm did not achieve to compete with the simplex algorithm in practice due to the expensive execution time per iteration. The first interior point method that outperformed simplex algorithm was proposed by Karmarkar [13]. Since Karmarkar's algorithm, many improvements have been made both in theory and in practice of interior point methods. Since then, many interior point methods have been proposed [10, 11, 17, 18]. Computational results showed that interior point methods outperform the simplex algorithm for large-scale LPs [1]. Nowadays, large-scale LPs can be solved either by a primal simplex algorithm or a dual simplex algorithm (or a combination of a primal and a dual algorithm) or an interior point algorithm. Interior point methods and simplex algorithm continue to be valuable in practice.

However, the implementation of pivoting algorithms that will be more efficient than simplex algorithm is still an active field of research. An approach to enhance the performance of simplex algorithm is to move in the exterior of the feasible solution and construct basic infeasible solutions instead of constructing feasible solutions like simplex algorithm does. Such an algorithm is called exterior point simplex algorithm and was proposed by Paparrizos initially for the assignment problem [19] and then for the solution of LPs [20, 21]. A more efficient approach is the Primal-Dual Exterior Point Simplex Algorithm (PDEPSA) proposed by Samaras [23] and Paparrizos et al. [22]. PDEPSA [9, 24] can deal with the problems of stalling and cycling more effectively and as a result improves the performance of the primal-dual exterior point algorithms. The advantage of PDEPSA stems from the fact that it uses an interior point in order to compute the leaving variable compared to the primal exterior point algorithm that uses a boundary point.

1.3 Why Another Book on Linear Programming? Why Linear Programming Using MATLAB?

LP is a significant area in the field of operations research. LP is used extensively in applications ranging from airline scheduling to logistics, transportation, decision making, and data mining. The simplex algorithm is one of the top ten algorithms with the greatest influence in the twentieth century [5] and the most widely used method for solving LPs. Nearly all Fortune 500 companies use the simplex algorithm to optimize several tasks. The simplex algorithm is the workhorse of algorithms for integer, stochastic, and nonlinear optimization. Hence, even small improvements in LP algorithms could result in considerable practical impact.

MATLAB is a high-level language for technical computing. It is widely used as a rapid prototyping tool in many scientific areas. Moreover, MATLAB includes sophisticated LP solvers in its Optimization Toolbox. The main feature of this book is the presentation of a variety of LP algorithms and methods and especially the revised simplex algorithm and its components. This book includes the thorough theoretical and computational presentation of four LP algorithmic families:

- the revised primal simplex algorithm,
- the revised dual simplex algorithm,
- the exterior point simplex algorithm, and
- Mehrotra's interior point method.

Furthermore, this book presents:

- 11 presolve techniques,
- 11 scaling techniques,

- 6 pivoting rules, and
- 4 basis inverse and update methods.

The novelty of this book is that the presentation of each LP algorithm or method is focused on three aspects:

- Initially, the theoretical background is presented for each algorithm or method including their mathematical formulation.
- Secondly, a thorough numerical example is presented for each algorithm or method.
- Finally, a MATLAB code is given to fully cover the presentation of each algorithm or method. All the codes were written in MATLAB, a widely used programming language, whose code is easy to understand. The MATLAB implementations that are presented in this book are sophisticated and allow the solution of large-scale benchmark LPs.

This book is addressed to students, scientists, and mathematical programmers. Students will learn different aspects about LP algorithms and especially the revised simplex algorithm through illustrative examples, while they will be able to solve them using the MATLAB codes given in this book. This book covers thoroughly a course on Linear Programming whether MATLAB is used or not. Scientists and mathematical programmers will have a book in their library that presents many different components of simplex-type methods, like presolve techniques, scaling techniques, pivoting rules, basis update methods, and sensitivity analysis. Moreover, the presentation of each component or algorithm is followed by a computational study to present the computational behavior of the presented methods or algorithms.

Even though there is an abundance of books on LP, none of them comprehensively contains both the theoretical and practical backgrounds of each method. Our dual goal is to fill this gap and provide a book to students and researchers that can be used to understand many widely used LP methods and algorithms and find codes to implement their own algorithms. All codes presented in this book are sophisticated implementations of LP algorithms and methods, aiming toward a balance between speed and ease of use. Experienced programmers can further optimize the presented implementations to achieve better execution times.

Our intention is to present many different LP methods and algorithms and especially present different techniques for the revised simplex algorithm. On the other hand, we decided not to include some other significant techniques, like LU basis update techniques and crash procedures for finding an initial basis. However, the inclusion of a part in this book presenting more advanced techniques in future editions is to be reconsidered.

We have selected to use MATLAB in order to implement the codes presented in this book for several reasons. First of all, MATLAB is a matrix language intended primarily for numerical computing. MATLAB is especially designed for matrix computations like solving systems of linear equations or factoring matrices. Secondly, MATLAB gives us the ability to create concise codes. Hence, readers will focus on the implementation of the different steps of the algorithms and not on how linear algebra operations (e.g., matrix operations, decompositions, etc.) are implemented. Finally, MATLAB provides sophisticated LP algorithms and users can use them to solve their LPs. Readers who are not familiar with programming can also use MATLAB's graphical user interface for solving LPs.

1.4 Our Approach: Organization of the Book

The scope of this book is threefold. Initially, the reader will learn the theoretical background of the most well-known LP methods and algorithms. Secondly, each method or algorithm is presented through a thorough illustrative example in order for the reader to understand better the presented method or algorithm. Finally, sophisticated LP implementations in MATLAB are presented that can help the reader understand the computational aspects of these algorithms and implement his/her own algorithms.

The main emphasis of this book is given on four aspects:

- The presentation (theoretical background, numerical example, and implementation in MATLAB) of all components of the revised primal simplex algorithm:

  1. presolve techniques,
  2. scaling techniques,
  3. pivoting rules, and
  4. basis inverse and update methods.

- The presentation (theoretical background, numerical example, and implementation in MATLAB) of the revised dual simplex algorithm.
- The presentation (theoretical background, numerical example, and implementation in MATLAB) of the exterior point simplex algorithm.
- The presentation (theoretical background, numerical example, and implementation in MATLAB) of Mehrotra's interior point method.

The main target group of this book is:

- Undergraduate and postgraduate students who take a course on Linear Programming. This book covers thoroughly a course on Linear Programming whether MATLAB is used or not.
- Scientists and mathematical programmers who want to learn about LP algorithms and especially the revised simplex algorithm. Computational studies on benchmark LPs are included in each section for the computational comparison of the presented methods and algorithms.

The readers of this book have basic knowledge of linear algebra and calculus. This book can be used by readers who want to learn about LP methods and algorithms whether they want to use MATLAB or not. The interested readers who want to utilize the MATLAB implementations of the presented methods and algorithms need to have some experience on MATLAB coding.

The rest of this book is organized as follows. Chapter 2 presents the theoretical background of LP. More specifically, the different formulations of the LP problem are presented. Moreover, detailed steps on how to formulate an LP problem are given. In addition, the geometry of the feasible region and the duality principle are also covered. Finally, a brief description of the LP algorithms that will be used in this book is also given.

Chapter 3 presents the Mathematical Programming System (MPS) format, a widely accepted standard for defining LPs. Two codes in MATLAB that can be used to convert an MPS file to MATLAB's matrix format (MAT) and vice versa are also presented. Furthermore, codes that can be used to create randomly generated sparse or dense LPs are also given. Finally, the most well-known public benchmark libraries for LPs are also presented.

Chapter 4 presents eleven presolve methods used prior to the execution of an LP algorithm: (i) eliminate zero rows, (ii) eliminate zero columns, (iii) eliminate singleton equality constraints, (iv) eliminate $k$ton equality constraints, (v) eliminate singleton inequality constraints, (vi) eliminate dual singleton inequality constraints, (vii) eliminate implied free singleton columns, (viii) eliminate redundant columns, (ix) eliminate implied bounds on rows, (x) eliminate redundant rows, and (xi) make coefficient matrix structurally full rank. Each method is presented with: (i) its mathematical formulation, (ii) a thorough illustrative numerical example, and (iii) its implementation in MATLAB. Finally, a computational study is performed. The aim of the computational study is twofold: (i) compare the execution time and the reduction to the problem size of the aforementioned presolve methods, and (ii) investigate the impact of preprocessing prior to the application of LP algorithms. The execution time and the number of iterations with and without preprocessing are presented.

Chapter 5 presents eleven scaling techniques used prior to the execution of an LP algorithm: (i) arithmetic mean, (ii) de Buchet for the case $p = 1$, (iii) de Buchet for the case $p = 2$, (iv) de Buchet for the case $p = \inf$, (v) entropy, (vi) equilibration, (vii) geometric mean, (viii) IBM MPSX, (ix) L_p-norm for the case $p = 1$, (x) L_p-norm for the case $p = 2$, and (xi) L_p-norm for the case $p = \infty$. Each technique is presented with: (i) its mathematical formulation, (ii) a thorough illustrative numerical example, and (iii) its implementation in MATLAB. Finally, a computational study is performed. The aim of the computational study is twofold: (i) compare the execution time of the aforementioned scaling techniques, and (ii) investigate the impact of scaling prior to the application of LP algorithms. The execution time and the number of iterations with and without scaling are presented.

Chapter 6 presents six pivoting rules used in each iteration of the simplex algorithm to determine the entering variable: (i) Bland's rule, (ii) Dantzig's rule, (iii) Greatest Increment Method, (iv) Least Recently Considered Method, (v) Partial Pricing rule, and (vi) Steepest Edge rule. Each technique is presented with: (i) its mathematical formulation, (ii) a thorough illustrative numerical example, and (iii) its implementation in MATLAB. Finally, a computational study is performed. The aim of the computational study is twofold: (i) compare the execution time of the aforementioned pivoting rules, and (ii) highlight the impact of the choice of the pivoting rule on the number of iterations and the execution time of the simplex algorithm.

Chapter 7 presents two basis inverse and two basis update methods used in simplex algorithm: (i) Gauss-Jordan elimination basis inverse method, (ii) LU Decomposition basis inverse method, (iii) Modification of the Product Form of the Inverse basis update method, and (iv) Product Form of the Inverse basis update method. Each technique is presented with: (i) its mathematical formulation, (ii) a thorough illustrative numerical example, and (iii) its implementation in MATLAB. Finally, a computational study is performed. The aim of the computational study is to compare the execution time of the aforementioned basis inverse and update methods and highlight the significance of the choice of the basis update method on simplex-type algorithms and the reduction that it can offer to the solution time.

Chapter 8 presents the revised primal simplex algorithm. Numerical examples are presented in order for the reader to understand better the algorithm. Furthermore, an implementation of the algorithm in MATLAB is given along with a computational study over benchmark LPs and randomly generated sparse LPs. The implementation is modular allowing the user to select which scaling technique, pivoting rule, and basis update method wants to use in order to solve LPs. The aim of the computational study is to compare the efficiency of the proposed implementation with MATLAB's simplex algorithm.

Chapter 9 presents the revised dual simplex algorithm. Numerical examples are presented in order for the reader to understand better the algorithm. Furthermore, an implementation of the algorithm in MATLAB is given along with a computational study over benchmark LPs and randomly generated sparse LPs. The implementation is modular allowing the user to select which scaling technique and basis update method wants to use in order to solve LPs. The aim of the computational study is to compare the efficiency of the proposed implementation with the revised primal simplex algorithm presented in Chapter 8.

Chapter 10 presents the exterior point simplex algorithm. Numerical examples are presented in order for the reader to understand better the algorithm. Furthermore, an implementation of the algorithm in MATLAB is given along with a computational study over benchmark LPs and randomly generated sparse LPs. The implementation is modular allowing the user to select which scaling technique and basis update method wants to use in order to solve LPs. The aim of the computational study is to compare the efficiency of the proposed implementation with the revised primal simplex algorithm presented in Chapter 8.

Chapter 11 provides a background on interior point methods. We also present Mehrotra's interior point method in detail. Numerical examples are presented in order for the reader to understand better the algorithm. Furthermore, an implementation of the algorithm in MATLAB is given along with a computational study over benchmark LPs and randomly generated sparse LPs. The aim of the computational study is to compare the efficiency of the proposed implementation with MATLAB's interior point method.

Chapter 12 presents topics about sensitivity analysis, i.e., how to deal with changes on an optimal LP problem that we have already solved with simplex algorithm. Sensitivity analysis is very useful in two situations: (i) when we wish to know how the solution will be affected if we perform a small change in the coefficients of an LP problem, and (ii) when we have already solved an LP problem and we want also to solve a second LP problem in which the data is only slightly different. Rather than restarting the simplex method from scratch for the modified LP problem, we want to solve the modified LP problem starting with the optimal basis of the original LP problem and perform only a few iterations to solve the modified LP problem (if necessary). We examine how the solution of an LP problem is affected when changes are made to the input data of the LP problem. More specifically, we examine changes in: (i) the cost vector, (ii) the right-hand side vector, and (iii) the coefficient of the constraints.

Appendix A presents the available functions of MATLAB's Optimization toolbox for solving LPs. Moreover, the Optimization App, a graphical user interface for solving optimization problems, is presented. In addition, LPs are solved using MATLAB in various ways. A computational study is performed. The aim of the computational study is to compare the efficiency of MATLAB's Optimization Toolbox solvers.

Appendix B presents two of the most powerful and state-of-the-art LP solvers, CLP and CPLEX. Moreover, we present codes to access these solvers from MATLAB. A computational study is performed. The aim of the computational study is to compare the efficiency of CLP and CPLEX.

1.5   Chapter Review

LP is a significant area in the field of operations research. LP is used extensively in many different applications and the simplex algorithm is one of the most widely used algorithms. In this chapter, we introduced some key features of LP and presented a brief history of LP algorithms. The three different families of algorithms that are used to solve an LP problem are: (i) simplex algorithm, (ii) exterior point simplex algorithm, and (iii) interior point methods. Finally, the reasons of the novelty of this book and its organization were also presented.

References

1. Bixby, E. R. (1992). Implementing the simplex method: The initial basis. *ORSA Journal on Computing, 4*, 267–284.
2. Borgwardt, H. K. (1982). The average number of pivot steps required by the simplex method is polynomial. *Zeitschrift fur Operational Research, 26*(1), 157–177.
3. Dantzig, G. B. (1949). Programming in linear structure. *Econometrica, 17*, 73–74.

4. Dantzig, G. B. (1963). *Linear programming and extensions*. Princeton, NJ: Princeton University Press.
5. Dongarra, J., & Sullivan, F. (2000). Guest editors' introduction: The top 10 algorithms. *Computing in Science & Engineering, 22–23.*
6. Forrest, J. J., & Goldfarb, D. (1992). Steepest-edge simplex algorithms for linear programming. *Mathematical Programming, 57*(1–3), 341–374.
7. Fourer, R. (1994). *Notes on the dual simplex method*. Draft report.
8. Frisch, K. F. (1955). *The logarithmic potential method of convex programming*. Technical report, University Institute of Economics, Oslo, Norway.
9. Glavelis, T., & Samaras, N. (2013). An experimental investigation of a primal–dual exterior point simplex algorithm. *Optimization, 62*(8), 1143–1152.
10. Gondzio, J. (1992). Splitting dense columns of constraint matrix in interior point methods for large-scale linear programming. *Optimization, 24*, 285–297.
11. Gondzio, J. (1996). Multiple centrality corrections in a primal-dual method for linear programming. *Computational Optimization and Applications, 6*, 137–156.
12. Hoffman, A. J., Mannos, M., Sokolowsky, D., & Wiegman, N. (1953). Computational experience in solving linear programs. *Journal of the Society for Industrial and Applied Mathematics, 1*, 17-33.
13. Karmarkar, N. K. (1984). A new polynomial-time algorithm for linear programming. *Combinatorica, 4*, 373–395.
14. Khachiyan, L. G. (1979). A polynomial algorithm in linear programming. *Soviet Mathematics Doklady, 20*, 191–194.
15. Klee, V., & Minty, G. J. (1972). How good is the simplex algorithm. In O. Shisha (Ed.), *Inequalities - III.* New York and London: Academic Press Inc.
16. Lemke, C. E. (1954). The dual method of solving the linear programming problem. *Naval Research Logistics Quarterly, 1*(1), 36–47.
17. Lustig, I. J., Marsten, R. E., & Shanno, D. F. (1992). On implementing Mehrotra's predictor corrector interior point method for linear programming. *SIAM Journal on Optimization, 2*, 435–449.
18. Mehrotra, S. (1992). On the implementation of a primal-dual interior point method. *SIAM Journal on Optimization, 2*, 575–601.
19. Paparrizos, K. (1991). An infeasible exterior point simplex algorithm for assignment problems. *Mathematical Programming, 51*(1–3), 45–54.
20. Paparrizos, K. (1993). An exterior point simplex algorithm for (general) linear programming problems. *Annals of Operations Research, 47*, 497–508.
21. Paparrizos, K., Samaras, N., & Stephanides, G. (2003). An efficient simplex type algorithm for sparse and dense linear programs. *European Journal of Operational Research, 148*(2), 323–334.
22. Paparrizos, K., Samaras, N., & Stephanides, G. (2003). A new efficient primal dual simplex algorithm. *Computers & Operations Research, 30*(9), 1383–1399.
23. Samaras, N. (2001). *Computational improvements and efficient implementation of two path pivoting algorithms*. Ph.D. dissertation, Department of Applied Informatics, University of Macedonia (in Greek).
24. Samaras, N., Sifelaras, A., & Triantafyllidis, C. (2009). A primal-dual exterior point algorithm for linear programming problems. *Yugoslav Journal of Operations Research, 19*(1), 123–132.
25. Von Neumann, J. (1947). *On a maximization problem*. Technical report, Institute for Advanced Study, Princeton, NJ, USA.

Chapter 2
Linear Programming Algorithms

Abstract LPs can be formulated in various forms. An LP problem consists of the objective function, the constraints, and the decision variables. This chapter presents the theoretical background of LP. More specifically, the different formulations of the LP problem are presented. Detailed steps on how to formulate an LP problem are given. In addition, the geometry of the feasible region and the duality principle are also covered. Finally, a brief description of LP algorithms that will be used in this book is also presented.

2.1 Chapter Objectives

- Present the different formulations of linear programming problems.
- Formulate linear programming problems.
- Solve linear programming problems graphically.
- Introduce the duality principle.
- Present the linear programming algorithms that will be used in this book.

2.2 Introduction

The LP problem is an optimization problem with a linear objective function and constraints. We can write the LP problem in its general form as shown in Equation (2.1):

$$
\begin{aligned}
\min z = \ & c_1 x_1 \ + \ c_2 x_2 \ + \cdots + \ c_n x_n \\
\text{s.t.} \quad & A_{11} x_1 + A_{12} x_2 + \cdots + A_{1n} x_n \oplus b_1 \\
& A_{21} x_1 + A_{22} x_2 + \cdots + A_{2n} x_n \oplus b_2 \\
& \ \vdots \qquad \vdots \qquad \vdots \qquad \vdots \quad \cdots \quad \vdots \qquad \vdots \\
& A_{m1} x_1 + A_{m2} x_2 + \cdots + A_{mn} x_n \oplus b_m
\end{aligned}
\tag{2.1}
$$

The original version of this chapter was revised. An erratum to this chapter can be found at
https://doi.org/10.1007/978-3-319-65919-0_13

© Springer International Publishing AG 2017
N. Ploskas, N. Samaras, *Linear Programming Using MATLAB®*,
Springer Optimization and Its Applications 127, DOI 10.1007/978-3-319-65919-0_2

Coefficients A_{ij}, c_j, and b_i, where $i = 1, 2, \cdots, m$ and $j = 1, 2, \cdots, n$, are known real numbers and are the input data of the problem. Symbol $\oplus$ defines the type of the constraints and can take the form of:

- $=$: equality constraint
- $\geq$: greater than or equal to constraint
- $\leq$: less than or equal to constraint

The word *min* shows that the problem is to minimize the objective function and *s.t.* is the shorthand for *subject to* that is followed by the constraints. Variables $x_1, x_2, \cdots, x_n$ are called decision variables and they are the unknown variables of the problem. When solving an LP problem, we want to find the values of the decision variables while minimizing the value of the objective function subject to the constraints.

Writing Equation (2.1) using matrix notation, the following format is extracted:

$$
\begin{aligned}
\min z &= c^T x \\
\text{s.t. } &Ax \oplus b
\end{aligned}
\tag{2.2}
$$

where $c, x \in \mathbb{R}^n$, $b \in \mathbb{R}^m$, and $A \in \mathbb{R}^{m \times n}$. x is the vector of the decision variables (size $n \times 1$), A is the matrix of coefficients of the constraints (size $m \times n$), c is the vector of coefficients of the objective function (size $n \times 1$), b is the vector of the right-hand side of the constraints (size $m \times 1$), and $\oplus$ (or *Eqin*) is the vector of the type of the constraints (size $m \times 1$) where:

- 0: defines an equality constraint ($=$)
- -1: defines a less than or equal to constraint ($\leq$)
- 1: defines greater than or equal to constraint ($\geq$)

Most LPs are applied problems and the decision variables are usually non-negative values. Hence, nonnegativity or natural constraints must be added to Equation (2.2):

$$
\begin{aligned}
\min z &= c^T x \\
\text{s.t. } &Ax \oplus b \\
&x \geq 0
\end{aligned}
\tag{2.3}
$$

All other constraints are called technological constraints. Equations (2.2) and (2.3) are called general LPs.

Let's transform a general problem written in the format shown in Equation (2.1) in matrix notation. The LP problem that will be transformed is the following:

$$\begin{aligned}
\min z = \;& 3x_1 \;-\; 4x_2 \;+\; 5x_3 \;-\; 2x_4 \\
\text{s.t.} \quad & -2x_1 \;-\; x_2 \;-\; 4x_3 \;-\; 2x_4 \;\le\; 4 \\
& 5x_1 \;+\; 3x_2 \;+\; x_3 \;+\; 2x_4 \;\le\; 18 \\
& 5x_1 \;+\; 3x_2 \qquad\qquad +\; x_4 \;\ge\; -13 \\
& 4x_1 \;+\; 6x_2 \;+\; 2x_3 \;+\; 5x_4 \;\ge\; -10 \\
& x_j \ge 0, \quad (j = 1, 2, 3, 4)
\end{aligned}$$

In matrix notation, the above LP problem is written as follows:

$$A = \begin{bmatrix} -2 & -1 & -4 & -2 \\ 5 & 3 & 1 & 2 \\ 5 & 3 & 0 & 1 \\ 4 & 6 & 2 & 5 \end{bmatrix}, c = \begin{bmatrix} 3 \\ -4 \\ 5 \\ -2 \end{bmatrix}, b = \begin{bmatrix} 4 \\ 18 \\ -13 \\ -10 \end{bmatrix}, Eqin = \begin{bmatrix} -1 \\ -1 \\ 1 \\ 1 \end{bmatrix}$$

This chapter presents the theoretical background of LP. More specifically, the different formulations of the LP problem are presented. Moreover, detailed steps on how to formulate an LP problem are given. In addition, the geometry of the feasible region and the duality principle are also covered. Finally, a brief description of LP algorithms that will be used in this book is also presented.

The structure of this chapter is as follows. In Section 2.3, the different formulations of the LP problem are presented. Section 2.4 presents the steps to formulate an LP problem. Section 2.5 presents the geometry of the feasible region, while Section 2.6 presents the duality principle. Section 2.7 provides a brief description of the LP algorithms that will be used in this book. Finally, conclusions and further discussion are presented in Section 2.8.

2.3   Linear Programming Problem

The general LP problem (Equation (2.2)) can contain both equality (=) and inequality ($\le, \ge$) constraints, ranges in the constraints, and the following types of variable constraints (bounds):

- lower bound: $x_j \ge l_j$
- upper bound: $x_j \le u_j$
- lower and upper bound: $l_j \le x_j \le u_j$
- fixed value (i.e., the lower and upper bounds are the same): $x_j = k, k \in \mathbb{R}$
- free variable: $-\infty \le x_j \le \infty$
- minus infinity: $x_j \ge -\infty$
- plus infinity: $x_j \le \infty$

This format cannot be used easily to describe algorithms, because there are many different forms of constraints that can be included. There are two other formulations of LPs that are more appropriate: (i) canonical form, and (ii) standard form.

An LP problem in its canonical form has all its technological constraints in the form 'less than or equal to' ($\leq$) and all variables are nonnegative, as shown in Equation (2.4):

$$
\begin{aligned}
\min z = \; & c_1 x_1 \; + \; c_2 x_2 \; + \cdots + \; c_n x_n \\
\text{s.t.} \quad & A_{11} x_1 + A_{12} x_2 + \cdots + A_{1n} x_n \leq b_1 \\
& A_{21} x_1 + A_{22} x_2 + \cdots + A_{2n} x_n \leq b_2 \\
& \; \vdots \qquad \vdots \qquad \vdots \qquad \vdots \quad \cdots \quad \vdots \qquad \vdots \qquad \vdots \\
& A_{m1} x_1 + A_{m2} x_2 + \cdots + A_{mn} x_n \leq b_m \\
x_j \geq 0, \quad & (j = 1, 2, \cdots, n)
\end{aligned}
\tag{2.4}
$$

Equation (2.4) can be written in matrix notation as shown in Equation (2.5):

$$
\begin{aligned}
\min z = \; & c^T x \\
\text{s.t.} \; & Ax \leq b \\
& x \geq 0
\end{aligned}
\tag{2.5}
$$

An LP problem in its standard form has all its technological constraints as equalities ($=$) and all variables are nonnegative, as shown in Equation (2.6):

$$
\begin{aligned}
\min z = \; & c_1 x_1 \; + \; c_2 x_2 \; + \cdots + \; c_n x_n \\
\text{s.t.} \quad & A_{11} x_1 + A_{12} x_2 + \cdots + A_{1n} x_n = b_1 \\
& A_{21} x_1 + A_{22} x_2 + \cdots + A_{2n} x_n = b_2 \\
& \; \vdots \qquad \vdots \qquad \vdots \qquad \vdots \quad \cdots \quad \vdots \qquad \vdots \qquad \vdots \\
& A_{m1} x_1 + A_{m2} x_2 + \cdots + A_{mn} x_n = b_m \\
x_j \geq 0, \quad & (j = 1, 2, \cdots, n)
\end{aligned}
\tag{2.6}
$$

Equation (2.6) can be written in matrix notation as shown in Equation (2.7):

$$
\begin{aligned}
\min z = \; & c^T x \\
\text{s.t.} \; & Ax = b \\
& x \geq 0
\end{aligned}
\tag{2.7}
$$

All LPs can be transformed to either the canonical form or the standard form. Below, we describe the steps that must be performed to transform:

- a general LP problem to its canonical form,
- a general LP to its standard form,
- an LP problem in its canonical form to its standard form, and
- an LP problem in its standard form to its canonical form.

Definition 2.1 Two LPs are equivalent, iff there is a one-to-one correspondence between their feasible points or solutions and their objective function values.

Or equivalently, if an LP problem is optimal, infeasible, or unbounded, then its equivalent LP problem is also optimal, infeasible, or unbounded, respectively.

2.3.1   General to Canonical Form

The steps that must be performed to transform a general LP problem to its canonical form are the following:

1. If the LP problem is a maximization problem, transform it to a minimization problem. Since maximizing a quantity is equivalent to minimizing its negative, any maximization objective function

$$\max z = c^T x = c_1 x_1 + c_2 x_2 + \cdots + c_n x_n \tag{2.8}$$

can be transformed to

$$\min z = -c^T x = -c_1 x_1 - c_2 x_2 - \cdots - c_n x_n \tag{2.9}$$

or in a more general form

$$\max \{c^T x : x \in \top\} = -\min \{-c^T x : x \in \top\} \tag{2.10}$$

where $\top$ is a set of numbers
2. An inequality constraint in the form 'greater than or equal to' ($\geq$)

$$a_1 x_1 + a_2 x_2 + \cdots + a_n x_n \geq b \tag{2.11}$$

can be transformed in the form 'less than or equal to' ($\leq$) if we multiply both the left-hand and the right-hand sides of that constraint by -1

$$-a_1 x_1 - a_2 x_2 - \cdots - a_n x_n \leq -b \tag{2.12}$$

3. An equality constraint

$$a_1 x_1 + a_2 x_2 + \cdots + a_n x_n = b \tag{2.13}$$

is equivalent with two inequality constraints

$$\begin{aligned} a_1 x_1 + a_2 x_2 + \cdots + a_n x_n \leq b \\ a_1 x_1 + a_2 x_2 + \cdots + a_n x_n \geq b \end{aligned} \tag{2.14}$$

Following the same procedure as previously, we can transform the second constraint of Equation (2.14) in the form 'greater than or equal to' ($\geq$) to an inequality constraint in the form 'less than or equal to' ($\leq$).

$$a_1x_1 + a_2x_2 + \cdots + a_nx_n \leq b$$
$$-a_1x_1 - a_2x_2 - \cdots - a_nx_n \leq -b \qquad (2.15)$$

However, this procedure increases the number of constraints (for each equality constraint we add two inequality constraints) and it is not efficient in practice. Another way to transform an equality to an inequality constraint is to solve the equation in terms of a variable a_jx_j with $a_j \neq 0$

$$x_1 = (b - a_2x_2 - \cdots - a_nx_n) \, / \, a_1, a_1 \neq 0 \qquad (2.16)$$

and extract the inequality (since $x_1 \geq 0$)

$$(a_2/a_1)x_2 + \cdots + (a_n/a_1)x_n \leq b/a_1, a_1 \neq 0 \qquad (2.17)$$

Finally, we remove variable x_1 from the objective function and the other constraints.

Let's transform a general problem to its canonical form. The LP problem that will be transformed is the following:

$$
\begin{aligned}
\max z = \ & 3x_1 \ - 4x_2 + 5x_3 \ - 2x_4 \\
\text{s.t.} \quad & -2x_1 \ - \ x_2 \ - 4x_3 \ - 2x_4 \leq \ \ 4 \\
& 5x_1 \ + 3x_2 + \ x_3 \ + 2x_4 \leq \ 18 \\
& 5x_1 \ + 3x_2 \qquad\quad + \ x_4 \geq -13 \\
& 4x_1 \ + 6x_2 + 2x_3 + 5x_4 = \ \ 10 \\
x_j \geq 0, \quad & (j = 1,2,3,4)
\end{aligned}
$$

Initially, we need to transform the objective function to minimization. According to the first step of the above procedure, we take its negative.

$$
\begin{aligned}
\min z = \ & -3x_1 + 4x_2 \ - 5x_3 + 2x_4 \\
\text{s.t.} \quad & -2x_1 \ - \ x_2 \ - 4x_3 \ - 2x_4 \leq \ \ 4 \quad (1) \\
& 5x_1 \ + 3x_2 + \ x_3 \ + 2x_4 \leq \ 18 \quad (2) \\
& 5x_1 \ + 3x_2 \qquad\quad + \ x_4 \geq -13 \quad (3) \\
& 4x_1 \ + 6x_2 + 2x_3 + 5x_4 = \ \ 10 \quad (4) \\
x_j \geq 0, \quad & (j = 1,2,3,4)
\end{aligned}
$$

Next, we transform all constraints in the form 'greater than or equal to' ($\geq$) to 'less than or equal to' ($\leq$). According to the second step of the aforementioned procedure, we multiply the third constraint by -1.

$$\min z = -3x_1 + 4x_2 - 5x_3 + 2x_4$$

$$
\begin{aligned}
\text{s.t.} \quad & -2x_1 - x_2 - 4x_3 - 2x_4 \le 4 && (1) \\
& 5x_1 + 3x_2 + x_3 + 2x_4 \le 18 && (2) \\
& -5x_1 - 3x_2 \qquad\quad - x_4 \le 13 && (3) \\
& 4x_1 + 6x_2 + 2x_3 + 5x_4 = 10 && (4)
\end{aligned}
$$

$$x_j \ge 0, \quad (j = 1, 2, 3, 4)$$

Finally, we transform all equalities (=) to inequality constraints in the form 'less than or equal to' ($\le$). According to the third step of the aforementioned procedure, we divide each term of the fourth constraint by 4 (the coefficient of the variable x_1) and remove variable x_1 from the objective function and other constraints.

$$\min z = 17/2x_2 - 7/2x_3 + 23/4x_4 - 15/2$$

$$
\begin{aligned}
\text{s.t.} \quad & 2x_2 - 3x_3 + 1/2x_4 \le 9 && (1) \\
& -9/2x_2 - 3/2x_3 - 17/4x_4 \le 11/2 && (2) \\
& 9/2x_2 + 5/2x_3 + 21/4x_4 \le 51/2 && (3) \\
& 3/2x_2 + 1/2x_3 + 5/4x_4 \le 5/2 && (4)
\end{aligned}
$$

$$x_j \ge 0, \quad (j = 2, 3, 4)$$

The following function presents the implementation in MATLAB of the transformation of a general LP problem to its canonical form (filename: general2canonical.m). Some necessary notations should be introduced before the presentation of the implementation for transforming a general LP problem to its canonical form. Let A be a $m \times n$ matrix with the coefficients of the constraints, c be a $n \times 1$ vector with the coefficients of the objective function, b be a $m \times 1$ vector of the right-hand side of the constraints, $Eqin$ be a $m \times 1$ vector of the type of the constraints (-1 for inequality constraints in the form 'less than or equal to' ($\le$), 0 for equalities, and 1 for inequality constraints in the form 'greater than or equal to' ($\ge$)), $MinMaxLP$ be a variable denoting the type of optimization (-1 for minimization and 1 for maximization), and $c0$ be the constant term of the objective function.

The function for the transformation of a general LP problem to its canonical form takes as input the matrix of coefficients of the constraints (matrix A), the vector of coefficients of the objective function (vector c), the vector of the right-hand side of the constraints (vector b), the vector of the type of the constraints (vector $Eqin$), a variable denoting the type of optimization (variable $MinMaxLP$), and the constant term of the objective function (variable $c0$), and returns as output the transformed matrix of coefficients of the constraints (matrix A), the transformed vector of coefficients of the objective function (vector c), the transformed vector of the right-hand side of the constraints (vector b), the transformed vector of the type of the constraints (vector $Eqin$), a variable denoting the type of optimization (variable $MinMaxLP$), and the updated constant term of the objective function (variable $c0$).

In lines 39–42, if the LP problem is a maximization problem, we transform it to a minimization problem. Finally, we find all constraints that are not in the form 'less than or equal to' and transform them in that form (lines 45–68).

```
1.    function [A, c, b, Eqin, MinMaxLP, c0] = ...
2.        general2canonical(A, c, b, Eqin, MinMaxLP, c0)
3.    % Filename: general2canonical.m
4.    % Description: the function is an implementation of the
5.    % transformation of a general LP problem to its
6.    % canonical form
7.    % Authors: Ploskas, N., & Samaras, N.
8.    %
9.    % Syntax: [A, c, b, Eqin, MinMaxLP, c0] = ...
10.   %   general2canonical(A, c, b, Eqin, MinMaxLP, c0)
11.   %
12.   % Input:
13.   % -- A: matrix of coefficients of the constraints
14.   %    (size m x n)
15.   % -- c: vector of coefficients of the objective function
16.   %    (size n x 1)
17.   % -- b: vector of the right-hand side of the constraints
18.   %    (size m x 1)
19.   % -- Eqin: vector of the type of the constraints
20.   %    (size m x 1)
21.   % -- MinMaxLP: the type of optimization
22.   % -- c0: constant term of the objective function
23.   %
24.   % Output:
25.   % -- A: transformed matrix of coefficients of the
26.   %    constraints (size m x n)
27.   % -- c: transformed vector of coefficients of the objective
28.   %    function (size n x 1)
29.   % -- b: transformed vector of the right-hand side of the
30.   %    constraints (size m x 1)
31.   % -- Eqin: transformed vector of the type of the
32.   %    constraints (size m x 1)
33.   % -- MinMaxLP: the type of optimization
34.   % -- c0: updated constant term of the objective function
35.
36.   [m, ~] = size(A); % size of matrix A
37.   % if the LP problem is a maximization problem, transform it
38.   % to a minimization problem
39.   if MinMaxLP == 1
40.       MinMaxLP = -1;
41.       c = -c;
42.   end
43.   % find all constraints that are not in the form 'less than
44.   % or equal to'
45.   for i = 1:m
46.       % transform constraints in the form 'greater than or
47.       % equal to'
48.       if Eqin(i) == 1
49.           A(i, :) = -A(i, :);
50.           b(i) = -b(i);
```

```
51.              Eqin(i) = -1;
52.        elseif Eqin(i) == 0 % transform equality constraints
53.              f = find(A(i, :) ~= 0);
54.              f = f(1);
55.              b(i) = b(i) / A(i, f);
56.              A(i, :) = A(i, :) / A(i, f);
57.              b([1:i - 1, i + 1:m]) = b([1:i - 1, i + 1:m]) - ...
58.                    A([1:i - 1, i + 1:m], f) .* b(i);
59.              A([1:i - 1, i + 1:m], :) = ...
60.                    A([1:i - 1, i + 1:m], :) - ...
61.                    A([1:i - 1, i + 1:m], f) * A(i, :);
62.              c0 = c0 + c(f) * b(i);
63.              c = c - c(f) * A(i, :);
64.              A(:, f) = [];
65.              c(f) = [];
66.              Eqin(i) = -1;
67.        end
68.   end
69.   end
```

2.3.2   General to Standard Form

The steps that must be performed to transform a general LP problem to its standard form are the following:

1. If the LP problem is a maximization problem, transform it to a minimization problem. Since maximizing a quantity is equivalent to minimizing its negative, any maximization objective function

$$\max z = c^T x = c_1 x_1 + c_2 x_2 + \cdots + c_n x_n \qquad (2.18)$$

can be transformed to

$$\min z = -c^T x = -c_1 x_1 - c_2 x_2 - \cdots - c_n x_n \qquad (2.19)$$

or in a more general form

$$\max \left\{ c^T x : x \in \top \right\} = -\min \left\{ -c^T x : x \in \top \right\} \qquad (2.20)$$

where $\top$ is a set of numbers

2. An inequality constraint in the form 'less than or equal to' ($\leq$)

$$a_1 x_1 + a_2 x_2 + \cdots + a_n x_n \leq b \qquad (2.21)$$

can be transformed to an equality ($=$) if we add a deficit slack variable to its left-hand side and add a corresponding nonnegativity constraint

$$a_1 x_1 + a_2 x_2 + \cdots + a_n x_n + x_{n+1} = b$$
$$\text{and } x_{n+1} \geq 0 \tag{2.22}$$

3. An inequality constraint in the form 'greater than or equal to' ($\geq$)

$$a_1 x_1 + a_2 x_2 + \cdots + a_n x_n \geq b \tag{2.23}$$

can be transformed to an equality ($=$) if we subtract a surplus slack variable to its left-hand side and add a corresponding nonnegativity constraint

$$a_1 x_1 + a_2 x_2 + \cdots + a_n x_n - x_{n+1} = b$$
$$\text{and } x_{n+1} \geq 0 \tag{2.24}$$

Let's transform a general problem to its standard form. The LP problem that will be transformed is the following:

$$
\begin{aligned}
\max z = \quad & 3x_1 - 4x_2 + 5x_3 - 2x_4 \\
\text{s.t.} \quad & -2x_1 - x_2 - 4x_3 - 2x_4 \leq 4 \\
& 5x_1 + 3x_2 + x_3 + 2x_4 \leq 18 \\
& 5x_1 + 3x_2 + x_4 \geq -13 \\
& 4x_1 + 6x_2 + 2x_3 + 5x_4 \geq -10 \\
& x_j \geq 0, \quad (j = 1, 2, 3, 4)
\end{aligned}
$$

Initially, we need to transform the objective function to minimization. According to the first step of the aforementioned procedure, we take its negative.

$$
\begin{aligned}
\min z = \quad & -3x_1 + 4x_2 - 5x_3 + 2x_4 \\
\text{s.t.} \quad & -2x_1 - x_2 - 4x_3 - 2x_4 \leq 4 \\
& 5x_1 + 3x_2 + x_3 + 2x_4 \leq 18 \\
& 5x_1 + 3x_2 + x_4 \geq -13 \\
& 4x_1 + 6x_2 + 2x_3 + 5x_4 \geq -10 \\
& x_j \geq 0, \quad (j = 1, 2, 3, 4)
\end{aligned}
$$

Next, we transform all constraints in the form 'less than or equal to' ($\leq$) to equalities ($=$). According to the second step of the above procedure, we add a deficit slack variable to the left-hand side and add a corresponding nonnegativity constraint.

$$\min z = -3x_1 + 4x_2 - 5x_3 + 2x_4$$

$$\text{s.t.} \quad -2x_1 - x_2 - 4x_3 - 2x_4 + x_5 \qquad\qquad = 4$$

$$5x_1 + 3x_2 + x_3 + 2x_4 \qquad + x_6 = 18$$

$$5x_1 + 3x_2 \qquad + x_4 \qquad\qquad \geq -13$$

$$4x_1 + 6x_2 + 2x_3 + 5x_4 \qquad\qquad \geq -10$$

$$x_j \geq 0, \quad (j = 1, 2, 3, 4, 5, 6)$$

Finally, we transform all constraints in the form 'greater than or equal to' ($\geq$) to equalities ($=$). According to the third step of the previously presented procedure, we subtract a surplus slack variable to the left-hand side and add a corresponding nonnegativity constraint.

$$\min z = -3x_1 + 4x_2 - 5x_3 + 2x_4$$

$$\text{s.t.} \quad -2x_1 - x_2 - 4x_3 - 2x_4 + x_5 \qquad\qquad = 4$$

$$5x_1 + 3x_2 + x_3 + 2x_4 \qquad + x_6 \qquad = 18$$

$$5x_1 + 3x_2 \qquad + x_4 \qquad - x_7 \qquad = -13$$

$$4x_1 + 6x_2 + 2x_3 + 5x_4 \qquad\qquad - x_8 = -10$$

$$x_j \geq 0, \quad (j = 1, 2, 3, 4, 5, 6, 7, 8)$$

The following function presents the implementation in MATLAB of the transformation of a general LP problem to its standard form (filename: general2standard.m). Some necessary notations should be introduced before the presentation of the implementation for transforming a general LP problem to its standard form. Let A be a $m \times n$ matrix with the coefficients of the constraints, c be a $n \times 1$ vector with the coefficients of the objective function, b be a $m \times 1$ vector of the right-hand side of the constraints, $Eqin$ be a $m \times 1$ vector of the type of the constraints (-1 for inequality constraints in the form 'less than or equal to' ($\leq$), 0 for equalities, and 1 for inequality constraints in the form 'greater than or equal to' ($\geq$)), and $MinMaxLP$ be a variable denoting the type of optimization (-1 for minimization and 1 for maximization).

The function for the transformation of a general LP problem to its standard form takes as input the matrix of coefficients of the constraints (matrix A), the vector of coefficients of the objective function (vector c), the vector of the right-hand side of the constraints (vector b), the vector of the type of the constraints (vector $Eqin$), and a variable denoting the type of optimization (variable $MinMaxLP$), and returns as output the transformed matrix of coefficients of the constraints (matrix A), the transformed vector of coefficients of the objective function (vector c), the transformed vector of the right-hand side of the constraints (vector b), the transformed vector of the type of the constraints (vector $Eqin$), and a variable denoting the type of optimization (variable $MinMaxLP$).

In order to improve the performance of memory bound code in MATLAB, in lines 35–36, we pre-allocate matrix A and vector c before accessing them within the for-loop. In lines 39–42, if the LP problem is a maximization problem, we transform it to a minimization problem. Finally, we find all constraints that are not equalities and transform them to equalities (lines 44–58).

```
1.    function [A, c, b, Eqin, MinMaxLP] = ...
2.        general2standard(A, c, b, Eqin, MinMaxLP)
3.    % Filename: general2standard.m
4.    % Description: the function is an implementation of the
5.    % transformation of a general LP problem to its
6.    % standard form
7.    % Authors: Ploskas, N., & Samaras, N.
8.    %
9.    % Syntax: [A, c, b, Eqin, MinMaxLP] = ...
10.   %    general2standard(A, c, b, Eqin, MinMaxLP)
11.   %
12.   % Input:
13.   % -- A: matrix of coefficients of the constraints
14.   %     (size m x n)
15.   % -- c: vector of coefficients of the objective function
16.   %     (size n x 1)
17.   % -- b: vector of the right-hand side of the constraints
18.   %     (size m x 1)
19.   % -- Eqin: vector of the type of the constraints
20.   %     (size m x 1)
21.   % -- MinMaxLP: the type of optimization
22.   %
23.   % Output:
24.   % -- A: transformed matrix of coefficients of the
25.   %     constraints (size m x n)
26.   % -- c: transformed vector of coefficients of the objective
27.   %     function (size n x 1)
28.   % -- b: transformed vector of the right-hand side of the
29.   %     constraints (size m x 1)
30.   % -- Eqin: transformed vector of the type of the
31.   %     constraints (size m x 1)
32.   % -- MinMaxLP: the type of optimization
33.
34.   [m, n] = size(A); % size of matrix A
35.   A = [A zeros(m, n)]; % preallocate matrix A
36.   c = [c; zeros(m, 1)]; % preallocate vector c
37.   % if the LP problem is a maximization problem, transform it
38.   % to a minimization problem
39.   if MinMaxLP == 1
40.       MinMaxLP = -1;
41.       c = -c;
42.   end
43.   % find all constraints that are not equalities
44.   for i = 1:m
45.       % transform constraints in the form 'less than or
46.       % equal to' to equality constraints
47.       if Eqin(i) == -1
```

```
48.         A(i, n + 1) = 1;
49.         Eqin(i) = 0;
50.         n = n + 1;
51.      % transform constraints in the form 'greater than or
52.      % equal to' to equality constraints
53.      elseif Eqin(i) == 1
54.         A(i, n + 1) = -1;
55.         Eqin(i) = 0;
56.         n = n + 1;
57.      end
58.   end
59.   end
```

2.3.3   Canonical to Standard Form

The steps that must be performed to transform an LP problem in its canonical form to its standard form are the same steps described previously in this section to transform a general LP problem to its standard form except that we do not have to check if the problem is a maximization problem and transform it to a minimization problem because an LP problem in its canonical form is already a minimization problem. Moreover, an LP problem in its canonical form contains only inequality constraints in the form 'less than or equal to'. The function with the implementation in MATLAB of the transformation of an LP problem in its canonical form to its standard form can be found in the function canonical2standard (filename: canonical2standard.m).

2.3.4   Standard to Canonical Form

The steps that must be performed to transform an LP problem in its standard form to its canonical form are the same steps described previously in this section to transform a general LP problem to its canonical form except that we do not have to check if the problem is a maximization problem and transform it to a minimization problem because an LP problem in its standard form is already a minimization problem. Moreover, an LP problem in its standard form contains only equality constraints. The function with the implementation in MATLAB of the transformation of an LP problem in its standard form to its canonical form can be found in the function standard2canonical (filename: standard2canonical.m).

2.3.5   Transformations of Constraint Ranges and Variable Constraints

As already mentioned in this section, a general LP problem can include ranges in the constraints and variable constraints (bounds).

If a constraint has ranges

$$l_b \leq a_1 x_1 + a_2 x_2 + \cdots + a_n x_n \leq u_b \tag{2.25}$$

then this restriction can be substituted by two constraints (one in the form of 'less than or equal to' and one in the form of 'greater than or equal to')

$$a_1 x_1 + a_2 x_2 + \cdots + a_n x_n \geq l_b$$
$$a_1 x_1 + a_2 x_2 + \cdots + a_n x_n \leq u_b$$

Constraints (bounds) can also be defined for variables:

1. **lower bound** $(x_j \geq l_j)$: We can add the bound to our constraints. Moreover, if we want to transform our LP problem to its standard form, then we subtract a slack variable to the left-hand side of this constraint $(x_j - y_j = l_j)$ and transform it to an equality constraint. However, that would add an extra constraint and an extra variable to the LP problem. Instead, we can substitute the initial decision variable (x_j) with the slack variable plus the lower bound $(x_j = y_j + l_j)$ in the constraints and the objective function. Using the latter procedure, the transformed LP problem has the same dimensions with the initial LP problem.

2. **upper bound** $(x_j \leq u_j)$: We can add the bound to our constraints. Moreover, if we want to transform our LP problem to its standard form, then we add a slack variable to the left-hand side of this constraint $(x_j + y_j = u_j)$ and transform it to an equality constraint. However, that would add an extra constraint and an extra variable to the LP problem. Instead, we can substitute the initial decision variable (x_j) with the upper bound minus the slack variable $(x_j = u_j - y_j)$ in the constraints and the objective function. Using the latter procedure, the transformed LP problem has the same dimensions with the initial LP problem.

3. **lower and upper bounds** $(l_j \leq x_j \leq u_j)$: We can add the bounds to our constraints as two separate constraints $(x_j \geq l_j$ and $x_j \leq u_j)$. Moreover, if we want to transform our LP problem to its standard form, then we subtract a slack variable to the left-hand side of the first constraint $(x_j - y_j = l_j)$ and add a slack variable to the left-hand side of the second constraint $(x_j + y_{j+1} = u_j)$ and transform them to equality constraints. However, that would add two extra constraints and two extra variables to the LP problem. Instead, we can write the compound inequality as $0 \leq x_j - l_j \leq u_j - l_j$. If we set $y_j = x_j - l_j \geq 0$, then we can substitute the initial decision variable (x_j) with the slack variable plus the lower bound $(x_j = y_j + l_j)$ in the constraints and the objective function. Finally, we add a constraint $y_j \leq u_j - l_j$ and add a slack variable to that constraint $(y_j + y_{j+1} = u_j - l_j)$ if we want to transform the LP problem to its standard form. Using the latter procedure, the transformed LP problem has one more constraint and one more variable than the initial LP problem.

4. **fixed variable** $(x_j = k)$: We can add the bound to our constraints. However, that would add an extra constraint to the LP problem. Instead, we can substitute the fixed bound (k) to the initial decision variable (x_j) in the constraints and the

objective function and eliminate that variable. Using the latter procedure, the transformed LP problem will have one variable less than the initial LP problem.

5. **free variable**: We can substitute a free variable x_j by $x_j^+ - x_j^-$, where x_j^+ and x_j^- are nonnegative, and eliminate that variable. However, that would add one more variable to the LP problem. Instead, we can find an equality constraint where we can express variable $A_{ij}x_j$, where $A_{ij} \neq 0$, as a function of the other variables and substitute x_j in the constraints and the objective function. Next, we can eliminate variable x_j from the constraints and the objective function and delete the equality constraint from which we expressed variable x_j as a function of the other variables. Using the latter procedure, the transformed LP problem has one less constraint and one less variable than the initial LP problem.

Let's transform a general problem to its standard form. The LP problem that will be transformed is the following:

$$
\begin{aligned}
\min z = -2x_1 &- 3x_2 + 4x_3 + x_4 + 5x_5 \\
\text{s.t.} \quad 2x_1 &- x_2 + 3x_3 + x_4 + x_5 = 5 \quad (1) \\
x_1 &+ 3x_2 + 2x_3 - 2x_4 \geq 4 \quad (2) \\
-x_1 &- x_2 + 3x_3 + x_4 \leq 3 \quad (3) \\
x_1 - free, x_2 &\geq -1, x_3 \leq 1, -1 \leq x_4 \leq 1, x_5 = 1
\end{aligned}
$$

Initially, we check if the problem is a maximization problem and we need to transform it to a minimization problem. This is not the case in that problem. Next, we transform all inequality constraints ($\leq, \geq$) to equalities ($=$). We subtract a surplus slack variable to the left-hand side of the second constraint and add a corresponding nonnegativity constraint. Moreover, we add a deficit slack variable to the left-hand side of the third constraint and add a corresponding nonnegativity constraint.

$$
\begin{aligned}
\min z = -2x_1 &- 3x_2 + 4x_3 + x_4 + 5x_5 \\
\text{s.t.} \quad 2x_1 &- x_2 + 3x_3 + x_4 + x_5 = 5 \quad (1) \\
x_1 &+ 3x_2 + 2x_3 - 2x_4 \quad - x_6 = 4 \quad (2) \\
-x_1 &- x_2 + 3x_3 + x_4 \quad + x_7 = 3 \quad (3) \\
x_1 - free, x_2 &\geq -1, x_3 \leq 1, -1 \leq x_4 \leq 1, x_5 = 1, x_6 \geq 0, x_7 \geq 0
\end{aligned}
$$

Next, we handle the lower bound of variable x_2. According to the first case of the aforementioned transformations, we can substitute the decision variable (x_2) with the slack variable plus the lower bound ($x_2 = y_2 - 1$).

$$
\begin{aligned}
\min z = -2x_1 &- 3(y_2 - 1) + 4x_3 + x_4 + 5x_5 \\
\text{s.t.} \quad 2x_1 &- (y_2 - 1) + 3x_3 + x_4 + x_5 = 5 \quad (1) \\
x_1 &+ 3(y_2 - 1) + 2x_3 - 2x_4 \quad - x_6 = 4 \quad (2) \\
-x_1 &- (y_2 - 1) + 3x_3 + x_4 \quad + x_7 = 3 \quad (3) \\
x_1 - free, y_2 &\geq 0, x_3 \leq 1, -1 \leq x_4 \leq 1, x_5 = 1, x_6 \geq 0, x_7 \geq 0
\end{aligned}
$$

Note, that the objective function also includes a constant term, $c_0 = 3$.

$$\min z = -2x_1 - 3x_2 + 4x_3 + x_4 + 5x_5 + 3$$
$$\text{s.t.} \quad 2x_1 - y_2 + 3x_3 + x_4 + x_5 \qquad\qquad\qquad = 4 \quad (1)$$
$$x_1 + 3y_2 + 2x_3 - 2x_4 \qquad\qquad - x_6 \qquad = 7 \quad (2)$$
$$-x_1 - y_2 + 3x_3 + x_4 \qquad\qquad\qquad + x_7 = 2 \quad (3)$$
$$x_1 - free, y_2 \geq 0, x_3 \leq 1, -1 \leq x_4 \leq 1, x_5 = 1, x_6 \geq 0, x_7 \geq 0$$

Next, we handle the upper bound of variable x_3. According to the second case of the aforementioned transformations, we can substitute the decision variable (x_3) with the upper bound minus the slack variable ($x_3 = 1 - y_3$).

$$\min z = -2x_1 - 3y_2 - 4y_3 + x_4 + 5x_5 + 7$$
$$\text{s.t.} \quad 2x_1 - y_2 - 3y_3 + x_4 + x_5 \qquad\qquad\qquad = 1 \quad (1)$$
$$x_1 + 3y_2 - 2y_3 - 2x_4 \qquad\qquad - x_6 \qquad = 5 \quad (2)$$
$$-x_1 - y_2 - 3y_3 + x_4 \qquad\qquad\qquad + x_7 = -1 \quad (3)$$
$$x_1 - free, y_2 \geq 0, y_3 \geq 0, -1 \leq x_4 \leq 1, x_5 = 1, x_6 \geq 0, x_7 \geq 0$$

Next, we handle the lower and upper bounds of variable x_4. According to the third case of the aforementioned transformations, we can substitute the initial decision variable (x_4) with the slack variable plus the lower bound ($x_4 = y_4 - 1$) in the constraints and the objective function. Finally, we add a new constraint $y_4 \leq 2$ and a slack variable y_5 to that constraint.

$$\min z = -2x_1 - 3y_2 - 4y_3 + y_4 + 5x_5 + 6$$
$$\text{s.t.} \quad 2x_1 - y_2 - 3y_3 + y_4 + x_5 \qquad\qquad\qquad = 2 \quad (1)$$
$$x_1 + 3y_2 - 2y_3 - 2y_4 \qquad\qquad - x_6 \qquad = 3 \quad (2)$$
$$-x_1 - y_2 - 3x_3 + y_4 \qquad\qquad\qquad + x_7 \qquad = 0 \quad (3)$$
$$y_4 \qquad\qquad\qquad\qquad + y_5 = 2 \quad (4)$$
$$x_1 - free, y_2 \geq 0, y_3 \geq 0, y_4 \geq 0, x_5 = 1, x_6 \geq 0, x_7 \geq 0, y_5 \geq 0$$

Next, we handle the fixed variable x_5. According to the fourth case of the aforementioned transformations, we can substitute the fixed bound (1) to the initial decision variable (x_5) and eliminate that variable.

$$\min z = -2x_1 - 3y_2 - 4y_3 + y_4 + 11$$
$$\text{s.t.} \quad 2x_1 - y_2 - 3y_3 + y_4 \qquad\qquad\qquad = 1 \quad (1)$$
$$x_1 + 3y_2 - 2y_3 - 2y_4 - x_6 \qquad\qquad = 3 \quad (2)$$
$$-x_1 - y_2 - 3x_3 + y_4 \qquad\qquad + x_7 \qquad = 0 \quad (3)$$
$$y_4 \qquad\qquad\qquad\qquad + y_5 = 2 \quad (4)$$
$$x_1 - free, y_2 \geq 0, y_3 \geq 0, y_4 \geq 0, x_6 \geq 0, x_7 \geq 0, y_5 \geq 0$$

Finally, we handle the free variable x_1. According to the fifth case of the aforementioned transformations, we can find an equality constraint where we can express variable x_1 as a function of the other variables. Let's choose the second constraint to express variable x_1 as a function of the other variables. Then, we substitute x_1 in the constraints and the objective function. Next, we eliminate variable x_1 from the constraints and the objective function and delete the second constraint. The final transformed LP problem is shown below:

$$
\begin{aligned}
\min z = \ & 3y_2 - 8y_3 - 3y_4 - 2x_6 + 5 \\
\text{s.t.} \quad & -7y_2 + y_3 + 5y_4 + 2x_6 && = -5 \quad (1) \\
& 2y_2 - 5x_3 - y_4 - x_6 + x_7 && = 3 \quad (2) \\
& y_4 \qquad\qquad\qquad\quad + y_5 && = 2 \quad (3) \\
& y_2 \geq 0, y_3 \geq 0, y_4 \geq 0, x_6 \geq 0, x_7 \geq 0, y_5 \geq 0
\end{aligned}
$$

2.4 Modeling Linear Programming Problems

Given a description of the problem, you should be able to formulate it as an LP problem following the steps below:

- Identify the decision variables,
- Identify the objective function,
- Identify the constraints, and
- Write down the entire problem adding nonnegativity constraints if necessary.

Let's follow these steps to formulate the associated LP problem in the following three examples.

Example 1 A company produces two products, A and B. One unit of product A is sold for $50, while one unit of product B is sold for $35. Each unit of product A requires 3 kg of raw material and 5 labor hours for processing, while each unit of product B requires 4 kg of raw material and 4 labor hours for processing. The company can buy 200 kg of raw material every week. Moreover, the company has currently 4 employees that work 8-hour shifts per day (Monday - Friday). The company wants to find the number of units of each product that should produce in order to maximize its revenue.

Initially, we identify the decision variables of the problem. The number of units of each product that the company should produce are the decision variables of this problem. Let x_1 and x_2 be the number of units of product A and B per week, respectively.

Next, we define the objective function. The company wants to maximize its revenue and we already know that the price for product A is $50 per unit and the price for product B is $35 per unit. Hence, the objective function is:

$$
\max z = 50x_1 + 35x_2
$$

Then, we identify the technological constraints of the given problem. First of all, there is a constraint about the raw material that should be used to produce the two products. Each unit of product A requires 3 kg of raw material and each unit of product B requires 4 kg of raw material, while the raw material used every week cannot exceed 200 kg. Hence, the first constraint is given by

$$3x_1 + 4x_2 \leq 200$$

There is another technological constraint about the labor hours that should be used to produce the two products. Each unit of product A requires 5 labor hours for processing and each unit of product B requires 4 labor hours for processing, while the available labor hours every week cannot exceed 160 h (4 employees × 5 days per week × 8-h shifts per day). Hence, the second constraint is given by

$$5x_1 + 4x_2 \leq 160$$

Moreover, we also add the nonnegativity constraints for variables x_1 and x_2. Hence, the LP problem is the following:

$$\max z = 50x_1 + 35x_2$$
$$\text{s.t.} \quad 3x_1 + 4x_2 \leq 200$$
$$5x_1 + 4x_2 \leq 160$$
$$x_1 \geq 0, x_2 \geq 0, \{x_1, x_2\} \in \mathbb{Z}$$

Example 2 A student is usually purchasing a snack every day from a small store close to his university. The store offer two choices of food: brownies and chocolate ice cream. One brownie costs $2 and one scoop of chocolate ice cream costs $1. The student purchases his snack from the same small store each day for many years now, so the owner of the store allows him to purchase a fraction of a product if he wishes. The student knows that each brownie contains 4 ounces of chocolate and 3 ounces of sugar, while each scoop of chocolate ice cream contains 3 ounces of chocolate and 2 ounces of sugar. The student has also decided that he needs at least 8 ounces of chocolate and 11 ounces of sugar per snack. He wants to find the amount of each product that should buy to meet his requirements by minimizing the cost.

Initially, we identify the decision variables of the problem. The amount of each product that the student should buy are the decision variables of this problem. Let x_1 and x_2 be the amount of brownies and the number of scoops of chocolate ice cream, respectively.

Next, we define the objective function. The student wants to minimize his cost and we already know that the cost for each brownie is $2 and the cost for a scoop of chocolate ice cream is $1. Hence, the objective function is:

$$\min z = 2x_1 + x_2$$

Then, we identify the technological constraints of the given problem. First of all, there is a constraint about the chocolate intake per snack. The student wants 8 ounces of chocolate per snack and each brownie contains 4 ounces of chocolate, while each scoop of chocolate ice cream contains 3 ounces of chocolate. Hence, the first constraint is given by

$$4x_1 + 3x_2 \geq 8$$

There is another technological constraint about the sugar intake per snack. The student wants 11 ounces of sugar per snack and each brownie contains 3 ounces of sugar, while each scoop of chocolate ice cream contains 2 ounces of sugar. Hence, the second constraint is given by

$$3x_1 + 2x_2 \geq 11$$

Moreover, we also add the nonnegativity constraints for variables x_1 and x_2. Hence, the LP problem is the following:

$$\min z = 2x_1 + x_2$$
$$\text{s.t.} \quad 4x_1 + 3x_2 \geq 8$$
$$3x_1 + 2x_2 \geq 11$$
$$x_1 \geq 0, x_2 \geq 0, \{x_1, x_2\} \in \mathbb{R}$$

Example 3 A 24/7 store has decided its minimal requirements for employees as follows:

1. *6 a.m.–10 a.m.–2 employees*
2. *10 a.m.–2 p.m.–5 employees*
3. *2 p.m.–6 p.m.–4 employees*
4. *6 p.m.–10 p.m.–3 employees*
5. *10 p.m.–2 a.m.–2 employees*
6. *2 a.m.–6 a.m.–1 employee*

Employees start working at the beginning of one of the above periods and work for 8 consecutive hours. The store wants to determine the minimum number of employees to be employed in order to have a sufficient number of employees available for each period.

Initially, we identify the decision variables of the problem. The number of employees start working at the beginning of each period are the decision variables of this problem. Let x_1, x_2, x_3, x_4, x_5, and x_6 be the number of employees start working at the beginning of each of the aforementioned 6 periods.

Next, we define the objective function. The store wants to minimize the number of employees to be employed. Hence, the objective function is:

$$\min z = x_1 + x_2 + x_3 + x_4 + x_5 + x_6$$

Then, we identify the technological constraints of the given problem. We take into account that an employee that starts working in a specific period will also work in the next period (8-h shift). So, the number of employees start working in the first period and the number of employees start working in the second period must be at least 5 (the maximum of 2 employees needed in the first period and 5 employees needed in the second period). Hence, the first constraint is given by

$$x_1 + x_2 \geq 5$$

The number of employees start working in the second period and the number of employees start working in the third period must be at least 5 (the maximum of 5 employees needed in the second period and 4 employees needed in the third period). Hence, the second constraint is given by

$$x_2 + x_3 \geq 5$$

The number of employees start working in the third period and the number of employees start working in the fourth period must be at least 4 (the maximum of 4 employees needed in the third period and 3 employees needed in the fourth period). Hence, the third constraint is given by

$$x_3 + x_4 \geq 4$$

The number of employees start working in the fourth period and the number of employees start working in the fifth period must be at least 3 (the maximum of 3 employees needed in the fourth period and 2 employees needed in the fifth period). Hence, the fourth constraint is given by

$$x_4 + x_5 \geq 3$$

The number of employees start working in the fifth period and the number of employees start working in the sixth period must be at least 2 (the maximum of 2 employees needed in the fifth period and 1 employee needed in the sixth period). Hence, the fifth constraint is given by

$$x_5 + x_6 \geq 2$$

The number of employees start working in the sixth period and the number of employees start working in the first period must be at least 2 (the maximum of 1 employee needed in the sixth period and 2 employees needed in the first period). Hence, the sixth constraint is given by

$$x_1 + x_6 \geq 2$$

Moreover, we also add the nonnegativity constraints for variables x_1, x_2, x_3, x_4, x_5, and x_6. Hence, the LP problem is the following:

$$
\begin{aligned}
\min z = x_1 &+ x_2 + x_3 + x_4 + x_5 + x_6 \\
\text{s.t.}\quad x_1 &+ x_2 & &\geq 5 \\
&\ \ x_2 + x_3 & &\geq 5 \\
&\qquad\ x_3 + x_4 & &\geq 4 \\
&\qquad\qquad x_4 + x_5 & &\geq 3 \\
&\qquad\qquad\qquad x_5 + x_6 &&\geq 2 \\
x_1 &\qquad\qquad\qquad\quad + x_6 &&\geq 2 \\
x_1 \geq 0, x_2 \geq 0,& x_3 \geq 0, x_4 \geq 0, x_5 \geq 0, x_6 \geq 0, \{x_1, x_2, x_3, x_4, x_5, x_6\} \in \mathbb{Z}
\end{aligned}
$$

2.5 Geometry of Linear Programming Problems

One way to solve LPs that involve two (or three) variables is the graphical solution of the LP problem. The steps that we must follow to solve an LP problem with this technique are the following:

- Draw the constraints and find the half-planes that represent them,
- Identify the feasible region that satisfies all constraints simultaneously, and
- Locate the corner points, evaluate the objective function at each point, and identify the optimum value of the objective function.

2.5.1 Drawing Systems of Linear Equalities

Definition 2.2 A set of points $x \in \mathbb{R}^n$ whose coordinates satisfy a linear equation of the form

$$
A_1 x_1 + A_2 x_2 + \cdots + A_n x_n = b
$$

where at least an $A_j \neq 0, j = 1, 2, \cdots, n$, is called an $(n-1)$ dimensional hyperplane.

Definition 2.3 Let $(x, y, z) \in \mathbb{R}^N$ and $t \geq 0$. A half line is the set of points $\{x | x = y + tz\}, t \in [0, \infty)$.

In order to graph linear inequalities with two variables, we need to:

- Draw the graph of the equation obtained for the given inequality by replacing the inequality sign with an equal sign, and
- Pick a point lying in one of the half-planes determined by the line sketched in the previous step and substitute the values of x and y into the given inequality. If the inequality is satisfied, the solution is the half-plane containing the point. Otherwise, the solution is the half-plane that does not contain the point.

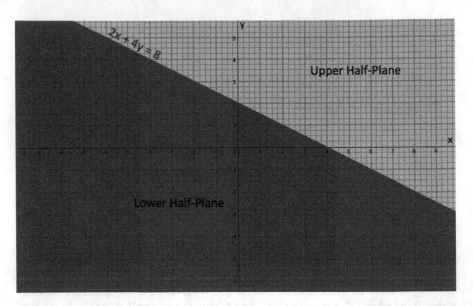

Fig. 2.1 Graph of the equation $2x + 4y = 8$

Let's consider the graph of the equation $2x + 4y \leq 8$. Initially, we draw the line $2x + 4y = 8$ (Figure 2.1).

Next, we determine which half-plane satisfies the inequality $2x + 4y \leq 8$. Pick any point lying in one of the half-planes and substitute the values of x and y into the inequality. Let's pick the origin $(0, 0)$, which lies in the lower half-plane. Substituting $x = 0$ and $y = 0$ into the given inequality, we find

$$2(0) + 4(0) \leq 8$$

or $0 \leq 8$, which is true. This tells us that the required half-plane is the one containing the point $(0, 0)$, i.e., the lower half-plane (Figure 2.2).

2.5.2 Identify the Feasible Region

Definition 2.4 Let $A \in \mathbb{R}^{m \times n}$, $b \in \mathbb{R}^n$, and let the polyhedron $P = \{x \in \mathbb{R}^n | Ax \leq b\}$. Then, polyhedron P is convex.

The above definition implies that the feasible region of any LP problem in canonical form is convex, or equivalently, the set of points corresponding to feasible or optimal solutions of the LP problem is a convex set.

A set of values of the variables of an LP problem that satisfies the constraints and the nonnegative restrictions is called a feasible solution. The feasible region is a convex polytope with polygonal faces. A feasible solution of an LP problem

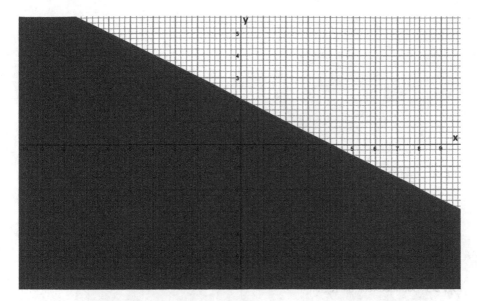

Fig. 2.2 Half-plane of the equation $2x + 4y \leq 8$

that optimizes its objective function is called the optimal solution. In order to solve geometrically an LP problem, we locate the half-planes determined by each constraint following the procedure that was described in the previous step (taking also into account the nonnegativity constraints of the variables). The solution set of a system of linear equalities is the feasible region. The feasible region can be bounded (Figure 2.3) or unbounded (Figure 2.4). In the latter case, if the optimal solution is also unbounded, then the LP problem is unbounded. Otherwise, the region can be unbounded but the LP can have an optimal solution. If the feasible region is empty, then the LP problem is infeasible (Figure 2.5).

2.5.3 Evaluate the Objective Function at Each Corner Point and Identify the Optimum Value of the Objective Function

Definition 2.5 A point $x \in \mathbb{R}^n$ in a convex set is said to be an extreme (corner) point if there are no two points x and y such that $z = tx + (1 - t)y$ for some $t \in (0, 1)$.

Definition 2.6 If the convex set corresponding to $\{Ax = b, x \geq 0\}$ is nonempty, then it has at least one extreme (corner) point.

Definition 2.7 The constraint set corresponding to $\{Ax = b, x \geq 0\}$ has a finite number of extreme (corner) points.

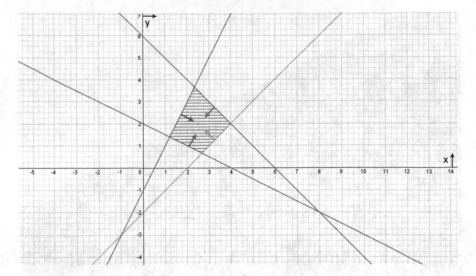

Fig. 2.3 Bounded feasible region

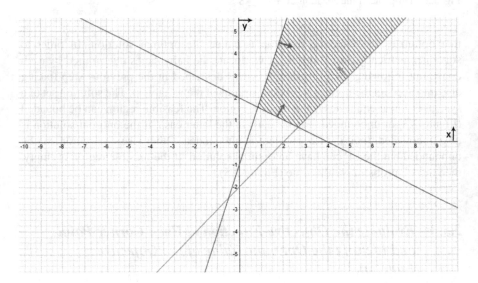

Fig. 2.4 Unbounded feasible region

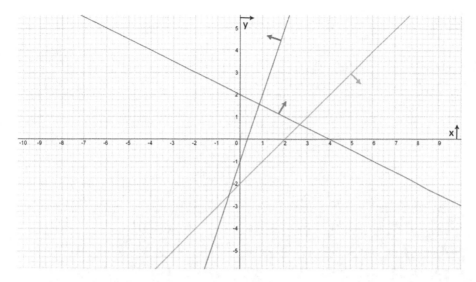

Fig. 2.5 Infeasible LP problem

Definition 2.8 A path in a polyhedral set is a sequence of adjacent extreme (corner) points.

If an LP problem has a solution, then it is located at a vertex (extreme point) of the feasible region. If an LP problem has multiple solutions, then at least one of them is located at a vertex of the set of feasible solutions. In order to find the optimum value of the objective function, we find all the points of intersection of the straight lines bounding the region, evaluate the objective function at each point, and identify the optimum value of the objective function.

Let's consider the bounded feasible region of the following LP problem (Figure 2.6).

We evaluate the objective function at each point of intersection of the straight lines bounding the region. Points 0, F, G, H, I, J, K, L, M, N, O, P, Q, and R are in the intersection of the straight lines of two constraints, but they are not in the feasible region. Hence, we evaluate the objective function only at points A, B, C, D, and E. Substituting x and y values of these points to the objective function, we find the value of the objective function at each point. The minimum or maximum value (depending if the problem is a minimization or a maximization problem) is the optimum value of the objective function and x and y values of the specific point are the optimum values for the LP problem.

The constraints in which the intersection lies the optimal point are called active because these constraints are satisfied as equalities. All other constraints are called inactive because these constraints are satisfied as inequalities.

Another way to find the optimal value of the objective function is to look at the contours of the objective function. The contours of the function $f(x_1, x_2) =$

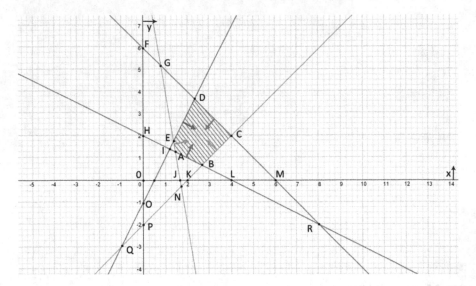

Fig. 2.6 Bounded feasible region of an LP problem

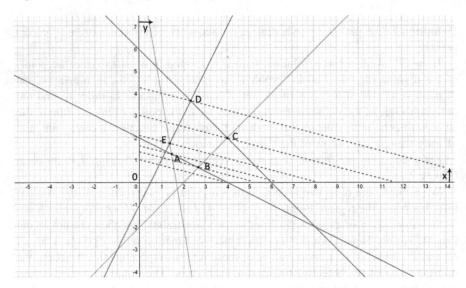

Fig. 2.7 Contours of the objective function

$a_1x_1 + a_2x_2$ are defined by $a_1x_1 + a_2x_2 = c$ for different values of c. The contours are all straight lines and for different values of c are all parallel. Let max $z = x + 4y$ be the objective function of the LP problem shown in Figure 2.6. Figure 2.7 includes the contours of the objective function for different values of c (dashed lines). We sketch a contour of the objective function and move perpendicular to this line in the direction in which the objective function increases until the boundary of the feasible region is reached. Hence, the optimum value of the objective function is at point D, because if we move further the contour then it will be outside of the feasible region boundary.

Example 1 Let's consider the following LP problem:

$$\max z = 2x + 3y$$
$$\text{s.t.} \qquad 2x + 4y \geq 8 \quad (1)$$
$$2x + 5y \leq 18 \quad (2)$$
$$3x + y \geq 5 \quad (3)$$
$$x - 2y \leq 2 \quad (4)$$
$$x, y \geq 0$$

Initially, we draw the constraints and find the half-planes that represent them. Graphing the first constraint

$$2x + 4y \geq 8$$

and taking also into account the nonnegativity constraints, we get the shaded region shown in Figure 2.8.
Adding the second constraint

$$2x + 5y \leq 18$$

into the previous graph, we now get the shaded region shown in Figure 2.9.

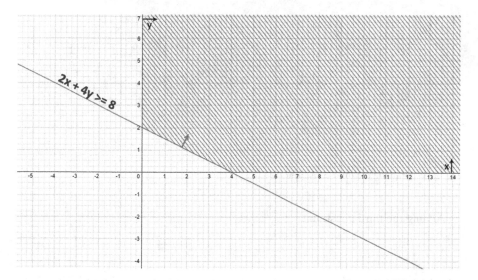

Fig. 2.8 Drawing the nonnegativity constraints and the first constraint

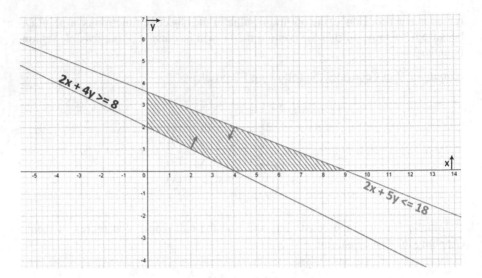

Fig. 2.9 Drawing the second constraint

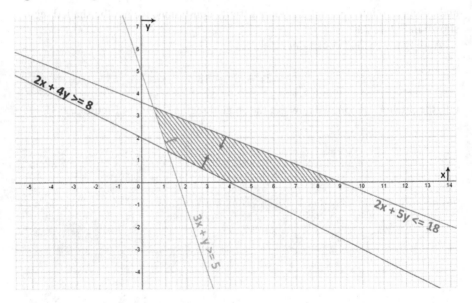

Fig. 2.10 Drawing the third constraint

Adding the third constraint

$$3x + y \geq 5$$

into the previous graph, we now get the shaded region shown in Figure 2.10.
Adding the fourth constraint

$$x - 2y \leq 2$$

into the previous graph, we now get the shaded region shown in Figure 2.11.

We have identified the feasible region (shaded region in Figure 2.11) which satisfies all the constraints simultaneously. Now, we locate the corner points, evaluate the objective function at each point, and identify the optimum value of the objective function. Figure 2.12 shows the points of intersection of the straight

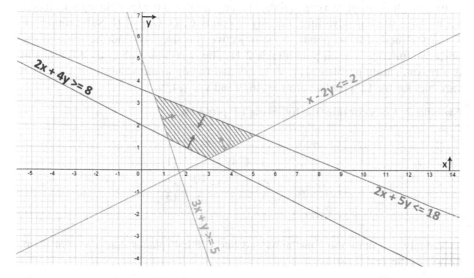

Fig. 2.11 Drawing the fourth constraint

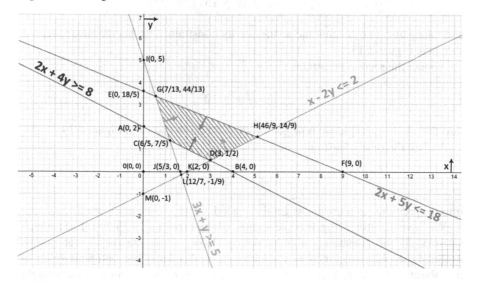

Fig. 2.12 Feasible region of the LP problem

Table 2.1 Evaluation of the objective function for the feasible boundary points

Point	x	y	z
C	6/5	7/5	33/5
D	3	1/2	15/2
G	7/13	44/13	146/13
H	46/9	14/9	134/9

lines of two constraints. Points $O, A, B, E, F, I, J, K, L$, and M are in the intersection of the straight lines of two constraints, but they are not in the feasible region. Hence, we evaluate the objective function only at points C, D, G, and H. Substituting the x and y values of these points to the objective function, we find the value of the objective function at each point (Table 2.1). The maximum value (the LP problem is a maximization problem) is $z = 134/9$ and the optimum values of the decision variables for the LP problem are $x = 46/9$ and $y = 14/9$.

Example 2 Let's consider the following LP problem:

$$\max z = x + 2y$$
$$\text{s.t.} \quad x + 4y \geq 15 \quad (1)$$
$$5x + y \geq 10 \quad (2)$$
$$x - y \leq 3 \quad (3)$$
$$x, y \geq 0$$

Initially, we draw the constraints and find the half-planes that represent them. Graphing the first constraint

$$x + 4y \geq 15$$

and taking also into account the nonnegativity constraints, we get the shaded region shown in Figure 2.13.

Adding the second constraint

$$5x + y \geq 10$$

into the previous graph, we now get the shaded region shown in Figure 2.14.

Adding the third constraint

$$x - y \leq 3$$

into the previous graph, we now get the shaded region shown in Figure 2.15.

We have identified the feasible region (shaded region in Figure 2.15) which satisfies all the constraints simultaneously. Now, we locate the corner points, evaluate the objective function at each point, and identify the optimum value of the objective function. Figure 2.16 shows the points of intersection of the straight

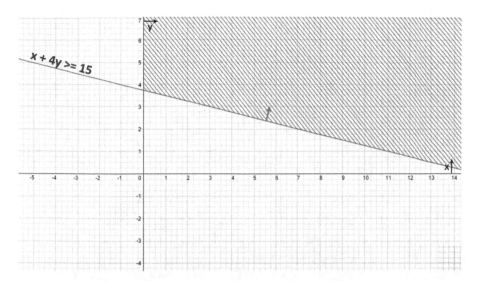

Fig. 2.13 Drawing the nonnegativity constraints and the first constraint

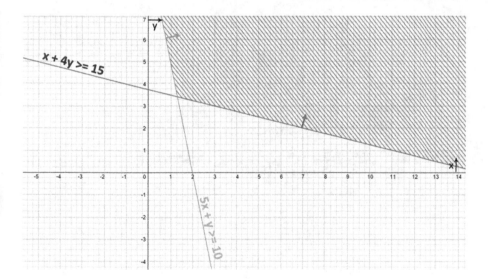

Fig. 2.14 Drawing the second constraint

lines of two constraints. Points O, A, B, F, G, H, and I are in the intersection of the straight lines of two constraints, but they are not in the feasible region. Hence, we evaluate the objective function only at points C, D, and E. Substituting the x and y values of these points to the objective function, we find the value of the objective function at each point (Table 2.2). However, the feasible region is unbounded and

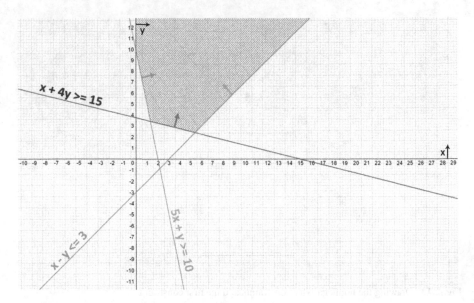

Fig. 2.15 Drawing the third constraint

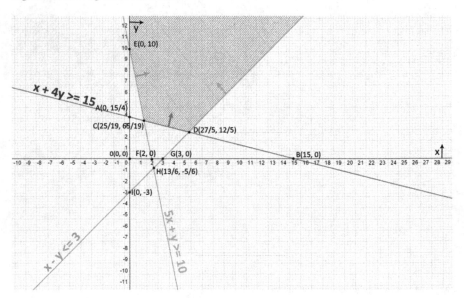

Fig. 2.16 Feasible region of the LP problem

the LP problem is a maximization problem, so we can find other points that have greater value than $z = 20$ (the value of the objective function if we substitute the x and y values of point E). Hence, the specific LP problem is unbounded.

Example 3 Let's consider the following LP problem:

Table 2.2 Evaluation of the objective function for the feasible boundary points

Point	x	y	z
C	25/19	65/19	155/19
D	27/5	12/5	51/5
E	0	10	20

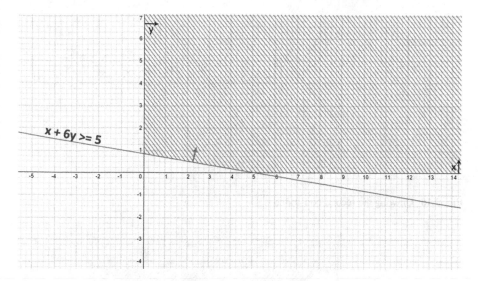

Fig. 2.17 Drawing the nonnegativity constraints and the first constraint

$$\min z = 3x + y$$
$$\text{s.t.} \quad x + 6y \geq 5 \quad (1)$$
$$3x + 3y \geq 5 \quad (2)$$
$$x - 2y \leq 1 \quad (3)$$
$$x, y \geq 0$$

Initially, we draw the constraints and find the half-planes that represent them. Graphing the first constraint

$$x + 6y \geq 5$$

and taking also into account the nonnegativity constraints, we get the shaded region shown in Figure 2.17.

Adding the second constraint

$$3x + 3y \geq 5$$

into the previous graph, we now get the shaded region shown in Figure 2.18.

Adding the third constraint

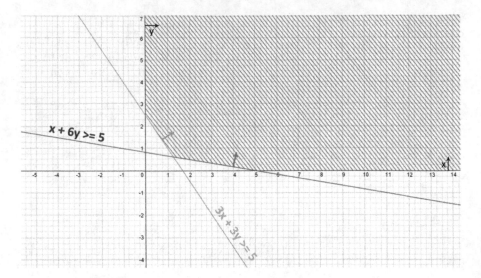

Fig. 2.18 Drawing the second constraint

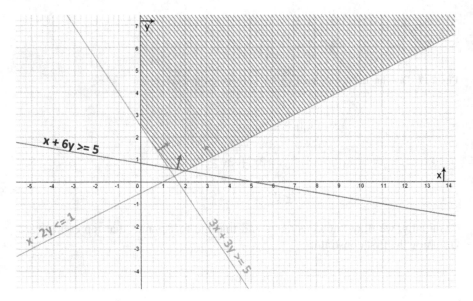

Fig. 2.19 Drawing the third constraint

$$x - 2y \leq 1$$

into the previous graph, we now get the shaded region shown in Figure 2.19.

We have identified the feasible region (shaded region in Figure 2.19) which satisfies all the constraints simultaneously. Now, we locate the corner points,

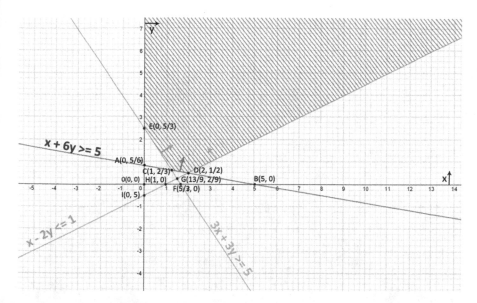

Fig. 2.20 Feasible region of the LP problem

Table 2.3 Evaluation of the objective function for the feasible boundary points

Point	x	y	z
C	1	2/3	11/3
D	2	1/2	13/2
E	0	5/3	5/3

evaluate the objective function at each point, and identify the optimum value of the objective function. Figure 2.20 shows the points of intersection of the straight lines of two constraints. Points O, A, B, F, G, H, and I are in the intersection of the straight lines of two constraints, but they are not in the feasible region. Hence, we evaluate the objective function only at points C, D, and E. Substituting the x and y values of these points to the objective function, we find the value of the objective function at each point (Table 2.3). The feasible region is unbounded, but the LP problem is a minimization problem. Hence, the minimum value is $z = 5/3$ and the optimum values of the decision variables for the LP problem are $x = 0$ and $y = 5/3$.

Example 4 Let's consider the following LP problem:

$$\min z = 2x - y$$
$$\text{s.t.} \quad x + 6y \leq 5 \quad (1)$$
$$-3x + y \geq 4 \quad (2)$$
$$2x - y \leq 3 \quad (3)$$
$$x, y \geq 0$$

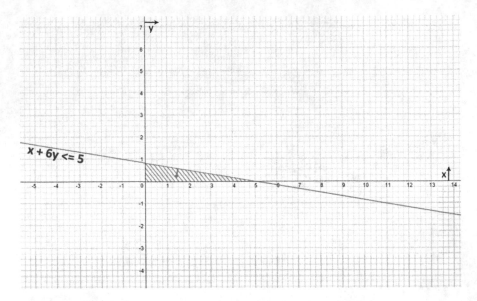

Fig. 2.21 Drawing the nonnegativity constraints and the first constraint

Initially, we draw the constraints and find the half-planes that represent them. Graphing the first constraint

$$x + 6y \leq 5$$

and taking also into account the nonnegativity constraints, we get the shaded region shown in Figure 2.21.

Adding the second constraint

$$-3x + y \geq 4$$

into the previous graph, we now see that there is not any feasible region, so the LP problem is infeasible (Figure 2.22).

For the sake of completeness, adding the third constraint

$$2x - y \leq 3$$

into the previous graph, we get the graph shown in Figure 2.23.

The LP problem is infeasible and there is no solution.

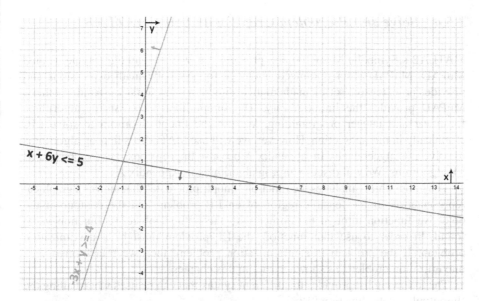

Fig. 2.22 Drawing the second constraint

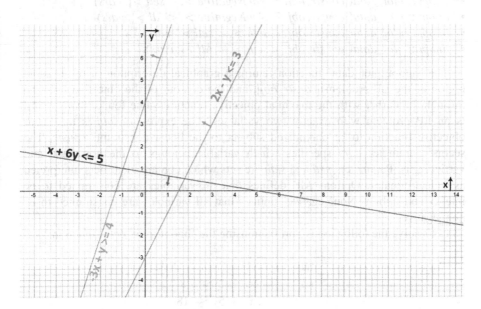

Fig. 2.23 Drawing the third constraint

2.5.4   Solve an LP Problem Graphically Using MATLAB

MATLAB's Symbolic Math Toolbox provides a powerful symbolic engine, named MuPAD. MuPAD is an optimized language for symbolic math expressions and provides an extensive set of mathematical functions and libraries. Among others, MuPAD provides the linopt library. The linopt library consists of algorithms for linear and integer programming. We use the linopt algorithm to solve LPs graphically in this subsection. However, we will use MATLAB's linprog algorithms to solve LPs algebraically in Appendix A because linprog consists of sophisticated LP algorithms. We will discuss linprog algorithms in Appendix A.

The linopt library includes the function 'linopt::plot_data' that returns a graphical representation of the feasible region of an LP problem and the contour of the objective function through the corner with the optimal objective function value found. In order to use that function, the LP problem must be a maximization problem. If the problem is a minimization problem, then we transform it to a maximization problem, as shown earlier in this chapter.

The following list provides the different syntax that can be used when calling the 'linopt::plot_data' function [3]:

- $linopt :: plot\_data([constr, obj, < NonNegative >, < seti >], vars)$
- $linopt :: plot\_data([constr, obj, < NonNegative >, < All >], vars)$
- $linopt :: plot\_data([constr, obj, < setn >, < seti >], vars)$
- $linopt :: plot\_data([constr, obj, < setn >, < All >], vars)$

The 'linopt::plot_data' function returns a graphical description of the feasible region of the LP problem $[constr, obj]$, and the contour of the objective function through the corner with the maximal objective function value found. $[constr, obj]$ is an LP problem with exactly two variables. The expression obj is the linear objective function to be maximized subject to the linear constraints $constr$. The second parameter $vars$ specifies which variable belongs to the horizontal and vertical axis. Parameters $seti$ and $setn$ are sets which contain identifiers interpreted as indeterminates. Option 'All' defines that all variables are constrained to be integer and option 'NonNegative' defines that all variables are constrained to be nonnegative.

Let's solve the following LP problem using the 'linopt::plot_data' function:

$$
\begin{aligned}
\max z = \quad & x \ + 2y \\
\text{s.t.} \quad & 2x \ + 2y \geq \ \ 8 \\
& 2x \ + 5y \leq \ 18 \\
& -3x \ - 4y \geq -17 \\
x, y \geq 0 \quad &
\end{aligned}
$$

To open the MuPAD Notebook, type *mupad* in the Command Window or start it from the MATLAB Apps tab (Figure 2.24).

Fig. 2.24 MuPAD notebook
in the MATLAB Apps tab

```
k := [{2*x + 2*y >= 8, 2*x + 5*y <= 18, -3*x - 4*y >= -17},
x + 2*y, NonNegative]:
g := linopt::plot_data(k, [x, y]):
plot(g):
linopt::maximize(k)
```

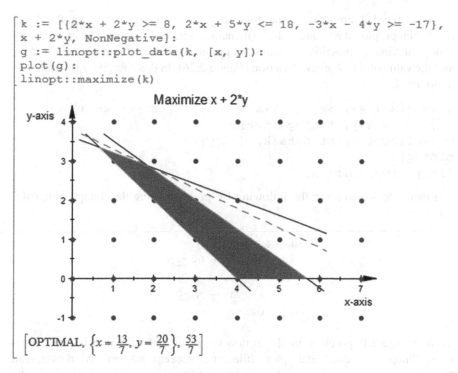

Fig. 2.25 Solve an LP problem graphically using MATLAB

Then, write the LP problem in the following form and plot it using the
functions 'linopt::plot_data' and 'plot' (filename: example1.mn). Moreover, call
'linopt:maximize' function to obtain the optimal values of the decision variables
and the value of the objective function (Figure 2.25). In this case, the LP problem is
optimal with an objective value $z = \frac{53}{7}$.

```
k := [{2*x + 2*y >= 8, 2*x + 5*y <= 18, -3*x - 4*y >= -17},
      x + 2*y, NonNegative]:
g := linopt::plot_data(k, [x, y]):
plot(g):
linopt::maximize(k)
```

Let's also solve the following LP problem using the 'linopt::plot_data' function:

$$\max z = x + 2y$$
$$\text{s.t.} \quad x + 4y \geq 15$$
$$5x + y \geq 10$$
$$x - y \leq 3$$
$$x, y \geq 0$$

Write the LP problem in the following form and plot it using the functions 'linopt::plot_data' and 'plot' (filename: example2.mn). Moreover, call 'linopt:maximize' function to obtain the optimal values of the decision variables and the value of the objective function (Figure 2.26). In this case, the LP problem is unbounded.

```
k := [{x + 4*y >= 15, 5*x + y >= 10, x - y <= 3},
      x + 2*y, NonNegative]:
g := linopt::plot_data(k, [x, y]):
plot(g):
linopt::maximize(k)
```

Finally, let's also solve the following LP problem using the 'linopt::plot_data' function:

$$\max z = 2x - y$$
$$\text{s.t.} \quad x + 6y \leq 5$$
$$-3x + y \geq 4$$
$$2x - y \leq 3$$
$$x, y \geq 0$$

Write the LP problem in the following form and plot it using the functions 'linopt::plot_data' and 'plot' (filename: example3.mn). Moreover, call 'linopt:maximize' function to solve the LP problem (Figure 2.27). In this case, the LP problem is infeasible.

```
k := [{x + 4*y >= 15, 5*x + y >= 10, x - y <= 3},
      x + 2*y, NonNegative]:
g := linopt::plot_data(k, [x, y]):
plot(g):
linopt::maximize(k)
```

```
k := [{x + 4*y >= 15, 5*x + y >= 10, x - y <= 3},
x + 2*y, NonNegative]:
g := linopt::plot_data(k, [x, y]):
plot(g):
linopt::maximize(k)
```

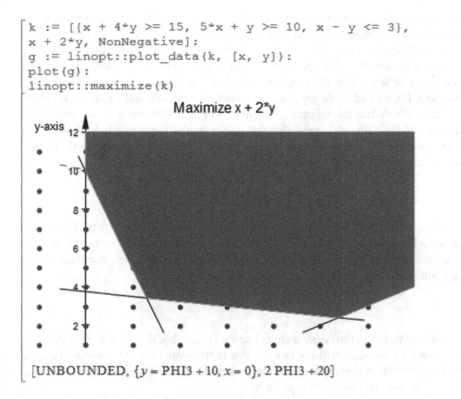

[UNBOUNDED, {y = PHI3 + 10, x = 0}, 2 PHI3 + 20]

Fig. 2.26 Solve an LP problem graphically using MATLAB

```
k := [{x + 6*y <= 5, -3*x + y >= 4, 2*x - y <= 3},
2*x - y, NonNegative]:
g := linopt::plot_data(k, [x, y]):
plot(g):
linopt::maximize(k)
Error: No feasible corners found. The feasible area is empty. [linopt::plot_data]
```

Fig. 2.27 Solve an LP problem graphically using MATLAB

2.6 Duality

Let's consider again the second example presented in Section 2.4 about the student's diet. The owner of the store informs his supplier that he needs at least 8 ounces of chocolate and 11 ounces of sugar to meet his client minimum requirements. He also provides the recipes that he uses to produce a brownie (4 ounces of chocolate and 3 ounces of sugar) and a scoop of chocolate ice cream (3 ounces of chocolate and 2 ounces of sugar). Moreover, he also informs the supplier that he sells each brownie for $2 and each scoop of chocolate ice cream for $1. The supplier now wants to determine the prices per ounce of chocolate and sugar so that he will maximize his revenue and the owner of that store will keep buying these products from him.

Let's formulate the problem that the supplier has to solve. Initially, we identify the decision variables of the problem. The price per ounce of chocolate and the price per ounce of sugar that the supplier should sell are the decision variables of this problem. Let w_1 and w_2 be the price per ounce of chocolate and sugar, respectively.

Next, we define the objective function. The supplier wants to maximize his revenue and we already know that the owner of the store will buy 8 ounces of chocolate and 11 ounces of sugar to meet his client minimum requirements. Hence, the objective function is:

$$\max z = 8w_1 + 11w_2$$

Then, we identify the technological constraints of the given problem. The cost to create a brownie should be below \$2, otherwise the owner of the shop will not buy from the supplier, because he runs the risk of making a loss if the student decides to buy brownies. Hence, the first constraint is given by

$$4w_1 + 3w_2 \leq 2$$

Similarly, the cost to create a chocolate ice cream should be below \$1, otherwise the owner of the shop will not buy the raw ingredients from the supplier, because he runs the risk of making a loss if the student decides to buy chocolate ice cream. Hence, the second constraint is given by

$$3w_1 + 2w_2 \leq 1$$

Moreover, we also add the nonnegativity constraints for variables w_1 and w_2. Hence, the LP problem is the following:

$$\max z = 8w_1 + 11w_2$$
$$\text{s.t.} \quad 4w_1 + 3w_2 \leq 2$$
$$3w_1 + 2w_2 \leq 1$$
$$w_1 \geq 0, w_2 \geq 0, \{w_1, w_2\} \in \mathbb{R}$$

Let's take a closer look in the LP problem we formulated in Section 2.4 and the LP problem we formulated in this section.

$$\min z = 2x_1 + x_2 \qquad\qquad \max z = 8w_1 + 11w_2$$
$$\text{s.t.} \quad 4x_1 + 3x_2 \geq 8 \qquad \text{s.t.} \quad 4w_1 + 3w_2 \leq 2$$
$$3x_1 + 2x_2 \geq 11 \qquad\qquad 3w_1 + 2w_2 \leq 1$$
$$x_1 \geq 0, x_2 \geq 0, \{x_1, x_2\} \in \mathbb{R} \quad w_1 \geq 0, w_2 \geq 0, \{w_1, w_2\} \in \mathbb{R}$$

The first problem (student's problem) is a minimization problem, while the second problem (supplier's problem) is a maximization problem. The coefficients of the objective function in the first problem are the same with the right-hand side in the second problem. The coefficients of the objective function in the second problem are the same with the right-hand side in the first problem. Moreover, the matrix of coefficients in the second problem is the transpose of the matrix of coefficients of the first problem. All these findings show that these two problems have a relation. Indeed, the second problem is the dual of the first problem. The first problem is called primal.

It turns out that LPs come in pairs. We can automatically derive two LPs from the same data. The relation of these problems is close. If the primal LP problem refers to the employer, then the dual LP problem refers to the employees. If the primal LP problem refers to the clients of a store (student's problem), then the dual LP problem refers to the suppliers of a store (supplier's problem).

Generalizing the previous findings, for a specific LP problem

$$\min z = c^T x$$
$$\text{s.t.}\ \ Ax \geq b \qquad\qquad (2.26)$$
$$x \geq 0$$

We can find its dual problem in its canonical form

$$\max z = b^T w$$
$$\text{s.t.}\ \ A^T w \leq c \qquad\qquad (2.27)$$
$$w \geq 0$$

Transforming an LP problem in the form shown in Equation (2.26), we can find its dual LP problem. For example, let's consider the following LP problem:

$$\min z = \ \ 3x_1 \ - \ 2x_2 + 2x_3$$
$$\text{s.t.} \qquad -2x_1 + 3x_2 + 5x_3 \geq \ 10$$
$$-x_1 \ - \ 2x_2 + 2x_3 \geq -5$$
$$x_1 \geq 0, x_2 \geq 0, x_3 \geq 0$$

The dual of the previous LP problem is the following:

$$\max z = \ \ 10w_1 \ - \ 5w_2$$
$$\text{s.t.} \qquad -2w_1 \ - \ w_2 \leq \ 3$$
$$3w_1 \ - \ 2w_2 \leq -2$$
$$5w_1 \ + \ 2w_2 \leq \ 2$$
$$w_1 \geq 0, w_2 \geq 0$$

The aforementioned procedure to transform a primal LP problem to its dual has a major drawback. The primal LP problem must be in the form shown in

Equation (2.26) or otherwise we convert it to that form. Let's describe a more general procedure to find the dual LP problem without the need to perform any transformations to the primal problem. The following steps will help us to find the dual LP problem:

- **Optimization type**: if the primal LP problem is a minimization problem, then the dual LP problem is a maximization problem. Similarly, if the primal LP problem is a maximization problem, then the dual LP problem is a minimization problem.
- **Decision variables**: the number of the decision variables of the dual LP problem is the same with the number of the constraints of the primal LP problem. The ith variable of the dual LP problem is related to the ith constraint of the primal LP problem. Hence:

  - if the ith constraint of the primal LP problem is an equality, then the ith variable of the dual LP problem is a free variable.
  - if the ith constraint of the primal LP problem is in the form 'greater than or equal to' ($\geq$), then the ith variable of the dual LP problem is greater than or equal to zero (nonnegativity constraint).
  - if the ith constraint of the primal LP problem is in the form 'less than or equal to' ($\leq$), then the ith variable of the dual LP problem is less than or equal to zero (upper bound).

- **Coefficients of the objective function**: The coefficients of the objective function of the dual LP problem are the right-hand side values of the primal LP problem.
- **Constraints**: the number of the constraints of the dual LP problem is the same with the number of the decision variables of the primal LP problem. The ith constraint of the dual LP problem is related to the ith variable of the primal LP problem. Hence:

  - if the ith variable (x_1) of the primal LP problem is a free variable, then the ith constraint of the dual LP problem is an equality.
  - if the ith variable of the primal LP problem is greater than or equal to zero (nonnegativity constraint), then the ith constraint of the dual LP problem is in the form of 'less than or equal to' ($\leq$).
  - if the ith variable of the primal LP problem is less than or equal to zero, then the ith constraint of the dual LP problem is in the form of 'greater than or equal to' ($\geq$).

- **Coefficient matrix**: The coefficient matrix of the dual LP problem is the transpose of the coefficient matrix of the primal LP problem.
- **Right-hand side values**: The right-hand side values of the dual LP problem are the coefficients of the objective function of the primal LP problem.

Table 2.4 summarizes the above procedure.

So, let's follow the aforementioned steps to find the dual LP problem of the following LP problem:

Table 2.4 Primal to dual transformations

min		↔	max	
Constraint	=	↔	Variable	Free
Constraint	≥	↔	Variable	≥ 0
Constraint	≤	↔	Variable	≤ 0
Variable	free	↔	Constraint	=
Variable	≥ 0	↔	Constraint	≤
variable	≤ 0	↔	Constraint	≥

$$\begin{aligned} \max z = \; & 2x_1 + 4x_2 \\ \text{s.t.} \quad & -3x_1 + 4x_2 - 3x_3 = -3 \quad (1) \\ & x_1 - 2x_2 + 4x_3 \geq 4 \quad (2) \\ & 2x_1 - x_2 - 2x_3 \leq 6 \quad (3) \\ & x_2 \geq 0, x_3 \leq 0 \end{aligned}$$

Since the primal LP problem is a maximization problem, the dual LP problem will be a minimization problem. The primal LP problem has 3 constraints, so the dual LP problem will have 3 decision variables. The first constraint of the primal LP problem is an equality, so the first variable of the dual LP problem will be a free variable. The second constraint of the primal LP is in the form of 'greater than or equal to' ($\geq$), so the second variable of the dual LP problem will be greater than or equal to zero. The third constraint of the primal LP problem is in the form of 'less than or equal to' ($\leq$), so the third variable of the dual LP problem will be less than or equal to zero. Moreover, the coefficients of the objective function of the dual LP problem will be the same as the right-hand side values of the primal LP problem. Hence, the objective function of the dual LP problem is:

$$\min z = -3w_1 + 4w_2 + 6w_3$$

The primal LP problem has 3 variables, so the dual LP problem will have 3 constraints. The first variable (x_1) of the primal LP problem is a free variable, so the first constraint of the dual LP problem will be an equality. The second variable (x_2) of the primal LP problem is greater than or equal to zero, so the second constraint of the dual LP problem will be in the form of 'less than or equal to' ($\leq$). The third variable (x_3) of the primal LP problem is less than or equal to zero, so the third constraint of the dual LP problem will be in the form of 'greater than or equal to' ($\geq$). The coefficient matrix of the dual LP problem will be the transpose of the coefficient matrix of the primal LP problem. The right-hand side values of the dual LP problem are the coefficients of the objective function of the primal LP problem. Hence, the constraints of the dual LP problem are:

$$-3w_1 + w_2 + 2w_3 = 2$$
$$4w_1 - 2w_2 - w_3 \leq 4$$
$$-3w_1 + 4w_2 - 2w_3 \geq 0$$

So, the dual LP problem is the following:

$$\min z = -3w_1 + 4w_2 + 6w_3$$
$$-3w_1 + w_2 + 2w_3 = 2$$
$$4w_1 - 2w_2 - w_3 \leq 4$$
$$-3w_1 + 4w_2 - 2w_3 \geq 0$$
$$w_2 \geq 0, w_3 \leq 0$$

A code that transforms a primal LP problem to its dual is presented (filename: `primal2dual.m`). Some necessary notations should be introduced before the presentation of this code. Let A be a $m \times n$ matrix with the coefficients of the constraints of the primal LP problem, c be a $n \times 1$ vector with the coefficients of the objective function of the primal LP problem, b be a $m \times 1$ vector with the right-hand side of the constraints of the primal LP problem, $Eqin$ be a $m \times 1$ vector with the type of the constraints of the primal LP problem, $MinMaxLP$ be a variable denoting the type of optimization of the primal LP problem (-1 for minimization and 1 for maximization), $VarConst$ be a $n \times 1$ vector of the variables' constraints of the primal LP problem (0—free variable, 1—variable ≤ 0, 2—variable ≥ 0), DA be a $n \times m$ matrix with the coefficients of the constraints of the dual LP problem, Dc be a $m \times 1$ vector with the coefficients of the objective function of the dual LP problem, Db be a $n \times 1$ vector with the right-hand side of the constraints of the dual LP problem, $DEqin$ be a $n \times 1$ vector with the type of the constraints of the dual LP problem, $DMinMaxLP$ be a variable denoting the type of optimization of the dual LP problem (-1 for minimization and 1 for maximization), and $DVarConst$ be a $m \times 1$ vector of the variables' constraints of the dual LP problem (0—free variable, 1—variable ≤ 0, 2—variable ≥ 0).

The function takes as input the matrix of coefficients of the constraints of the primal LP problem (matrix A), the vector of the coefficients of the objective function of the primal LP problem (vector c), the vector of the right-hand side of the constraints of the primal LP problem (vector b), the vector of the type of the constraints of the primal LP problem (vector $Eqin$), a variable denoting the type of optimization of the primal LP problem (variable $MinMaxLP$), and the vector of the variables' constraints of the primal LP problem (vector $VarConst$), and returns as output the matrix of coefficients of the constraints of the dual LP problem (matrix DA), the vector of the coefficients of the objective function of the dual LP problem (vector Dc), the vector of the right-hand side of the constraints of the dual LP problem (vector Db), the vector of the type of the constraints of the dual LP problem (vector $DEqin$), a variable denoting the type of optimization of the dual LP problem (variable $DMinMaxLP$), and the vector of the variables' constraints of the dual LP problem (vector $DVarConst$).

If the primal LP problem is a minimization problem, then the dual LP problem is a maximization one and vice versa (line 43). The coefficients of the objective function of the dual LP problem are the right-hand side values of the primal LP problem (line 47). The right-hand side values of the dual LP problem are the coefficients of the objective function of the primal LP problem (line 51). The coefficient matrix of the dual LP problem is the transpose of the coefficient matrix of the primal LP problem (line 55). Then, we calculate the vector of the type of the constraints (lines 58–72). Finally, we calculate the vector of the variables' constraints (lines 74–88).

```
1.    function [DA, Dc, Db, DEqin, DMinMaxLP, DVarConst] = ...
2.        primal2dual(A, c, b, Eqin, MinMaxLP, VarConst)
3.    % Filename: primal2dual.m
4.    % Description: the function is an implementation of the
5.    % transformation of a primal LP problem to its dual
6.    % Authors: Ploskas, N., & Samaras, N.
7.    %
8.    % Syntax: [DA, Dc, Db, DEqin, DMinMaxLP, DVarConst] = ...
9.    %    primal2dual(A, c, b, Eqin, MinMaxLP, VarConst)
10.   %
11.   % Input:
12.   % -- A: matrix of coefficients of the constraints of the
13.   %    primal LP problem (size m x n)
14.   % -- c: vector of coefficients of the objective function
15.   %    of the primal LP problem (size n x 1)
16.   % -- b: vector of the right-hand side of the constraints
17.   %    of the primal LP problem (size m x 1)
18.   % -- Eqin: vector of the type of the constraints of the
19.   %    primal LP problem (size m x 1)
20.   % -- MinMaxLP: the type of optimization of the primal
21.   %    LP problem
22.   % -- VarConst: vector of the variables' constraints of the
23.   %    primal LP problem (0 - free variable, 1 - variable
24.   %    <= 0, 2 - variable >= 0) (size n x 1)
25.   %
26.   % Output:
27.   % -- DA: matrix of coefficients of the constraints of the
28.   %    dual LP problem (size n x m)
29.   % -- Dc: vector of coefficients of the objective function
30.   %    of the dual LP problem (size m x 1)
31.   % -- Db: vector of the right-hand side of the constraints
32.   %    of the dual LP problem (size n x 1)
33.   % -- DEqin: vector of the type of the constraints of the
34.   %    dual LP problem (size n x 1)
35.   % -- DMinMaxLP: the type of optimization of the dual
36.   %    LP problem
37.   % -- DVarConst: vector of the variables' constraints of the
38.   %    dual LP problem (0 - free variable, 1 - variable
39.   %    <= 0, 2 - variable >= 0) (size m x 1)
40.
41.   % if the primal is a minimization problem, then the
42.   % dual is a maximization one and vice versa
```

```
43.    DMinMaxLP = -MinMaxLP;
44.    % the coefficients of the objective function of the
45.    % dual LP problem are the right-hand side values of
46.    % the primal LP problem
47.    Dc = b;
48.    % the right-hand side values of the dual LP problem
49.    % are the coefficients of the objective function of the
50.    % primal LP problem
51.    Db = c;
52.    % the coefficient matrix of the dual LP problem is the
53.    % transpose of the coefficient matrix of the primal LP
54.    % problem
55.    DA = A';
56.    [m, n] = size(A); % size of matrix A
57.    % find the vector of the type of the constraints
58.    DEqin = zeros(n, 1);
59.    for i = 1:n
60.        % free variable -> equality constraint
61.        if VarConst(i) == 0
62.            DEqin(i) = 0;
63.        % variable <= 0 -> 'greater than or equal to'
64.        % constraint
65.        elseif VarConst(i) == 1
66.            DEqin(i) = 1;
67.        % variable >= 0 -> 'less than or equal to'
68.        % constraint
69.        elseif VarConst(i) == 2
70.            DEqin(i) = -1;
71.        end
72.    end
73.    % find the vector of variables' constraints
74.    DVarConst = zeros(m, 1);
75.    for j = 1:m
76.        % equality constraint -> free variable
77.        if Eqin(j) == 0
78.            DVarConst(j) = 0;
79.        % 'less than or equal to' constraint ->
80.        % variable <= 0
81.        elseif Eqin(j) == -1
82.            DVarConst(j) = 1;
83.        % 'greater than or equal to' constraint ->
84.        % variable >= 0
85.        elseif Eqin(j) == 1
86.            DVarConst(j) = 2;
87.        end
88.    end
89. end
```

As we already know, an LP problem can be infeasible, unbounded, or have a finite optimal solution or infinite optimal solutions. Suppose that x is a feasible solution of the primal and w is a feasible solution of the dual. Then, $Ax \leq b$, $w^T A \geq c^T$, $x \geq 0$, and $w \geq 0$. It follows that $w^T Ax \geq c^T x$ and $w^T Ax \leq w^T b$. Hence, $c^T x \leq b^T w$. This theorem is known as the weak duality theorem and we can conclude that if x

Table 2.5 Relationship between the primal and dual LPs

Primal \ Dual	Finite optimum	Unbounded	Infeasible
Finite optimum	Yes; values equal	Impossible	Impossible
Unbounded	Impossible	Impossible	Possible
Infeasible	Impossible	Possible	Possible

and w are feasible solutions of the primal and dual problems and $c^T x = b^T w$, then x and w must be optimal solutions to the primal and dual LP problems, respectively. However, this does not imply that there are feasible solutions x and w such that $c^T x = b^T w$.

The strong duality theorem guarantees that the dual has an optimal solution if and only if the primal does. If x and w are optimal solutions to the primal and dual LP problems, then $c^T x = b^T w$. As a result, the objective values of the primal and dual LP problems are the same. Moreover, the primal and dual LP problems are related. We can conclude to the following relationship between the primal and dual LP problems (Table 2.5):

- If the primal LP problem has a finite solution, then the dual LP problem has also a solution. The values of the objective function of the primal and the dual LPs are equal.
- If the primal LP problem is unbounded, then the dual LP problem is infeasible.
- If the primal LP problem is infeasible, then the dual LP problem may be either unbounded or infeasible.

Similarly:

- If the dual LP problem has a finite solution, then the primal LP problem has also a solution. The values of the objective function of the primal and the dual LPs are equal.
- If the dual LP problem is unbounded, then the primal LP problem is infeasible.
- If the dual LP problem is infeasible, then the primal LP problem may be either unbounded or infeasible.

The duality theorem implies a relationship between the primal and dual LPs that is known as complementary slackness. As you already know, the number of the variables in the dual LP problem is equal to the number of the constraints in the primal LP problem. Moreover, the number of the constraints in the dual LP problem is equal to the number of the variables in the primal LP problem. This relationship suggests that the variables in one problem are complementary to the constraints in the other. A constraint is binding if changing it also changes the optimal solution. A constraint has slack if it is not binding. An inequality constraint has slack if the slack variable is positive.

Assume that the primal LP problem has a solution x and the dual LP problem has a solution w. The complementary slackness theorem states that:

- If $x_j > 0$, then the jth constraint in the dual LP problem is binding.
- If the jth constraint in the dual LP problem is not binding, then $x_j = 0$.
- If $w_i > 0$, then the ith constraint in the primal LP problem is binding.
- If the ith constraint in the primal LP problem is not binding, then $w_i = 0$.

The complementary slackness theorem is useful, because it helps us to interpret the dual variables. If we already know the solution to the primal LP problem, then we can easily find the solution to the dual LP problem.

2.7   Linear Programming Algorithms

As already discussed in Chapter 1, there are two type of LP algorithms: (i) simplex-type algorithms, and (ii) interior point algorithms. The most widely used simplex-type algorithm is the revised simplex algorithm proposed by Dantzig [1]. The revised simplex algorithm begins with a feasible basis and uses pricing operations until an optimum solution is computed. The revised simplex algorithm moves from extreme point to extreme point on the boundary of the feasible region. In addition, the revised dual simplex algorithm, which was initially proposed by Lemke [2], is another alternative for solving LPs. The revised dual simplex algorithm begins with a dual feasible basis and uses pricing operations until an optimum solution is computed. Another efficient simplex-type algorithm is the exterior point simplex algorithm that was proposed by Paparrizos [5, 6]. The main idea of the exterior point simplex algorithm is that it moves in the exterior of the feasible region and constructs basic infeasible solutions instead of feasible solutions calculated by the simplex algorithm. This book describes in detail these three algorithms. Moreover, we also present another family of LP algorithms, the interior point algorithms. Interior point algorithms traverse across the interior of the feasible region.

We use four LP algorithms in the following chapters of this book:

- Revised primal simplex algorithm (presented in detail in Chapter 8).
- Revised dual simplex algorithm (presented in detail in Chapter 9).
- Exterior point simplex algorithm (presented in detail in Chapter 10).
- Mehrotra's predictor-corrector method (presented in detail in Chapter 11).

A brief description of these algorithms follows.

2.7.1   Revised Primal Simplex Algorithm

A formal description of the revised primal simplex algorithm is given in Table 2.6 and a flow diagram of its major steps in Figure 2.28. Initially, the LP problem is presolved. Presolve methods are important in solving LPs, as they reduce LPs' size and discover whether the problem is unbounded or infeasible. Presolve methods are

Table 2.6 Revised primal simplex algorithm

Step 0. *(Initialization).*

Presolve the LP problem.

Scale the LP problem.

Select an initial basic solution (B, N).

if the initial basic solution is feasible then proceed to step 2.

Step 1. *(Phase I).*

Construct an auxiliary problem by adding an artificial variable x_{n+1} with a

coefficient vector equal to $-A_B e$, where $e = (1, 1, \cdots, 1)^T \in \mathbb{R}^m$.

Apply the revised simplex algorithm. If the final basic solution (B, N) is feasible,

then proceed to step 2 in order to solve the initial problem. The auxiliary LP

problem can be either optimal or infeasible.

Step 2. *(Phase II).*

Step 2.0. *(Initialization).*

Compute $(A_B)^{-1}$ and vectors x_B, w, and s_N.

Step 2.1. *(Test of Optimality).*

if $s_N \geq 0$ then STOP. The LP problem is optimal.

else

    Choose the index l of the entering variable using a pivoting rule.

    Variable x_l enters the basis.

Step 2.2. *(Pivoting).*

Compute the pivot column $h_l = (A_B)^{-1} A_l$.

if $h_l \leq 0$ then STOP. The LP problem is unbounded.

else

Choose the leaving variable $x_{B[r]} = x_k$ using the following relation:

$$x_k = x_{B[r]} = \frac{x_{B[r]}}{h_{il}} = \min\left\{\frac{x_{B[i]}}{h_{il}} : h_{il} > 0\right\}$$

Step 2.3. *(Update).*

Swap indices k and l. Update the new basis inverse $(A_{\overline{B}})^{-1}$, using a basis

update scheme. Update vectors x_B, w, and s_N.

Go to Step 2.1.

---

used prior to the application of an LP algorithm in order to: (i) eliminate redundant constraints, (ii) fix variables, (iii) transform bounds of single structural variables, and (iv) reduce the number of variables and constraints by eliminations. A detailed analysis of presolve methods is presented in Chapter 4.

Next, the LP problem is scaled. Scaling is used prior to the application of an LP algorithm in order to: (i) produce a compact representation of the variable bounds, (ii) reduce the condition number of the constraint matrix, (iii) improve the numerical behavior of the algorithms, (iv) reduce the number of iterations required to solve LPs, and (v) simplify the setup of the tolerances. A detailed analysis of scaling techniques is presented in Chapter 5.

Then, we calculate an initial basis. If the basis is feasible then we proceed to apply the revised primal simplex algorithm to the original problem (Phase II). If the

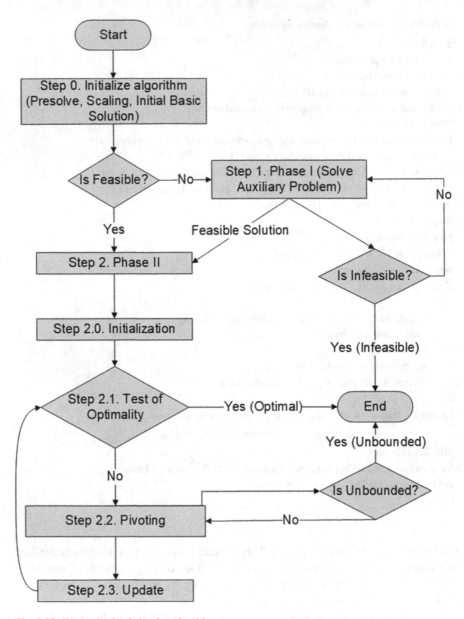

Fig. 2.28 Revised primal simplex algorithm

initial basis is not feasible, then we construct an auxiliary problem to find an initial basis (Phase I). In Phase I, it is possible to find out that the LP problem is infeasible. More details about this step will be given in Chapter 8.

Next, we perform the test of optimality to identify if the current basis is optimal and terminate the algorithm. If not, then we apply a pivoting rule to select the entering variable. The choice of the pivot element at each iteration is one of the most critical steps in simplex-type algorithms. A detailed analysis of pivoting rules is presented in Chapter 6.

Then, we perform the pivoting step where we select the leaving variable. At this step, it is possible to identify that the LP problem is unbounded. More details about this step will be given in Chapter 8.

Next, we update the appropriate variables. At this step, we need to update the basis inverse using a basis update scheme. A detailed analysis of basis update schemes is presented in Chapter 7. Finally, we continue with the next iteration of the algorithm.

Initially, we will present the different methods available for each step of the algorithm in Chapters 4–7 (presolve methods, scaling techniques, pivoting rules, and basis update schemes) and then present the whole algorithm in Chapter 8.

2.7.2 Revised Dual Simplex Algorithm

A formal description of the revised dual simplex algorithm is given in Table 2.7 and a flow diagram of its major steps in Figure 2.29. Initially, the LP problem is presolved. Next, the LP problem is scaled. Then, we calculate an initial basis. If the basis is dual feasible then we proceed to apply the revised dual simplex algorithm to the original problem. If the initial basis is not dual feasible, then we construct an auxiliary problem using the big-M method and solve this problem with the revised dual simplex algorithm. More details about this step will be given in Chapter 9.

At each iteration of the revised dual simplex algorithm, we perform the test of optimality to identify if the current basis is optimal and terminate the algorithm. If not, then we select the leaving variable. Then, we perform the pivoting step where we select the entering variable using the minimum ratio test. At this step, it is possible to identify that the LP problem is infeasible. More details about this step will be given in Chapter 9.

Next, we update the appropriate variables. At this step, we need to update the basis inverse using a basis update scheme. Finally, we continue with the next iteration of the algorithm.

Initially, we will present the different methods available for each step of the algorithm in Chapters 4, 5, and 7 (presolve methods, scaling techniques, and basis update schemes) and then present the whole algorithm in Chapter 9.

2.7.3 Exterior Point Simplex Algorithm

A formal description of the exterior point simplex algorithm is given in Table 2.8 and a flow diagram of its major steps in Figure 2.30. Similar to the revised simplex algorithm, the LP problem is presolved and scaled.

Table 2.7 Revised dual simplex algorithm

Step 0. *(Initialization).*

Presolve the LP problem.

Scale the LP problem.

Select an initial basic solution (B, N).

if the initial basic solution is dual feasible then proceed to step 2.

Step 1. *(Dual Algorithm with big-M Method).*

Add an artificial constraint $e^T x_N + x_{n+1} = M, x_{n+1} \geqq 0$. Construct the

vector of the coefficients of M in the right-hand side, $\bar{\bar{b}}$.

Set $s_p = \min \{s_j : j \in N\}, \bar{B} = B \cup p$, and $\bar{N} = N \cup \{n + 1\}$.

Find $\bar{x}_B, \bar{\bar{x}}_B, w,$ and s_N. Now, apply the algorithm of Step 2

to the modified big-M problem. The original LP problem can be either

optimal or infeasible or unbounded.

Step 2. *(Dual Simplex Algorithm).*

Step 2.0. *(Initialization).*

Compute $(A_B)^{-1}$ and vectors $x_B, w,$ and s_N.

Step 2.1. *(Test of Optimality).*

if $x_B \geq 0$ then STOP. The primal LP problem is optimal.

else

    Choose the leaving variable k such that $x_{B[r]} = x_k = \min \{x_{B[i]} : x_{B[i]} \leq 0\}$.

    Variable x_k leaves the basis.

Step 2.2. *(Pivoting).*

Compute vector $H_{rN} = (A_B^{-1})_r A_N$.

if $H_{rN} \geq 0$ then STOP. The primal LP problem is infeasible.

else

    Choose the entering variable $x_{N[t]} = x_l$ using the minimum ratio test:

$$x_l = x_{N[t]} = \frac{-s_{N[t]}}{H_{rN}} = \min \left\{ \frac{-s_{N[i]}}{H_{iN}} : H_{iN} < 0 \right\}$$

Step 2.3. *(Update).*

Swap indices k and l. Update the new basis inverse $(A_{\bar{B}})^{-1}$, using a basis

update scheme. Update vectors $x_B, w,$ and s_N.

Go to Step 2.1.

Then, we calculate an initial basis. If $P \neq \emptyset$ and the improving direction crosses the feasible region then we proceed to apply the exterior point simplex algorithm to the original problem (Phase II). Otherwise, we construct an auxiliary problem and apply the revised simplex algorithm to find an initial basis (Phase I). In Phase I, it is possible to find out that the LP problem is infeasible. More details about this step will be given in Chapter 10.

Next, we perform the test of optimality to identify if the current basis is optimal and terminate the algorithm. If not, then we choose the leaving variable. At this step, it is possible to identify that the LP problem is unbounded. Then, we perform the pivoting step to find the entering variable. More details about these steps will be given in Chapter 10.

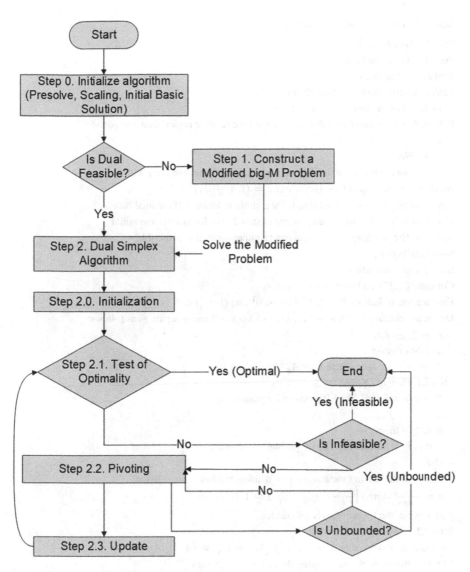

Fig. 2.29 Revised dual simplex algorithm

Next, we update the appropriate variables. At this step, we need to update the basis inverse using a basis update scheme. Finally, we continue with the next iteration of the algorithm.

Initially, we will present the different methods available for each step of the algorithm in Chapters 4, 5, and 7 (presolve methods, scaling techniques and basis update schemes) and then present the whole algorithm in Chapter 10.

Table 2.8 Exterior point simplex algorithm

Step 0. *(Initialization).*

Presolve the LP problem.

Scale the LP problem.

Select an initial basic solution (B, N).

Find the set of indices $P = \{j \in N : s_j < 0\}$.

If $P \neq \emptyset$ and the improving direction crosses the feasible region then proceed to step 2.

Step 1. *(Phase I).*

Construct an auxiliary problem by adding an artificial variable x_{n+1} with a coefficient vector equal to $-A_B e$, where $e = (1, 1, \cdots, 1)^T \in \mathbb{R}^m$.

Apply the steps of the revised simplex algorithm in Phase I. If the final basic solution (B, N) is feasible, then proceed to step 2 in order to solve the initial problem. The auxiliary LP problem can be either optimal or infeasible.

Step 2. *(Phase II).*

Step 2.0. *(Initialization).*

Compute $(A_B)^{-1}$ and vectors x_B, w, and s_N.

Find the sets of indices $P = \{j \in N : s_j < 0\}$ and $Q = \{j \in N : s_j \geq 0\}$.

Define an arbitrary vector $\lambda = (\lambda_1, \lambda_2, \cdots, \lambda_{|P|}) > 0$ and compute s_0 as follows:

$s_0 = \sum_{j \in P} \lambda_j s_j$

and the direction

$d_B = -\sum_{j \in P} \lambda_j h_j$, where $h_j = A_B^{-1} A_j$.

Step 2.1. *(Test of Optimality).*

if $P = \emptyset$ then STOP. The LP problem is optimal.

else

    if $d_B \geq 0$ then

        if $s_0 = 0$ then STOP. The LP problem is optimal.

    else

        choose the leaving variable $x_{B[r]} = x_k$ using the following relation:

    $a = \frac{x_{B[r]}}{-d_{B[r]}} = \min\left\{\frac{x_{B[i]}}{-d_{B[i]}} : d_{B[i]} < 0\right\}, i = 1, 2, \cdots, m$

    if $a = \infty$, the LP problem is unbounded.

Step 2.2. *(Pivoting).*

Compute the row vectors: $H_{rP} = \left(A_B^{-1}\right)_{r.} A_P$ and $H_{rQ} = \left(A_B^{-1}\right)_{r.} A_Q$.

Compute the ratios θ_1 and θ_2 using the following relations:

$\theta_1 = \frac{-s_P}{H_{rP}} = \min\left\{\frac{-s_j}{H_{rj}} : H_{rj} > 0 \wedge j \in P\right\}$ and

$\theta_2 = \frac{-s_Q}{H_{rQ}} = \min\left\{\frac{-s_j}{H_{rj}} : H_{rj} < 0 \wedge j \in Q\right\}$

Determine the indices t_1 and t_2 such that $P[t_1] = p$ and $Q[t_2] = q$.

if $\theta_1 \leq \theta_2$ then set $l = p$

else set $l = q$.

Step 2.3. *(Update).*

Swap indices k and l. Update the new basis inverse $\left(A_{\overline{B}}\right)^{-1}$, using a basis update scheme. Set $B(r) = l$. If $\theta_1 \leq \theta_2$, set $P = P \setminus \{l\}$ and $Q = Q \cup \{k\}$.

Otherwise, set $Q(t_2) = k$. Update vectors x_B, w, and s_N.

Update the new improving direction $\overline{d}_B$ using the relation

$d_{\overline{B}} = E^{-1} d_B$. If $l \in P$, set $d_{\overline{B}[r]} = d_{\overline{B}[r]} + \lambda_l$.

Go to Step 2.1.

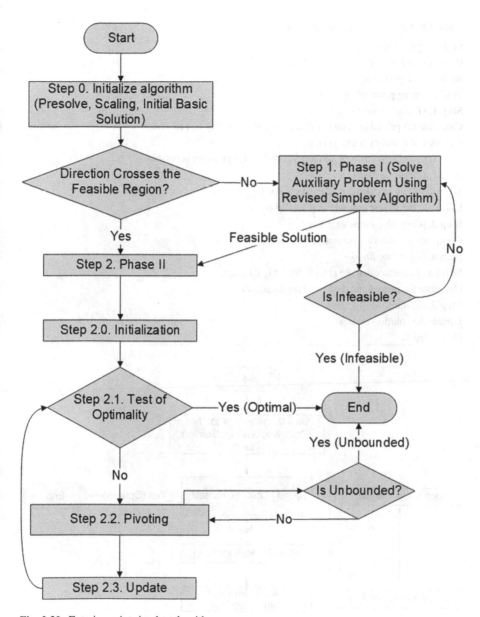

Fig. 2.30 Exterior point simplex algorithm

2.7.4 Mehrotra's Predictor-Corrector Method

A formal description of Mehrotra's Predictor-Corrector (MPC) method [4] is given in Table 2.9 and a flow diagram of its major steps in Figure 2.31. Similar to the revised simplex algorithm, the LP problem is presolved and scaled.

Table 2.9 Mehrotra's predictor-corrector method

Step 0. *(Initialization).*
Presolve the LP problem.
Scale the LP problem.
Find a starting point (x^0, w^0, s^0).
Step 1. *(Test of Optimality).*
Calculate the primal (r_p), dual (r_d), and complementarity (r_c) residuals.
Calculate the duality measure (μ).
if $\max(\mu, ||r_p||, ||r_d||) <= tol$ then STOP. The LP problem is optimal.
Step 2. *(Predictor Step).*
Solve a system to calculate $(\Delta x^p, \Delta w^p, \Delta s^p)$.
Calculate the largest possible step lengths α_p^p, α_d^p.
Step 3. *(Centering Parameter Step).*
Compute the centering parameter σ.
Step 4. *(Corrector Step).*
Solve a system to calculate (11.25) for $(\Delta x, \Delta w, \Delta s)$.
Calculate the primal and dual step lengths α_p, α_d.
Step 5. *(Update Step).*
Update the solution (x, w, s).
Go to Step 1.

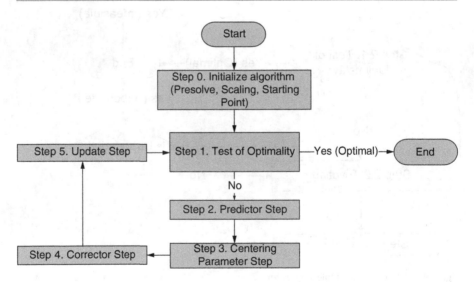

Fig. 2.31 Mehrotra's predictor-corrector method

Then, we find a starting point. Next, we perform the test of optimality to identify if the current solution is optimal and terminate the algorithm. In the predictor step, we solve a system to calculate $(\Delta x^p, \Delta w^p, \Delta s^p)$ and obtain the largest possible step lengths α_p^p and α_d^p. Next, we compute the centering parameter σ. In the corrector step, we solve a system to calculate $(\Delta x, \Delta w, \Delta s)$ and obtain the primal and dual step lengths α_p and α_d. Finally, we update the solution (x, w, s) and continue to apply these steps until the solution is optimal.

This algorithm will be presented in detail in Chapter 11, because it requires the introduction of some basic principles of the interior point methods.

2.8 Chapter Review

In this chapter, we presented the theoretical background of LP. More specifically, the different formulations of the LP problem were presented. Moreover, detailed steps on how to model an LP problem were given. In addition, the geometry of the feasible region and the duality principle were also covered. Finally, a brief description of LP algorithms that will be used in this book was also given.

References

1. Dantzig, G. B. (1953). *Computational algorithm of the revised simplex method*. RAND Report RM-1266, The RAND Corporation, Santa Monica, CA.
2. Lemke, C. E. (1954). The dual method of solving the linear programming problem. *Naval Research Logistics Quarterly, 1*(1), 36–47.
3. MathWorks. (2017). *MuPAD: user's guide*. Available online at: http://cn.mathworks.com/help/pdf_doc/symbolic/mupad_ug.pdf (Last access on March 31, 2017).
4. Mehrotra, S. (1992). On the implementation of a primal-dual interior point method. *SIAM Journal on Optimization, 2*, 575–601.
5. Paparrizos, K. (1991). An infeasible exterior point simplex algorithm for assignment problems. *Mathematical Programming, 51*(1–3), 45–54.
6. Paparrizos, K. (1993). An exterior point simplex algorithm for (general) linear programming problems. *Annals of Operations Research, 47*, 497–508.

Chapter 3
Linear Programming Benchmark and Random Problems

Abstract Mathematical Programming System (MPS) format is a widely accepted standard for defining LPs. The vast majority of solvers takes as input an LP problem in MPS format. The given LP problem can be either a benchmark problem, i.e., a problem that is publicly available, or a randomly generated LP problem. This chapter presents the MPS format and two codes in MATLAB that can be used to convert an MPS file to MATLAB's matrix format (MAT) and vice versa. Moreover, codes that can be used to create randomly generated sparse or dense LPs are also given. Finally, the most well-known benchmark libraries for LPs are also presented.

3.1 Chapter Objectives

- Present the Mathematical Programming System format.
- Implement a code that converts an MPS file to MATLAB's matrix format.
- Implement a code that converts MATLAB's matrix format to an MPS file.
- Implement a code to randomly generate sparse or dense LPs.
- Implement a code to randomly generate optimal sparse or dense LPs.
- Present the statistics of the most well-known benchmark libraries for LPs.

3.2 Introduction

There are a few file formats for defining LPs. In addition, each LP solver may take as input solver-specific file formats. However, the Mathematical Programming System (MPS) format is the most widely accepted standard for defining LPs and the vast majority of LP solvers takes as input an LP problem in MPS format. An MPS file consists of seven sections; two of them are optional. Moreover, MPS is a column-oriented format and it is not free format, i.e., each field starts in a specific column.

The LPs that are given as input to LP solvers are either benchmark problems or randomly generated problems. Benchmark problems are publicly available LPs that are hard to solve. Most of them are degenerate LPs. Many libraries containing LPs exist. The most widely used benchmark libraries are:

© Springer International Publishing AG 2017
N. Ploskas, N. Samaras, *Linear Programming Using MATLAB®*,
Springer Optimization and Its Applications 127, DOI 10.1007/978-3-319-65919-0_3

- The Netlib LP test set [1]
- The Kennington LP test set [2]
- The infeas section of the Netlib LP test set [3]
- The misc section of the Mészáros LP test set [4]
- The problematic section of the Mészáros LP test set [5]
- The new section of the Mészáros LP test set [6]
- The stochlp section of the Mészáros LP test set [7]
- The Mittelmann LP test set [8]

This chapter presents the MPS format and two codes in MATLAB that can be used to convert an MPS file to MATLAB's matrix format (MAT) and vice versa. Moreover, codes that can be used to create randomly generated sparse or dense LPs are also given. In addition, the most well-known benchmark libraries for LPs are also presented.

The structure of this chapter is as follows. In Section 3.3, the MPS format is presented. Section 3.4 presents two codes in MATLAB that can be used to convert an MPS file to MATLAB's matrix format (MAT) and vice versa. In Section 3.5, codes that can be used to create randomly generated sparse or dense LPs are also given. Section 3.6 provides the statistics of the aforementioned benchmark libraries. Finally, conclusions and further discussion are presented in Section 3.7.

3.3   Mathematical Programming System Format

3.3.1   Sections and Fields

The MPS format is a widely accepted standard for defining LPs. MPS format is a column-oriented format, i.e., problems are specified by column rather than by row. MPS is an old format and it is not free format. Fields start in columns 1, 5, 15, 25, 40, and 50. Sections are marked by headers and are distinguished by starting in the first column. It is typical to use uppercase letters throughout the MPS file, but most MPS readers also accept lowercase or mixed-case letters in all parts of an MPS file except the headers.

Headers separate the sections of the MPS file. Each header consists of a single word that begins in the first column. There are seven types of headers, each corresponding to a section of the MPS file:

- *NAME*: specifies the name of the problem. It is the only header that is followed by data, i.e., the name of the problem.
- *ROWS*: specifies the name and the type of each constraint.
- *COLUMNS*: specifies the name assigned to each variable along with the coefficients of the constraints and the objective function.
- *RHS*: specifies the names of the right-hand side vector and the values for each constraint.

- *RANGES*: specifies the constraints that are restricted to be in the interval of two specific values.
- *BOUNDS*: specifies the lower and/or upper bounds of each variable.
- *ENDDATA*: specifies the end of transmission of the MPS file.

Each section of the MPS file is mandatory, except the *RANGES* and *BOUNDS* sections. If no *BOUNDS* section is found in the MPS file, then all variables x_j have their bounds set to $l_j = 0$ and $u_j = \infty$. The following template is the skeleton of an MPS file:

```
---------------------------------------------------------------
Field:     1        2         3         4         5         6
Columns:   1-3      5-12      15-22     25-36     40-47     50-61

           NAME                problem name
           ROWS
              type   name
           COLUMNS
                     column    row       value     row       value
                     name      name                name
           RHS
                     rhs       row       value     row       value
                     name      name                name
           RANGES
                     range     row       value     row       value
                     name      name                name
           BOUNDS
              type   bound     column    value
                     name      name
           ENDATA
---------------------------------------------------------------
```

There is not any field in the MPS file than specifies the direction of optimization, i.e., minimization or maximization. Most LP solvers minimize the defined LPs.

Data records include information of an LP problem. Each data record consists of six fields:

- Field1: indicator
- Field2: name
- Field3: name
- Field4: value
- Field5: name
- Field6: value

The fields must be separated by white space, i.e., blank space, tab, etc., and the first field must start at the second column. Some sections are not using specific fields, as will be described later in this subsection. Any ASCII character (32 through 126) is allowed, but names must not contain any blank space. If a dollar sign is found in the first character of the third or the fifth field, then the remaining characters of this record are treated as a comment. If an asterisk is found in the first column of a line, then everything on that line is treated as a comment. Values can be defined with either decimal or exponential notation and can utilize 25 characters. If a value is defined in exponential notation, then a plus or minus sign must precede the value.

In the *ROWS* section, each row of the problem is specified with its name and type, one row per record. The first field contains a single letter that defines the type of each constraint:

- *N*: defines a free constraint (objective function)
- *G*: defines a 'greater than or equal to' constraint
- *L*: defines a 'less than or equal to' constraint
- *E*: defines an equality constraint

Only one row must be defined as a free row, indicating the objective function. The second field contains the name of the row. Fields 3–6 are not used in the *ROWS* section.

In the *COLUMNS* section, all the columns of the constraint matrix are defined with their name and only their nonzero elements are shown in the MPS file. Multiple records may be included to completely define a column of the constraint matrix. The first field is blank. The second field contains the column name, while the third field contains the row name. The value of the specific column and row in the constraint matrix is given in the fourth field. The fifth and sixth fields are optional; the fifth field contains a row name, while the value of the column, which was specified in the second field, and the row, which was specified in the fifth field, is given in the sixth field.

In the *RHS* section, the nonzero right-hand side values of the constraints are specified. The first field is blank. The second field contains the right-hand side name, while the third field contains the row name. The value of the specific right-hand side and row is specified by the fourth field. The fifth and sixth fields are optional; the fifth field contains a row name, while the value of the right-hand side, which was specified in the second field, and the row, which was specified in the fifth field, is given in the sixth field.

In the *RANGES* section, right-hand side values that are applied to constraints may be specified. The first field is blank. The second field contains the right-hand side range vector name, while the third field contains the row name. The value of the specific right-hand side range vector and row is given in the fourth field. The fifth and sixth fields are optional; the fifth field contains a row name, while the value of the right-hand side range vector, which was specified in the second field, and the row, which was specified in the fifth field, is given in the sixth field. The following cases can be distinguished depending on the type of the specified row and sign of the range value:

- Type of row *G* and sign of the range value + or −: the right-hand side lower limit is *rhs* and the upper limit is *rhs* + |*range*|.
- Type of row *L* and sign of the range value + or −: the right-hand side lower limit is *rhs* − |*range*| and the upper limit is *rhs*.

- Type of row E and sign of the range value $+$: the right-hand side lower limit is *rhs* and the upper limit is *rhs* + |*range*|.
- Type of row E and sign of the range value $-$: the right-hand side lower limit is *rhs* + |*range*| and the upper limit is *rhs*.

In the *BOUNDS* section, the bound values of the variables may be specified. The first field defines the type of the bound:

- *LO*: defines a lower bound
- *UP*: defines an upper bound
- *FX*: defines a fixed value, i.e., the lower and upper bounds are the same
- *FR*: defines a free variable, i.e., the lower bound of that variable is $-\infty$ and the upper bound of that variable is ∞
- *MI*: defines a minus infinity, i.e., the lower bound of that variable is $-\infty$
- *PL*: defines a plus infinity, i.e., the upper bound of that variable is ∞

The second field contains the bound name, while the third field contains the column name. The value of the specific bound is given in the fourth field. The fifth and sixth fields are not used in the *BOUNDS* section.

If no bounds are specified for some variables, then it is assumed that the lower bound of these variables is 0 and the upper bound of these variables is ∞ ($x_j \geq 0$).

3.3.2   Examples

In order to better understand the MPS format, two examples are presented in this subsection. The first one converts an LP problem from its mathematical notation to the MPS format, while the second converts an MPS file to an LP problem in mathematical notation.

1st example

The LP problem that will be converted to MPS format is the following:

$$
\begin{aligned}
\min z = {}& -2x_1 + 4x_2 - 2x_3 + 2x_4 \\
\text{s.t.} \quad & 2x_2 + 3x_3 && \geq && 6 \\
& 4x_1 - 3x_2 + 8x_3 - x_4 && = && 20 \\
& -3x_1 + 2x_2 - 4x_4 && \leq && -8 \\
& 4x_1 - x_3 + 4x_4 && = && 18 \\
& x_1 && \geq && 1 \\
& x_3 && \geq && 2 \\
& x_3 && \leq && 10 \\
& x_1 - free, x_3 - free, x_j \geq 0, \quad (j = 2, 4)
\end{aligned}
$$

The equivalent MPS file is shown below. In *ROWS* section, the objective function and the name and type of the constraints are defined. The *COLUMNS* section includes the values of the constraint matrix, while *RHS* section includes the values of the right-hand side vector.

```
NAME                    EXAMPLE1_1
ROWS
 N   OBJ
 G   R1
 E   R2
 L   R3
 E   R4
 G   R5
 G   R6
 L   R7
COLUMNS
     X1          R2          4
     X1          R3          -3
     X1          R4          4
     X1          R5          1
     X1          OBJ         -2
     X2          R1          2
     X2          R2          -3
     X2          R3          2
     X2          OBJ         4
     X3          R1          3
     X3          R2          8
     X3          R4          -1
     X3          R6          1
     X3          R7          1
     X3          OBJ         -2
     X4          R2          -1
     X4          R3          -4
     X4          R4          4
     X4          OBJ         2
RHS
     RHS1        R1          6
     RHS1        R2          20
     RHS1        R3          -8
     RHS1        R4          18
     RHS1        R5          1
     RHS1        R6          2
     RHS1        R7          10
BOUNDS
 FR  BND1        X1
```

```
 FR BND1        X3
ENDATA
```

Another way is to treat the three latter constraints as bounds:

```
NAME               EXAMPLE1_2
ROWS
 N   OBJ
 G   R1
 E   R2
 L   R3
 E   R4
COLUMNS
     X1        R2          4
     X1        R3         -3
     X1        R4          4
     X1        OBJ        -2
     X2        R1          2
     X2        R2         -3
     X2        R3          2
     X2        OBJ         4
     X3        R1          3
     X3        R2          8
     X3        R4         -1
     X3        OBJ        -2
     X4        R2         -1
     X4        R3         -4
     X4        R4          4
     X4        OBJ         2
RHS
     RHS1      R1          6
     RHS1      R2         20
     RHS1      R3         -8
     RHS1      R4         18
BOUNDS
 LO BND1       X1                    1
 LO BND1       X3                    2
 UP BND1       X3                   10
ENDATA
```

Moreover, we can also use the fifth and sixth field in *COLUMNS* and *RHS* sections to reduce the number of records:

```
NAME                 EXAMPLE1_3
ROWS
 N   OBJ
 G   R1
 E   R2
 L   R3
 E   R4
COLUMNS
     X1         R2          4          R3            -3
     X1         R4          4          OBJ           -2
     X2         R1          2          R2            -3
     X2         R3          2          OBJ            4
     X3         R1          3          R2             8
     X3         R4         -1          OBJ           -2
     X4         R2         -1          R3            -4
     X4         R4          4          OBJ            2
RHS
     B1         R1          6          R2            20
     B1         R3         -8          R4            18
BOUNDS
 LO  BND1       X1          1
 LO  BND1       X3          2
 UP  BND1       X3         10
ENDATA
```

2nd example

The MPS file that will be converted to its mathematical notation is the following:

```
NAME                 EXAMPLE2
ROWS
 N   OBJ
 L   R1
 G   R2
 E   R3
COLUMNS
     X1         R1          1          R2             1
     X1         OBJ         1
     X2         R1          1          R3            -1
     X2         OBJ         4
     X3         R2          1          R3             1
     X3         OBJ         9
RHS
     B1         R1          5          R2            10
```

```
      B1            R3            7
BOUNDS
  UP  BND1         X1            4
  LO  BND1         X2           -1
  UP  BND1         X2            1
ENDATA
```

The equivalent LP problem in its mathematical notation is the following:

$$
\begin{aligned}
\min z = x_1 &+ 4x_2 + 9x_3 \\
\text{s.t.} \quad x_1 &+ x_2 && \leq 5 \\
x_1 & && + x_3 \geq 10 \\
&- x_2 && + x_3 = 7 \\
x_1 & && \leq 4 \\
& x_2 && \leq 1 \\
& x_2 && \geq -1 \\
x_j \geq 0, \quad (j = 3)
\end{aligned}
$$

3.4 Convert MAT2MPS and MPS2MAT Format

3.4.1 Convert MAT2MPS Format

This subsection presents the implementation in MATLAB of a code to convert a MAT file (file format that MATLAB uses) to an MPS file (filename: `mat2mps.m`). One necessary notation should be introduced before the presentation of the implementation of the code that converts a MAT file to an MPS file. Let *filename* be the name of the MAT file (without the .mat extension) that will be converted.

The function that converts a MAT file to an MPS file takes as input the filename of the MAT file without the .mat extension (variable *filename*), and outputs a file named filename.mps that contains the equivalent LP problem in MPS format. The input MAT file should contain the following variables:

- A: the constraint matrix (size $m \times n$)
- b: the right-hand side vector (size $m \times 1$)
- c: the vector with the coefficients of the objective function (size $n \times 1$)
- *Eqin*: the vector with the type of the constraints (size $m \times 1$), where its elements have three possible values:

  - -1: 'less than or equal to' type of constraint
  - 0: equality constraint
  - 1: 'greater than or equal to' type of constraint

- *MinMaxLP*: the optimization direction where -1 defines minimization and 1 maximization. If no such variable exists, then the default direction is minimization (optional)
- *Name*: the name of the LP problem. If no such variable exists, then the name of the LP problem is the input filename (optional)
- *R*: the matrix with the constraint ranges (optional)
- *BS*: the matrix with the variable bounds (optional)
- *NonZeros*: the number of the nonzero elements of the constraint matrix (optional)
- $c0$ and $c00$: the objective function constant terms. If no such variables exist, then the objective constant is equal to 0. If both variables exist, then the objective function constant is equal to $c00-c0$. If only variable $c00$ exists, then the objective constant is equal to $c00$. Finally, if only variable $c0$ exists, the objective constant is equal to $-c0$ (optional)

In lines 43–56, we create an MPS file with the same name as the LP problem and write the MPS filename to it. Then, we write the type of optimization to the MPS file (lines 57–64). In lines 65–91, we create the *ROWS* section of the MPS file. Then, we create the *COLUMNS* section of the MPS file (lines 92–151). In lines 152–195, we create the *RHS* section of the MPS file. Then, we create the *RANGES* section (lines 196–239) and the *BOUNDS* section (lines 240–279). Finally, we write the *ENDATA* header and close the file (lines 280–282).

```
1.    function mat2mps(filename)
2.    % Filename: mat2mps.m
3.    % Description: Converts a MAT file to an MPS file
4.    % Authors: Ploskas, N., & Samaras, N., Triantafyllidis, Ch.
5.    %
6.    % Syntax: mat2mps(filename)
7.    %
8.    % Input:
9.    % -- filename: the name of the MAT file (without the .mat
10.   %    extension)
11.   %
12.   % Output: an MPS file named filename.mps
13.
14.   % copy the specified file to matlab.mat
15.   copyfile([filename '.mat'], 'matlab.mat');
16.   load; % load file
17.   A = full(A); % make A full
18.   b = full(b); % make b full
19.   c = full(c); % make c full
20.   Eqin = full(Eqin); % make Eqin full
21.   C = c; % copy c to C
22.   B = b; % copy b to B
23.   % if MinMaxLP does not exist, the LP problem is minimization
24.   if exist('MinMaxLP', 'var') == 0
25.       MinMaxLP = -1;
26.   end
27.   % if the Name does not exist, use the filename
```

```
28.  if exist('Name', 'var') == 0
29.      Name = filename;
30.  end
31.  % if R does not exist, set to empty
32.  if exist('R', 'var') == 0
33.      R = [];
34.  else
35.      R = full(R);
36.  end
37.  % if BS does not exist, set to empty
38.  if exist('BS','var') == 0
39.      BS = [];
40.  else
41.      BS = full(BS);
42.  end
43.  % create an mps file with the same name of the LP and write
44.  % MPS filename to it
45.  y = length(Name);
46.  Name2 = Name;
47.  Name(1, y + 1) = '.';
48.  Name(1, y + 2) = 'm';
49.  Name(1, y + 3) = 'p';
50.  Name(1, y + 4) = 's';
51.  fid = fopen(Name, 'wt');
52.  id = ['NAME          ',Name2];
53.  fprintf(fid, '%s', id);
54.  for i = 1:50 + (8 - y)
55.      nullname(i) = ' ';
56.  end
57.  % write the type of optimization to the MPS file
58.  if MinMaxLP == -1 % minimization
59.      fprintf(fid, '%s', nullname);
60.      fprintf(fid, '%s', '( MIN)');
61.  elseif MinMaxLP == 1 % maximization
62.      fprintf(fid, '%s', nullname);
63.      fprintf(fid, '%s', '( MAX)');
64.  end
65.  % create ROWS section
66.  fprintf(fid, '\n');
67.  fprintf(fid, '%s', 'ROWS'); % print the ROWS header
68.  fprintf(fid, '\n');
69.  obj = [' N  OBJ'];
70.  fprintf(fid, '%s', obj); % print the objective function
71.  fprintf(fid, '\n');
72.  a = length(Eqin);
73.  Eqin2 = {};
74.  for i = 1:a
75.      if Eqin(i) == -1 % less than or equal to constraint
76.          Eqin2{i} = 'L';
77.      elseif Eqin(i) == 0 % equality constraint
78.          Eqin2{i} = 'E';
```

```
79.      elseif Eqin(i) == 1 % greater than or equal to constraint
80.          Eqin2{i} = 'G';
81.      end
82.  end
83.  Eqin2 = cell2mat(Eqin2); % convert cell to matrix
84.  for i = 1:a % print constraints
85.      fprintf(fid, '%s', ' ');
86.      fprintf(fid, '%s', Eqin2(i));
87.      fprintf(fid, '%s', ' ');
88.      fprintf(fid, '%s', 'R');
89.      fprintf(fid, '%i', i);
90.      fprintf(fid, '\n');
91.  end
92.  % create COLUMNS section
93.  fprintf(fid, '%s', 'COLUMNS'); % print COLUMNS header
94.  [a, b] = size(A);
95.  fprintf(fid, '\n');
96.  for i = 1:b % column-oriented
97.      if i > 1
98.          if C(i - 1) ~= 0 % new line after each record
99.              fprintf(fid, '\n');
100.         end
101.     end
102.     j = 1;
103.     while j <= a
104.         if A(j, i) ~= 0 % print the constraint matrix A
105.             fprintf(fid, '%s', '    X');
106.             fprintf(fid, '%i', i);
107.             szcount = num2str(i);
108.             y = length(szcount);
109.             k = 5 - y;
110.             for m = 1:k
111.                 fprintf(fid, '%s', ' ');
112.             end
113.             fprintf(fid, '%s', '    ');
114.             fprintf(fid, '%s', 'R');
115.             fprintf(fid, '%i', j);
116.             szcount = num2str(j);
117.             y = length(szcount);
118.             k = 5 - y;
119.             for m = 1:k
120.                 fprintf(fid, '%s', ' ');
121.             end
122.             fprintf(fid, '%s', '    ');
123.             fprintf(fid, '%f', A(j, i));
124.             fprintf(fid, '\n');
125.         end
126.         if j < a
127.             j = j + 1;
128.         else
129.             if C(i) ~= 0 % print the objective function
```

```
130.                    fprintf(fid, '%s', '     X');
131.                    fprintf(fid, '%i', i);
132.                    szcount = num2str(i);
133.                    y = length(szcount);
134.                    k = 5 - y;
135.                    for m = 1:k
136.                        fprintf(fid, '%s', ' ');
137.                    end
138.                    fprintf(fid, '%s', '    ');
139.                    fprintf(fid, '%s', 'OBJ');
140.                    fprintf(fid, '%s', '        ');
141.                    fprintf(fid, '%f', C(i));
142.                    if i == b
143.                        if j == a
144.                            fprintf(fid, '\n');
145.                        end
146.                    end
147.                end
148.                break
149.            end
150.        end
151. end
152. % create RHS section
153. fprintf(fid, '%s', 'RHS'); % print the RHS header
154. fprintf(fid, '\n');
155. a = length(B);
156. for i = 1:a % print each right-hand side
157.     fprintf(fid, '%s', '    RHS1');
158.     fprintf(fid, '%s', '       ');
159.     fprintf(fid, '%s', 'R');
160.     fprintf(fid, '%i', i);
161.     szcount = num2str(i);
162.     y = length(szcount);
163.     k = 5 - y;
164.     for m = 1:k
165.         fprintf(fid, '%s', ' ');
166.     end
167.     fprintf(fid, '%s', '    ');
168.     fprintf(fid, '%f', B(i));
169.     fprintf(fid, '\n');
170. end
171. % write the objective constants if exist
172. objconst = [];
173. if exist('c00', 'var') == 1
174.     if exist('c0', 'var') == 1
175.         objconst = c00 - c0;
176.     else
177.         objconst = c00;
178.     end
179. else
180.     if exist('c0', 'var') == 1
```

```
181.          objconst = -c0;
182.     else
183.          objconst = [];
184.     end
185. end
186. if ~isempty(objconst)
187.     if objconst ~= 0
188.          fprintf(fid, '%s', '     RHS1');
189.          fprintf(fid, '%s', '        ');
190.          fprintf(fid, '%s', 'OBJ');
191.          fprintf(fid, '%s', '          ');
192.          fprintf(fid, '%f', objconst);
193.          fprintf(fid, '\n');
194.     end
195. end
196. % create RANGES section
197. if isempty(R) ~= 1
198.     fprintf(fid, '%s', 'RANGES'); % print the RANGES header
199.     fprintf(fid, '\n');
200.     [a, ~] = size(R);
201.     for i = 1:a % print all ranges
202.          fprintf(fid, '%s', '     RANGE1');
203.          fprintf(fid, '%s', '       ');
204.          fprintf(fid, '%s', 'R');
205.          fprintf(fid, '%i', R(i, 1));
206.          szcount = num2str(R(i, 1));
207.          y = length(szcount);
208.          k = 5 - y;
209.          for m = 1:k
210.              fprintf(fid, '%s', ' ');
211.          end
212.          fprintf(fid, '%s', '  ');
213.          if Eqin(R(i, 1)) == -1
214.              if R(i, 4) == 1
215.                  fprintf(fid, '%f', -(R(i, 2) - B(R(i, 1))));
216.                  fprintf(fid, '\n');
217.              else
218.                  fprintf(fid, '%f', R(i, 2) - B(R(i, 1)));
219.                  fprintf(fid, '\n');
220.              end
221.          elseif Eqin(R(i, 1)) == 0
222.              if R(i, 4) == 1
223.                  fprintf(fid, '%f', R(i, 3) - B(R(i, 1)));
224.                  fprintf(fid, '\n');
225.              else
226.                  fprintf(fid, '%f', R(i, 2) - B(R(i, 1)));
227.                  fprintf(fid, '\n');
228.              end
229.          elseif Eqin(R(i, 1)) == 1
230.              if R(i, 4) == 1
231.                  fprintf(fid, '%f', R(i, 3) - R(i, 2));
```

```
232.                    fprintf(fid, '\n');
233.              else
234.                    fprintf(fid, '%f', -(R(i, 3) - R(i, 2)));
235.                    fprintf(fid, '\n');
236.              end
237.           end
238.        end
239. end
240. % create BOUNDS section
241. if isempty(BS) ~= 1
242.      fprintf(fid, '%s', 'BOUNDS'); % print the BOUNDS header
243.      fprintf(fid, '\n');
244.      a = length(BS);
245.      for i = 1:a % print each bound
246.          if BS(i, 2) == -1
247.                fprintf(fid, '%s', ' UP');
248.          elseif BS(i, 2) == 1
249.                fprintf(fid, '%s', ' LO');
250.          elseif BS(i, 2) == 0
251.                fprintf(fid, '%s', ' FX');
252.          elseif BS(i, 2) == 11
253.                fprintf(fid, '%s', ' FR');
254.          elseif BS(i, 2) == 111
255.                fprintf(fid, '%s', ' MI');
256.          elseif BS(i, 2) == 1111
257.                fprintf(fid, '%s', ' PL');
258.          end
259.          fprintf(fid, '%s', ' ');
260.          fprintf(fid, '%s', 'BOUND1    ');
261.          fprintf(fid, '%s', 'X');
262.          fprintf(fid, '%i', BS(i, 1));
263.          % write the value field for LO, UP, and FX type of
264.          % constraints
265.          if BS(i, 2) == -1 || BS(i, 2) == 1 || BS(i, 2) == 0
266.                szcount = num2str(i);
267.                y = length(szcount);
268.                k = 7 - y;
269.                for m = 1:k
270.                      fprintf(fid, '%s', ' ');
271.                end
272.                fprintf(fid, '%s', ' ');
273.                fprintf(fid, '%f', BS(i,3));
274.                fprintf(fid, '\n');
275.          else
276.                fprintf(fid, '\n');
277.          end
278.      end
279. end
280. % write the ENDATA header
281. fprintf(fid, '%s', 'ENDATA');
282. fclose(fid); % close the file
```

```
283. delete matlab.mat % delete matlab.mat
284. end
```

3.4.2   Convert MPS2MAT Format

This subsection presents the implementation in MATLAB of a code to convert an MPS file to a MAT file (filename: `mps2mat.m`). One necessary notation should be introduced before the presentation of the implementation of the code that converts an MPS file to a MAT file. Let *filename* be the name of the MPS file (without the .mps extension) that will be converted.

The function that converts an MPS file to a MAT file takes as input the filename of the MPS file without the .mps extension (variable *filename*), and outputs a file named filename.mat that contains the equivalent LP problem in MATLAB's matrix format. The MPS file should contain at least the *NAME*, *ROWS*, *COLUMNS*, and *RHS* sections and may also contain the *RANGES* and *BOUNDS* sections. The output MAT file contains the following variables:

- A: the constraint matrix (size $m \times n$)
- b: the right-hand side vector (size $m \times 1$)
- c: the vector with the coefficients of the objective function (size $n \times 1$)
- *Eqin*: the vector with the type of the constraints (size $m \times 1$), where its elements have three possible values:

  - -1: 'less than or equal to' type of constraint
  - 0: equality constraint
  - 1: 'greater than or equal to' type of constraint

- *MinMaxLP*: the optimization direction where -1 defines minimization and 1 maximization
- *Name*: the name of the LP problem
- R: the matrix with the constraint ranges
- *BS*: the matrix with the variable bounds
- *NonZeros*: the number of the nonzero elements of the constraint matrix
- $c0$ and $c00$: the objective constant terms

In lines 29–54, we read the *NAME* section from the MPS file and we create variables *Name* and *MinMaxLP*. Then, we read the *ROWS* section from the MPS file and populate the *Eqin* vector (lines 60–89). In lines 90–374, we read the *COLUMNS* section from the MPS file and populate vector c and matrix A. Then, we read the *RHS* section from the MPS file and populate vector b (lines 387–429). In lines 430–530, we read the *RANGES* section from the MPS file and populate matrix R. Then, we read the *BOUNDS* section and populate matrix *BS* (lines 531–579). Finally, we make the vectors and matrices sparse and write the variables to the corresponding MAT file (lines 580–595).

```
1.    function mps2mat(filename)
2.    % Filename: mps2mat.m
3.    % Description: Converts an MPS file to a MAT file
4.    % Authors: Ploskas, N., & Samaras, N., Triantafyllidis, Ch.
5.    %
6.    % Syntax: mps2mat(filename)
7.    %
8.    % Input:
9.    % -- filename: the name of the MPS file (without the .mps
10.   %    extension)
11.   %
12.   % Output: a MAT file named filename.mat
13.
14.   % set the search starting point to 1
15.   searchStartingPoint = 1;
16.   % create an empty vector for the objective constant
17.   c0 = [];
18.   % add the extension to the mps file
19.   mpsFilename = [filename '.mps'];
20.   % find a rough estimation for the number of variables and
21.   % the nonzero elements
22.   s = dir(mpsFilename);
23.   filesize = s.bytes;
24.   estimation = floor(filesize / 220);
25.   nnzmax = floor((filesize / 30));
26.   fid = fopen(mpsFilename, 'rt'); % open the file
27.   Eqin = []; % create an empty vector for Eqin
28.   MinMaxLP = []; % create an empty vector for MinMaxLP
29.   % find the name of the problem and the optimization
30.   % direction
31.   line = fgets(fid);
32.   % read the NAME header
33.   useless = sscanf(line(1, 1:4), '%s%', 1);
34.   % if the optimization direction is included
35.   if length(line) > 23
36.       % read the Name
37.       Name = sscanf(line(1, 15:23), '%s%', 1);
38.       % read the optimization direction
39.       MinMaxLP = sscanf(line(1, 75:length(line)), '%s%', 1);
40.   else % if the optimization direction is not included
41.       % read the name
42.       Name = sscanf(line(1, 15:length(line)), '%s%', 1);
43.       MinMaxLP = []; % optimization direction does not exist
44.   end
45.   % if the optimization direction is not empty
46.   if isempty(MinMaxLP) ~= 1
47.       if isequal(MinMaxLP, 'MIN') == 1 % minimization
48.           MinMaxLP = -1;
49.       elseif isequal(MinMaxLP, 'MAX') == 1 % maximization
50.           MinMaxLP = 1;
51.       end
52.   else % if it is empty, the default is minimization
53.       MinMaxLP = -1;
54.   end
```

```
55.  fgetl(fid);
56.  constraintType = '';
57.  objPos = 1;
58.  k = 1;
59.  lineE = 1;
60.  % read the ROWS section and populate vector Eqin
61.  while isequal(constraintType, 'COLUMNS') == 0
62.      % get the name and type of the constraints
63.      constraintType = fscanf(fid, '%s%', 1);
64.      if isequal(constraintType, 'COLUMNS') ~= 1
65.          constraintName = fscanf(fid, '%s%', 1);
66.      else % break when you read the COLUMNS header
67.          break
68.      end
69.      X{k, 1} = constraintType;
70.      X{k, 2} = constraintName;
71.      if isequal(X{k, 1}, 'N') == 1 % objective function
72.          objPos = k;
73.          ant = X{k, 2};
74.      % equality constraint
75.      elseif isequal(constraintType, 'E' ) == 1
76.          Eqin(lineE) = 0;
77.          lineE = lineE + 1;
78.      % less than or equal to constraint
79.      elseif isequal(constraintType, 'L') == 1
80.          Eqin(lineE) = -1;
81.          lineE = lineE + 1;
82.      % greater than or equal to constraint
83.      elseif isequal(constraintType, 'G') == 1
84.          Eqin(lineE) = 1;
85.          lineE = lineE + 1;
86.      end
87.      k = k + 1;
88.  end
89.  Eqin = Eqin'; % transpose Eqin
90.  % read COLUMNS section and populate vector c and matrix A
91.  X(objPos, :) = [];
92.  [m, ~] = size(X);
93.  A = spalloc(m, estimation, nnzmax); % preallocate matrix A
94.  B = zeros(m, 1); %create vector B
95.  C = {}; % create cell C
96.  R = []; % create vector R
97.  BS = {}; % create vector BS
98.  clmn = 1;
99.  variables = {};
100. objectivity = 0;
101. fgets(fid);
102. D = X(:, 2);
103. flag = 0;
104. [a, ~] = size(D);
105. D{a + 1, :} = ant;
106. D = char(D);
107. mas = length(D);
108. objInd = a + 1;
```

```
109. % read the first line in COLUMNS header
110. line = fgets(fid);
111. b = length(line);
112. % read the variable name, the constraint name and the value
113. varname = sscanf(line(1, 5:12), '%s%', 1);
114. varpos = sscanf(line(1, 15:22), '%s%', 1);
115. value = sscanf(line(1, 25:b), '%s%', 1);
116. value = str2double(value);
117. if b > 40 % read the fifth and the sixth fields if exist
118.     flag = 1;
119.     varpos2 = sscanf(line(1, 40:b), '%s%', 1);
120.     value2 = str2double(line(1, 50:b));
121. end
122. % check if during the reading we changed the current variable
123. if objectivity == 0
124.     C{clmn} = 0;
125. else
126.     objectivity = 0;
127. end
128. currentvar = varname;
129. clmn = clmn + 1;
130. variables{clmn, 1} = currentvar;
131. k = strmatch(varpos, D, 'exact');
132. % store the value to vector C if the variable refers to the
133. % objective function
134. if k == objInd
135.     C{clmn} = value;
136.     objectivity = 1;
137. else
138.     A(k, clmn) = value; % store to matrix A
139. end
140. if b > 40 % if exists the fifth and sixth fields
141.     k2 = strmatch(varpos2, D, 'exact');
142.     % store the value to vector C if the variable refers to
143.     % the objective function
144.     if k2 == objInd
145.         C{clmn} = value2;
146.         objectivity = 1;
147.     else
148.         A(k2, clmn) = value2; % store to matrix A
149.     end
150. end
151. % read the second line in COLUMNS header
152. varname = '';
153. varpos = '';
154. value = '';
155. varpos2 = '';
156. value2 = '';
157. line = fgets(fid);
158. b = length(line);
159. % if we reached the RHS section
160. if isequal(line(1, 1:3), 'RHS')
161.     if objectivity == 0
162.         C{clmn} = 0;
```

```
163.      end
164. end
165. % read the variable name, the constraint name and the value
166. varname = sscanf(line(1, 5:12), '%s%', 1);
167. varpos = sscanf(line(1, 15:22), '%s%', 1);
168. value = sscanf(line(1, 25:b), '%s%', 1);
169. value = str2double(value);
170. if b > 40 % read the fifth and the sixth field if exist
171.      varpos2 = sscanf(line(1, 40:48), '%s%', 1);
172.      value2 = str2double(line(1, 50:b));
173.      flag = 1;
174. end
175. % if the variable changes, then we must reset the starting
176. % point of the search
177. if isequal(varname, currentvar) == 0
178.      searchStartingPoint = 1;
179.      % check if during the reading we changed the current
180.      % variable
181.      if objectivity == 0
182.          C{clmn} = 0;
183.      else
184.          objectivity = 0;
185.      end
186.      currentvar = varname;
187.      clmn = clmn + 1;
188.      variables{clmn, 1} = currentvar;
189. end
190. k = strmatch(varpos, D(searchStartingPoint:mas, :), ...
191.      'exact');
192. if searchStartingPoint ~= 1
193.      k = k + searchStartingPoint - 1;
194. end
195. if isempty(k)
196.      k = strmatch(varpos, D(1:searchStartingPoint - 1, ...
197.          :),'exact');
198. end
199. searchStartingPoint = k + 1;
200. % store the value to vector C if the variable refers to
201. % the objective function
202. if k == objInd
203.      C{clmn} = value;
204.      objectivity = 1;
205. else
206.      % store to matrix A
207.      A(k, clmn) = value;
208. end
209. if b > 40 % if exists the fifth and sixth fields
210.      k2 = strmatch(varpos2, D(searchStartingPoint:mas, ...
211.          :), 'exact');
212.      if searchStartingPoint ~= 1
213.          k2 = k2 + searchStartingPoint - 1;
214.      end
215.      if isempty(k2)
216.          k2 = strmatch(varpos2, D(1:searchStartingPoint ...
```

```
217.                    - 1, :), ...
218.                    'exact');
219.        end
220.        searchStartingPoint = k2 + 1;
221.        % store the value to vector C if the variable refers to
222.        % the objective function
223.        if k2 == objInd
224.            C{clmn} = value2;
225.            objectivity = 1;
226.        else
227.            A(k2, clmn) = value2; % store to matrix A
228.        end
229. end
230. % read the rest of the records of COLUMNS section
231. % set the first found index to use later for faster search
232. searchStartingPoint = k + 1;
233. flag = 1;
234. if flag == 1
235.        % stop when you reach the RHS section
236.        while isequal(varname, 'RHS') == 0
237.            varname = '';
238.            varpos = '';
239.            value = '';
240.            varpos2 = '';
241.            value2 = '';
242.            line = fgets(fid);
243.            b = length(line);
244.            % if we reach the RHS section, then break the loop
245.            if isequal(line(1, 1:3),'RHS')
246.                if objectivity == 0
247.                    C{clmn} = 0;
248.                end
249.                break
250.            end
251.            % read the variable name, the constraint name and
252.            % the value
253.            varname = sscanf(line(1, 5:12), '%s%', 1);
254.            varpos = sscanf(line(1, 15:22), '%s%', 1);
255.            value = sscanf(line(1, 25:b), '%s%', 1);
256.            value = str2double(value);
257.            % read the fifth and the sixth fields if exist
258.            if b > 40
259.                varpos2 = sscanf(line(1, 40:b), '%s%', 1);
260.                value2 = str2double(line(1, 50:b));
261.            end
262.            if isequal(varname, currentvar) == 0
263.                % if the variable changes, then we must reset
264.                % the starting point of the search
265.                searchStartingPoint = 1;
266.                % check if during the reading we changed the
267.                % current variable
268.                if objectivity == 0
269.                    C{clmn} = 0;
270.                else
```

```
271.                     objectivity = 0;
272.                 end
273.                 currentvar = varname;
274.                 clmn = clmn + 1;
275.                 variables{clmn, 1} = currentvar;
276.             end
277.             k = strmatch(varpos, D(searchStartingPoint:mas, ...
278.                 :), 'exact');
279.             if searchStartingPoint ~= 1
280.                 k = k + searchStartingPoint - 1;
281.             end
282.             if isempty(k)
283.                 k = strmatch(varpos, D(1:searchStartingPoint ...
284.                     - 1, :), 'exact');
285.             end
286.             searchStartingPoint = k + 1;
287.             % store the value to vector C if the variable refers
288.             % to the objective function
289.             if k == objInd
290.                 C{clmn} = value;
291.                 objectivity = 1;
292.             else
293.                 % store to matrix A
294.                 A(k, clmn) = value;
295.             end
296.             if b > 40 % if exists the fifth and sixth fields
297.                 k2 = strmatch(varpos2, ...
298.                     D(searchStartingPoint:mas, :), 'exact');
299.                 if searchStartingPoint ~= 1
300.                     k2 = k2 + searchStartingPoint - 1;
301.                 end
302.                 if isempty(k2)
303.                     k2 = strmatch(varpos2, ...
304.                         D(1:searchStartingPoint - 1, :), ...
305.                         'exact');
306.                 end
307.                 searchStartingPoint = k2 + 1;
308.                 % store the value to vector C if the variable
309.                 % refers to the objective function
310.                 if k2 == objInd
311.                     C{clmn} = value2;
312.                     objectivity = 1;
313.                 else
314.                     A(k2, clmn) = value2; % store to matrix A
315.                 end
316.             end
317.         end
318. end
319. % process the final record of COLUMNS section
320. if flag == 0
321.     % stop when the RHS section is reached
322.     while isequal(varname, 'RHS') == 0
323.         varname = '';
324.         varpos = '';
```

```
325.            value = '';
326.            line = fgets(fid);
327.            b = length(line);
328.            if isequal(line(1, 1:3), 'RHS')
329.                if objectivity == 0
330.                    C{clmn} = 0;
331.                end
332.                break
333.            end
334.            % read the variable name, the constraint name and
335.            % the value
336.            varname = sscanf(line(1, 5:12), '%s%', 1);
337.            varpos = sscanf(line(1, 15:22), '%s%', 1);
338.            value = sscanf(line(1, 25:b), '%s%', 1);
339.            value = str2double(value);
340.            if isequal(varname, currentvar) == 0
341.                % if the variable changes, then we must reset
342.                % the starting point of the search
343.                searchStartingPoint = 1;
344.                % check if during the reading we changed the
345.                % current variable
346.                if objectivity == 0
347.                    C{clmn} = 0;
348.                else
349.                    objectivity = 0;
350.                end
351.                currentvar = varname;
352.                clmn = clmn + 1;
353.                variables{clmn, 1} = currentvar;
354.            end
355.            k = strmatch(varpos, ...
356.                D(searchStartingPoint:mas, :), 'exact');
357.            if searchStartingPoint ~= 1
358.                k = k + searchStartingPoint - 1;
359.            end
360.            if isempty(k)
361.                k = strmatch(varpos, ...
362.                    D(1:searchStartingPoint - 1, :), 'exact');
363.            end
364.            searchStartingPoint = k + 1;
365.            % store the value to vector C if the variable
366.            % refers to the objective function
367.            if k == objInd
368.                C{clmn} = value;
369.                objectivity = 1;
370.            else
371.                A(k, clmn) = value; % store to matrix A
372.            end
373.        end
374. end
375. A(:, 1) = [];
376. variables(1, :) = [];
377. [x, ~] = size(variables);
378. % checking for any mistakes made during the preallocation
```

```
379.  % of matrix A
380.  if estimation > x
381.      A(:, x + 1:(estimation - 1)) = [];
382.  end
383.  C(:, 1) = [];
384.  [fr, ~] = size(A);
385.  rflag = 0;
386.  bflag = 0;
387.  % read the RHS section and populate vector b
388.  % stop if the end of the MPS file reached
389.  while isequal(varname, 'ENDATA') == 0
390.      varname = '';
391.      varpos = '';
392.      value = '';
393.      varpos2 = '';
394.      value2 = '';
395.      line = fgets(fid);
396.      b = length(line);
397.      % stop if the end of the MPS file reached
398.      if isequal(line(1, 1:6), 'ENDATA')
399.          rflag = 0;
400.          bflag = 0;
401.          break
402.          % stop if we reached the RANGES section
403.      elseif isequal(line(1, 1:6), 'RANGES')
404.          rflag = 1;
405.          break
406.          % stop if we reached the BOUNDS section
407.      elseif isequal(line(1, 1:6), 'BOUNDS')
408.          bflag = 1;
409.          break
410.      end
411.      % read the right-hand side name, the constraint name
412.      % and the value
413.      varname = sscanf(line(1, 5:12), '%s%', 1);
414.      varpos = sscanf(line(1, 15:22), '%s%', 1);
415.      value = sscanf(line(1, 25:b), '%s%', 1);
416.      value = str2double(value);
417.      if b > 40 % if exists the fifth and sixth fields
418.          varpos2 = sscanf(line(1, 40:b), '%s%', 1);
419.          value2 = str2double(line(1, 50:b));
420.          k2 = strmatch(varpos2, D, 'exact');
421.          B(k2) = value2; % store the value to vector B
422.      end
423.      k = strmatch(varpos, D, 'exact');
424.      B(k) = value; % store the value to vector B
425.  end
426.  frb = length(B);
427.  if frb > fr % check if the objective has a constant
428.      B(frb) = [];
429.  end
430.  % read the RANGES section and populate matrix R
431.  if rflag == 1
432.      range_ind = 1;
```

```
433.        % stop if the end of the MPS file reached
434.        while isequal(varname, 'ENDATA') == 0
435.            varname = '';
436.            varpos = '';
437.            value = '';
438.            varpos2 = '';
439.            value2 = '';
440.            line = fgets(fid);
441.            b = length(line);
442.            % stop if the end of the MPS file reached
443.            if isequal(line(1, 1:6), 'ENDATA')
444.                bflag = 0;
445.                break
446.                % stop if we reached the BOUNDS section
447.            elseif isequal(line(1, 1:6), 'BOUNDS')
448.                bflag = 1;
449.                break
450.            end
451.            % read the range name, the constraint name and
452.            % the value
453.            varname = sscanf(line(1, 5:12), '%s%', 1);
454.            varpos = sscanf(line(1, 15:22), '%s%', 1);
455.            value = sscanf(line(1, 25:b), '%s%', 1);
456.            value = str2double(value);
457.            if b > 40 % if exists the fifth and sixth fields
458.                varpos2 = sscanf(line(1, 40:b), '%s%', 1);
459.                value2 = str2double(line(1, 50:b));
460.            end
461.            k = strmatch(varpos, D, 'exact');
462.            R(range_ind, 1) = k;
463.            % store range to matrix R
464.            if isequal(X{k, 1}, 'E')
465.                if value > 0 % Type of row E and sign of the
466.                    % range value +
467.                    R(range_ind, 2) = B(k);
468.                    R(range_ind, 3) = (B(k) + abs(value));
469.                    R(range_ind, 4) = 1;
470.                elseif value < 0 % Type of row E and sign of
471.                    % the range value -
472.                    R(range_ind, 2) = B(k) - abs(value);
473.                    R(range_ind, 3) = B(k);
474.                    R(range_ind, 4) = -1;
475.                end
476.            elseif isequal(X{k, 1}, 'L') % Type of row L
477.                R(range_ind, 2) = B(k) - abs(value);
478.                R(range_ind, 3) = B(k);
479.                if value > 0
480.                    R(range_ind, 4) = 1;
481.                else
482.                    R(range_ind, 4) = -1;
483.                end
484.            elseif isequal(X{k, 1}, 'G') % Type of row G
485.                R(range_ind, 2) = B(k);
486.                R(range_ind, 3) = B(k) + abs(value);
```

```
487.            if value > 0
488.                R(range_ind, 4) = 1;
489.            else
490.                R(range_ind, 4) = -1;
491.            end
492.        end
493.        range_ind = range_ind + 1;
494.        k = strmatch(varpos2, D, 'exact');
495.        if isempty(k) ~= 1
496.            R(range_ind, 1) = k;
497.            % store range to matrix R
498.            if isequal(X{k, 1}, 'E')
499.                if value2 > 0 % Type of row E and sign of
500.                    % the range value +
501.                    R(range_ind, 2) = B(k);
502.                    R(range_ind, 3) = B(k) + abs(value2);
503.                    R(range_ind, 4) = 1;
504.                elseif value2 < 0 % Type of row E and sign
505.                    % of the range value -
506.                    R(range_ind, 2) = B(k) - abs(value2);
507.                    R(range_ind, 3) = B(k);
508.                    R(range_ind, 4) = -1;
509.                end
510.            elseif isequal(X{k, 1}, 'L') % Type of row L
511.                R(range_ind, 2) = B(k) - abs(value2);
512.                R(range_ind, 3) = B(k);
513.                if value2 > 0
514.                    R(range_ind, 4) = 1;
515.                else
516.                    R(range_ind, 4) = -1;
517.                end
518.            elseif isequal(X{k, 1}, 'G') % Type of row G
519.                R(range_ind, 2) = B(k);
520.                R(range_ind, 3) = B(k) + abs(value2);
521.                if value2 > 0
522.                    R(range_ind, 4) = 1;
523.                else
524.                    R(range_ind, 4) = -1;
525.                end
526.            end
527.            range_ind = range_ind + 1;
528.        end
529.    end
530. end
531. % read the BOUNDS section and populate matrix BS
532. if bflag == 1
533.    D = variables(:, 1);
534.    % possible bound types
535.    Types = {'LO' 'UP' 'FX' 'FR' 'MI' 'PL'};
536.    bound_ind = 1;
537.    while isequal(varname, 'ENDATA') == 0
538.        boundtype = '';
539.        varname = '';
540.        value = '';
```

```
541.          line = fgets(fid);
542.          b = length(line);
543.          % stop if the end of the MPS file reached
544.          if isequal(line(1, 1:6), 'ENDATA')
545.              break
546.          end
547.          % read the bound type, the bound name, the
548.          % variable name and the value
549.          boundtype = sscanf(line(1, 2:4), '%s%', 1);
550.          if b > 22 % % LO, UP and FX type of constraints
551.              % have a value field
552.              varname = sscanf(line(1, 15:22), '%s%', 1);
553.              value = sscanf(line(1, 25:b), '%s%', 1);
554.              value = str2double(value);
555.              BS{bound_ind, 3} = value;
556.          else % FR, MI, PL type of constraints do not
557.              % have a value field
558.              varname = sscanf(line(1, 15:b), '%s%', 1);
559.              BS{bound_ind, 3} = NaN;
560.          end
561.          k = strmatch(varname, D, 'exact');
562.          BS{bound_ind, 1} = k;
563.          k2 = strmatch(boundtype, Types, 'exact');
564.          if k2 == 1 % LO bound
565.              BS{bound_ind, 2} = 1;
566.          elseif k2 == 2 % UP bound
567.              BS{bound_ind, 2} = -1;
568.          elseif k2 == 3 % FX bound
569.              BS{bound_ind, 2} = 0;
570.          elseif k2 == 4 % FR bound
571.              BS{bound_ind, 2} = 11;
572.          elseif k2 == 5 % MI bound
573.              BS{bound_ind, 2} = 111;
574.          elseif k2 == 6 % MV bound
575.              BS{bound_ind, 2} = 1111;
576.          end
577.          bound_ind = bound_ind + 1;
578.      end
579. end
580. c = cell2mat(C); % convert cell to matrix
581. c = sparse(c); % make vector c sparse
582. % calculate the nonzero elements
583. NonZeros = nnz(A);
584. b = sparse(B); % make vector b sparse
585. A = sparse(A); % make matrix A sparse
586. R = sparse(R); % make matrix R sparse
587. BS = cell2mat(BS); % convert cell to mat
588. BS = sparse(BS); % make matrix BS sparse
589. Eqin = sparse(Eqin); % make vector Eqin sparse
590. Name = filename;
591. % save to file
592. c0 = 0;
593. save matlab.mat Name c A b R BS Eqin NonZeros ...
594.      MinMaxLP c0
```

```
595. copyfile('matlab.mat', [filename '.mat']); % copy file
596. fclose(fid); % close the file
597. delete matlab.mat % delete matlab.mat
598. end
```

3.5 Generate Random Sparse and Dense Linear Programming Problems

This section presents the implementation in MATLAB of a code that generates random sparse and dense LPs. Some necessary notations should be introduced before the presentation of the implementation of the code that generates random sparse and dense LPs. Let m be the number of the constraints of the generated random LPs, n be the number of the variables of the generated random LPs, A be a $m \times n$ matrix with the coefficients of the constraints, c be a $n \times 1$ vector with the coefficients of the objective function, b be a $m \times 1$ vector with the right-hand side of the constraints, $Eqin$ be a $m \times 1$ vector with the type of the constraints, $optType$ be the type of optimization (0—min, 1—max, 2—randomly selected), Alu be a 1×2 vector of the ranges of the values for matrix A, clu be a 1×2 vector of the ranges of the values for vector c, blu be a 1×2 vector of the ranges of the values for vector b, $Eqinlu$ be a 1×2 vector of the ranges of the values for vector $Eqin$, $MinMaxLP$ be the type of optimization (-1—min, 1—max), $center$ be the center of the circle, r be the radius of the circle, $dense$ be the density of matrix A, $t1$ be the lower bound of the polyhedron, and $t2$ be the upper bound of the polyhedron.

The code that generates random sparse and dense LPs is implemented with seven functions: (i) one function that displays a custom menu for creating random LPs (filename: `GenerateRandomLPs.m`), (ii) one function that displays a custom menu for creating dense random LPs (filename: `denseMenu.m`), (iii) one function that creates dense random LPs (filename: `denseRandom.m`), (iv) one function that creates dense random optimal LPs (filename: `denseRandomOptimal.m`), (v) one function that displays a custom menu for creating sparse random LPs (filename: `sparseMenu.m`), (vi) one function that creates sparse random LPs (filename: `sparseRandom.m`), and (vii) one function that creates sparse random optimal LPs (filename: `sparseRandomOptimal.m`).

The function `GenerateRandomLPs` that displays a custom menu for creating random LPs returns as output the files (mat files) that store the generated random LPs.

In lines 15–17, a menu is presented to the user and the user selects one option. Then, depending on the user's selected option, the random number generator is initialized (lines 18–21), the function to display the menu for creating random dense LPs is called (lines 22–25), the function to display the menu for creating random sparse LPs is called (lines 26–29), or the function is terminated (lines 30–32). Built-in function *rand* generates a pseudo-random number uniformly distributed on the interval $[0, 1]$. Built-in function *randn* generates a pseudo-random number using a Gaussian distribution with mean value equal to 0 and standard deviation equal to 1.

```
1.    function GenerateRandomLPs
2.    % Filename: GenerateRandomLPs.m
3.    % Description: Custom menu for creating random LPs
4.    % Authors: Ploskas, N., & Samaras, N.
5.    %
6.    % Syntax: GenerateRandomLPs
7.    %
8.    % Input: None
9.    %
10.   % Output: files that store the generated random LPs
11.
12.   choice = 0;
13.   while choice ~= 4
14.       % show menu
15.       choice = menu('Generate Random LPs', ...
16.           'Generator Initialization', 'Dense Random LPs', ...
17.           'Sparse Random LPs', 'Exit');
18.       if choice == 1 % initialize generator
19.           seedNumber = input('Please give the seed number: ');
20.           rand('state', seedNumber);
21.           sprand('state');
22.       elseif choice == 2 % create dense random LPs
23.           m = input('Please give the number of constraints: ');
24.           n = input('Please give the number of variables: ');
25.           denseMenu(m, n);
26.       elseif choice == 3 % create sparse random LPs
27.           m = input('Please give the number of constraints: ');
28.           n = input('Please give the number of variables: ');
29.           sparseMenu(m, n);
30.       elseif choice == 4 % exit
31.           return;
32.       end
33.   end
34.   end
```

The function denseMenu that displays a custom menu for creating dense random LPs takes as input the number of the constraints of the generated random LPs (variable m) and the number of the variables of the generated random LPs (variable n), and returns as output the files that store the generated random LPs. The files that are generated start with the word 'fdata', then a number that shows the id of the problem follows and the dimension of the problem is also added. For example, a filename 'fdata3_1000x1500.mat' refers to the third file of the files that were generated; the LP problem has $1,000$ constraints and $1,500$ variables. The output MAT files contain the following variables:

- A: the constraint matrix (size $m \times n$)
- b: the right-hand side vector (size $m \times 1$)
- c: the vector with the coefficients of the objective function (size $n \times 1$)
- $Eqin$: the vector with the type of the constraints (size $m \times 1$), where its elements have three possible values:

- − −1: 'less than or equal to' type of constraint
- − 0: equality constraint
- − 1: 'greater than or equal to' type of constraint

- *MinMaxLP*: the optimization direction where −1 defines minimization and 1 maximization
- *Name*: the name of the LP problem
- *R*: the matrix with the constraint ranges
- *BS*: the matrix with the variable bounds
- *NonZeros*: the number of the nonzero elements of the constraint matrix
- *c0* and *c00*: the objective constant terms

In lines 15–31, the information needed to generate the LPs is given by the user. If the user selects to create only randomly generated dense optimal LPs, then the function that generates only optimal dense LPs is called and the returned LP problem is stored to a MAT file (lines 32–59). Otherwise, the function that generates random dense LPs (not only optimal) is called from the script and the returned LP problem is also stored to a MAT file (lines 60–87).

```
1.   function denseMenu(m, n)
2.   % Filename: denseMenu.m
3.   % Description: Custom menu for creating dense random LPs
4.   % Authors: Ploskas, N., & Samaras, N.
5.   %
6.   % Syntax: denseMenu(m, n)
7.   %
8.   % Input:
9.   % -- m: the number of constraints of the generated random LPs
10.  % -- n: the number of variables of the generated random LPs
11.  %
12.  % Output: files that store the generated dense random LPs
13.
14.  % read the directory to store the LPs
15.  pathName = input(['Please give the path where you want ' ...
16.      'to store the generated LPs: ']);
17.  % read the number of the LPs to generate
18.  nOfProblems = input(['Please give the number of the LPs ' ...
19.      'that you want to create:  ']);
20.  % read the ranges of the values for matrix A
21.  Alu = input(['Please give the range of the values for ' ...
22.      'matrix A: ']);
23.  % read the ranges of the values for vector c
24.  clu = input(['Please give the range of the values for ' ...
25.      'vector c: ']);
26.  % read the type of optimization
27.  optType = input(['Please give the type of optimization ' ...
28.      '(0 min, 1 max, 2 random): ']);
29.  % read if all LPs will be optimal
30.  optimality = input(['Do you want all the LPs to be ' ...
31.      'optimal (1 yes, 2 no): ']);
32.  if optimality == 1 % optimal LPs
33.      % read the center of the circle
34.      center = input(['Please give the center of the circle: ']);
35.      % read the radius of the circle
```

```
36.        R = input(['Please give the radius of the circle ' ...
37.            '(must be less than the center of the circle: ']);
38.        while R >= center % read R
39.            R = input(['Please give the radius of the circle ' ...
40.                '(must be less than the center of the circle: ']);
41.        end
42.        % create nOfProblems dense random optimal LPs
43.        for i = 1:nOfProblems
44.            [A, c, b, Eqin, MinMaxLP] = denseRandomOptimal(m, n, ...
45.                optType, Alu, clu, center, R);
46.            s1 = num2str(m);
47.            s2 = num2str(n);
48.            k = num2str(i);
49.            fname = [pathName '/' 'fdata' k '_' s1 'x' s2];
50.            Name = ['fdata' k '_' s1 'x' s2];
51.            R = [];
52.            BS = [];
53.            NonZeros = nnz(A);
54.            c0 = 0;
55.            c00 = 0;
56.            % save variables in a MAT file
57.            eval(['save ' fname '.mat A c b Eqin MinMaxLP Name ' ...
58.                'R BS NonZeros c0 c00']);
59.        end
60.    else % not optimal LPs
61.        % read the ranges of the values for vector b
62.        blu = input(['Please give the range of the values for ' ...
63.            'vector b: ']);
64.        % read the ranges of the values for vector Eqin
65.        Eqinlu = input(['Please give the range of the values ' ...
66.            'for vector Eqin (0 - equality constraints, 1 - ' ...
67.            ' less than or equal to inequality constraints, ' ...
68.            ' 2 - greater than or equal to inequality ' ...
69.            'constraints: ']);
70.        for i = 1:nOfProblems % create nOfProblems dense random LPs
71.            [A, c, b, Eqin, MinMaxLP] = denseRandom(m, n, ...
72.                optType, Alu, clu, blu, Eqinlu);
73.            s1 = num2str(m);
74.            s2 = num2str(n);
75.            k = num2str(i);
76.            fname = [pathName '/' 'fdata' k '_' s1 'x' s2];
77.            Name = ['fdata' k '_' s1 'x' s2];
78.            R = [];
79.            BS = [];
80.            NonZeros = nnz(A);
81.            c0 = 0;
82.            c00 = 0;
83.            % save variables in a MAT file
84.            eval(['save ' fname '.mat A c b Eqin MinMaxLP Name ' ...
85.                'R BS NonZeros c0 c00']);
86.        end
87.    end
88. end
```

The function `denseRandom` is used to create dense random LPs that are not guaranteed to be optimal. The function that creates dense random LPs takes as input the number of the constraints of the generated random LPs (variable m), the number of the variables of the generated random LPs (variable n), the type of optimization (variable *optType*), the ranges of the values for matrix A (vector *Alu*), the ranges of the values for vector c (vector *clu*), the ranges of the values for vector b (vector *blu*), and the ranges of the values for vector *Eqin* (vector *Eqinlu*), and returns as output the matrix of the coefficients of the constraints (matrix A), the vector of the coefficients of the objective function (vector c), the vector of the right-hand side of the constraints (vector b), the vector of the type of the constraints (vector *Eqin*), and the type of optimization (variable *MinMaxLP*).

In lines 37–40, matrix A is created. Then, vectors c (lines 41–44), b (lines 45–48), and *Eqin* (lines 49–56) are also created. Finally, the optimization type is determined (lines 57–66).

```
1.    function [A, c, b, Eqin, MinMaxLP]= denseRandom(m, n, ...
2.         optType, Alu, clu, blu, Eqinlu)
3.    % Filename: denseRandom.m
4.    % Description: Creates dense random LPs
5.    % Authors: Ploskas, N., & Samaras, N.
6.    %
7.    % Syntax: [A, c, b, MinMaxLP, Eqin]= denseRandom(m, n, ...
8.    %    minCard, Alu, clu, blu, Eqinlu)
9.
10.   % Input:
11.   % -- m: the number of constraints of the generated random
12.   %    LPs
13.   % -- n: the number of variables of the generated random
14.   %    LPs
15.   % -- optType: the type of optimization (0 min, 1 max, 2
16.   %    random)
17.   % -- Alu: the ranges of the values for matrix A
18.   %    (size 1 x 2)
19.   % -- clu: the ranges of the values for vector c
20.   %    (size 1 x 2)
21.   % -- blu: the ranges of the values for vector b
22.   %    (size 1 x 2)
23.   % -- Eqinlu: the ranges of the values for vector Eqin
24.   %    (size 1 x 2)
25.   %
26.   % Output:
27.   % -- A: matrix of coefficients of the constraints
28.   %    (size m x n)
29.   % -- c: vector of coefficients of the objective function
30.   %    (size n x 1)
31.   % -- b: vector of the right-hand side of the constraints
32.   %    (size m x 1)
33.   % -- Eqin: vector of the type of the constraints
34.   %    (size m x 1)
35.   % -- MinMaxLP: the type of optimization
```

```
36.    %
37.    Al = Alu(1); % get the lower bound of A
38.    Au = Alu(2); % get the upper bound of A
39.    % create matrix A
40.    A = floor((Au - Al + 1) * rand(m, n)) + Al;
41.    cl = clu(1); % get the lower bound of c
42.    cu = clu(2); % get the upper bound of c
43.    % create vector c
44.    c = floor((cu - cl + 1) * rand(n, 1)) + cl;
45.    bl = blu(1); % get the upper bound of b
46.    bu = blu(2); % get the upper bound of c
47.    % create vector b
48.    b = floor((bu - bl + 1) * rand(m, 1)) + bl;
49.    eqinl = Eqinlu(1); % get the lower bound of Eqin
50.    eqinu = Eqinlu(2); % get the upper bound of Eqin
51.    % create vector Eqin
52.    Eqin = floor((eqinu - eqinl + 1) * rand(m, 1)) + eqinl;
53.    % less than or equal to inequality constraints
54.    Eqin(Eqin == 1) = -1;
55.    % greater than or equal to inequality constraints
56.    Eqin(Eqin == 2) = 1;
57.    if optType == 0 % minimization
58.        MinMaxLP = -1;
59.    elseif optType == 1 % maximization
60.        MinMaxLP = 1;
61.    else % pick random optimization type
62.        MinMaxLP = -1;
63.        if randi(2) < 2
64.            MinMaxLP = 1;
65.        end
66.    end
67.    end
```

The function denseRandomOptimal is used to create dense random LPs that are guaranteed to be optimal [9]. The generated optimal random LPs have the following form:

$$\min \quad c^T x$$
$$\text{s.t. } Ax \le b$$
$$x \ge 0$$

Every constraint of the form $A_i x = b_i$, $i = 1, 2, \cdots, m$, corresponds to a hyperplane. Hence, a random hyperplane must be created in order to generate a random LP problem. The hyperplanes are created so as to be tangent to a sphere, $B\left(x^0, R\right)$, where x^0 is the center and R the radius of the sphere. As a result, a random unit vector d is created and the intersection of the radius $(x^0 + td : t \ge 0)$ to the sphere is found. Let x^1 be the intersection of the radius to the sphere (Figure 3.1):

Fig. 3.1 Tangent hyperplane
to a sphere

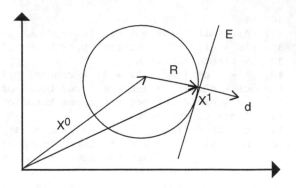

$$\left\{ \begin{array}{l} x^1 = x^0 + td, \|d\| = 1, t \geq 0 \\ \|x^1 - x^0\| = R, R > 0 \end{array} \right\} \Leftrightarrow \left\{ \begin{array}{l} x^1 - x^0 = td \\ \left(x^1 - x^0\right)^2 = R^2 \end{array} \right\}$$

$$\Leftrightarrow \left\{ \begin{array}{l} x^1 - x^0 = td \\ (td)^2 = R^2 \end{array} \right\} \Leftrightarrow \left\{ \begin{array}{l} x^1 - x^0 = td \\ t^2 = R^2 \end{array} \right\} \Leftrightarrow \left\{ \begin{array}{l} x^1 - x^0 = td \\ t = R \vee -R \end{array} \right\}$$

Hence, $x^1 = x^0 + Rd$, $\|d\| = 1$.

Then, we create a hyperplane that passes from x^1 and is tangent to the sphere (a vector perpendicular to vector d). We then define an inequality $a^T x \leq b$ so as to be satisfied by the center of the sphere. Regarding the hyperplane that passes from x^1 and is tangent to the sphere: if x is a point of the hyperplane, vector v from x^1 to x is $v = x - x^1$ and v is perpendicular to vector d, then:

$$d^T v = 0 \Rightarrow d^T \left(x - x^0 - Rd\right) = 0 \Rightarrow d^T x = d^T x^0 + Rd^T d \Rightarrow d^T x = d^T x^0 + R$$

Hence, the equation of the hyperplane that is tangent in a random point of the sphere $B\left(x^0, R\right)$ is:

$$d^T x = R + d^T x^0$$

So, the corresponding inequality is $d^T x \leq R + d^T x^0$, i.e., $a = d$ and $b = R + d^T x^0$.

The function `denseRandomOptimal` is used to create dense random optimal LPs. The function that creates dense random optimal LPs takes as input the number of the constraints of the generated random LPs (variable m), the number of the variables of the generated random LPs (variable n), the type of optimization (variable *optType*), the ranges of the values for matrix A (vector *Alu*), the ranges of the values for vector c (vector *clu*), the center of the circle (variable *circle*), and the radius of the circle (variable R), and returns as output the matrix of the coefficients of the constraints (matrix A), the vector of the coefficients of the objective function (vector c), the vector of the right-hand side of the constraints (vector b), the vector of the type of the constraints (vector *Eqin*), and the type of optimization (variable *MinMaxLP*).

In lines 35–38, vector c is created. Then, matrix A and vector b are also created (lines 39–53). In line 55, vector *Eqin* is created. Finally, the optimization type is determined (lines 56–65).

```
1.    function [A, c, b, Eqin, MinMaxLP] = ...
2.        denseRandomOptimal(m, n, optType, Alu, clu, center, R)
3.    % Filename: denseRandomOptimal.m
4.    % Description: Creates dense random optimal LPs
5.    % Authors: Ploskas, N., & Samaras, N.
6.    %
7.    % Syntax: [A, c, b, Eqin, MinMaxLP] = ...
8.    %       denseRandomOptimal(m, n, optType, Alu, clu, center, R)
9.    %
10.   % Input:
11.   % -- m: the number of constraints of the generated random
12.   %       LPs
13.   % -- n: the number of variables of the generated random
14.   %       LPs
15.   % -- optType: the type of optimization (0 min, 1 max, 2
16.   %       random)
17.   % -- Alu: the ranges of the values for matrix A
18.   %       (size 1 x 2)
19.   % -- clu: the ranges of the values for vector c
20.   %       (size 1 x 2)
21.   % -- center: the center of the circle
22.   % -- R: the radius of the circle
23.   %
24.   % Output:
25.   % -- A: matrix of coefficients of the constraints
26.   %       (size m x n)
27.   % -- c: vector of coefficients of the objective function
28.   %       (size n x 1)
29.   % -- b: vector of the right-hand side of the constraints
30.   %       (size m x 1)
31.   % -- Eqin: vector of the type of the constraints
32.   %       (size m x 1)
33.   % -- MinMaxLP: the type of optimization
34.
35.   cl = clu(1); % get the lower bound of c
36.   cu = clu(2); % get the upper bound of c
37.   % create vector c
38.   c = round((cu - cl + 1) * randn(n, 1)) + cl;
39.   Al = Alu(1); % get the lower bound of A
40.   Au = Alu(2); % get the upper bound of A
41.   A = zeros(m, n);
42.   b = zeros(m, 1);
43.   k(1:n) = center;
44.   for i = 1:m % create matrix A and vector b
45.       % create a row of matrix A
46.       a = round((Au - Al + 1) * rand(1, n)) + Al;
47.       % calculate the hyperplane that is tangent
```

```
48.      % in a random point of the sphere
49.      y = k + R * (a / norm(a));
50.      b0 = a * y';
51.      b(i, 1) = b0; % add the point to vector b
52.      A(i, :) = a; % add the row to matrix A
53.   end
54.   % create vector Eqin
55.   Eqin(1:m, 1) = -1;
56.   if optType == 0 % minimization
57.      MinMaxLP = -1;
58.   elseif optType == 1 % maximization
59.      MinMaxLP = 1;
60.   else % pick random optimization type
61.      MinMaxLP = -1;
62.      if randi(2) < 2
63.            MinMaxLP = 1;
64.      end
65.   end
66.   end
```

The function sparseMenu that displays a custom menu for creating sparse random LPs takes as input the number of the constraints of the generated random LPs (variable m) and the number of the variables of the generated random LPs (variable n), and returns as output the files that store the generated random LPs. The files that are generated start with the word 'sdata', then a number that shows the id of the problem follows and the dimension of the problem is also added. For example, a filename 'sdata3_1000x1500.mat' refers to the third file of the files that were generated; the LP problem has $1,000$ constraints and $1,500$ variables. The output MAT files contain the following variables:

- A: the constraint matrix (size $m \times n$)
- b: the right-hand side vector (size $m \times 1$)
- c: the vector with the coefficients of the objective function (size $n \times 1$)
- $Eqin$: the vector with the type of the constraints (size $m \times 1$), where its elements have three possible values:

    - -1: 'less than or equal to' type of constraint
    - 0: equality constraint
    - 1: 'greater than or equal to' type of constraint

- $MinMaxLP$: the optimization direction where -1 defines minimization and 1 maximization
- $Name$: the name of the LP problem
- R: the matrix with the constraint ranges
- BS: the matrix with the variable bounds
- $NonZeros$: the number of the nonzero elements of the constraint matrix
- $c0$ and $c00$: the objective constant terms

In lines 16–44, the information needed to generate the LPs is given by the user. If the user selects to create only randomly generated sparse optimal LPs, then

the function that generates only optimal sparse LPs is called and the returned LP
problem is stored to a MAT file (lines 45–71). Otherwise, the function that generates
random sparse LPs (not only optimal) is called and the returned LP problem is also
stored to a MAT file (lines 72–94).

```
1.    function sparseMenu(m, n)
2.    % Filename: sparseMenu.m
3.    % Description: Custom menu for creating sparse random LPs
4.    % Authors: Ploskas, N., & Samaras, N.
5.    %
6.    % Syntax: sparseMenu(m, n)
7.    %
8.    % Input:
9.    % -- m: the number of constraints of the generated random
10.   %    LPs
11.   % -- n: the number of variables of the generated random
12.   %    LPs
13.   %
14.   % Output: files that store the generated sparse random LPs
15.
16.   % read the directory to store the LPs
17.   pathName = input(['Please give the path where you ' ...
18.       'want to store the generated LPs: ']);
19.   % read the number of the LPs to generate
20.   nOfProblems = input(['Please give the number of the ' ...
21.       'LPs that you want to create: ']);
22.   % read the density of the LPs to generate
23.   dense = input('Please enter the density of the LP: ');
24.   % read the ranges of the values for matrix A
25.   Alu = input(['Please give the range of the values for ' ...
26.       'matrix A: ']);
27.   % read the ranges of the values for vector c
28.   clu = input(['Please give the range of the values for ' ...
29.       'vector c: ']);
30.   % read the ranges of the values for vector b
31.   blu = input(['Please give the range of the values for ' ...
32.       'vector b: ']);
33.   % read the ranges of the values for vector Eqin
34.   Eqinlu = input(['Please give the range of the values ' ...
35.       'for vector Eqin (0 - equality constraints, 1 - ' ...
36.       'less than or equal to inequality constraints, ' ...
37.       '2 - greater than or equal to inequality ' ...
38.       'constraints: ']);
39.   % read the type of optimization
40.   optType = input(['Please give the type of ' ...
41.       'optimization (0 min, 1 max, 2 random): ']);
42.   % read if all LPs will be optimal
43.   optimality = input(['Do you want all the LPs to be ' ...
44.       'optimal (1 yes, 2 no): ']);
45.   if optimality == 1 % optimal LPs
46.       % read the lower bound of b1
```

```
47.     t1 = input('Please give the lower bound of b1: ');
48.     % read the upper bound of b1
49.     t2 = input('Please give the upper bound of b1: ');
50.     % create nOfProblems sparse random optimal LPs
51.     for i = 1:nOfProblems
52.         [A, c, b, Eqin, MinMaxLP] = ...
53.             sparseRandomOptimal(m, n, optType, Alu, ...
54.             clu, blu, Eqinlu, dense, t1, t2);
55.         s1 = num2str(m);
56.         s2 = num2str(n);
57.         k = num2str(i);
58.         fname = [pathName '/' 'sdata' k '_' s1 'x' s2];
59.         Name = ['fdata' k '_' s1 'x' s2];
60.         A = sparse(A);
61.         c = sparse(c);
62.         b = sparse(b);
63.         Eqin = sparse(Eqin);
64.         R = [];
65.         BS = [];
66.         NonZeros = nnz(A);
67.         c0 = 0;
68.         c00 = 0;
69.         eval (['save ' fname '.mat A c b Eqin MinMaxLP ' ...
70.             'Name R BS NonZeros c0 c00']);
71.     end
72. else % not optimal LPs
73.     % create nOfProblems sparse random LPs
74.     for i = 1:nOfProblems
75.         [A, c, b, Eqin, MinMaxLP] = sparseRandom(m, n, ...
76.             optType, Alu, clu, blu, Eqinlu, dense);
77.         s1 = num2str(m);
78.         s2 = num2str(n);
79.         k = num2str(i);
80.         fname = [pathName '/' 'sdata' k '_' s1 'x' s2];
81.         Name = ['fdata' k '_' s1 'x' s2];
82.         A = sparse(A);
83.         c = sparse(c);
84.         b = sparse(b);
85.         Eqin = sparse(Eqin);
86.         R = [];
87.         BS = [];
88.         NonZeros = nnz(A);
89.         c0 = 0;
90.         c00 = 0;
91.         eval (['save ' fname '.mat A c b Eqin MinMaxLP ' ...
92.             ' Name R BS NonZeros c0 c00']);
93.     end
94. end
95. end
```

The function `sparseRandom` is used to create sparse random LPs that are not guaranteed to be optimal. The function that creates sparse random LPs takes as input the number of the constraints of the generated random LPs (variable m), the number of the variables of the generated random LPs (variable n), the type of optimization (variable *optType*), the ranges of the values for matrix A (vector *Alu*), the ranges of the values for vector c (vector *clu*), the ranges of the values for vector b (vector *blu*), the ranges of the values for vector *Eqin* (vector *Eqinlu*), and the density of matrix A (variable *dense*), and returns as output the matrix of the coefficients of the constraints (matrix A), the vector of the coefficients of the objective function (vector c), the vector of the right-hand side of the constraints (vector b), the vector of the type of the constraints (vector *Eqin*), and the type of optimization (variable *MinMaxLP*).

In lines 38–45, matrix A is created. Then, vectors c (lines 46–49), b (lines 50–53), and *Eqin* (lines 54–61) are also created. Finally, the optimization type is determined (lines 62–71).

```
1.    function [A, c, b, Eqin, MinMaxLP] = sparseRandom(m, n, ...
2.        optType, Alu, clu, blu, Eqinlu, dense)
3.    % Filename: sparseRandom.m
4.    % Description: Creates sparse random LPs
5.    % Authors: Ploskas, N., & Samaras, N.
6.    %
7.    % Syntax: [A, c, b, Eqin, MinMaxLP] = sparseRandom(m, n, ...
8.    %        optType, Alu, clu, blu, dense)
9.    %
10.   % Input:
11.   % -- m: the number of constraints of the generated random
12.   %    LPs
13.   % -- n: the number of variables of the generated random
14.   %    LPs
15.   % -- optType: the type of optimization (0 min, 1 max, 2
16.   %    random)
17.   % -- Alu: the ranges of the values for matrix A
18.   %    (size 1 x 2)
19.   % -- clu: the ranges of the values for vector c
20.   %    (size 1 x 2)
21.   % -- blu: the ranges of the values for vector b
22.   %    (size 1 x 2)
23.   % -- Eqinlu: the ranges of the values for vector Eqin
24.   %    (size 1 x 2)
25.   % -- dense: the density of matrix A
26.   %
27.   % Output:
28.   % -- A: matrix of coefficients of the constraints
29.   %    (size m x n)
30.   % -- c: vector of coefficients of the objective function
31.   %    (size n x 1)
32.   % -- b: vector of the right-hand side of the constraints
33.   %    (size m x 1)
34.   % -- Eqin: vector of the type of the constraints
```

```
35.  %      (size m x 1)
36.  % -- MinMaxLP: the type of optimization
37.
38.  Al = Alu(1); % get the lower bound of A
39.  Au = Alu(2); % get the upper bound of A
40.  % create matrix A
41.  A = round((Au - Al + 1) * sprand(m, n, dense));
42.  [k, l, s] = find(A);
43.  for i = 1:length(k)
44.      A(k(i), l(i)) = s(i) + Al;
45.  end
46.  cl = clu(1); % get the lower bound of c
47.  cu = clu(2); % get the upper bound of c
48.  % create vector c
49.  c = floor((cu - cl + 1) * rand(n, 1)) + cl;
50.  bl = blu(1); % get the upper bound of b
51.  bu = blu(2); % get the upper bound of b
52.  % create vector b
53.  b = floor((bu - bl + 1) * rand(m, 1)) + bl;
54.  eqinl = Eqinlu(1); % get the lower bound of Eqin
55.  eqinu = Eqinlu(2); % get the upper bound of Eqin
56.  % create vector Eqin
57.  Eqin = floor((eqinu - eqinl + 1) * rand(m, 1)) + eqinl;
58.  % less than or equal to inequality constraints
59.  Eqin(Eqin == 1) = -1;
60.  % greater than or equal to inequality constraints
61.  Eqin(Eqin == 2) = 1;
62.  if optType == 0 % minimization
63.      MinMaxLP = -1;
64.  elseif optType == 1 % maximization
65.      MinMaxLP = 1;
66.  else % pick random optimization type
67.      MinMaxLP = -1;
68.      if randi(2) < 2
69.          MinMaxLP = 1;
70.      end
71.  end
72.  end
```

The function sparseRandomOptimal is used to create sparse random LPs that are guaranteed to be optimal. This is achieved by increasing the right-hand side value of the 'less than or equal to' constraints in order to create artificially a closed polyhedron. The function that creates sparse random optimal LPs takes as input the number of the constraints of the generated random LPs (variable m), the number of the variables of the generated random LPs (variable n), the type of optimization (variable *optType*), the ranges of the values for matrix A (vector *Alu*), the ranges of the values for vector c (vector *clu*), the ranges of the values for vector b (vector *blu*), the ranges of the values for vector *Eqin* (vector *Eqinlu*), the density of matrix A (variable *dense*), the lower bound of $b1$ (variable *t1*), and the upper bound of $b1$ (variable *t2*), and returns as output the matrix of the coefficients of the constraints (matrix *A*), the vector of the coefficients of the objective function (vector *c*), the vector of the right-hand side of the constraints (vector *b*), the vector of the type of the constraints (vector *Eqin*), and the type of optimization (variable *MinMaxLP*).

In lines 42–49, matrix *A* is created. Then, vectors *c* (lines 50–53), *b* (lines 54–57), and *Eqin* (lines 58–72) are also created. Finally, the optimization type is determined (lines 73–82).

```
1.   function [A, c, b, Eqin, MinMaxLP] = ...
2.       sparseRandomOptimal(m, n, optType, Alu, clu, blu, ...
3.       Eqinlu, dense, t1, t2)
4.   % Filename: sparseRandomOptimal.m
5.   % Description: Creates sparse random optimal LPs
6.   % Authors: Ploskas, N., & Samaras, N.
7.   %
8.   % Syntax: [A, c, b, Eqin, MinMaxLP] = ...
9.   %       sparseRandomOptimal(m, n, optType, Alu, clu, blu, ...
10.  %       Eqinlu, dense, t1, t2)
11.  %
12.  % Input:
13.  % -- m: the number of constraints of the generated random
14.  %    LPs
15.  % -- n: the number of variables of the generated random
16.  %    LPs
17.  % -- optType: the type of optimization (0 min, 1 max, 2
18.  %    random)
19.  % -- Alu: the ranges of the values for matrix A
20.  %    (size 1 x 2)
21.  % -- clu: the ranges of the values for vector c
22.  %    (size 1 x 2)
23.  % -- blu: the ranges of the values for vector b
24.  %    (size 1 x 2)
25.  % -- Eqinlu: the ranges of the values for vector Eqin
26.  %    (size 1 x 2)
27.  % -- dense: the density of matrix A
28.  % -- t1: the lower bound of b1
29.  % -- t2: the upper bound of b1
30.  %
31.  % Output:
32.  % -- A: matrix of coefficients of the constraints
33.  %    (size m x n)
34.  % -- c: vector of coefficients of the objective function
35.  %    (size n x 1)
36.  % -- b: vector of the right-hand side of the constraints
37.  %    (size m x 1)
38.  % -- Eqin: vector of the type of the constraints
39.  %    (size m x 1)
40.  % -- MinMaxLP: the type of optimization
41.
42.  Al = Alu(1); % get the lower bound of A
43.  Au = Alu(2); % get the upper bound of A
44.  % create matrix A
```

```
45.   A = round((Au - Al + 1) * sprand(m, n, dense));
46.   [k, l, s] = find(A);
47.   for i = 1:length(k)
48.       A(k(i), l(i)) = s(i) + Al;
49.   end
50.   cl = clu(1); % get the lower bound of c
51.   cu = clu(2); % get the upper bound of c
52.   % create vector c
53.   c = floor((cu - cl + 1) * rand(n, 1)) + cl;
54.   bl = blu(1); % get the upper bound of b
55.   bu = blu(2); % get the upper bound of c
56.   % create vector b
57.   b = floor((bu - bl + 1) * rand(m, 1)) + bl;
58.   eqinl = Eqinlu(1); % get the lower bound of Eqin
59.   eqinu = Eqinlu(2); % get the upper bound of Eqin
60.   % create vector Eqin
61.   Eqin = floor((eqinu - eqinl + 1) * rand(m, 1)) + eqinl;
62.   if any(Eqin == 1) % increase the value of the less than or
63.       % equal to constraints in order to create artificially
64.       % a closed polyhedron
65.       m_1 = find(Eqin == 1);
66.       m_11 = length(m_1);
67.       b(m_1) = floor((t2 - t1 + 1) * rand(m_11, 1)) + t1;
68.   end
69.   % less than or equal to inequality constraints
70.   Eqin(Eqin == 1) = -1;
71.   % greater than or equal to inequality constraints
72.   Eqin(Eqin == 2) = 1;
73.   if optType == 0 % minimization
74.       MinMaxLP = -1;
75.   elseif optType == 1 % maximization
76.       MinMaxLP = 1;
77.   else % pick random optimization type
78.       MinMaxLP = -1;
79.       if randi(2) < 2
80.            MinMaxLP = 1;
81.       end
82.   end
83.   end
```

In the remaining of the section, we will demonstrate the procedure on how to create sparse and dense randomly generated LPs using the above codes. Initially, run the GenerateRandomLPs function (Figure 3.2). You can select the 'Generator Initialization' option to set the seed number and control the random number generation (Figure 3.3).

Select the 'Dense Random LPs' option in order to generate dense random LPs. If you choose to create only optimal LPs, then you will be asked to enter the number of constraints, the number of variables, the path to store the generated LPs, the number of the LPs that you want to create, the range of the values for matrix A, the range of the values for vector c, the type of optimization, the center of the circle, and the

Fig. 3.2 Menu for generating random LPs

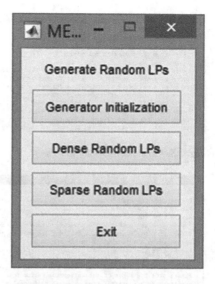

Fig. 3.3 Control random number generation

```
Command Window

  >> GenerateRandomLPs
  Please give the seed number: 4
fx |
```

```
Command Window
  >> GenerateRandomLPs
  Please give the seed number: 4
  Please give the number of rows: 100
  Please give the number of columns: 100
  Please give the path where you want to store the generated LPs: 'test'
  Please give the number of the LPs that you want to create:  10
  Please give the range of the values for matrix A: [-200 200]
  Please give the range of the values for vector c: [-100 100]
  Please give the type of optimization (0 min, 1 max, 2 random): 0
  Do you want all the LPs to be optimal (1 yes, 2 no): 1
  Please give the center of the circle: 8
  Please give the radius of the circle (must be less than the center of the circle: 5
```

Fig. 3.4 Create dense random optimal LPs

radius of the circle (Figure 3.4). Otherwise, you will be asked to enter the number of constraints, the number of variables, the path to store the generated LPs, the number of the LPs that you want to create, the range of the values for matrix A, the range of the values for vector c, the range of the values for vector b, the range of the values for vector $Eqin$, and the type of optimization (Figure 3.5).

Select the 'Sparse Random LPs' option in order to generate sparse random LPs. If you choose to create only optimal LPs, then you will be asked to enter the number of

```
Command Window
>> GenerateRandomLPs
Please give the seed number: 4
Please give the number of rows: 100
Please give the number of columns: 100
Please give the path where you want to store the generated LPs: 'test'
Please give the number of the LPs that you want to create:  10
Please give the range of the values for matrix A: [-200 200]
Please give the range of the values for vector c: [-100 100]
Please give the type of optimization (0 min, 1 max, 2 random): 0
Do you want all the LPs to be optimal (1 yes, 2 no): 2
Please give the range of the values for vector b: [10 100]
Please give the range of the values for vector Eqin (0 - equality constraints, 1 - less than or equal to inequality
constraints, 2 - greater than or equal to inequality constraints: [1 2]
```

Fig. 3.5 Create dense random LPs

```
Command Window
>> GenerateRandomLPs
Please give the seed number: 4
Please give the number of rows: 100
Please give the number of columns: 100
Please give the path where you want to store the generated LPs: 'test'
Please give the number of the LPs that you want to create:  10
Please enter the density of the LP:  0.05
Please give the range of the values for matrix A: [-200 200]
Please give the range of the values for vector c: [-100 100]
Please give the range of the values for vector b: [10 100]
Please give the range of the values for vector Eqin (0 - equality constraints, 1 - less than or equal to inequality
constraints, 2 - greater than or equal to inequality constraints: [1 2]
Please give the type of optimization (0 min, 1 max, 2 random): 0
Do you want all the LPs to be optimal (1 yes, 2 no): 1
Please give the lower bound of b1: 500
Please give the upper bound of b1: 10000
```

Fig. 3.6 Create sparse random optimal LPs

```
Command Window
>> GenerateRandomLPs
Please give the seed number: 4
Please give the number of rows: 100
Please give the number of columns: 100
Please give the path where you want to store the generated LPs: 'test'
Please give the number of the LPs that you want to create:  10
Please enter the density of the LP:  0.05
Please give the range of the values for matrix A: [-200 200]
Please give the range of the values for vector c: [-100 100]
Please give the range of the values for vector b: [10 100]
Please give the range of the values for vector Eqin (0 - equality constraints, 1 - less than or equal to inequality
constraints, 2 - greater than or equal to inequality constraints: [1 2]
Please give the type of optimization (0 min, 1 max, 2 random): 0
Do you want all the LPs to be optimal (1 yes, 2 no): 2
```

Fig. 3.7 Create sparse random LPs

constraints, the number of variables, the path to store the generated LPs, the number of the LPs that you want to create, the range of the values for matrix A, the range of the values for vector c, the range of the values for vector b, the range of the values for vector *Eqin*, the type of optimization, the lower bound of $t1$, and the upper bound of $t1$ (Figure 3.6). Otherwise, you will be asked to enter the number of constraints, the number of variables, the path to store the generated LPs, the number of the LPs that you want to create, the range of the values for matrix A, the range of the values for vector c, the range of the values for vector b, the range of the values for vector *Eqin*, and the type of optimization (Figure 3.7).

3.6 Linear Programming Benchmark Problems

Benchmark problems are publicly available LPs that are hard to solve. Many libraries containing LPs exist. The most widely used benchmark libraries are:

- The Netlib LP test set [1]
- The Kennington LP test set [2]
- The infeas section of the Netlib LP test set [3]
- The misc section of the Mészáros LP test set [4]
- The problematic section of the Mészáros LP test set [5]
- The new section of the Mészáros LP test set [6]
- The stochlp section of the Mészáros LP test set [7]
- The Mittelmann LP test set [8]

The Netlib test set has become the standard set of test problems for testing and comparing algorithms and software for LP. The Netlib library is a well-known suite containing 98 real world LPs. Ordonez and Freund [10] have shown that 71% of the Netlib LPs are ill-conditioned. Hence, numerical difficulties may occur. In addition, Mészáros and Mittelmann benchmark libraries include many LPs; some of them are very degenerate and hard to solve.

Tables 3.1–3.8 provide the statistics for the LPs of the aforementioned libraries. Table 3.1 presents the statistics of the LPs for the Netlib LP test set, Table 3.2 for the Kennington LP test set, Table 3.3 for the infeas section of the Netlib LP test set, Table 3.4 presents the statistics of the LPs for the misc section of the Mészáros LP test set, Table 3.5 presents the statistics of the LPs for the problematic section of the Mészáros LP test set, Table 3.6 presents the statistics of the LPs for the new section of the Mészáros LP test set, Table 3.7 presents the statistics of the LPs for the stochlp section of the Mészáros LP test set, while Table 3.8 presents the statistics of the LPs for the Mittelmann LP test set. The first column of the tables includes the name of the problem, the second the number of the constraints, the third the number of the variables, the fourth the nonzero elements of matrix A, and the fifth the optimal objective value. The whole test set includes 98 LPs from the Netlib LP test set, 16 LPs from the Kennington LP test set, 29 LPs from the infeas section of the Netlib LP test set, 242 LPs from the misc section of the Mészáros LP test set, 9 LPs from the problematic section of the Mészáros LP test set, 4 LPs from the new section of the Mészáros LP test set, 120 LPs from the stochlp section of the Mészáros LP test set, and 50 LPs from the Mittelmann LP test set.

Table 3.1 Statistics of the Netlib LP test set

Name	Constraints	Variables	Nonzeros	Optimal value
25fv47	821	1,571	10,400	5.50E+03
80bau3b	2,262	9,799	21,002	9.87E+05
adlittle	56	97	383	2.25E+05
afiro	27	32	83	−4.65E+02
agg	488	163	2,410	−3.60E+07
agg2	516	302	4,284	−2.02E+07
agg3	516	302	4,300	1.03E+07
bandm	305	472	2,494	−1.59E+02
beaconfd	173	262	3,375	3.36E+04
blend	74	83	491	−3.08E+01
bnl1	643	1,175	5,121	1.98E+03
bnl2	2,324	3,489	13,999	1.81E+03
boeing1	351	384	3,485	−3.35E+02
boeing2	166	143	1,196	−3.15E+02
bore3d	233	315	1,429	1.37E+03
brandy	220	249	2,148	1.52E+03
capri	271	353	1,767	2.69E+03
cycle	1,903	2,857	20,720	−5.23E+00
czprob	929	3,523	10,669	2.19E+06
d2q06c	2,171	5,167	32,417	1.23E+05
d6cube	415	6,184	37,704	3.15E+02
degen2	444	534	3,978	−1.44E+03
degen3	1,503	1,818	24,646	−9.87E+02
dfl001	6,071	12,230	35,632	1.13E+07
e226	223	282	2,578	−1.88E+01
etamacro	400	688	2,409	−7.56E+02
fffff800	524	854	6,227	5.56E+05
finnis	497	614	2,310	1.73E+05
fit1d	24	1,026	13,404	−9.15E+03
fit1p	627	1,677	9,868	9.15E+03
fit2d	25	10,500	129,018	−6.85E+04
fit2p	3,000	13,525	50,284	6.85E+04
forplan	161	421	4,563	−6.64E+02
ganges	1,309	1,681	6,912	−1.10E+05
gfrd-pnc	616	1,092	2,377	6.90E+06
greenbea	2,392	5,405	30,877	−7.25E+07
greenbeb	2,392	5,405	30,877	−4.30E+06
grow15	300	645	5,620	−1.07E+08
grow22	440	946	8,252	−1.61E+08
grow7	140	301	2,612	−4.78E+07

(continued)

Table 3.1 (continued)

Name	Constraints	Variables	Nonzeros	Optimal value
israel	174	142	2,269	$-8.97E+05$
kb2	43	41	286	$-1.75E+03$
lotfi	153	308	1,078	$-2.53E+01$
maros	846	1,443	9,614	$-5.81E+04$
maros-r7	3,136	9,408	144,848	$1.50E+06$
modszk1	687	1,620	3,168	$3.21E+02$
nesm	662	2,923	13,288	$1.41E+07$
perold	625	1,376	6,018	$-9.38E+03$
pilot	410	1,000	5,141	$-5.57E+02$
pilot.ja	2,030	4,883	73,152	$-6.11E+03$
pilot.we	940	1,988	14,698	$-2.72E+06$
pilot4	1,441	3,652	43,167	$-2.58E+03$
pilot87	975	2,172	13,057	$3.02E+02$
pilotnov	722	2,789	9,126	$-4.50E+03$
qap8	912	1,632	7,296	$2.04E+02$
qap12	3,192	8,856	38,304	$5.23E+02$
qap15	6,330	22,275	94,950	$1.04E+03$
recipe	91	180	663	$-2.67E+02$
sc105	105	103	280	$-5.22E+01$
sc205	205	203	551	$-5.22E+01$
sc50a	50	48	130	$-6.46E+01$
sc50b	50	48	118	$-7.00E+01$
scagr25	471	500	1,554	$-1.48E+07$
scagr7	129	140	420	$-2.33E+06$
scfxm1	330	457	2,589	$1.84E+04$
scfxm2	660	914	5,183	$3.67E+04$
scfxm3	990	1,371	7,777	$5.49E+04$
scorpion	388	358	1,426	$1.88E+03$
scrs8	490	1,169	3,182	$9.04E+02$
scsd1	77	760	2,388	$8.67E+00$
scsd6	147	1,350	4,316	$5.05E+01$
scsd8	397	2,750	8,584	$9.05E+02$
sctap1	300	480	1,692	$1.41E+03$
sctap2	1,090	1,880	6,714	$1.72E+03$
sctap3	1,480	2,480	8,874	$1.42E+03$
seba	515	1,028	4,352	$1.57E+04$
share1b	117	225	1,151	$-7.66E+04$
share2b	96	79	694	$-4.16E+02$

(continued)

Table 3.1 (continued)

Name	Constraints	Variables	Nonzeros	Optimal value
shell	536	1,775	3,556	1.21E+09
ship04l	402	2,118	6,332	1.79E+06
ship04s	402	1,458	4,352	1.80E+06
ship08l	778	4,283	12,802	1.91E+06
ship08s	778	2,387	7,114	1.92E+06
ship12l	1,151	5,427	16,170	1.47E+06
ship12s	1,151	2,763	8,178	1.49E+06
sierra	1,227	2,036	7,302	1.54E+07
stair	356	467	3,856	−2.51E+02
standata	359	1,075	3,031	1.26E+03
standgub	361	1,184	3,139	1.26E+03
standmps	467	1,075	3,679	1.41E+03
stocfor1	117	111	447	−4.11E+04
stocfor2	2,157	2,031	8,343	−3.90E+04
stocfor3	16,675	15,695	64,875	−4.00E+04
truss	1,000	8,806	27,836	4.59E+05
tuff	333	587	4,520	2.92E−01
vtp.base	198	203	908	1.30E+05
wood1p	244	2,594	70,215	1.44E+00
woodw	1,098	8,405	37,474	1.30E+00

Table 3.2 Statistics of the Kennington LP test set

Name	Constraints	Variables	Nonzeros	Optimal value
cre-a	3,516	4,067	14,987	2.36E+07
cre-b	9,648	72,447	256,095	2.31E+07
cre-c	3,068	3,678	13,244	2.53E+07
cre-d	8,926	69,980	242,646	2.45E+07
ken-07	2,426	3,602	8,404	−6.80E+08
ken-11	14,694	21,349	49,058	−6.97E+09
ken-13	28,632	42,659	97,246	−1.03E+10
ken-18	105,127	154,699	358,171	−5.22E+10
osa-07	1,118	23,949	143,694	5.36E+05
osa-14	2,337	52,460	314,760	1.11E+06
osa-30	4,350	100,024	600,138	2.14E+06
osa-60	10,280	232,966	1,397,793	4.04E+06
pds-02	2,953	7,535	16,390	2.89E+10
pds-06	9,881	28,655	62,524	2.78E+10
pds-10	16,558	48,763	106,436	2.67E+10
pds-20	33,874	105,728	230,200	2.38E+10

Table 3.3 Statistics of the infeas section of the Netlib LP test set

Name	Constraints	Variables	Nonzeros	Optimal value
bgdbg1	348	407	1,440	Infeasible
bgetam	400	688	2,409	Infeasible
bgindy	2,671	10,116	65,502	Infeasible
bgprtr	20	34	64	Infeasible
box1	231	261	651	Infeasible
ceria3d	3,576	824	17,602	Infeasible
chemcom	288	720	1,566	Infeasible
cplex1	3,005	3,221	8,944	Infeasible
cplex2	224	221	1,058	Infeasible
ex72a	197	215	467	Infeasible
ex73a	193	211	457	Infeasible
forest6	66	95	210	Infeasible
galenet	8	8	16	Infeasible
gosh	3,792	10,733	97,231	Infeasible
gran	2,658	2,520	20,106	Infeasible
greenbea	2,393	5,405	30,883	Infeasible
itest2	9	4	17	Infeasible
itest6	11	8	20	Infeasible
klein1	54	54	696	Infeasible
klein2	477	54	4,585	Infeasible
klein3	994	88	12,107	Infeasible
mondou2	312	604	1,208	Infeasible
pang	361	460	2,652	Infeasible
pilot4i	410	1,000	5,141	Infeasible
qual	323	464	1,646	Infeasible
reactor	318	637	2,420	Infeasible
refinery	323	464	1,626	Infeasible
vol1	323	464	1,646	Infeasible
woodinfe	35	89	140	Infeasible

Table 3.4 Statistics of the misc section of the Mészáros LP test set

Name	Constraints	Variables	Nonzeros	Optimal value
aa01	823	8,904	72,965	5.55E+04
aa03	825	8,627	70,806	4.96E+04
aa3	825	8,627	70,806	4.96E+04
aa4	426	7,195	52,121	2.59E+04
aa5	801	8,308	65,953	5.37E+04
aa6	646	7,292	51,728	2.70E+04

(continued)

Table 3.4 (continued)

Name	Constraints	Variables	Nonzeros	Optimal value
air02	50	6,774	61,555	7.64E+03
air03	124	10,757	91,028	3.39E+05
air04	823	8,904	72,965	5.55E+04
air05	426	7,195	52,121	2.59E+04
air06	825	8,627	70,806	4.96E+04
aircraft	3,754	7,517	20,267	1.57E+03
bas1lp	5,411	4,461	582,411	3.63E+02
baxter	27,441	15,128	95,971	5.60E+07
car4	16,384	33,052	63,724	3.55E+01
cari	400	1,200	152,800	5.82E+02
ch	3,700	5,062	20,873	9.26E+05
co5	5,774	7,993	53,661	7.14E+05
co9	10,789	14,851	101,578	9.47E+05
complex	1,023	1,408	46,463	−9.97E+01
cq5	5,048	7,530	47,353	4.00E+05
cq9	9,278	13,778	88,897	5.06E+05
cr42	905	1,513	6,614	2.80E+01
crew1	135	6,469	46,950	2.06E+01
dano3mip	3,202	13,873	79,655	5.76E+02
dbic1	43,200	183,235	1,038,761	−9.77E+06
dbir1	18,804	27,355	1,058,605	−8.11E+06
dbir2	18,906	27,355	1,139,637	−6.12E+06
delf000	3,128	5,464	12,606	3.08E+00
delf001	3,098	5,462	13,214	2.36E+03
delf002	3,135	5,460	13,287	2.83E+00
delf003	3,065	5,460	13,269	9.12E+03
delf004	3,142	5,464	13,546	1.59E+02
delf005	3,103	5,464	13,494	2.29E+03
delf006	3,147	5,469	13,604	2.28E+02
delf007	3,137	5,471	13,758	3.80E+02
delf008	3,148	5,472	13,821	2.41E+02
delf009	3,135	5,472	13,750	4.84E+02
delf010	3,147	5,472	13,802	2.29E+02
delf011	3,134	5,471	13,777	4.73E+02
delf012	3,151	5,471	13,793	1.79E+02
delf013	3,116	5,472	13,809	2.62E+03
delf014	3,170	5,472	13,866	1.53E+02
delf015	3,161	5,471	13,793	7.37E+02

(continued)

Table 3.4 (continued)

Name	Constraints	Variables	Nonzeros	Optimal value
delf017	3,176	5,471	13,732	4.62E+02
delf018	3,196	5,471	13,774	1.42E+02
delf019	3,185	5,471	13,762	2.34E+03
delf020	3,213	5,472	14,070	3.52E+02
delf021	3,208	5,471	14,068	3.95E+02
delf022	3,214	5,472	14,060	3.65E+02
delf023	3,214	5,472	14,098	3.54E+02
delf024	3,207	5,466	14,456	3.51E+02
delf025	3,197	5,464	14,447	3.50E+02
delf026	3,190	5,462	14,220	3.52E+02
delf027	3,187	5,457	14,200	3.19E+02
delf028	3,177	5,452	14,402	2.97E+02
delf029	3,179	5,454	14,402	2.75E+02
delf030	3,199	5,469	14,262	2.54E+02
delf031	3,176	5,455	14,205	2.34E+02
delf032	3,196	5,467	14,251	2.16E+02
delf033	3,173	5,456	14,205	2.00E+02
delf034	3,175	5,455	14,208	1.89E+02
delf035	3,193	5,468	14,284	1.77E+02
delf036	3,170	5,459	14,202	1.64E+02
df2177	630	9,728	21,706	9.00E+01
disp3	2,182	1,856	6,407	2.02E+05
dsbmip	1,182	1,886	7,366	−3.05E+02
e18	24,617	14,231	132,095	5.63E+02
ex3sta1	17,443	8,156	59,419	−6.31E+01
farm	7	12	36	1.75E+04
gams10a	114	61	297	1.00E+00
gams30a	354	181	937	1.00E+00
ge	10,099	11,098	39,554	5.58E+06
iiasa	669	2,970	6,648	2.63E+08
jendrec1	2,109	4,228	89,608	7.03E+03
kent	31,300	16,620	184,710	3.46E+08
kl02	71	36,699	212,536	2.15E+02
kleemin3	3	3	6	−1.00E+04
kleemin4	4	4	10	−1.00E+06
kleemin5	5	5	15	−1.00E+08
kleemin6	6	6	21	−1.00E+10
kleemin7	7	7	28	−1.00E+12
kleemin8	8	8	36	−1.00E+14
l9	244	1,401	4,577	9.98E−01
large000	4,239	6,833	16,573	7.26E+01
large001	4,162	6,834	17,225	3.32E+04
large002	4,249	6,835	18,330	2.88E+00

(continued)

Table 3.4 (continued)

Name	Constraints	Variables	Nonzeros	Optimal value
large003	4,200	6,835	18,016	2.46E+04
large004	4,250	6,836	17,739	1.75E+02
large005	4,237	6,837	17,575	4.12E+02
large006	4,249	6,837	17,887	2.47E+02
large007	4,236	6,836	17,856	5.03E+02
large008	4,248	6,837	17,898	2.68E+02
large009	4,237	6,837	17,878	5.03E+02
large010	4,247	6,837	17,887	2.56E+02
large011	4,236	6,837	17,878	4.94E+02
large012	4,253	6,838	17,919	2.00E+02
large013	4,248	6,838	17,941	5.71E+02
large014	4,271	6,838	17,979	1.72E+02
large015	4,265	6,838	17,957	6.73E+02
large016	4,287	6,838	18,029	1.63E+02
large017	4,277	6,837	17,983	6.26E+02
large018	4,297	6,837	17,791	2.15E+02
large019	4,300	6,836	17,786	5.26E+02
large020	4,315	6,837	18,136	3.78E+02
large021	4,311	6,838	18,157	4.22E+02
large022	4,312	6,834	18,104	3.77E+02
large023	4,302	6,835	18,123	3.62E+02
large024	4,292	6,831	18,599	3.56E+02
large025	4,297	6,832	18,743	3.55E+02
large026	4,284	6,824	18,631	3.56E+02
large027	4,275	6,821	18,562	3.34E+02
large028	4,302	6,833	18,886	3.11E+02
large029	4,301	6,832	18,952	2.87E+02
large030	4,285	6,823	18,843	2.65E+02
large031	4,294	6,826	18,867	2.45E+02
large032	4,292	6,827	18,850	2.26E+02
large033	4,273	6,817	18,791	2.10E+02
large034	4,294	6,831	18,855	1.99E+02
large035	4,293	6,829	18,881	1.86E+02
large036	4,282	6,822	18,840	1.71E+02
lp22	2,958	13,434	65,560	4.59E+02
lpl1	39,951	125,000	381,259	6.72E+10
lpl2	3,294	10,755	32,106	4.47E+03
lpl3	10,828	33,538	100,377	2.73E+04
mod2	34,774	31,728	165,129	4.36E+07
model1	4,400	15,447	149,000	−1.55E+05
model2	7,056	18,288	55,859	0.00E+00
model3	362	798	3,028	0.00E+00
model4	379	1,212	7,498	−7.40E+03

(continued)

Table 3.4 (continued)

Name	Constraints	Variables	Nonzeros	Optimal value
model5	1,609	3,840	23,236	1.75E+04
model6	1,337	4,549	45,340	1.11E+06
model7	1,888	11,360	89,483	−8.58E+05
model8	2,096	5,001	27,340	1.18E+05
model9	3,358	8,007	49,452	4.94E+04
model10	2,896	6,464	25,277	0.00E+00
model11	2,879	10,257	55,274	−1.41E+05
multi	61	102	961	4.44E+04
nemsafm	334	2,252	2,730	−6.79E+03
nemscem	651	1,570	3,698	8.98E+04
nemsemm1	3,945	71,413	1,050,047	5.13E+05
nemsemm2	6,943	42,133	175,267	5.81E+05
nemspmm1	2,372	8,622	55,586	−3.27E+05
nemspmm2	2,301	8,413	67,904	−2.92E+05
nemswrld	7,138	27,174	190,907	2.60E+02
nl	7,039	9,718	41,428	1.23E+06
nsct1	22,901	14,981	656,259	−3.89E+07
nsct2	23,003	14,981	675,156	−3.72E+07
nsic1	451	463	2,853	−9.17E+06
nsic2	465	463	3,015	−8.20E+06
nsir1	4,407	5,717	138,955	−2.89E+07
nsir2	4,453	5,717	150,599	−2.72E+07
nug05	210	225	1,050	5.00E+01
nug06	372	486	2,232	8.60E+01
nug07	602	931	4,214	1.48E+02
nug08	912	1,632	7,296	2.04E+02
nug12	3,192	8,856	38,304	5.23E+02
nug15	6,330	22,275	94,950	1.04E+03
nw14	73	123,409	904,910	6.18E+04
orna1	882	882	3,108	−5.04E+08
orna2	882	882	3,108	−5.86E+08
orna3	882	882	3,108	−5.84E+08
orna4	882	882	3,108	6.72E+08
orna7	882	882	3,108	−5.84E+08
orswq2	80	80	264	4.85E−01
p0033	15	33	98	2.52E+03
p0040	23	40	110	6.18E+04
p010	10,090	19,000	117,910	1.12E+06
p0201	133	201	1,923	6.88E+03
p0282	241	282	1,966	1.77E+05
p0291	252	291	2,031	1.71E+03
p05	176	548	1,711	3.15E+02
p0548	5,090	9,500	58,955	5.56E+05

(continued)

Table 3.4 (continued)

Name	Constraints	Variables	Nonzeros	Optimal value
p19	284	586	5,305	2.54E+05
p2756	755	2,756	8,937	2.69E+03
p6000	2,095	5,872	17,731	−2.35E+06
pcb1000	1,565	2,428	20,071	5.68E+04
pcb3000	3,960	6,810	56,557	1.37E+05
pf2177	9,728	900	21,706	9.00E+01
pldd000b	3,069	3,267	8,980	2.74E+00
pldd001b	3,069	3,267	8,981	3.81E+00
pldd002b	3,069	3,267	8,982	4.07E+00
pldd003b	3,069	3,267	8,983	4.03E+00
pldd004b	3,069	3,267	8,984	4.18E+00
pldd005b	3,069	3,267	8,985	4.15E+00
pldd006b	3,069	3,267	8,986	4.18E+00
pldd007b	3,069	3,267	8,987	4.04E+00
pldd008b	3,069	3,267	9,047	4.17E+00
pldd009b	3,069	3,267	9,050	4.54E+00
pldd010b	3,069	3,267	9,053	4.83E+00
pldd011b	3,069	3,267	9,055	4.97E+00
pldd012b	3,069	3,267	9,057	4.61E+00
primagaz	1,554	10,836	21,665	1.07E+09
problem	12	46	86	0.00E+00
progas	1,650	1,425	8,422	7.61E+05
qiulp	1,192	840	3,432	−9.32E+02
r05	5,190	9,500	103,955	5.58E+05
rat1	3,136	9,408	88,267	2.00E+06
rat5	3,136	9,408	137,413	3.08E+06
rat7a	3,136	9,408	268,908	2.07E+06
refine	29	33	124	−3.93E+05
rlfddd	4,050	57,471	260,577	−1.30E+01
rlfdual	8,052	66,918	273,979	−1.00E+00
rlfprim	58,866	8,052	265,927	1.00E+00
rosen1	2,056	4,096	62,136	−1.74E+05
rosen2	1,032	2,048	46,504	−5.44E+04
rosen7	264	512	7,770	−2.03E+04
rosen8	520	1,024	15,538	−4.21E+04
rosen10	2,056	4,096	62,136	−1.74E+05
route	20,894	23,923	187,686	5.94E+03
seymourl	4,944	1,372	33,549	4.04E+02
slptsk	2,861	3,347	72,465	2.99E+01
small000	709	1,140	2,749	2.13E+01
small001	687	1,140	2,871	2.01E+03
small002	713	1,140	2,946	3.77E+01

(continued)

Table 3.4 (continued)

Name	Constraints	Variables	Nonzeros	Optimal value
small003	711	1,140	2,945	1.85E+02
small004	717	1,140	2,983	5.47E+01
small005	717	1,140	3,017	3.42E+01
small006	710	1,138	3,024	2.44E+01
small007	711	1,137	3,079	1.39E+01
small008	712	1,134	3,042	8.05E+00
small009	710	1,135	3,030	4.91E+00
small010	711	1,138	3,027	2.26E+00
small011	705	1,133	3,005	8.87E−01
small012	706	1,134	3,014	6.23E−01
small013	701	1,131	2,989	9.63E−01
small014	687	1,130	2,927	1.23E+00
small015	683	1,130	2,967	1.68E+00
small016	677	1,130	2,937	2.20E+00
south31	18,425	35,421	111,498	1.82E+09
stat96v1	5,995	197,472	588,798	3.74E+00
stat96v2	29,089	957,432	2,852,184	3.17E+00
stat96v3	33,841	1,113,780	3,317,736	3.74E+00
stat96v4	3,173	62,212	490,472	−1.16E+00
stat96v5	2,307	75,779	233,921	1.32E+00
sws	14,310	12,465	93,015	2.60E+08
t0331-4l	664	46,915	430,982	2.97E+04
testbig	17,613	31,223	61,639	−6.04E+01
ulevimin	6,590	44,605	162,206	−5.19E+09
us04	163	28,016	297,538	1.77E+04
world	34,506	32,734	164,470	6.91E+07
zed	116	43	567	−1.51E+04

Table 3.5 Statistics of the problematic section of the Mészáros LP test set

Name	Constraints	Variables	Nonzeros	Optimal value
de063155	852	1,487	4,553	9.88E+09
de063157	936	1,487	4,699	2.15E+07
de080285	936	1,487	4,662	1.39E+01
gen	769	2,560	63,085	0.00E+00
gen1	769	2,560	63,085	0.00E+00
gen2	1,121	3,264	81,855	3.29E+00
gen4	1,537	4,297	107,102	0.00E+00
iprob	3,001	3,001	9,000	Infeasible
l30	2,701	15,380	51,169	9.53E−01

Table 3.6 Statistics of the new section of the Mészáros LP test set

Name	Constraints	Variables	Nonzeros	Optimal value
degme	185,501	659,415	8,127,528	−4.69E+06
karted	46,502	133,115	1,770,349	−4.73E+05
tp-6	142,752	1,014,301	11,537,419	−1.32E+07
ts-palko	22,002	47,235	1,076,903	−1.49E+04

Table 3.7 Statistics of the stochlp section of the Mészáros LP test set

Name	Constraints	Variables	Nonzeros	Optimal value
aircraft	3,754	7,517	20,267	1.57E+03
cep1	1,521	3,248	6,712	3.55E+05
deter0	1,923	5,468	11,173	−2.05E+00
deter1	5,527	15,737	32,187	−2.56E+00
deter2	6,095	17,313	35,731	−1.83E+00
deter3	7,647	21,777	44,547	−2.11E+00
deter4	3,235	9,133	19,231	−1.43E+00
deter5	5,103	14,529	29,715	−2.26E+00
deter6	4,255	12,113	24,771	−2.31E+00
deter7	6,375	18,153	37,131	−2.23E+00
deter8	3,831	10,905	22,299	−2.79E+00
fxm2-6	1,520	2,172	12,139	1.84E+04
fxm2-16	3,900	5,602	32,239	1.84E+04
fxm3_6	6,200	9,492	54,589	1.86E+04
fxm3_16	41,340	64,162	370,839	1.84E+04
fxm4_6	22,400	30,732	248,989	1.86E+04
pgp2	4,034	9,220	18,440	4.47E+02
pltexpa2-6	686	1,820	3,708	−9.48E+00
pltexpa2-16	1,726	4,540	9,223	−9.66E+00
pltexpa3-6	4,430	11,612	23,611	−1.40E+01
pltexpa3-16	28,350	74,172	150,801	−1.42E+01
pltexpa4-6	26,894	70,364	143,059	−1.96E+01
sc205-2r-4	101	102	270	−6.04E+01
sc205-2r-8	189	190	510	−6.04E+01
sc205-2r-16	365	366	990	−5.54E+01
sc205-2r-27	607	608	1,650	−1.51E+01
sc205-2r-32	717	718	1,950	−5.54E+01
sc205-2r-50	1,113	1,114	3,030	−3.08E+01
sc205-2r-64	1,421	1,422	3,870	−5.54E+01
sc205-2r-100	2,213	2,214	6,030	−1.01E+01
sc205-2r-200	4,413	4,414	12,030	−1.01E+01
sc205-2r-400	8,813	8,814	24,030	−1.01E+01
sc205-2r-800	17,613	17,614	48,030	−1.01E+01
sc205-2r-1600	35,213	35,214	96,030	0.00E+00

(continued)

Table 3.7 (continued)

Name	Constraints	Variables	Nonzeros	Optimal value
scagr7-2b-4	167	180	546	$-8.33E+05$
scagr7-2b-16	623	660	2,058	$-8.33E+05$
scagr7-2b-64	9,743	10,260	32,298	$-8.33E+05$
scagr7-2c-4	167	180	546	$-8.32E+05$
scagr7-2c-16	623	660	2,058	$-8.32E+05$
scagr7-2c-64	2,447	2,580	8,106	$-8.19E+05$
scagr7-2r-4	167	180	546	$-8.33E+05$
scagr7-2r-8	319	340	1,050	$-8.33E+05$
scagr7-2r-16	623	660	2,058	$-8.33E+05$
scagr7-2r-27	1,041	1,100	3,444	$-8.34E+05$
scagr7-2r-32	1,231	1,300	4,074	$-8.33E+05$
scagr7-2r-54	2,067	2,180	6,846	$-8.34E+05$
scagr7-2r-64	2,447	2,580	8,106	$-8.33E+05$
scagr7-2r-108	4,119	4,340	13,542	$-8.34E+05$
scagr7-2r-216	8,223	8,660	27,042	$-8.34E+05$
scagr7-2r-432	16,431	17,300	54,042	$-8.34E+05$
scagr7-2r-864	32,847	34,580	108,042	$-8.34E+05$
scfxm1-2b-4	684	1,014	3,999	$2.88E+03$
scfxm1-2b-16	2,460	3,714	13,959	$2.88E+03$
scfxm1-2b-64	19,036	28,914	106,919	$2.88E+03$
scfxm1-2c-4	684	1,014	3,999	$2.88E+03$
scfxm1-2r-4	684	1,014	3,999	$2.88E+03$
scfxm1-2r-8	1,276	1,914	7,319	$2.88E+03$
scfxm1-2r-16	2,460	3,714	13,959	$2.88E+03$
scfxm1-2r-27	4,088	6,189	23,089	$2.89E+03$
scfxm1-2r-32	4,828	7,314	27,239	$2.88E+03$
scfxm1-2r-64	9,564	14,514	53,799	$2.88E+03$
scfxm1-2r-96	14,300	21,714	80,359	$2.89E+03$
scfxm1-2r-128	19,036	28,914	106,919	$2.89E+03$
scfxm1-2r-256	37,980	57,714	213,159	$2.89E+03$
scrs8-2b-4	140	189	457	$1.12E+02$
scrs8-2b-16	476	645	1,633	$1.12E+02$
scrs8-2b-64	1,820	2,469	6,337	$1.12E+03$
scrs8-2c-4	140	189	457	$1.12E+02$
scrs8-2c-8	252	341	849	$1.12E+02$
scrs8-2c-16	476	645	1,633	$1.12E+02$
scrs8-2c-32	924	1,253	3,201	$1.12E+02$
scrs8-2c-64	1,820	2,469	6,337	$1.12E+02$
scrs8-2r-4	140	189	457	$1.23E+02$
scrs8-2r-8	252	341	849	$1.12E+03$
scrs8-2r-16	476	645	1,633	$1.23E+02$
scrs8-2r-27	784	1,063	2,711	$5.87E+02$
scrs8-2r-32	924	1,253	3,201	$1.23E+02$

(continued)

Table 3.7 (continued)

Name	Constraints	Variables	Nonzeros	Optimal value
scrs8-2r-64	1,820	2,469	6,337	1.23E+02
scrs8-2r-64b	1,820	2,469	6,337	1.35E+03
scrs8-2r-128	3,612	4,901	12,609	1.18E+03
scrs8-2r-256	7,196	9,765	25,153	1.14E+03
scrs8-2r-512	14,364	19,493	50,241	1.07E+03
scsd8-2b-4	90	630	1,890	1.53E+01
scsd8-2b-16	330	2,310	7,170	2.75E+01
scsd8-2b-64	5,130	35,910	112,770	1.60E+01
scsd8-2c-4	90	630	1,890	1.50E+01
scsd8-2c-16	330	2,310	7,170	1.50E+01
scsd8-2c-64	5,130	35,910	112,770	1.50E+01
scsd8-2r-4	90	630	1,890	1.55E+01
scsd8-2r-8	170	1,190	3,650	1.60E+01
scsd8-2r-8b	170	1,190	3,650	1.60E+01
scsd8-2r-16	330	2,310	7,170	1.60E+01
scsd8-2r-27	550	3,850	12,010	2.40E+01
scsd8-2r-32	650	4,550	14,210	1.60E+01
scsd8-2r-54	1,090	7,630	23,890	2.39E+01
scsd8-2r-64	1,290	9,030	28,290	1.58E+01
scsd8-2r-108	2,170	15,190	47,650	2.37E+01
scsd8-2r-216	4,330	30,310	95,170	2.34E+01
scsd8-2r-432	8,650	60,550	190,210	2.58E+01
sctap1-2b-4	270	432	1,516	2.39E+02
sctap1-2b-16	990	1,584	5,740	2.81E+02
sctap1-2b-64	15,390	24,624	90,220	4.85E+02
sctap1-2c-4	270	432	1,516	2.36E+02
sctap1-2c-16	990	1,584	5,740	3.26E+02
sctap1-2c-64	3,390	5,424	19,820	2.00E+02
sctap1-2r-4	270	432	1,516	2.81E+02
sctap1-2r-8	510	816	2,924	3.61E+02
sctap1-2r-8b	510	816	2,924	2.50E+02
sctap1-2r-16	990	1,584	5,740	3.59E+02
sctap1-2r-27	1,650	2,640	9,612	2.48E+02
sctap1-2r-32	1,950	3,120	11,372	3.54E+02
sctap1-2r-54	3,270	5,232	19,116	2.49E+02
sctap1-2r-64	3,870	6,192	22,636	3.44E+02
sctap1-2r-108	6,510	10,416	38,124	2.48E+02
sctap1-2r-216	12,990	20,784	76,140	2.49E+02
sctap1-2r-480	28,830	46,128	169,068	2.49E+02
stormg2_1000	528,185	1,259,121	3,341,696	1.58E+07
stormg2-8	4,409	10,193	27,424	1.55E+07
stormg2-27	14,441	34,114	90,903	1.55E+07
stormg2-125	66,185	157,496	418,321	1.55E+07

Table 3.8 Statistics of the Mittelmann LP test set

Name	Constraints	Variables	Nonzeros	Optimal value
16_n14	16,384	262,144	524,288	3.24E+11
cont1	160,792	40,398	399,990	0.0088
cont1_l	1,918,399	641,598	5,752,001	2.77E−03
cont4	160,792	40,398	398,398	6.25E−03
cont11	160,792	80,396	399,990	6.10E+01
cont11_l	1,468,599	981,396	4,403,001	2.06E+02
fome11	12,142	24,460	71,264	2.25E+07
fome12	24,284	48,920	142,528	4.51E+07
fome13	48,568	97,840	285,056	9.01E+07
fome20	33,874	105,728	230,200	2.38E+10
fome21	67,748	211,456	460,400	4.73E+10
i_n13	8,192	741,455	1,482,910	8.48E+11
L1_sixm250obs	986,069	428,032	4,280,320	4.63E+03
Linf_520c	93,326	69,004	566,193	1.99E−01
lo10	46,341	406,225	812,450	1.45E+12
long15	32,769	753,687	1,507,374	6.33E+10
neos	479,119	36,786	1,047,675	2.25E+08
neos1	131,581	1,892	468,009	1.90E+01
neos2	132,568	1,560	552,519	4.55E+02
neos3	512,209	6,624	1,542,816	2.78E+04
netlarge1	47,700	8,001,358	16,002,716	2.33E+08
netlarge2	40,000	1,160,000	2,320,000	5.74E+08
netlarge3	40,000	4,676,000	9,352,000	1.52E+08
netlarge6	8,000	15,000,000	30,000,000	1.23E+09
ns1687037	50,622	43,749	1,406,739	3.66E+00
ns1688926	32,768	16,587	1,712,128	2.62E+02
nug08-3rd	19,728	20,448	139,008	2.14E+02
nug20	15,240	72,600	304,800	2.18E+03
nug30	52,260	379,350	1,567,800	4.80E+03
pds-20	33,874	105,728	230,200	2.38E+10
pds-30	49,944	154,998	337,144	2.14E+10
pds-40	66,844	212,859	462,128	1.89E+10
pds-50	83,060	270,095	585,114	1.66E+10
pds-60	99,431	329,643	712,779	1.43E+10
pds-70	114,944	382,311	825,771	1.22E+10
pds-80	12,918	426,278	919,524	1.15E+10
pds-90	142,823	466,671	1,005,359	1.11E+10
pds-100	156,243	505,360	1,086,785	1.09E+10
rail507	507	63,009	409,349	1.72E+02
rail516	516	47,311	314,896	1.82E+02
rail582	582	55,515	401,708	2.10E+02
rail2586	2,586	920,683	8,008,776	9.36E+02

(continued)

Table 3.8 (continued)

Name	Constraints	Variables	Nonzeros	Optimal value
rail4284	4,284	1,092,610	11,279,748	1.05E+03
sgpf5y6	246,077	308,634	828,070	−6.44E+03
spal_004	10,203	321,696	46,168,124	−4.98E−01
square15	32,762	753,526	1,507,052	6.15E+10
stormG2_1000	528,185	1,259,121	3,341,696	1.58E+07
watson_1	201,155	383,927	1,052,028	−1.69E+03
watson_2	352,013	671,861	1,841,028	−7.54E+03
wide15	32769	753687	1507374	6.33E+10

Table 3.9 Statistics of the benchmark set used in the computational studies of this book

Name	Constraints	Variables	Nonzeros	Optimal value
beaconfd	173	262	3,375	3.36E+04
cari	400	1,200	152,800	5.82E+02
farm	7	12	36	1.75E+04
itest6	11	8	20	Infeasible
klein2	477	54	4,585	Infeasible
nsic1	451	463	2,853	−9.17E+06
nsic2	465	463	3,015	−8.20E+06
osa-07	1,118	23,949	143,694	5.36E+05
osa-14	2,337	52,460	314,760	1.11E+06
osa-30	4,350	100,024	600,138	2.14E+06
rosen1	520	1,024	23,274	−2.76E+04
rosen10	2,056	4,096	62,136	−1.74E+05
rosen2	1,032	2,048	46,504	−5.44E+04
rosen7	264	512	7,770	−2.03E+04
rosen8	520	1,024	15,538	−4.21E+04
sc205	205	203	551	−5.22E+01
scfxm1	330	457	2,589	1.84E+04
sctap2	1,090	1,880	6,714	1.72E+03
sctap3	1,480	2,480	8,874	1.42E+03

In the next chapters, we use the LPs presented in Table 3.9 to experiment with different algorithms and techniques. Our aim is to present the computational performance of the presented methods and algorithms, and not to perform a thorough computational study. Hence, we have selected only a limited number of representative LPs. All computational studies have been performed on a quad-processor Intel Core i7 3.4 GHz with 32 GB of main memory and 8 cores, a clock of 3700 MHz, an L1 code cache of 32 KB per core, an L1 data cache of 32 KB per core, an L2 cache of 256 KB per core, an L3 cache of 8 MB and a memory bandwidth of 21 GB/s, running under Microsoft Windows 7 64-bit. The MATLAB version used in this book is R2015a. In Appendix B, we use CLP 1.16.9 and CPLEX 12.6.1.

3.7 Chapter Review

The MPS format is the most widely accepted standard for defining LPs. The vast majority of solvers takes as input an LP problem in MPS format. The given LP problem can be either a benchmark problem, i.e., a hard problem to solve that is publicly available, or a randomly generated LP problem. In this chapter, we presented the MPS format and two codes in MATLAB that can be used to convert an MPS file to MATLAB's matrix format and vice versa. Moreover, codes that can be used to create randomly generated sparse or dense LPs were also given. Finally, the most well-known benchmark libraries for LPs were also presented.

References

1. Netlib Repository. (2017). *Netlib LP test set.* Available online at: http://www.netlib.org/lp/data/ (Last access on March 31, 2017).
2. Netlib Repository. (2017). *Kennington LP test set.* Available online at: http://www.netlib.org/lp/data/kennington/ (Last access on March 31, 2017).
3. Netlib Repository. (2017). *infeas section of the Netlib LP test set.* Available online at: http://www.netlib.org/lp/infeas/ (Last access on March 31, 2017).
4. Mészáros, C. (2017). *The misc section of the Mészáros LP test set.* Available online at: http://www.sztaki.hu/~meszaros/public_ftp/lptestset/misc/ (Last access on March 31, 2017).
5. Mészáros, C. (2017). *The problematic section of the Mészáros LP test set.* Available online at: http://www.sztaki.hu/~meszaros/public_ftp/lptestset/problematic/ (Last access on March 31, 2017).
6. Mészáros, C. (2017). *The new section of the Mészáros LP test set.* Available online at: http://www.sztaki.hu/~meszaros/public_ftp/lptestset/New/ (Last access on March 31, 2017).
7. Mészáros, C. (2017). *The stochlp section of the Mészáros LP test set.* Available online at: http://www.sztaki.hu/~meszaros/public_ftp/lptestset/stochlp/ (Last access on March 31, 2017).

8. Mittelmann, H. (2017). *The Mittelmann LP test set*. Available online at: http://plato.asu.edu/ ftp/lptestset/ (Last access on March 31, 2017).
9. Paparrizos, K., Samaras, N., & Stephanides, G. (2000). A method for generating random optimal linear problems and a comparative computational study. In *Proceedings of the 13th National Conference Hellenic Operational Research Society*, 30 November–2 December, University of Piraeus, Greece, 785–794 (In Greek).
10. Ordóñez, F., & Freund, R. M. (2003). Computational experience and the explanatory value of condition measures for linear optimization. *SIAM Journal on Optimization, 14*(2), 307–333.

Chapter 4
Presolve Methods

Abstract Presolve methods are important in solving LPs, as they reduce the size of the problem and discover whether an LP is unbounded or infeasible. Presolve methods are used prior to the application of an LP algorithm in order to: (i) eliminate redundant constraints, (ii) fix variables, (iii) transform bounds of single structural variables, and (iv) reduce the number of variables and constraints by eliminations. This chapter presents eleven presolve methods used prior to the execution of an LP algorithm: (i) eliminate zero rows, (ii) eliminate zero columns, (iii) eliminate singleton equality constraints, (iv) eliminate $k$ton equality constraints, (v) eliminate singleton inequality constraints, (vi) eliminate dual singleton inequality constraints, (vii) eliminate implied free singleton columns, (viii) eliminate redundant columns, (ix) eliminate implied bounds on rows, (x) eliminate redundant rows, and (xi) make coefficient matrix structurally full rank. Each method is presented with: (i) its mathematical formulation, (ii) a thorough numerical example, and (iii) its implementation in MATLAB. In addition, we discuss how to transform a solution back in terms of the original variables and constraints of the problem (postsolve). Finally, a computational study on benchmark LPs is performed. The aim of the computational study is twofold: (i) compare the execution time and the reduction to the problem size of the aforementioned presolve methods, and (ii) investigate the impact of preprocessing prior to the application of LP algorithms. The execution time and the number of iterations with and without preprocessing are presented.

4.1 Chapter Objectives

- Present the importance of preprocessing prior to the execution of a linear programming algorithm.
- Provide the mathematical formulation of the presolve methods.
- Present the presolve methods through an illustrative numerical example.
- Implement the presolve methods using MATLAB.
- Compare the computational performance of the presolve methods.

The original version of this chapter was revised. An erratum to this chapter can be found at https://doi.org/10.1007/978-3-319-65919-0_13

© Springer International Publishing AG 2017   135
N. Ploskas, N. Samaras, *Linear Programming Using MATLAB®*,
Springer Optimization and Its Applications 127, DOI 10.1007/978-3-319-65919-0_4

4.2 Introduction

Preprocessing is a procedure where a number of presolve methods are applied on an LP problem prior to the application of an LP algorithm. A presolve method attempts to identify and remove as many redundancies as possible. The number of constraints and variables of the LPs are these that determine the increase or decrease in the computational complexity [3]. The purpose of the presolve analysis is to improve problem numerical properties and computational characteristics. Using presolve methods prior to the application of an LP algorithm, we can achieve significant reductions to the problem size. This reduction can result to better execution times in order to solve the problem with an LP solver. Presolve methods are used in order to [16]: (i) eliminate redundant constraints and variables, (ii) fix variables, (iii) transform bounds of single structural variables, (iv) reduce the number of variables and constraints by eliminations, and (v) improve sparsity. The reductions to the problem size can be achieved by eliminating redundant constraints and variables. Moreover, presolve methods are able to spot possible infeasibility and unboundedness of the problem prior to the application of an LP algorithm. Hence, presolve analysis became a fundamental component of all commercial and noncommercial LP solvers.

Preprocessing is not a new research area. Researchers have previously discussed and studied thoroughly the use of presolve methods in LP [4, 13, 16, 20]. Brearly et al. [4] proposed many of the presolve methods presented in this chapter, i.e., eliminate zero rows, zero columns, singleton rows, singleton columns, primal and dual forcing constraints, bound compatibility, redundant constraints, and implied free constraints. Tomlin and Welch [18] presented the preprocessing for various types of problem reductions. Other approaches included the idea of using general linear transformations on the constraints in order to reduce the number of nonzero elements in the LPs can be found in [1, 5, 15]. Andersen and Andersen [2] performed different combinations of presolve methods presenting the reduction in the number of variables and the execution time. Andersen and Andersen [2] presented a thorough review of many presolve methods. A new set of presolve methods for LPs with box constraints was presented by Ioslovich [12]. Solving large-scale LPs efficiently has always been a challenge for Operations Research community. The wide range of industrial and scientific applications that requires fast and efficient LP solvers is the reason for the existence of presolve methods. These presolve methods can be combined with simplex-type algorithms and interior-point algorithms [9, 11, 19, 21].

Gould and Toint [10] proposed new presolve methods for both quadratic and linear optimization problems. They proposed a set of presolve methods for LPs showing a reduction of 20% in the number of variables and a reduction of 10% in the execution time. Elble [8] presented an out-of-order preprocessor randomizing the order in which the techniques are executed. Swietanowski [17] proposed a set of

presolve methods for LPs showing a reduction of 5.72% in the number of iterations
and a reduction of 19.36% in the execution time.

This chapter presents eleven presolve methods used prior to the execution of an
LP algorithm: (i) eliminate zero rows, (ii) eliminate zero columns, (iii) eliminate
singleton equality constraints, (iv) eliminate $k$ton equality constraints, (v) eliminate
singleton inequality constraints, (vi) eliminate dual singleton inequality constraints,
(vii) eliminate implied free singleton columns, (viii) eliminate redundant columns,
(ix) eliminate implied bounds on rows, (x) eliminate redundant rows, and (xi)
make coefficient matrix structurally full rank. Each method is presented with:
(i) its mathematical formulation, (ii) a thorough numerical example, and (iii)
its implementation in MATLAB. Finally, a computational study is performed. A
presolve session is repeated until the constraint matrix A has been passed twice
with no reduction. The aim of the computational study is twofold: (i) compare the
execution time and the reduction to the problem size of the previously presented
presolve methods, and (ii) investigate the impact of preprocessing prior to the
application of LP algorithms. The execution time and the number of iterations with
and without preprocessing are presented.

The structure of this chapter is as follows. Section 4.3 presents the mathematical
background. In Sections 4.4–4.14, eleven presolve methods are presented. More
specifically, Section 4.4 presents the presolve method that eliminates zero rows,
Section 4.5 presents the presolve method that eliminates zero columns, Section 4.6
presents the presolve method that eliminates singleton equality constraints, Section 4.7 presents the presolve method that eliminates $k$ton equality constraints,
Section 4.8 presents the presolve method that eliminates singleton inequality
constraints, Section 4.9 presents the presolve method that eliminates dual singleton
inequality constraints, Section 4.10 presents the presolve method that eliminates
implied free singleton columns, Section 4.11 presents the presolve method that
eliminates redundant columns, Section 4.12 presents the presolve method that
eliminates implied bounds on rows, Section 4.13 presents the presolve method
that eliminates redundant rows, while Section 4.14 presents the presolve method
that makes the coefficient matrix structurally full rank. Section 4.15 presents the
MATLAB code for the whole presolve analysis. Section 4.16 presents guidelines on
how to transform a solution back in terms of the original variables and constraints of
the problem (postsolve). Section 4.17 presents a computational study that compares
the execution time and the reduction to the problem size of the presented presolve
methods and investigates the impact of preprocessing prior to the application
of LP algorithms. Finally, conclusions and further discussion are presented in
Section 4.18.

4.3   Background

The following LP problem in canonical form is considered throughout this chapter:

$$\min \quad z = c^T x$$
$$\text{s.t.} \quad \underline{b} \le Ax \le \overline{b}$$
$$\underline{x} \le x \le \overline{x}$$

where $c, x \in \mathbb{R}^n, A \in \mathbb{R}^{m \times n}, \underline{x} = 0, \overline{x} = \infty, \underline{b} = (\mathbb{R} \cup \{-\infty\})^m, \overline{b} = (\mathbb{R} \cup \{+\infty\})^m$, and T denotes transposition. Let $A_{i.}$ be the ith row of matrix A and $A_{.j}$ be the jth column of matrix A.

4.4 Eliminate Zero Rows

4.4.1 Mathematical Formulation

A row of the coefficient matrix A is an empty row iff all the coefficients in that row are equal to zero. A zero row can be formulated as:

$$\underline{b}_i \le A_{i1}x_1 + A_{i2}x_2 + \cdots + A_{in}x_n \le \overline{b}_i \tag{4.1}$$

where $A_{ij} = 0, i = 1, 2, \cdots, m$, and $j = 1, 2, \cdots, n$. A constraint of this type may be redundant or may state that the LP problem is infeasible. All possible cases are distinguished in the following theorem.

Theorem 4.1 *For each empty row, we distinguish the following cases:*

1. $A_{i1}x_1 + A_{i2}x_2 + \cdots + A_{in}x_n \le \overline{b}_i$ and $\overline{b}_i \ge 0$: The constraint is redundant and can be deleted.
2. $A_{i1}x_1 + A_{i2}x_2 + \cdots + A_{in}x_n \le \overline{b}_i$ and $\overline{b}_i < 0$: The LP problem is infeasible.
3. $A_{i1}x_1 + A_{i2}x_2 + \cdots + A_{in}x_n \ge \underline{b}_i$ and $\underline{b}_i \le 0$: The constraint is redundant and can be deleted.
4. $A_{i1}x_1 + A_{i2}x_2 + \cdots + A_{in}x_n \ge \underline{b}_i$ and $\underline{b}_i > 0$: The LP problem is infeasible.
5. $A_{i1}x_1 + A_{i2}x_2 + \cdots + A_{in}x_n \ge \underline{b}_i = \overline{b}_i = b_i$ and $b_i = 0$: The constraint is redundant and can be deleted.
6. $A_{i1}x_1 + A_{i2}x_2 + \cdots + A_{in}x_n \ge \underline{b}_i = \overline{b}_i = b_i$ and $b_i \ne 0$: The LP problem is infeasible.

Proof In order to prove the different cases of this theorem, we use the standard form of the LP problem.

1. Adding the slack variable x_{n+1}, the empty constraint can be written as:

$$0x_1 + 0x_2 + \cdots + 0x_n + x_{n+1} = \overline{b}_i \ge 0$$

where $x_{n+1} \ge 0$. It is obvious that this row includes only one nonzero element, $A_{i,n+1} = 1$. This is a singleton equality constraint (more information in Section 4.6) and can be written as:

$$x_{n+1} = \overline{b}_i \geq 0$$

Hence, the value of x_{n+1} is fixed at b_i. We replace variable x_{n+1} to all constraints. The new right-hand side is:

$$\overline{\overline{b}} = \overline{b} - x_{n+1}$$

After that replacement, row i is deleted from matrix A, element i is deleted from vectors $\overline{\overline{b}}$ and $Eqin$, column $n+1$ is deleted from matrix A, and element $n+1$ is deleted from vector c.

2. Adding the slack variable x_{n+1}, the empty constraint can be written as:

$$0x_1 + 0x_2 + \cdots + 0x_n + x_{n+1} = \overline{b}_i \leq 0$$

where $x_{n+1} \geq 0$. However, solving the constraint for x_{n+1}, we find out that $x_{n+1} = \overline{b}_i \leq 0$. Hence, the LP problem is infeasible.

3. Subtracting the slack variable x_{n+1}, the empty constraint can be written as:

$$0x_1 + 0x_2 + \cdots + 0x_n - x_{n+1} = \underline{b}_i \leq 0$$

where $x_{n+1} \geq 0$. The empty constraint can be written as:

$$x_{n+1} = \underline{b}_i \geq 0$$

and we continue as in case 1.

4. Subtracting the slack variable x_{n+1}, the empty constraint can be written as:

$$0x_1 + 0x_2 + \cdots + 0x_n - x_{n+1} = \underline{b}_i \geq 0$$

where $x_{n+1} \geq 0$. However, solving the constraint for x_{n+1}, we find out that $x_{n+1} = \underline{b}_i \leq 0$. Hence, the LP problem is infeasible.

5. The constraint is already an equality constraint, so we do not add a slack variable:

$$0x_1 + 0x_2 + \cdots + 0x_n = b_i = 0$$

It is obvious that this constraint is redundant and can be deleted.

6. The constraint is already an equality constraint, so we do not add a slack variable:

$$0x_1 + 0x_2 + \cdots + 0x_n = b_i \neq 0$$

It is obvious that the LP is infeasible.

$\square$

4.4.2   Numerical Example

This subsection demonstrates the presolve method that eliminates zero rows through two illustrative examples.

   1st example

The LP problem that will be presolved is the following:

$$
\begin{aligned}
\min z = -x_1 + 4x_2 + 5x_3 &- 2x_4 - 8x_5 + 2x_6 \\
\text{s.t.} \quad 2x_1 - 3x_2 \qquad\qquad &+ 3x_5 + x_6 \le \quad 9 \quad (1) \\
-x_1 + 3x_2 + 2x_3 \qquad &- x_5 - 2x_6 \ge \quad 1 \quad (2) \\
0x_1 + 0x_2 + 0x_3 + 0x_4 &+ 0x_5 + 0x_6 \ge -5 \quad (3) \\
7x_1 + 5x_2 + 2x_3 \qquad &- 2x_5 + 4x_6 = \quad 7 \quad (4) \\
0x_1 + 0x_2 + 0x_3 + 0x_4 &+ 0x_5 + 0x_6 \ge -10 \quad (5) \\
x_j \ge 0, \quad (j = 1,2,3,4,5,6) &
\end{aligned}
$$

In matrix notation the above LP problem is written as follows:

$$
A = \begin{bmatrix} 2 & -3 & 0 & 0 & 3 & 1 \\ -1 & 3 & 2 & 0 & -1 & -2 \\ 0 & 0 & 0 & 0 & 0 & 0 \\ 7 & 5 & 2 & 0 & -2 & 4 \\ 0 & 0 & 0 & 0 & 0 & 0 \end{bmatrix}, c = \begin{bmatrix} -1 \\ 4 \\ 5 \\ -2 \\ -8 \\ 2 \end{bmatrix}, b = \begin{bmatrix} 9 \\ 1 \\ -5 \\ 7 \\ -10 \end{bmatrix}, Eqin = \begin{bmatrix} -1 \\ 1 \\ 1 \\ 0 \\ 1 \end{bmatrix}
$$

Initially, we start searching for rows that have all their elements equal to zero. We observe that all elements of the third row are equal to zero:

$$
0x_1 + 0x_2 + 0x_3 + 0x_4 + 0x_5 + 0x_6 \ge -5
$$

According to the third case of the previous subsection ($A_{i1}x_1 + A_{i2}x_2 + \cdots + A_{in}x_n \ge \underline{b_i}$ and $\underline{b_i} \le 0$), the constraint is redundant and can be deleted. Hence, we can delete the third row from matrix A and the third element from vectors b and $Eqin$. The presolved LP problem is now:

$$
A = \begin{bmatrix} 2 & -3 & 0 & 0 & 3 & 1 \\ -1 & 3 & 2 & 0 & -1 & -2 \\ 7 & 5 & 2 & 0 & -2 & 4 \\ 0 & 0 & 0 & 0 & 0 & 0 \end{bmatrix}, c = \begin{bmatrix} -1 \\ 4 \\ 5 \\ -2 \\ -8 \\ 2 \end{bmatrix}, b = \begin{bmatrix} 9 \\ 1 \\ 7 \\ -10 \end{bmatrix}, Eqin = \begin{bmatrix} -1 \\ 1 \\ 0 \\ 1 \end{bmatrix}
$$

Similarly, we observe that all elements of the fourth row (initially fifth row) are equal to zero:

$$0x_1 + 0x_2 + 0x_3 + 0x_4 + 0x_5 + 0x_6 \geq -10$$

According to the third case of the previous subsection ($A_{i1}x_1 + A_{i2}x_2 + \cdots + A_{in}x_n \geq \underline{b}_i$ and $\underline{b}_i \leq 0$), the constraint is redundant and can be deleted. Hence, we can delete the fourth row from matrix A and the fourth element from vectors b and $Eqin$:

$$A = \begin{bmatrix} 2 & -3 & 0 & 0 & 3 & 1 \\ -1 & 3 & 2 & 0 & -1 & -2 \\ 7 & 5 & 2 & 0 & -2 & 4 \end{bmatrix}, c = \begin{bmatrix} -1 \\ 4 \\ 5 \\ -2 \\ -8 \\ 2 \end{bmatrix}, b = \begin{bmatrix} 9 \\ 1 \\ 7 \end{bmatrix}, Eqin = \begin{bmatrix} -1 \\ 1 \\ 0 \end{bmatrix}$$

Finally, the equivalent LP problem after presolve is the following:

$$\begin{aligned} \min z = -x_1 &+ 4x_2 + 5x_3 - 2x_4 - 8x_5 + 2x_6 \\ \text{s.t.} \quad 2x_1 &- 3x_2 \qquad\qquad\quad + 3x_5 + x_6 \leq 9 \\ -x_1 &+ 3x_2 + 2x_3 \qquad\quad - x_5 - 2x_6 \geq 1 \\ 7x_1 &+ 5x_2 + 2x_3 \qquad\quad - 2x_5 + 4x_6 = 7 \\ x_j \geq 0, &\quad (j = 1, 2, 3, 4, 5, 6) \end{aligned}$$

2nd example

The LP problem that will be presolved is the following:

$$\begin{aligned} \min z = -x_1 &+ 3x_2 + 4x_3 - 2x_4 - 8x_5 \\ \text{s.t.} \quad x_1 &- 4x_2 \qquad\qquad + 3x_5 \leq 12 \quad (1) \\ -x_1 &+ 2x_2 + 5x_3 \qquad - 2x_5 \geq 2 \quad (2) \\ 0x_1 &+ 0x_2 + 0x_3 + 0x_4 + 0x_5 = -5 \quad (3) \\ 5x_1 &+ 4x_2 + 3x_3 \qquad - 2x_5 = 7 \quad (4) \\ x_j \geq 0, &\quad (j = 1, 2, 3, 4, 5) \end{aligned}$$

In matrix notation the above LP problem is written as follows:

$$A = \begin{bmatrix} 1 & -4 & 0 & 0 & 3 \\ -1 & 2 & 5 & 0 & -2 \\ 0 & 0 & 0 & 0 & 0 \\ 5 & 4 & 3 & 0 & -2 \end{bmatrix}, c = \begin{bmatrix} -1 \\ 3 \\ 4 \\ -2 \\ -8 \end{bmatrix}, b = \begin{bmatrix} 12 \\ 2 \\ -5 \\ 7 \end{bmatrix}, Eqin = \begin{bmatrix} -1 \\ 1 \\ 0 \\ 0 \end{bmatrix}$$

Initially, we start searching for rows that have all their elements equal to zero. We observe that all elements of the third row are equal to zero:

$$0x_1 + 0x_2 + 0x_3 + 0x_4 + 0x_5 + 0x_6 = -5$$

According to the second case of the previous subsection ($A_{i1}x_1 + A_{i2}x_2 + \cdots + A_{in}x_n \leq \overline{b}_i$ and $\overline{b}_i < 0$), the LP problem is infeasible.

4.4.3   Implementation in MATLAB

This subsection presents the implementation in MATLAB of the presolve method that eliminates zero rows (filename: `eliminateZeroRows.m`). Some necessary notations should be introduced before the presentation of the implementation of the presolve method that eliminates zero rows. Let A be a $m \times n$ matrix with the coefficients of the constraints, b be a $m \times 1$ vector with the right-hand side of the constraints, $Eqin$ be a $m \times 1$ vector with the type of the constraints, and *infeasible* be a flag variable showing if the LP problem is infeasible or not.

The function for the presolve method that eliminates zero rows takes as input the matrix of coefficients of the constraints (matrix A), the vector of the right-hand side of the constraints (vector b), and the vector of the type of the constraints (vector $Eqin$), and returns as output the presolved matrix of coefficients of the constraints (matrix A), the presolved vector of the right-hand side of the constraints (vector b), the presolved vector of the type of the constraints (vector $Eqin$), and a flag variable showing if the LP problem is infeasible or not (variable *infeasible*).

In line 31, we compute the number of nonzero elements in each row. If any row has all its elements equal to zero, then we eliminate it or declare infeasibility (lines 33–66). Initially, we examine if the LP problem is infeasible according to cases 2, 4, and 6 that were presented in subsection 4.4.1 (lines 37–56). If the LP problem is not infeasible, then we eliminate zero rows and update matrix A and vectors b and $Eqin$ (lines 57–63). Finally, the number of the eliminated rows is printed (lines 64–65).

```
1.    function [A, b, Eqin, infeasible] = ...
2.        eliminateZeroRows(A, b, Eqin)
3.    % Filename: eliminateZeroRows.m
4.    % Description: the function is an implementation of the
5.    % presolve method that eliminates zero rows
6.    % Authors: Ploskas, N., & Samaras, N.
7.    %
8.    % Syntax: [A, b, Eqin, infeasible] = ...
9.    %    eliminateZeroRows(A, b, Eqin)
10.   %
11.   % Input:
12.   % -- A: matrix of coefficients of the constraints
13.   %    (size m x n)
14.   % -- b: vector of the right-hand side of the constraints
15.   %    (size m x 1)
```

```
16.   % -- Eqin: vector of the type of the constraints
17.   %      (size m x 1)
18.   %
19.   % Output:
20.   % -- A: presolved matrix of coefficients of the constraints
21.   %      (size m x n)
22.   % -- b: presolved vector of the right-hand side of the
23.   %      constraints (size m x 1)
24.   % -- Eqin: presolved vector of the type of the constraints
25.   %      (size m x 1)
26.   % -- infeasible: flag variable showing if the LP is
27.   %      infeasible or not
28.
29.   infeasible = 0;
30.   % compute the number of nonzero elements in each row
31.   delrows = sum(spones(A'));
32.   % if any row has all its element equal to zero
33.   if any(delrows == 0)
34.       % get the indices of all rows that have all their
35.       % elements equal to zero
36.       id = int16(find(delrows == 0));
37.       % check if the LP problem is infeasible (Case 2)
38.       k = int16(find(Eqin(id) == -1));
39.       if any(b(id(k)) < 0) && ~isempty(k)
40.           infeasible = 1;
41.           return;
42.       else
43.           % check if the LP problem is infeasible (Case 4)
44.           k = int16(find(Eqin(id) == 1));
45.           if any(b(id(k)) > 0) && ~isempty(k)
46.               infeasible = 1;
47.               return;
48.           else
49.               % check if the LP problem is infeasible (Case 6)
50.               k = int16(find(Eqin(id) == 0));
51.               if any(b(id(k)) ~= 0) && ~isempty(k)
52.                   infeasible = 1;
53.                   return;
54.               end
55.           end
56.       end
57.       % find the indices of the nonzero rows
58.       idnnz = int16(find(delrows > 0));
59.       % update matrix A and vectors b and Eqin by deleting
60.       % zero rows
61.       A = A(idnnz, :);
62.       b = b(idnnz);
63.       Eqin = Eqin(idnnz);
64.       fprintf(['ELIMINATE ZERO ROWS: %i zero-rows were ' ...
65.           'eliminated\n'], length(id))
66.   end
67.   end
```

4.5 Eliminate Zero Columns

4.5.1 Mathematical Formulation

A column of the coefficient matrix A is an empty column iff all the coefficients in that column are equal to zero. A variable of this type may be redundant or may state that the LP problem is unbounded. The two cases are distinguished in the following theorem:

Theorem 4.2 *For each empty column, we distinguish the following cases:*

1. $c_j \geq 0$: *The variable is redundant and can be deleted.*
2. $c_j < 0$: *The LP problem is unbounded.*

Proof An LP problem is optimal iff the reduced costs (s_N) are greater than or equal to zero, that means:

$$s_N = c_N - w^T A \geq 0$$

1. If $c_j \geq 0$, then the nonbasic variable $x_j = 0$ satisfies the above equation. Hence, x_j will not become a basic variable, meaning that variable x_j will be equal to zero in the optimal point $x^* = (x_B, x_N)^T$ and will belong to the set of nonbasic variables, i.e., $j \in N$. Therefore, this variable is redundant and can be deleted from the LP problem.
2. If $c_j \leq 0$, then the nonbasic variable $x_j = 0$ does not satisfy the above equation. Hence, x_j may become a basic variable. However, all elements A_j in this column are equal to zero. Therefore, the LP problem is unbounded. In addition, the dual LP problem is infeasible.

$\square$

4.5.2 Numerical Example

This subsection demonstrates the presolve method that eliminates zero columns through two illustrative examples.

1st example

The LP problem that will be presolved is the following:

$$\min z = -x_1 + 4x_2 + 5x_3 + 2x_4 - 8x_5 + 2x_6$$

$$
\begin{aligned}
\text{s.t.} \quad 2x_1 - 3x_2 \qquad\qquad & + 3x_5 + x_6 \le 9 \\
-x_1 + 3x_2 \qquad\qquad & - x_5 - 2x_6 \ge 1 \\
x_1 \qquad\qquad\qquad\qquad\quad & \ge -5 \\
6x_1 + 5x_2 \qquad\qquad & - 2x_5 + 4x_6 = 7 \\
& 3x_5 + 4x_6 \ge -10
\end{aligned}
$$

$$x_j \ge 0, \quad (j = 1, 2, 3, 4, 5, 6)$$

In matrix notation the above LP problem is written as follows:

$$
A = \begin{bmatrix} 2 & -3 & 0 & 0 & 3 & 1 \\ -1 & 3 & 0 & 0 & -1 & -2 \\ 1 & 0 & 0 & 0 & 0 & 0 \\ 6 & 5 & 0 & 0 & -2 & 4 \\ 0 & 0 & 0 & 0 & 3 & 4 \end{bmatrix}, c = \begin{bmatrix} -1 \\ 4 \\ 5 \\ 2 \\ -8 \\ 2 \end{bmatrix}, b = \begin{bmatrix} 9 \\ 1 \\ -5 \\ 7 \\ -10 \end{bmatrix}, Eqin = \begin{bmatrix} -1 \\ 1 \\ 1 \\ 0 \\ 1 \end{bmatrix}
$$

Initially, we start searching for columns that have all their elements equal to zero. We observe that all elements of the third column are equal to zero. According to the first case of the previous subsection ($c_j \ge 0$), the variable is redundant and can be deleted. Hence, we can delete the third column from matrix A and the third element from vector c. The presolved LP problem is now:

$$
A = \begin{bmatrix} 2 & -3 & 0 & 3 & 1 \\ -1 & 3 & 0 & -1 & -2 \\ 1 & 0 & 0 & 0 & 0 \\ 6 & 5 & 0 & -2 & 4 \\ 0 & 0 & 0 & 3 & 4 \end{bmatrix}, c = \begin{bmatrix} -1 \\ 4 \\ 2 \\ -8 \\ 2 \end{bmatrix}, b = \begin{bmatrix} 9 \\ 1 \\ -5 \\ 7 \\ -10 \end{bmatrix}, Eqin = \begin{bmatrix} -1 \\ 1 \\ 1 \\ 0 \\ 1 \end{bmatrix}
$$

Similarly, we observe that all elements of the new third column (initially fourth column) are equal to zero. According to the first case of the previous subsection ($c_j \ge 0$), the variable is redundant and can be deleted. Hence, we can delete the third column from matrix A and the third element from vector c:

$$
A = \begin{bmatrix} 2 & -3 & 3 & 1 \\ -1 & 3 & -1 & -2 \\ 1 & 0 & 0 & 0 \\ 6 & 5 & -2 & 4 \\ 0 & 0 & 3 & 4 \end{bmatrix}, c = \begin{bmatrix} -1 \\ 4 \\ -8 \\ 2 \end{bmatrix}, b = \begin{bmatrix} 9 \\ 1 \\ -5 \\ 7 \\ -10 \end{bmatrix}, Eqin = \begin{bmatrix} -1 \\ 1 \\ 1 \\ 0 \\ 1 \end{bmatrix}
$$

Finally, the equivalent LP problem after presolve is the following:

$$\min z = -x_1 + 4x_2 - 8x_5 + 2x_6$$
$$\text{s.t.} \quad 2x_1 - 3x_2 + 3x_5 + \ x_6 \le \ 9$$
$$-x_1 + 3x_2 - \ x_5 - 2x_6 \ge \ 1$$
$$x_1 \qquad\qquad\qquad \ge -5$$
$$6x_1 + 5x_2 - 2x_5 + 4x_6 = \ 7$$
$$3x_5 + 4x_6 \ge -10$$
$$x_j \ge 0, \quad (j = 1, 2, 5, 6)$$

2nd example

The LP problem that will be presolved is the following:

$$\min z = -x_1 + 4x_2 + 5x_3 - 2x_4 - 8x_5 + 2x_6$$
$$\text{s.t.} \quad 2x_1 - 3x_2 \qquad\qquad + 3x_5 + \ x_6 \le \ 9$$
$$-x_1 + 3x_2 \qquad\qquad - \ x_5 - 2x_6 \ge \ 1$$
$$x_1 \qquad\qquad\qquad\qquad\qquad \ge -5$$
$$6x_1 + 5x_2 \qquad\qquad - 2x_5 + 4x_6 = \ 7$$
$$3x_5 + 4x_6 \ge -10$$
$$x_j \ge 0, \quad (j = 1, 2, 3, 4, 5, 6)$$

In matrix notation the above LP problem is written as follows:

$$A = \begin{bmatrix} 2 & -3 & 0 & 0 & 3 & 1 \\ -1 & 3 & 0 & 0 & -1 & -2 \\ 1 & 0 & 0 & 0 & 0 & 0 \\ 6 & 5 & 0 & 0 & -2 & 4 \\ 0 & 0 & 0 & 0 & 3 & 4 \end{bmatrix}, c = \begin{bmatrix} -1 \\ 4 \\ 5 \\ -2 \\ -8 \\ 2 \end{bmatrix}, b = \begin{bmatrix} 9 \\ 1 \\ -5 \\ 7 \\ -10 \end{bmatrix}, Eqin = \begin{bmatrix} -1 \\ 1 \\ 1 \\ 0 \\ 1 \end{bmatrix}$$

Initially, we start searching for columns that have all their elements equal to zero. We observe that all elements of the third column are equal to zero. According to the first case of the previous subsection ($c_j \ge 0$), the variable is redundant and can be deleted. Hence, we can delete the third column from matrix A and the third element from vector c. The presolved LP problem is now:

$$A = \begin{bmatrix} 2 & -3 & 0 & 3 & 1 \\ -1 & 3 & 0 & -1 & -2 \\ 1 & 0 & 0 & 0 & 0 \\ 6 & 5 & 0 & -2 & 4 \\ 0 & 0 & 0 & 3 & 4 \end{bmatrix}, c = \begin{bmatrix} -1 \\ 4 \\ -2 \\ -8 \\ 2 \end{bmatrix}, b = \begin{bmatrix} 9 \\ 1 \\ -5 \\ 7 \\ -10 \end{bmatrix}, Eqin = \begin{bmatrix} -1 \\ 1 \\ 1 \\ 0 \\ 1 \end{bmatrix}$$

Similarly, we observe that all elements of the new third column (initially fourth column) are equal to zero. According to the second case of the previous subsection ($c_j < 0$), the LP problem is unbounded.

4.5.3 Implementation in MATLAB

This subsection presents the implementation in MATLAB of the presolve method that eliminates zero columns (filename: `eliminateZeroColumns.m`). Some necessary notations should be introduced before the presentation of the implementation of the presolve method that eliminates zero columns. Let A be a $m \times n$ matrix with the coefficients of the constraints, c be a $n \times 1$ vector with the coefficients of the objective function, and *unbounded* be a flag variable showing if the LP problem is unbounded or not.

The function for the presolve method that eliminates zero columns takes as input the matrix of coefficients of the constraints (matrix A) and the vector of the coefficients of the objective function (vector c), and returns as output the presolved matrix of coefficients of the constraints (matrix A), the presolved vector of the coefficients of the objective function (vector c), and a flag variable showing if the LP problem is unbounded or not (variable *unbounded*).

In line 25, we find the columns that have all their elements equal to zero. If any column has all its elements equal to zero, then we eliminate it or declare unboundedness (lines 27–44). Initially, we get the indices of all columns that have all their elements equal to zero (line 30) and examine if the LP problem is unbounded according to case 2 that was presented in subsection 4.5.1 (lines 32–35). If the LP problem is not unbounded, then we eliminate the zero columns and update matrix A and vector c (lines 40–41). Finally, the number of the eliminated columns is printed (lines 42–43).

```
1.   function [A, c, unbounded] = eliminateZeroColumns(A, c)
2.   % Filename: eliminateZeroColumns.m
3.   % Description: the function is an implementation of the
4.   % presolve method that eliminates zero columns
5.   % Authors: Ploskas, N., & Samaras, N.
6.   %
7.   % Syntax: [A, c, unbounded] = eliminateZeroColumns(A, c)
8.   %
9.   % Input:
10.  % -- A: matrix of coefficients of the constraints
11.  %    (size m x n)
12.  % -- c: vector of coefficients of the objective function
13.  %    (size n x 1)
14.  %
15.  % Output:
16.  % -- A: presolved matrix of coefficients of the constraints
17.  %    (size m x n)
18.  % -- c: presolved vector of coefficients of the objective
```

```
19.    %      function (size 1 x n)
20.    % -- unbounded: flag variable showing if the LP is unbounded
21.    %      or not
22.
23.    unbounded = 0;
24.    % find the columns that have all their elements equal to zero
25.    delcols = (max(abs(A)) == 0)';
26.    % if any column has all its element equal to zero
27.    if any(delcols == 1)
28.        % get the indices of all columns that have all their
29.        % elements equal to zero
30.        idelcols = int16(find(delcols));
31.        % check if the LP problem is infeasible (Case 2)
32.        if any(c(idelcols) < 0)
33.            unbounded = 1;
34.            return;
35.        end
36.        % get the indices of all columns that have at least one
37.        % nonzero element
38.        idnnz = int16(find(1 - delcols));
39.        % update matrix A and vector c by deleting zero columns
40.        A = A(:, idnnz);
41.        c = c(idnnz);
42.        fprintf(['ELIMINATE ZERO COLUMNS: %i zero columns ' ...
43.                'were eliminated\n'], nnz(delcols))
44.    end
45.    end
```

4.6   Eliminate Singleton Equality Constraints

4.6.1   Mathematical Formulation

An equality row of the coefficient matrix A is a singleton row iff one coefficient in that row is nonzero. A singleton equality row can be formulated as:

$$A_{i1}x_1 + A_{i2}x_2 + \cdots + A_{in}x_n = b_i \qquad (4.2)$$

where $A_{ik} \neq 0 \wedge A_{ij} = 0$, $i = 1, 2, \cdots, m$, $j = 1, 2, \cdots, n$, and $j \neq k$. The previous constraint can be rewritten as:

$$A_{ik}x_k = b_i \qquad (4.3)$$

Hence, the value of x_k is fixed at b_i/A_{ik}. A constraint of this type may be redundant or may state that the LP problem is infeasible. We can distinguish the following cases:

1. $x_k \geq 0$: Row i and column k are redundant and can be deleted.
2. $x_k < 0$: The LP problem is infeasible.

If $x_k \geq 0$, we replace variable x_k to all constraints:

$$\bar{b} = b - x_k A_{.k} \tag{4.4}$$

If $c_k \neq 0$, then a constant term of the objective function is computed as:

$$c_0 = c_0 - c_k * (b_i / A_{ik}) \tag{4.5}$$

After that replacement, row i is deleted from matrix A, element i is deleted from vectors b and $Eqin$, column k is deleted from matrix A, and element k is deleted from vector c.

It is very often to arise a new singleton equality row after the elimination of the previous one. Hence, the current presolve method continues until no other singleton equality row appears.

4.6.2 Numerical Example

This subsection demonstrates the presolve method that eliminates singleton equality constraints through two illustrative examples.

1st example

The LP problem that will be presolved is the following:

$$
\begin{aligned}
\min z = -2x_1 &+ 4x_2 - 2x_3 + 2x_4 \\
\text{s.t.} \qquad\qquad 3x_3 \quad &= \;\; 6 \quad (1) \\
4x_1 - 3x_2 + 8x_3 - x_4 &= 20 \quad (2) \\
-3x_1 + 2x_2 \qquad - 4x_4 &= -8 \quad (3) \\
4x_1 \qquad - x_3 \quad &= 18 \quad (4) \\
x_j \geq 0, \quad (j = 1, 2, 3, 4)
\end{aligned}
$$

In matrix notation the above LP problem is written as follows:

$$
A = \begin{bmatrix} 0 & 0 & 3 & 0 \\ 4 & -3 & 8 & -1 \\ -3 & 2 & 0 & -4 \\ 4 & 0 & -1 & 0 \end{bmatrix}, c = \begin{bmatrix} -2 \\ 4 \\ -2 \\ 2 \end{bmatrix}, b = \begin{bmatrix} 6 \\ 20 \\ -8 \\ 18 \end{bmatrix}, Eqin = \begin{bmatrix} 0 \\ 0 \\ 0 \\ 0 \end{bmatrix}
$$

Initially, we start searching for equality constraints that have only one nonzero element. We observe that all elements of the first equality constraint are equal to zero except the third element:

$$0x_1 + 0x_2 + 3x_3 + 0x_4 = 6$$

So, variable x_3 equals:

$$x_3 = \frac{6}{3} = 2$$

According to the first case of the previous subsection ($x_k \geq 0$), the first row and the third column are redundant and can be deleted. We update vector b:

$$\bar{b} = b - x_3 A_{.3} = \begin{bmatrix} 6 \\ 20 \\ -8 \\ 18 \end{bmatrix} - 2 \begin{bmatrix} 3 \\ 8 \\ 0 \\ -1 \end{bmatrix} = \begin{bmatrix} 0 \\ 4 \\ -8 \\ 20 \end{bmatrix}$$

$c_3 \neq 0$, so a constant term of the objective function is computed as:

$$c_0 = c_0 - (-2) * (6/3) = 0 + 4 = 4$$

Then, we can delete the first row and the third column from matrix A, the first element from vectors b and $Eqin$, and the third element from vector c. The presolved LP problem is now:

$$A = \begin{bmatrix} 4 & -3 & -1 \\ -3 & 2 & -4 \\ 4 & 0 & 0 \end{bmatrix}, c = \begin{bmatrix} -2 \\ 4 \\ 2 \end{bmatrix}, b = \begin{bmatrix} 4 \\ -8 \\ 20 \end{bmatrix}, Eqin = \begin{bmatrix} 0 \\ 0 \\ 0 \end{bmatrix}, c_0 = 4$$

Similarly, we observe that all elements of the third equality constraint (initially fourth) are equal to zero except the first element:

$$4x_1 + 0x_2 + 0x_3 = 20$$

So, variable x_1 equals:

$$x_1 = \frac{20}{4} = 5$$

According to the first case of the previous subsection ($x_k \geq 0$), the third row and the first column are redundant and can be deleted. We update vector b:

$$\bar{b} = b - x_1 A_{.1} = \begin{bmatrix} 4 \\ -8 \\ 20 \end{bmatrix} - 5 \begin{bmatrix} 4 \\ -3 \\ 4 \end{bmatrix} = \begin{bmatrix} -16 \\ 7 \\ 0 \end{bmatrix}$$

$c_3 \neq 0$, so a constant term of the objective function is computed as:

$$c_0 = c_0 - (-2) * (20/4) = 4 + 10 = 14$$

Then, we can delete the third row and the first column from matrix A, the third element from vectors b and $Eqin$, and the first element from vector c:

$$A = \begin{bmatrix} -3 & -1 \\ 2 & -4 \end{bmatrix}, c = \begin{bmatrix} 4 \\ 2 \end{bmatrix}, b = \begin{bmatrix} -16 \\ 7 \end{bmatrix}, Eqin = \begin{bmatrix} 0 \\ 0 \end{bmatrix}, c_0 = 14$$

Finally, the equivalent LP problem after presolve is the following:

$$
\begin{aligned}
\min z = \quad & 4x_2 + 2x_4 + \quad 14 \\
\text{s.t.} \quad & -3x_2 - x_4 = -16 \\
& 2x_2 - 4x_4 = \quad 7 \\
& x_j \geq 0, \quad (j = 2, 4)
\end{aligned}
$$

2nd example

The LP problem that will be presolved is the following:

$$
\begin{aligned}
\min z = -2x_1 & + 4x_2 - 2x_3 + 2x_4 \\
\text{s.t.} \quad & 3x_3 \qquad\qquad = -6 \quad (1) \\
4x_1 & - 3x_2 + 8x_3 - x_4 = 20 \quad (2) \\
-3x_1 & + 2x_2 \qquad - 4x_4 = -8 \quad (3) \\
4x_1 & \qquad - x_3 \qquad = 18 \quad (4) \\
x_j \geq 0, \quad & (j = 1, 2, 3, 4)
\end{aligned}
$$

In matrix notation the above LP problem is written as follows:

$$A = \begin{bmatrix} 0 & 0 & 3 & 0 \\ 4 & -3 & 8 & -1 \\ -3 & 2 & 0 & -4 \\ 4 & 0 & -1 & 0 \end{bmatrix}, c = \begin{bmatrix} -2 \\ 4 \\ -2 \\ 2 \end{bmatrix}, b = \begin{bmatrix} -6 \\ 20 \\ -8 \\ 18 \end{bmatrix}, Eqin = \begin{bmatrix} 0 \\ 0 \\ 0 \\ 0 \end{bmatrix}$$

Initially, we start searching for equality constraints that have only one nonzero element. We observe that all elements of the first equality constraint are equal to zero except the third element:

$$0x_1 + 0x_2 + 3x_3 + 0x_4 = -6$$

So, variable x_3 equals:

$$x_3 = \frac{-6}{3} = -2$$

According to the second case of the previous subsection ($x_k < 0$), the LP problem is infeasible.

4.6.3   Implementation in MATLAB

This subsection presents the implementation in MATLAB of the presolve method that eliminates singleton equality constraints (filename: eliminateSingleton EqualityConstraints.m). Some necessary notations should be introduced before the presentation of the implementation of the presolve method that eliminates singleton equality constraints. Let A be a $m \times n$ matrix with the coefficients of the constraints, c be a $n \times 1$ vector with the coefficients of the objective function, b be a $m \times 1$ vector with the right-hand side of the constraints, $Eqin$ be a $m \times 1$ vector with the type of the constraints, $c0$ be the constant term of the objective function, and *infeasible* be a flag variable showing if the LP problem is infeasible or not.

The function for the presolve method that eliminates singleton equality constraints takes as input the matrix of coefficients of the constraints (matrix A), the vector of the coefficients of the objective function (vector c), the vector of the right-hand side of the constraints (vector b), the vector of the type of the constraints (vector $Eqin$), and the constant term of the objective function (variable $c0$), and returns as output the presolved matrix of coefficients of the constraints (matrix A), the presolved vector of the coefficients of the objective function (vector c), the presolved vector of the right-hand side of the constraints (vector b), the presolved vector of the type of the constraints (vector $Eqin$), the updated constant term of the objective function (variable $c0$), and a flag variable showing if the LP problem is infeasible or not (variable *infeasible*).

In lines 38–41, we find the rows that have only one nonzero element and set all the other rows equal to zero. Moreover, we find the number of the rows that have only one nonzero element (line 44). If any singleton equality constraint exists, then we eliminate it or declare infeasibility (lines 45–86). Initially, we compute the indices of the singleton equality constraints (lines 46–52). For each singleton equality constraint, we update matrix A and vectors c, b, and $Eqin$ (lines 56–82). In addition, we examine if the LP problem is infeasible according to case 2 that was presented in subsection 4.6.1 (lines 77–80). Finally, the number of the eliminated singleton equality constraints is printed (lines 83–85).

```
1.    function [A, c, b, Eqin, c0, infeasible] = ...
2.        eliminateSingletonEqualityConstraints(A, c, b, Eqin, c0)
3.    % Filename: eliminateSingletonEqualityConstraints.m
4.    % Description: the function is an implementation of the
5.    % presolve technique that eliminates singleton equality
6.    % constraints
7.    % Authors: Ploskas, N., & Samaras, N.
8.    %
9.    % Syntax: [A, c, b, Eqin, c0, infeasible] = ...
10.   %   eliminateSingletonEqualityConstraints(A, c, b, Eqin, c0)
11.   %
12.   % Input:
13.   % -- A: matrix of coefficients of the constraints
14.   %    (size m x n)
15.   % -- c: vector of coefficients of the objective function
16.   %    (size n x 1)
17.   % -- b: vector of the right-hand side of the constraints
18.   %    (size m x 1)
19.   % -- Eqin: vector of the type of the constraints
20.   %    (size m x 1)
21.   % -- c0: constant term of the objective function
22.   %
23.   % Output:
24.   % -- A: presolved matrix of coefficients of the constraints
25.   %    (size m x n)
26.   % -- c: presolved vector of coefficients of the objective
27.   %    function (size n x 1)
28.   % -- b: presolved vector of the right-hand side of the
29.   %    constraints (size m x 1)
30.   % -- Eqin: presolved vector of the type of the constraints
31.   %    (size m x 1)
32.   % -- c0: updated constant term of the objective function
33.   % -- infeasible: flag variable showing if the LP is
34.   %    infeasible or not
35.
36.   infeasible = 0;
37.   % compute the number of nonzero elements in each row
38.   delrows = sum(spones(A'));
39.   % find the rows that have only one nonzero element and set
40.   % all the other rows equal to zero
41.   singleton = (delrows == 1);
42.   % find the number of the rows that have only one nonzero
43.   % element
44.   cardrows = nnz(singleton);
45.   if cardrows > 0 % if singleton rows exist
46.       % get the indices of the singleton rows
47.       idsrows = int16(find(singleton));
48.       % find equality constraints
49.       [row, ~] = find(Eqin == 0);
50.       % compute the indices of the singleton equality
51.       % constraints
52.       idsrows = intersect(idsrows, row);
53.       % compute the number of the singleton equality
54.       % constraints
```

```
55.        cardrows = length(idsrows);
56.        % for each singleton equality constraint
57.        for i = cardrows:-1:1
58.            ii = idsrows(i); % get the index of the constraint
59.            % find the nonzero elements of this row
60.            [row1, j] = find(A(ii, :));
61.            if ~isempty(row1) % if there are nonzero elements
62.                b(ii) = b(ii) / A(ii, j); % update b
63.                % if b is greater than or equal to zero
64.                if b(ii) >= 0
65.                    b_temp = b(ii) .* A(:, j);
66.                    A(:, j) = []; % delete column from A
67.                    if c(j) ~= 0 % update c and c0
68.                        c(j) = c(j) * b(ii);
69.                        c0 = c0 - c(j);
70.                    end
71.                    % update matrix A and vectors c, b, and Eqin
72.                    c(j) = [];
73.                    b = b - b_temp;
74.                    A(ii, :) = [];
75.                    b(ii) = [];
76.                    Eqin(ii) = [];
77                 else % the LP problem is infeasible (Case 2)
78.                    infeasible = 1;
79.                    return;
80.                end
81.            end
82.        end
83.        fprintf(['ELIMINATE SINGLETON EQUALITY CONSTRAINTS: ' ...
84.            '%i singleton equality constraints were ' ....
85.            'eliminated\n'], cardrows);
86.    end
87. end
```

4.7 Eliminate *k*ton Equality Constraints

4.7.1 Mathematical Formulation

The elimination of singleton equality constraints was initially proposed by Brearley et al. [4]. Lustig et al. [14] and later Mészáros and Suhl [16] and Gould and Toint [10] discussed the elimination of doubleton equality constraints. Similar to singleton constraints, we can extend the previous presolve method in order to identify and eliminate doubleton, tripleton, and more general k-ton equality constraints, where $k \geq 1$. A row of the coefficient matrix A is a $k$ton row iff k coefficients in that row are not equal to zero. A $k$ton equality row can be formulated as:

$$A_{i1}x_1 + A_{i2}x_2 + \cdots + A_{in}x_n = b_i \tag{4.6}$$

where $A_{ik} \neq 0 \wedge A_{ij} = 0$, $i = 1, 2, \cdots, m$, $j = 1, 2, \cdots, n$, and $k \in \mathbb{N}, 1 \leq k_p \leq n$, $1 \leq p \leq k, j \neq k$. The previous constraint can be rewritten as:

$$A_{ik_1} x_{k_1} + A_{ik_2} x_{k_2} + \cdots + A_{ik_p} x_{k_p} = b_i \tag{4.7}$$

If $k > 1$, we find the column j that the last nonzero element exists in row i. Then, we update matrix A and vector b:

$$
\begin{aligned}
b_i &= b_i / A_{ij} \\
A_{i.} &= A_{i.} / A_{ij}
\end{aligned}
\tag{4.8}
$$

If $c_j \neq 0$, then:

$$
\begin{aligned}
c_0 &= c_0 + c_j b_i \\
c &= c - c_j A_{i.}^T
\end{aligned}
\tag{4.9}
$$

Moreover, the equality constraint is transformed to an inequality constraint in the form of 'less than or equal to' ($Eqin_i = -1$).

Next, we can eliminate variable x_j. However, we first update the constraints where variable x_j is nonzero except constraint i (the $k$ton constraint). Let's assume that vector t holds the indices of the constraints where variable x_j is nonzero except constraint i and $b_i \neq 0$:

$$
\begin{aligned}
b_t &= b_t - A_{tj} b_i \\
A_{t.} &= A_{t.} - A_{tj} A_{i.}
\end{aligned}
\tag{4.10}
$$

Finally, we eliminate variable x_j from matrix A and vector c.

If $k = 1$, then the constraint may be redundant or may state that the LP problem is infeasible. Similar to the singleton equality constraints, we can distinguish the following cases:

1. $x_k \geq 0$: Row i and column k are redundant and can be deleted.
2. $x_k < 0$: The LP problem is infeasible.

If $x_k \geq 0$, we replace variable x_k to all constraints:

$$\bar{b} = b - x_k A_{.k} \tag{4.11}$$

If $c_j \neq 0$, then:

$$c_0 = c_0 + c_j (b_i / A_{ik}) \tag{4.12}$$

After that replacement, row i is deleted from matrix A, element i is deleted from vectors b and $Eqin$, column k is deleted from matrix A, and element k is deleted from vector c.

It is very often to arise a new $k$ton equality row after the elimination of the previous one. Hence, the current presolve method continues until no other $k$ton equality row appears.

4.7.2 Numerical Example

This subsection demonstrates the presolve method that eliminates $k$ton equality constraints through two illustrative examples.

1st example

The LP problem that will be presolved is the following:

$$
\begin{aligned}
\min z = {} & -2x_1 + 4x_2 - 2x_3 + 2x_4 \\
\text{s.t.} \quad & 3x_2 && = 6 && (1) \\
& 4x_1 - 3x_2 + 8x_3 - x_4 = 20 && (2) \\
& -3x_1 + 2x_2 \quad\ \ - 4x_4 = -8 && (3) \\
& 4x_1 \quad\ \ - x_3 \quad\ = 18 && (4) \\
& x_j \ge 0, \quad (j = 1, 2, 3, 4)
\end{aligned}
$$

In matrix notation the above LP problem is written as follows:

$$
A = \begin{bmatrix} 0 & 3 & 0 & 0 \\ 4 & -3 & 8 & -1 \\ -3 & 2 & 0 & -4 \\ 4 & 0 & -1 & 0 \end{bmatrix}, c = \begin{bmatrix} -2 \\ 4 \\ -2 \\ 2 \end{bmatrix}, b = \begin{bmatrix} 6 \\ 20 \\ -8 \\ 18 \end{bmatrix}, Eqin = \begin{bmatrix} 0 \\ 0 \\ 0 \\ 0 \end{bmatrix}
$$

Let's assume that $k = 2$. Initially, we start searching for rows that have only two nonzero elements. We observe that the fourth row ($i = 4$) has only two nonzero elements (the first and the third elements):

$$4x_1 + 0x_2 - 1x_3 + 0x_4 = 18$$

Then, we find the column j that the last nonzero element exists in the fourth row. That column is the third ($j = 3$). So, we update matrix A and vector b:

$$b_4 = b_4/A_{43} = 18/(-1) = -18$$
$$A_{4.} = A_{4.}/A_{4j} = \begin{bmatrix} 4 & 0 & -1 & 0 \end{bmatrix} /(-1) = \begin{bmatrix} -4 & 0 & 1 & 0 \end{bmatrix}$$

Moreover, the equality constraint is transformed to an inequality constraint in the form of 'less than or equal to' ($Eqin_i = -1$).

Hence, the updated variables of the problem are the following:

$$A = \begin{bmatrix} 0 & 3 & 0 & 0 \\ 4 & -3 & 8 & -1 \\ -3 & 2 & 0 & -4 \\ -4 & 0 & 1 & 0 \end{bmatrix}, c = \begin{bmatrix} -2 \\ 4 \\ -2 \\ 2 \end{bmatrix}, b = \begin{bmatrix} 6 \\ 20 \\ -8 \\ -18 \end{bmatrix}, Eqin = \begin{bmatrix} 0 \\ 0 \\ 0 \\ -1 \end{bmatrix}$$

Next, we can eliminate variable x_3. However, we first update the constraints where variable x_3 is nonzero except the fourth constraint (the $k$ton constraint). Variable x_3 is nonzero on the second constraint.

$$b_2 = b_2 - A_{23} * b_4 = 20 - 8 * (-18) = 164$$

$$A_{2.} = A_{2.} - A_{23} * A_{4.} = \begin{bmatrix} 4 & -3 & 8 & -1 \end{bmatrix} - 8 * \begin{bmatrix} -4 & 0 & 1 & 0 \end{bmatrix} = \begin{bmatrix} 36 & -3 & 0 & -1 \end{bmatrix}$$

$c_3 \neq 0$, so:

$$c_0 = c_0 + c_3 * b_4 = 0 + (-2) * (-18) = 36$$

$$c = c - c_3 * A_{4.}^T = \begin{bmatrix} -2 \\ 4 \\ -2 \\ 2 \end{bmatrix} - (-2) * \begin{bmatrix} -4 \\ 0 \\ 1 \\ 0 \end{bmatrix} = \begin{bmatrix} -10 \\ 4 \\ 0 \\ 2 \end{bmatrix}$$

We also delete the third variable. Hence, the updated variables of the problem are the following:

$$A = \begin{bmatrix} 0 & 3 & 0 \\ 36 & -3 & -1 \\ -3 & 2 & -4 \\ -4 & 0 & 0 \end{bmatrix}, c = \begin{bmatrix} -10 \\ 4 \\ 2 \end{bmatrix}, b = \begin{bmatrix} 6 \\ 164 \\ -8 \\ -18 \end{bmatrix}, Eqin = \begin{bmatrix} 0 \\ 0 \\ 0 \\ -1 \end{bmatrix}, c_0 = 36$$

So, we mark variable x_3 to be deleted after the completion of this presolve method and we proceed to eliminate singleton equality constraints ($k = 1$). Initially, we start searching for rows that have only one nonzero element. We observe that all elements of the first row are equal to zero except the second element:

$$0x_1 + 3x_2 + 0x_4 = 6$$

So, variable x_2 equals:

$$x_2 = \frac{6}{3} = 2$$

According to the first case of the previous subsection ($x_k \geq 0$), the first row and the second column are redundant and can be deleted. We update vector b:

$$\bar{b} = b - x_2 A_{.2} = \begin{bmatrix} 6 \\ 164 \\ -8 \\ -18 \end{bmatrix} - 2 \begin{bmatrix} 3 \\ -3 \\ 2 \\ 0 \end{bmatrix} = \begin{bmatrix} 0 \\ 170 \\ -12 \\ -18 \end{bmatrix}$$

$c_2 \neq 0$, so a constant term of the objective function is computed as:

$$c_0 = 36 + (4) * (6/3) = 36 + 8 = 44$$

Then, we can delete the first row and the second column from matrix A, the first element from vectors b and $Eqin$, and the second element from vector c. The presolved LP problem is now:

$$A = \begin{bmatrix} 36 & -1 \\ -3 & -4 \\ -4 & 0 \end{bmatrix}, c = \begin{bmatrix} -10 \\ 2 \end{bmatrix}, b = \begin{bmatrix} 170 \\ -12 \\ -18 \end{bmatrix}, Eqin = \begin{bmatrix} 0 \\ 0 \\ -1 \end{bmatrix}, c_0 = 44$$

Similarly, we observe that all elements of the third row (initially fourth) are equal to zero except the first element, but it is not an equality constraint.

Finally, the equivalent LP problem after presolve is the following:

$$\begin{aligned} \min z = -10x_2 + 2x_4 + \quad 44 \\ \text{s.t.} \qquad 36x_2 - \quad x_4 = \quad 170 \\ -3x_2 - 4x_4 = -12 \\ -4x_2 \qquad \leq -18 \\ x_j \geq 0, \quad (j = 2, 4) \end{aligned}$$

2nd example

The LP problem that will be presolved is the following:

$$\begin{aligned} \min z = -2x_1 + 4x_2 - 2x_3 + 2x_4 \\ \text{s.t.} \qquad 3x_2 \qquad\qquad = -6 \quad (1) \\ 4x_1 - 3x_2 + 8x_3 - \quad x_4 = 20 \quad (2) \\ -3x_1 + 2x_2 \qquad - 4x_4 = -8 \quad (3) \\ 4x_1 \qquad - x_3 \qquad = 18 \quad (4) \\ x_j \geq 0, \quad (j = 1, 2, 3, 4) \end{aligned}$$

In matrix notation the above LP problem is written as follows:

$$A = \begin{bmatrix} 0 & 3 & 0 & 0 \\ 4 & -3 & 8 & -1 \\ -3 & 2 & 0 & -4 \\ 4 & 0 & -1 & 0 \end{bmatrix}, c = \begin{bmatrix} -2 \\ 4 \\ -2 \\ 2 \end{bmatrix}, b = \begin{bmatrix} -6 \\ 20 \\ -8 \\ 18 \end{bmatrix}, Eqin = \begin{bmatrix} 0 \\ 0 \\ 0 \\ 0 \end{bmatrix}$$

Let's assume that $k = 2$. Initially, we start searching for rows that have only two nonzero elements. We observe that the fourth row ($i = 4$) has only two nonzero elements (the first and the third elements):

$$4x_1 + 0x_2 - 1x_3 + 0x_4 = 18$$

Then, we find the column j that the last nonzero element exists in the fourth row. That column is the third ($j = 3$). So, we update matrix A and vector b:

$$b_4 = b_4/A_{43} = 18/(-1) = -18$$

$$A_4. = A_4./A_{4j} = \begin{bmatrix} 4 & 0 & -1 & 0 \end{bmatrix}/(-1) = \begin{bmatrix} -4 & 0 & 1 & 0 \end{bmatrix}$$

Moreover, the equality constraint is transformed to an inequality constraint in the form of 'less than or equal to' ($Eqin_i = -1$).

Hence, the updated variables of the problem are the following:

$$A = \begin{bmatrix} 0 & 3 & 0 & 0 \\ 4 & -3 & 8 & -1 \\ -3 & 2 & 0 & -4 \\ -4 & 0 & 1 & 0 \end{bmatrix}, c = \begin{bmatrix} -2 \\ 4 \\ -2 \\ 2 \end{bmatrix}, b = \begin{bmatrix} -6 \\ 20 \\ -8 \\ -18 \end{bmatrix}, Eqin = \begin{bmatrix} 0 \\ 0 \\ 0 \\ -1 \end{bmatrix}$$

Next, we can eliminate variable x_3. However, we first update the constraints where variable x_3 is nonzero except the fourth constraint (the *k*ton constraint). Variable x_3 is nonzero on the second constraint.

$$b_2 = b_2 - A_{23} * b_4 = 20 - 8 * (-18) = 164$$

$$A_2. = A_2. - A_{23} * A_4. = \begin{bmatrix} 4 & -3 & 8 & -1 \end{bmatrix} - 8 * \begin{bmatrix} -4 & 0 & 1 & 0 \end{bmatrix} = \begin{bmatrix} 36 & -3 & 0 & -1 \end{bmatrix}$$

$c_3 \neq 0$, so:

$$c_0 = c_0 + c_3 * b_4 = 0 + (-2) * (-18) = 36$$

$$c = c - c_3 * A_4.^T = \begin{bmatrix} -2 \\ 4 \\ -2 \\ 2 \end{bmatrix} - (-2) * \begin{bmatrix} -4 \\ 0 \\ 1 \\ 0 \end{bmatrix} = \begin{bmatrix} -10 \\ 4 \\ 0 \\ 2 \end{bmatrix}$$

We also delete the third variable. Hence, the updated variables of the problem are the following:

$$A = \begin{bmatrix} 0 & 3 & 0 \\ 36 & -3 & -1 \\ -3 & 2 & -4 \\ -4 & 0 & 0 \end{bmatrix}, c = \begin{bmatrix} -10 \\ 4 \\ 2 \end{bmatrix}, b = \begin{bmatrix} -6 \\ 164 \\ -8 \\ -18 \end{bmatrix}, Eqin = \begin{bmatrix} 0 \\ 0 \\ 0 \\ -1 \end{bmatrix}$$

So, we proceed to eliminate the singleton equality constraints ($k = 1$). Initially, we start searching for rows that have only one nonzero element. We observe that all elements of the first row are equal to zero except the second element:

$$0x_1 + 3x_2 + 0x_4 = -6$$

So, variable x_2 equals:

$$x_2 = \frac{-6}{3} = -2$$

According to the second case of the previous subsection ($x_k < 0$), the LP problem is infeasible.

4.7.3 Implementation in MATLAB

This subsection presents the implementation in MATLAB of the presolve method that eliminates $k$ton equality constraints (filename: eliminateKtonEquality Constraints.m). Some necessary notations should be introduced before the presentation of the implementation of the presolve method that eliminates $k$ton equality constraints. Let A be a $m \times n$ matrix with the coefficients of the constraints, c be a $n \times 1$ vector with the coefficients of the objective function, b be a $m \times 1$ vector with the right-hand side of the constraints, $Eqin$ be a $m \times 1$ vector with the type of the constraints, $c0$ be the constant term of the objective function, $kton$ be the type of equality constraints to eliminate (1: singleton, 2: doubleton, etc.), and $infeasible$ be a flag variable showing if the LP problem is infeasible or not.

The function for the presolve method that eliminates $k$ton equality constraints takes as input the matrix of coefficients of the constraints (matrix A), the vector of the coefficients of the objective function (vector c), the vector of the right-hand side of the constraints (vector b), the vector of the type of the constraints (vector $Eqin$), the constant term of the objective function (variable $c0$), and the type of equality constraints to eliminate (variable $kton$), and returns as output the presolved matrix of coefficients of the constraints (matrix A), the presolved vector of the coefficients of the objective function (vector c), the presolved vector of the right-hand side of the constraints (vector b), the presolved vector of the type of the constraints (vector $Eqin$), the updated constant term of the objective function (variable $c0$), and a flag variable showing if the LP problem is infeasible or not (variable $infeasible$).

In lines 40–42, if $k = 1$ then we call the function to eliminate the singleton equality constraints (Section 4.7.1). We perform the same procedure for all k, $1 \leq k \leq kton$ (lines 50–129). In lines 51–55, we find the rows that have only k nonzero elements and set all the other rows equal to zero. Moreover, we find the number of the rows that have only k nonzero elements (line 58). If any $k$ton equality constraint exists, then we find the column of the last nonzero element in that constraint and update matrix A and vectors c, b, and *Eqin* (lines 59–128). If $k = 1$ then we eliminate the constraint or declare infeasibility (lines 114–125). We also eliminate the adequate variables (lines 130–137). Finally, the number of the eliminated $k$ton equality constraints is printed (lines 138–143).

```
1.    function [A, c, b, Eqin, c0, infeasible] = ...
2.        eliminateKtonEqualityConstraints(A, c, b, Eqin, ...
3.        c0, kton)
4.    % Filename: eliminateKtonEqualityConstraints.m
5.    % Description: the function is an implementation of the
6.    % presolve technique that eliminates kton equality
7.    % constraints
8.    % Authors: Ploskas, N., & Samaras, N.
9.    %
10.   % Syntax: [A, c, b, Eqin, c0, infeasible] = ...
11.   % eliminateKtonEqualityConstraints(A, c, b, Eqin, ...
12.   % c0, kton)
13.   %
14.   % Input:
15.   % -- A: matrix of coefficients of the constraints
16.   %    (size m x n)
17.   % -- c: vector of coefficients of the objective function
18.   %    (size n x 1)
19.   % -- b: vector of the right-hand side of the constraints
20.   %    (size m x 1)
21.   % -- Eqin: vector of the type of the constraints
22.   %    (size m x 1)
23.   % -- c0: constant term of the objective function
24.   % -- kton: the type of equality constraints to eliminate (1:
25.   %    singleton, 2: doubleton, etc.)
26.   %
27.   % Output:
28.   % -- A: presolved matrix of coefficients of the constraints
29.   %    (size m x n)
30.   % -- c: presolved vector of coefficients of the objective
31.   %    function (size n x 1)
32.   % -- b: presolved vector of the right-hand side of the
33.   %    constraints (size m x 1)
34.   % -- Eqin: presolved vector of the type of the constraints
35.   %    (size m x 1)
36.   % -- c0: updated constant term of the objective function
37.   % -- infeasible: flag variable showing if the LP is
38.   %    infeasible or not
39.
40.   if kton == 1
41.       [A, c, b, Eqin, c0, infeasible] = ...
```

```
42.                eliminateSingletonEqualityConstraints(A, c, b, Eqin, c0);
43.  else
44.      infeasible = 0;
45.      [m, n] = size(A);
46.      tol = max(m, n) * eps * norm(A, 'inf'); % compute tolerance
47.      colindex = sparse(1, n);
48.      ci = 1;
49.      eliminatedRows = 0;
50.      for k = kton:-1:1 % for each k, 1 <= k <= kton
51.          % compute the number of nonzero elements in each row
52.          delrows = sum(spones(A'));
53.          % find the rows that have only k nonzero elements and
54.          % set all the other rows equal to zero
55.          singleton = (delrows == k);
56.          % find the number of the rows that have only k
57.          % nonzero elements
58.          cardrows = nnz(singleton);
59.          % if kton rows exist
60.          if cardrows >= max(1, .01 * size(A, 1))
61.              % get the indices of the kton rows
62.              idsrows = int16(find(singleton));
63.              % find equality constraints
64.              [row , ~] = find(Eqin == 0);
65.              % compute the indices of the kton equality
66.              % constraints
67.              idsrows = intersect(idsrows, row);
68.              % compute the number of the kton equality
69.              % constraints
70.              cardrows = length(idsrows);
71.              % for each kton equality constraint
72.              for i = cardrows:-1:1
73.                  % get the index of the constraint
74.                  ii = idsrows(i);
75.                  % find the nonzero elements of this row
76.                  [~, j] = find(A(ii, :));
77.                  % find the column of the last nonzero element
78.                  if length(j) > 1
79.                      j = j(length(j));
80.                  elseif isempty(j)
81.                      continue;
82.                  end
83.                  % set to zero if the value is less than or
84.                  % equal to tol and continue to the next
85.                  % constraint
86.                  if abs(A(ii, j)) <= tol
87.                      A(ii, j) = 0;
88.                      continue;
89.                  else % update matrix A and vectors c, b,
90.                       % and Eqin
91.                      b(ii) = b(ii) / A(ii, j);
92.                      A(ii, :) = A(ii, :) / A(ii, j);
93.                      i_nz = find(A(:, j));
94.                      i_nz = setdiff(i_nz, ii);
95.                      if ~isempty(i_nz)
```

```
96.                          for t = i_nz
97.                             if b(ii) ~= 0
98.                                 b(t) = b(t) - A(t, j) * b(ii);
99.                             end
100.                            A(t, :) = A(t, :) - A(t, j) * A(ii, :);
101.                            Eqin(ii) = -1;
102.                            colindex(ci) = j;
103.                            ci = ci + 1;
104.                         end
105.                     else
106.                         Eqin(ii) = -1;
107.                         colindex(ci) = j;
108.                         ci = ci + 1;
109.                     end
110.                     if c(j) ~= 0
111.                         c0 = c0 + c(j) * b(ii);
112.                         c = c - c(j) * A(ii, :)';
113.                     end
114.                     % the LP problem is infeasible (Case 2)
115.                     if (k == 1) && (b(ii) < 0)
116.                         infeasible = 1;
117.                         return;
118.                     end
119.                     if (k == 1) && (b(ii) >= 0)
120.                         % delete the iith constraint
121.                         A(ii, :) = [];
122.                         b(ii) = [];
123.                         Eqin(ii) = [];
124.                         eliminatedRows = eliminatedRows + 1;
125.                     end
126.                 end
127.             end
128.         end
129.     end
130.     colindex = sort(colindex, 'descend');
131.     cardcols = nnz(colindex);
132.     if cardcols > 0 % delete variables
133.         for j = 1:cardcols
134.             A(:, colindex(j)) = [];
135.             c(colindex(j)) = [];
136.         end
137.     end
138.     if (eliminatedRows > 0) || (cardcols > 0)
139.         fprintf(['ELIMINATE KTON EQUALITY CONSTRAINTS: ' ...
140.             '%i kton equality constraints were ' ...
141.             'found. %i constraints and %i variables were ' ...
142.             'eliminated\n'], cardcols, eliminatedRows, cardcols);
143.     end
144. end
145. end
```

Table 4.1 shows the results before and after presolving a set of benchmark LPs by applying only the *k*ton elimination method with various values of *k*. Using $k = 2$, the LP problem after presolve is sparser and has fewer constraints compared to other

Table 4.1 Results of the elimination of *k*ton equality constraints on a benchmark set using different values of *k*

Problem	Before presolve			After presolve k = 2			After presolve k = 3			After presolve k = 4			After presolve k = 5			After presolve k = 10		
	Rows	Columns	Nonzeros	Rows	Columns	Nonzeros	Rows	Columns	Nonzeros	Rows	Columns	Nonzeros	Rows	Columns	Nonzeros	Rows	Columns	Nonzeros
25fv47	821	1,571	10,400	799	1,494	10,188	799	1,452	10,214	799	1,397	10,482	799	1,362	10,896	801	1,286	12,625
aa3	825	8,627	70,806	825	8,627	70,806	825	8,627	70,806	825	8,627	70,806	825	8,627	70,806	825	8,627	70,806
adlittle	56	97	383	55	96	376	55	95	374	55	95	374	55	92	372	55	91	375
agg	488	163	2,410	488	162	2,408	488	159	2,403	488	158	2,403	488	152	2,402	488	141	2,425
agg3	516	302	4,300	516	301	4,298	516	297	4,291	516	297	4,292	516	289	4,292	516	268	4,344
aircraft	3,754	7,517	20,267	3,754	7,517	20,267	3,754	7,517	20,267	3,754	6,766	21,016	3,754	6,014	24,764	3,754	3,764	22,514
bandm	305	472	2,494	268	380	1,938	269	361	1,923	270	347	1,921	273	341	2,002	277	314	2,525
beaconfd	173	262	3,375	146	206	2,471	144	184	1,657	143	172	1,024	143	171	1,022	145	170	1,086
bgindy	2,671	10,116	65,502	2,671	10,049	65,435	2,671	9,746	67,406	2,671	9,368	71,590	2,671	8,799	80,327	2,671	8,262	88,596
blend	74	83	491	74	75	467	74	70	488	74	68	537	74	63	525	74	51	589
bnl1	643	1,175	5,121	643	1,089	5,015	643	1,061	5,055	643	1,057	5,125	643	1,057	5,125	643	1,032	5,298
bnl2	2,324	3,489	13,999	2,312	2,628	13,061	2,313	2,440	13,173	2,313	2,431	13,820	2,313	2,423	13,866	2,313	2,385	14,142
brandy	220	249	2,148	187	208	1,932	187	201	1,928	188	193	1,944	192	178	2,001	194	173	2,383
car4	16,384	33,052	63,724	16,384	17,368	48,040	16,384	17,368	48,040	16,384	17,368	48,040	16,384	17,368	48,040	16,384	17,368	48,040
cre-b	9,648	72,447	256,095	9,648	72,353	255,721	9,648	72,295	255,673	9,648	72,263	255,815	9,648	72,240	255,920	9,648	71,748	257,775
d2q06c	2,171	5,167	32,417	2,161	5,057	32,280	2,161	4,895	32,454	2,162	4,729	32,597	2,163	4,601	32,888	2,163	4,357	34,487
degen3	1,503	1,818	24,646	1,503	1,727	24,555	1,503	1,251	30,563	1,503	1,234	30,275	1,503	1,234	30,158	1,503	1,213	31,063
fffff800	524	854	6,227	501	808	5,940	501	733	5,895	501	719	5,687	501	661	5,701	501	576	5,801
kent	31,300	16,620	184,710	26,260	10,800	136,050	26,260	10,440	138,750	26,260	10,440	138,750	26,260	10,440	138,750	26,260	10,440	138,750
lotfi	153	308	1,078	151	294	1,022	151	289	1,022	151	275	1,000	151	269	1,012	151	228	1,138
nug05	210	225	1,050	210	225	1,050	210	225	1,050	210	225	1,050	210	77	2,265	210	77	2,265
nug06	372	486	2,232	372	486	2,232	372	486	2,232	372	486	2,232	372	486	2,232	372	206	6,416
nug07	602	931	4,214	602	931	4,214	602	931	4,214	602	931	4,214	602	931	4,214	602	457	14,637
nug08	912	1,632	7,296	912	1,632	7,296	912	1,632	7,296	912	1,632	7,296	912	1,632	7,296	912	890	28,980
nug12	3,192	8,856	38,304	3,192	8,856	38,304	3,192	8,856	38,304	3,192	8,856	38,304	3,192	8,856	38,304	3,192	8,856	38,304
p010	10,090	19,000	117,910	10,090	18,010	116,920	10,090	17,020	127,360	10,090	10,000	221,050	10,090	10,000	221,050	10,090	10,000	221,050
p05	5,090	9,500	58,955	5,090	9,005	58,460	5,090	8,510	63,680	5,090	5,000	110,525	5,090	5,000	110,525	5,090	5,000	110,525
qap08	912	1,632	7,296	912	1,632	7,296	912	1,632	7,296	912	1,632	7,296	912	1,632	7,296	912	890	28,980
r05	5,190	9,500	103,955	5,190	9,005	103,460	5,190	8,510	114,080	5,190	5,000	204,875	5,190	5,000	204,875	5,190	5,000	204,875
sc205	205	203	551	205	181	528	205	113	1,311	205	113	1,311	205	113	1,311	205	113	1,311

scagr25	471	500	1,554	470	421	1,368	470	348	6,150	470	323	7,185	470	298	7,235	470	226	11,624
scagr7	129	140	420	128	115	360	128	96	588	128	89	633	128	82	647	128	62	903
scfxm1	330	457	2,589	322	438	2,548	322	414	2,539	322	388	2,700	322	372	2,882	322	351	3,136
scfxm2	660	914	5,183	644	876	5,101	644	828	5,083	644	776	5,407	644	744	5,771	644	702	6,279
scfxm3	990	1,371	7,777	966	1,314	7,654	966	1,242	7,627	966	1,164	8,114	966	1,116	8,660	966	1,053	9,422
scorpion	388	358	1,426	375	295	1,302	375	282	1,302	375	150	678	381	150	678	381	126	870
scrs8	490	1,169	3,182	485	1,128	3,056	485	857	3,056	485	842	3,265	485	838	3,277	485	827	3,491
sctap1	300	480	1,692	300	480	1,692	300	360	1,788	300	360	1,788	300	360	1,788	300	360	1,788
sctap2	1,090	1,880	6,714	1,090	1,880	6,714	1,090	1,410	7,090	1,090	1,410	7,090	1,090	1,410	7,090	1,090	1,410	7,090
sctap3	1,480	2,480	8,874	1,480	2,480	8,874	1,480	1,860	9,370	1,480	1,860	9,370	1,480	1,860	9,370	1,480	1,860	9,370
share1b	117	225	1,151	112	208	1,108	112	198	1,130	112	192	1,210	112	192	1,210	112	179	1,273
shp04l	402	2,118	6,332	402	2,110	6,324	402	2,110	6,324	402	2,110	6,324	402	2,110	6,324	402	1,846	8,660
ship04s	402	1,458	4,352	310	1,326	4,036	310	1,326	4,036	310	1,326	4,036	310	1,326	4,036	310	1,178	5,376
ship08l	778	4,283	12,802	754	4,259	12,730	754	4,259	12,730	754	4,259	12,730	754	4,251	12,770	754	3,715	17,530
ship08s	778	2,387	7,114	482	2,027	6,162	482	2,027	6,162	482	2,027	6,162	482	2,027	6,162	482	1,819	8,066
ship12l	1,151	5,427	16,170	947	5,223	15,558	947	5,223	15,558	947	5,223	15,558	947	5,223	15,558	947	4,587	21,378
ship12s	1,151	2,763	8,178	575	2,091	6,354	575	2,091	6,354	575	2,091	6,354	575	2,091	6,354	575	1,923	8,058
stocfor1	117	111	447	109	91	397	109	67	451	109	67	451	109	61	605	109	54	658
stocfor2	2,157	2,031	8,343	2,157	1,779	8,091	2,157	1,275	10,579	2,157	1,275	10,579	2,157	1,149	17,373	2,157	1,021	18,598
stocfor3	16,675	15,695	57,378	16,032	12,940	49,291	16,189	10,782	65,141	16,146	10,550	64,254	16,189	10,089	98,549	16,212	9,378	86,103
sws	14,310	12,465	93,015	10,530	8,100	62,190	10,530	7,830	63,810	10,530	7,830	63,810	10,530	7,830	63,810	10,530	7,830	63,810
testbig	17,613	31,223	61,639	17,613	30,419	60,035	17,613	20,015	54,431	17,613	16,813	51,229	17,613	13,611	48,027	17,613	13,611	48,027
zed	116	43	567	112	34	554	114	31	551	115	30	551	115	30	551	115	29	638
Shifted geometric mean	2,408	3,340	23,076	2,348	3,154	22,401	2,348	3,022	22,913	2,348	2,907	23,636	2,349	2,864	24,113	2,349	2,723	25,961
Reduction	—	—	—	−2.49%	−5.58%	−2.93%	−2.48%	−9.52%	−0.71%	−2.48%	−12.98%	2.42%	−2.45%	−14.26%	4.49%	−2.44%	−18.49%	12.50

values of k. Using $k = 10$, the LP problems have fewer variables compared to other values of k but they also become denser. The average value of multiple instances is computed using the shifted geometric means of the individual instances. We use a shift of 1, 000 for the rows and columns and 10, 000 for the nonzeros.

4.8   Eliminate Singleton Inequality Constraints

4.8.1   Mathematical Formulation

An inequality row of the coefficient matrix A is a singleton row iff one coefficient in that row is not equal to zero. A singleton inequality row can be formulated as:

$$\underline{b}_i \leq A_{i1}x_1 + A_{i2}x_2 + \cdots + A_{in}x_n \leq \overline{b}_i \tag{4.13}$$

where $A_{ik} \neq 0 \wedge A_{ij} = 0$, $i = 1, 2, \cdots, m$, $j = 1, 2, \cdots, n$, and $j \neq k$. A constraint of this type may be redundant or may state that the LP problem is infeasible. All possible cases are distinguished in the following theorem.

Theorem 4.3 *For each singleton inequality constraint, we distinguish the following cases:*

1. *Constraint type $\leq \overline{b}_i$, $A_{ik} > 0$ and $\overline{b}_i < 0$: The LP problem is infeasible.*
2. *Constraint type $\leq \overline{b}_i$, $A_{ik} < 0$ and $\overline{b}_i > 0$: Row i is redundant and can be deleted.*
3. *Constraint type $\leq \overline{b}_i$, $A_{ik} > 0$ and $\overline{b}_i = 0$: Row i and column k are redundant and can be deleted.*
4. *Constraint type $\leq \overline{b}_i$, $A_{ik} < 0$ and $\overline{b}_i = 0$: Row i is redundant and can be deleted.*
5. *Constraint type $\geq \underline{b}_i$, $A_{ik} > 0$ and $\underline{b}_i < 0$: Row i is redundant and can be deleted.*
6. *Constraint type $\geq \underline{b}_i$, $A_{ik} < 0$ and $\underline{b}_i > 0$: The LP problem is infeasible.*
7. *Constraint type $\geq \underline{b}_i$, $A_{ik} > 0$ and $\underline{b}_i = 0$: Row i is redundant and can be deleted.*
8. *Constraint type $\geq \underline{b}_i$, $A_{ik} < 0$ and $\underline{b}_i = 0$: Row i and column k are redundant and can be deleted.*

Proof

1. If $A_{ik} > 0$ and $\overline{b}_i < 0$, the constraint can be written as:

$$A_{ik}x_k \leq \overline{b}_i \Rightarrow x_k \leq \frac{\overline{b}_i}{A_{ik}}$$

where $\frac{\overline{b}_i}{A_{ik}} \leq 0$. However, x_k must be greater than or equal to zero, so the LP problem is infeasible.

2. If $A_{ik} < 0$ and $\overline{b}_i > 0$, the constraint can be written as:

$$A_{ik}x_k \leq \overline{b}_i \Rightarrow x_k \geq \frac{\overline{b}_i}{A_{ik}}$$

where $\frac{\bar{b}_i}{A_{ik}} \leq 0$. Taking also into account that $x_k \geq 0$, then:

$$x_k = [0, \infty) \wedge \left[\frac{\bar{b}_i}{A_{ik}}, \infty \right) = [0, \infty)$$

The bounds of variable x_k did not alter, so constraint i can be eliminated.

3. If $A_{ik} > 0$ and $\bar{b}_i = 0$, the constraint can be written as:

$$A_{ik}x_k \leq 0 \Rightarrow x_k \leq 0$$

Taking also into account that $x_k \geq 0$, then:

$$x_k = [0, \infty) \wedge (-\infty, 0] = 0$$

According to Section 4.6, this constraint is a singleton constraint and can be eliminated. Since $x_k = 0$, there is no need to replace variable x_k to the other constraints; we just eliminate it from the problem.

4. If $A_{ik} < 0$ and $\bar{b}_i = 0$, the constraint can be written as:

$$A_{ik}x_k \leq 0 \Rightarrow x_k \geq 0$$

Taking also into account that $x_k \geq 0$, then:

$$x_k = [0, \infty) \wedge [0, \infty) = [0, \infty)$$

The bounds of variable x_k did not alter, so constraint i can be eliminated.

5. If $A_{ik} > 0$ and $\underline{b}_i < 0$, the constraint can be written as:

$$A_{ik}x_k \geq \underline{b}_i \Rightarrow x_k \geq \frac{\underline{b}_i}{A_{ik}}$$

where $\frac{\underline{b}_i}{A_{ik}} \leq 0$. We continue as in case 2.

6. If $A_{ik} < 0$ and $\bar{b}_i > 0$, the constraint can be written as:

$$A_{ik}x_k \geq \underline{b}_i \Rightarrow x_k \leq \frac{\underline{b}_i}{A_{ik}}$$

where $\frac{\underline{b}_i}{A_{ik}} \leq 0$. We continue as in case 1.

7. If $A_{ik} > 0$ and $\underline{b}_i = 0$, the constraint can be written as:

$$A_{ik}x_k \geq 0 \Rightarrow x_k \geq 0$$

We continue as in case 4.

8. If $A_{ik} < 0$ and $\underline{b}_i = 0$, the constraint can be written as:

$$A_{ik}x_k \geq 0 \Rightarrow x_k \leq 0$$

We continue as in case 3.

$\square$

4.8.2   Numerical Example

This subsection demonstrates the presolve method that eliminates singleton inequality constraints through two illustrative examples.

1st example

The LP problem that will be presolved is the following:

$$
\begin{array}{rrcl}
\min z = -2x_1 + 4x_2 - 2x_3 + 2x_4 \\
\text{s.t.} \qquad\qquad - 3x_3 \qquad\quad &\leq& 2 & (1)\\
4x_1 - 3x_2 + 8x_3 - x_4 &=& 20 & (2)\\
-3x_1 + 2x_2 \qquad\quad - 4x_4 &\geq& -8 & (3)\\
- x_3 \qquad &=& 18 & (4)\\
\end{array}
$$
$$x_j \geq 0, \quad (j = 1,2,3,4)$$

In matrix notation the above LP problem is written as follows:

$$
A = \begin{bmatrix} 0 & 0 & -3 & 0 \\ 4 & -3 & 8 & -1 \\ -3 & 2 & 0 & -4 \\ 0 & 0 & -1 & 0 \end{bmatrix}, c = \begin{bmatrix} -2 \\ 4 \\ -2 \\ 2 \end{bmatrix}, b = \begin{bmatrix} 2 \\ 20 \\ -8 \\ 18 \end{bmatrix}, Eqin = \begin{bmatrix} -1 \\ 0 \\ 1 \\ 0 \end{bmatrix}
$$

Initially, we start searching for inequality constraints that have only one nonzero element. We observe that all elements of the first inequality constraint are equal to zero except the third element:

$$0x_1 + 0x_2 - 3x_3 + 0x_4 \leq 2$$

According to the second case of the previous subsection (constraint type $\leq \overline{b}_i$, $A_{ik} < 0$, and $\overline{b}_i > 0$), the first row is redundant and can be deleted. So, we can delete the first row from matrix A and the first element from vectors b and $Eqin$:

$$
A = \begin{bmatrix} 4 & -3 & 8 & -1 \\ -3 & 2 & 0 & -4 \\ 0 & 0 & -1 & 0 \end{bmatrix}, c = \begin{bmatrix} -2 \\ 4 \\ -2 \\ 2 \end{bmatrix}, b = \begin{bmatrix} 20 \\ -8 \\ 18 \end{bmatrix}, Eqin = \begin{bmatrix} 0 \\ 1 \\ 0 \end{bmatrix}
$$

Finally, the equivalent LP problem after presolve is the following:

$$\min z = -2x_1 + 4x_2 - 2x_3 + 2x_4$$

$$\text{s.t.} \quad 4x_1 - 3x_2 + 8x_3 - x_4 = 20 \quad (1)$$

$$-3x_1 + 2x_2 \quad\quad\quad - 4x_4 \geq -8 \quad (2)$$

$$- x_3 \quad\quad\quad = 18 \quad (3)$$

$$x_j \geq 0, \quad (j = 1, 2, 3, 4)$$

Now, we observe that all coefficients of the third equality constraint are equal to 0 except the third element:

$$-x_3 = 18$$

Hence, we can eliminate this constraint by applying the singleton equality presolve method.

2nd example

The LP problem that will be presolved is the following:

$$\min z = -2x_1 + 4x_2 - 2x_3 + 2x_4$$

$$\text{s.t.} \quad - 3x_3 \quad\quad\quad \geq 2 \quad (1)$$

$$-4x_1 - 3x_2 + 8x_3 - x_4 = 20 \quad (2)$$

$$-3x_1 + 2x_2 \quad\quad\quad - 4x_4 \geq -8 \quad (3)$$

$$- x_3 \quad\quad\quad = 18 \quad (4)$$

$$x_j \geq 0, \quad (j = 1, 2, 3, 4)$$

In matrix notation the above LP problem is written as follows:

$$A = \begin{bmatrix} 0 & 0 & -3 & 0 \\ -4 & -3 & 8 & -1 \\ -3 & 2 & 0 & -4 \\ 0 & 0 & -1 & 0 \end{bmatrix}, c = \begin{bmatrix} -2 \\ 4 \\ -2 \\ 2 \end{bmatrix}, b = \begin{bmatrix} 2 \\ 20 \\ -8 \\ 18 \end{bmatrix}, Eqin = \begin{bmatrix} 1 \\ 0 \\ 1 \\ 0 \end{bmatrix}$$

Initially, we start searching for inequality constraints that have only one nonzero element. We observe that all elements of the first row are equal to zero except the third element:

$$0x_1 + 0x_2 - 3x_3 + 0x_4 \geq 2$$

According to the sixth case of the previous subsection (constraint type $\geq \underline{b}_i$, $A_{ik} < 0$, and $\underline{b}_i > 0$), the LP problem is infeasible.

4.8.3   Implementation in MATLAB

This subsection presents the implementation in MATLAB of the presolve method
that eliminates singleton inequality constraints (filename: `eliminateSing
leton InequalityConstraints.m`). Some necessary notations should be
introduced before the presentation of the implementation of the presolve method
that eliminates singleton inequality constraints. Let A be a $m \times n$ matrix with the
coefficients of the constraints, c be a $n \times 1$ vector with the coefficients of the
objective function, b be a $m \times 1$ vector with the right-hand side of the constraints,
$Eqin$ be a $m \times 1$ vector with the type of the constraints, and $infeasible$ be a flag
variable showing if the LP problem is infeasible or not.

The function for the presolve method that eliminates singleton inequality
constraints takes as input the matrix of coefficients of the constraints (matrix A),
the vector of the coefficients of the objective function (vector c), the vector of
the right-hand side of the constraints (vector b), and the vector of the type of the
constraints (vector $Eqin$), and returns as output the presolved matrix of coefficients
of the constraints (matrix A), the presolved vector of the coefficients of the objective
function (vector c), the presolved vector of the right-hand side of the constraints
(vector b), the presolved vector of the type of the constraints (vector $Eqin$), and a
flag variable showing if the LP problem is infeasible or not (variable $infeasible$).

In lines 36–39, we find the rows that have only one nonzero element and set all
the other rows equal to zero. Moreover, we find the number of the rows that have
only one nonzero element (line 42). If any singleton inequality constraint exists,
then we eliminate it or declare infeasibility (lines 45–133). Initially, we compute the
indices of the singleton inequality constraints (line 52). For each singleton inequality
constraint, we examine the eight cases that were presented in subsection 4.8.1.
According to cases 2 (lines 70–74), 3 (lines 77–84), 4 (lines 87–92), 5 (lines 96–
100), 7 (lines 108–112), and 8 (lines 115–123), we eliminate the singleton inequality
constraints and update the adequate variables. According to cases 1 (lines 65–67)
and 6 (lines 103–105), the LP problem is infeasible. Finally, the number of the
eliminated singleton inequality constraints is printed (lines 127–132).

```
1.    function [A, c, b, Eqin, infeasible] = ...
2.        eliminateSingletonInequalityConstraints(A, c, b, Eqin)
3.    % Filename: eliminateSingletonInequalityConstraints.m
4.    % Description: the function is an implementation of the
5.    % presolve technique that eliminates singleton
6.    % inequality constraints
7.    % Authors: Ploskas, N., & Samaras, N.
8.    %
9.    % Syntax: [A, c, b, Eqin, infeasible] = ...
10.   %   eliminateSingletonInequalityConstraints(A, c, b, Eqin)
11.   %
12.   % Input:
13.   % -- A: matrix of coefficients of the constraints
14.   %    (size m x n)
15.   % -- c: vector of coefficients of the objective function
```

```
16.  %      (size n x 1)
17.  % -- b: vector of the right-hand side of the constraints
18.  %      (size m x 1)
19.  % -- Eqin: vector of the type of the constraints
20.  %      (size m x 1)
21.  %
22.  % Output:
23.  % -- A: presolved matrix of coefficients of the constraints
24.  %      (size m x n)
25.  % -- c: presolved vector of coefficients of the objective
26.  %      function (size n x 1)
27.  % -- b: presolved vector of the right-hand side of the
28.  %      constraints (size m x 1)
29.  % -- Eqin: presolved vector of the type of the constraints
30.  %      (size m x 1)
31.  % -- infeasible: flag variable showing if the LP is
32.  %      infeasible or not
33.
34.  infeasible = 0;
35.  % compute the number of nonzero elements in each row
36.  delrows = sum(spones(A'));
37.  % find the rows that have only one nonzero element and set
38.  % all the other rows equal to zero
39.  singleton = (delrows == 1);
40.  % find the number of the rows that have only one nonzero
41.  % element
42.  cardrows = nnz(singleton);
43.  counterrows = 0;
44.  countercols = 0;
45.  if cardrows > 0 % if singleton rows exist
46.      % get the indices of the singleton rows
47.      idsrows = int16(find(singleton));
48.      % find inequality constraints
49.      [row, ~] = find(Eqin ~= 0);
50.      % compute the indices of the singleton inequality
51.      % constraints
52.      idsrows = intersect(idsrows, row);
53.      % compute the number of the singleton inequality
54.      % constraints
55.      cardrows = length(idsrows);
56.      % for each singleton inequality constraint
57.      for i = cardrows:-1:1
58.          ii = idsrows(i); % get the index of the constraint
59.          % find the nonzero elements of this row
60.          [row1, j] = find(A(ii, :));
61.          if ~isempty(row1) % if there are nonzero elements
62.              if Eqin(ii) == -1 % constraint type <=
63.                  % check if the LP problem is infeasible
64.                  % (Case 1)
65.                  if (A(ii, j) > 0) && (b(ii) < 0)
66.                      infeasible = 1;
67.                      return;
68.                  % eliminate the singleton inequality constraints
69.                  % (Case 2)
```

```
70.                    elseif (A(ii, j) < 0) && (b(ii) > 0)
71.                       A(ii, :) = [];
72.                       b(ii) = [];
73.                       Eqin(ii) = [];
74.                        counterrows = counterrows + 1;
75.                    % eliminate the singleton inequality constraints
76.                    % (Case 3)
77.                    elseif (A(ii, j) > 0) && (b(ii) == 0)
78.                       A(ii, :) = [];
79.                       b(ii) = [];
80.                       Eqin(ii) = [];
81.                       c(j) = [];
82.                       A(:, j) = [];
83.                        counterrows = counterrows + 1;
84.                        countercols = countercols + 1;
85.                    % eliminate the singleton inequality constraints
86.                    % (Case 4)
87.                    elseif (A(ii, j) < 0) && (b(ii) == 0)
88.                       A(ii, :) = [];
89.                       b(ii) = [];
90.                       Eqin(ii) = [];
91.                        counterrows = counterrows + 1;
92.                    end
93.                elseif Eqin(ii) == 1 % constraint type >=
94.                    % eliminate the singleton inequality constraints
95.                    % (Case 5)
96.                    if((A(ii, j) > 0) && (b(ii) < 0))
97.                       A(ii, :) = [];
98.                       b(ii) = [];
99.                       Eqin(ii) = [];
100.                       counterrows = counterrows + 1;
101.                    % check if the LP problem is infeasible
102.                    % (Case 6)
103.                    elseif (A(ii, j) < 0) && (b(ii) > 0)
104.                        infeasible = 1;
105.                        return;
106.                    % eliminate the singleton inequality constraints
107.                    % (Case 7)
108.                    elseif (A(ii, j) > 0) && (b(ii) == 0)
109.                       A(ii, :) = [];
110.                       b(ii) = [];
111.                       Eqin(ii) = [];
112.                       counterrows = counterrows + 1;
113.                    % eliminate the singleton inequality constraints
114.                    % (Case 8)
115.                    elseif (A(ii,j) < 0) && (b(ii) == 0)
116.                       A(ii, :) = [];
117.                       b(ii) = [];
118.                       Eqin(ii) = [];
119.                       c(j) = [];
120.                       A(:, j) = [];
```

```
121.                        counterrows = counterrows + 1;
122.                        countercols = countercols + 1;
123.                    end
124.                end
125.            end
126.        end
127.        if (counterrows ~= 0) || (countercols ~= 0)
128.            fprintf(['ELIMINATE SINGLETON INEQUALITY CONSTRAINTS: ' ...
129.                '%i singleton inequality constraints were found. ' ...
130.                '%i constraints and %i variables were eliminated ' ...
131.                '\n'], cardrows, counterrows, countercols);
132.        end
133. end
134. end
```

4.9   Eliminate Dual Singleton Inequality Constraints

4.9.1   Mathematical Formulation

This method is similar to the previous one, but it is applied to the dual LP problem. A column of the coefficient matrix A is a singleton column iff only one coefficient in that column is not equal to zero. Transforming the primal LP problem to its dual LP problem, a singleton column will form a dual singleton constraint in the dual LP problem. A dual singleton inequality row can be formulated as:

$$A_{j1}w_1 + A_{j2}w_2 + \cdots + A_{jm}w_m \leq c_j \qquad (4.14)$$

where $A_{jk} \neq 0 \wedge A_{ji} = 0, i = 1, 2, \cdots, m, j = 1, 2, \cdots, n$, and $i \neq k$. A constraint of this type may be redundant or may state that the dual LP problem is unbounded and the primal LP problem is infeasible. We present this method and the eliminations that it performs without the need to transform the initial problem to its dual, i.e., instead of eliminating a row of the dual LP problem, we can eliminate a column of the primal LP problem. All possible cases are distinguished in the following theorem.

Theorem 4.4 *For each dual singleton inequality constraint, we distinguish the following cases:*

1. *Constraint type $\leq$, $A_{kj} > 0$, and $c_j > 0$: Column j is redundant and can be deleted.*
2. *Constraint type $\leq$, $A_{kj} < 0$, and $c_j < 0$: The LP problem is infeasible.*
3. *Constraint type $\leq$, $A_{kj} > 0$, and $c_j = 0$: Column j is redundant and can be deleted.*

4. *Constraint type $\leq$, $A_{kj} < 0$, and $c_j = 0$: Row k and column j are redundant and can be deleted.*
5. *Constraint type $\geq$, $A_{kj} > 0$, and $c_j < 0$: The LP problem is infeasible.*
6. *Constraint type $\geq$, $A_{kj} < 0$, and $c_j > 0$: Column j is redundant and can be deleted.*
7. *Constraint type $\geq$, $A_{kj} > 0$, and $c_j = 0$: Row k and column j are redundant and can be deleted.*
8. *Constraint type $\geq$, $A_{kj} < 0$, and $c_j = 0$: Column j is redundant and can be deleted.*

Proof A singleton column of the primal LP problem forms a dual singleton constraint in the dual LP problem. Depending on the type of the constraint i of the primal LP problem, the bound of the dual variable w_i will be either $(-\infty, 0]$ (for inequality constraints of the form less than or equal to) or $[0, \infty)$ (for inequality constraints of the form greater than or equal to). We can prove all cases in a similar way as in Theorem 4.3.   □

4.9.2 Numerical Example

This subsection demonstrates the presolve method that eliminates dual singleton inequality constraints through two illustrative examples.

1st example

The LP problem that will be presolved is the following:

$$
\begin{aligned}
\min z = \quad & 4x_1 + \ x_2 - 2x_3 + 7x_4 \\
\text{s.t.} \quad & 3x_1 \qquad\ \ - \ x_3 - 6x_4 \leq 0 \quad (1) \\
& -3x_1 - 2x_2 + 5x_3 - \ x_4 \geq 0 \quad (2) \\
& 4x_1 \qquad\quad\ + 3x_3 + 4x_4 \leq 5 \quad (3) \\
& x_j \geq 0, \quad (j = 1, 2, 3, 4)
\end{aligned}
$$

In matrix notation the above LP problem is written as follows:

$$
A = \begin{bmatrix} 3 & 0 & -1 & -6 \\ -3 & -2 & 5 & -1 \\ 4 & 0 & 3 & 4 \end{bmatrix}, c = \begin{bmatrix} 4 \\ 1 \\ -2 \\ 7 \end{bmatrix}, b = \begin{bmatrix} 0 \\ 0 \\ 5 \end{bmatrix}, Eqin = \begin{bmatrix} -1 \\ 1 \\ -1 \end{bmatrix}
$$

Initially, we start searching for columns that have only one nonzero element. We observe that all elements of the second column are equal to zero except the second element. According to the sixth case of the previous subsection (constraint type $\geq$, $A_{kj} < 0$, and $c_j > 0$), the second column is redundant and can be deleted. So, we can delete the second column from matrix A and the second element from vector c:

$$A = \begin{bmatrix} 3 & -1 & -6 \\ -3 & 5 & -1 \\ 4 & 3 & 4 \end{bmatrix}, c = \begin{bmatrix} 4 \\ -2 \\ 7 \end{bmatrix}, b = \begin{bmatrix} 0 \\ 0 \\ 5 \end{bmatrix}, Eqin = \begin{bmatrix} -1 \\ 1 \\ -1 \end{bmatrix}$$

Finally, the equivalent LP problem after presolve is the following:

$$
\begin{aligned}
\min z = \quad & 4x_1 - 2x_3 + 7x_4 \\
\text{s.t.} \quad & 3x_1 - x_3 - 6x_4 \leq 0 \\
& -3x_1 + 5x_3 - x_4 \geq 0 \\
& 4x_1 + 3x_3 + 4x_4 \leq 5 \\
& x_j \geq 0, \quad (j = 1, 3, 4)
\end{aligned}
$$

2nd example

The LP problem that will be presolved is the following:

$$
\begin{aligned}
\min z = \quad & 4x_1 - x_2 - 2x_3 + 7x_4 \\
\text{s.t.} \quad & 3x_1 \quad\quad - x_3 - 6x_4 \leq 0 \quad (1) \\
& -3x_1 + 2x_2 + 5x_3 - x_4 \geq 0 \quad (2) \\
& 4x_1 \quad\quad + 3x_3 + 4x_4 \leq 5 \quad (3) \\
& x_j \geq 0, \quad (j = 1, 2, 3, 4)
\end{aligned}
$$

In matrix notation the above LP problem is written as follows:

$$A = \begin{bmatrix} 3 & 0 & -1 & -6 \\ -3 & 2 & 5 & -1 \\ 4 & 0 & 3 & 4 \end{bmatrix}, c = \begin{bmatrix} 4 \\ -1 \\ -2 \\ 7 \end{bmatrix}, b = \begin{bmatrix} 0 \\ 0 \\ 5 \end{bmatrix}, Eqin = \begin{bmatrix} -1 \\ 1 \\ -1 \end{bmatrix}$$

Initially, we start searching for columns that have only one nonzero element. We observe that all elements of the second column are equal to zero except the second element. According to the fifth case of the previous subsection (constraint type $\geq$, $A_{kj} > 0$, and $c_j < 0$), the LP problem is infeasible.

4.9.3 Implementation in MATLAB

This subsection presents the implementation in MATLAB of the presolve method that eliminates dual singleton inequality constraints (filename: `eliminateDual SingletonInequalityConstraints.m`). Some necessary notations should be introduced before the presentation of the implementation of the presolve method that eliminates dual singleton inequality constraints. Let A be a $m \times n$ matrix with the coefficients of the constraints, c be a $n \times 1$ vector with the coefficients of the objective function, b be a $m \times 1$ vector with the right-hand side of the constraints, *Eqin* be a $m \times 1$ vector with the type of the constraints, and *infeasible* be a flag variable showing if the LP problem is infeasible or not.

The function for the presolve method that eliminates dual singleton inequality constraints takes as input the matrix of coefficients of the constraints (matrix A), the vector of the coefficients of the objective function (vector c), the vector of the right-hand side of the constraints (vector b), and the vector of the type of the constraints (vector *Eqin*), and returns as output the presolved matrix of coefficients of the constraints (matrix A), the presolved vector of the coefficients of the objective function (vector c), the presolved vector of the right-hand side of the constraints (vector b), the presolved vector of the type of the constraints (vector *Eqin*), and a flag variable showing if the LP problem is infeasible or not (variable *infeasible*).

In lines 36–39, we find the columns that have only one nonzero element and set all the other columns equal to zero. Moreover, we find the number of the columns that have only one nonzero element (line 42). If any dual singleton inequality constraint exists, then we eliminate it or declare infeasibility (lines 45–123). Initially, we find the indices of the dual singleton inequality constraints (line 47). For each dual singleton inequality constraint, we examine the eight cases that were presented in subsection 4.9.1. According to cases 1 (lines 59–62), 3 (lines 70–73), 4 (lines 76–84), 6 (lines 93–96), 7 (lines 99–106), and 8 (lines 109–113), we eliminate the dual singleton inequality constraints and update the adequate variables. According to cases 2 (lines 65–67) and 5 (lines 88–90), the LP problem is infeasible. Finally, the number of the eliminated dual singleton inequality constraints is printed (lines 117–122).

```
1.    function [A, c, b, Eqin, infeasible] = ...
2.        eliminateDualSingletonInequalityConstraints(A, c, b, Eqin)
3.    % Filename: eliminateDualSingletonInequalityConstraints.m
4.    % Description: the function is an implementation of the
5.    % presolve technique that eliminates dual singleton
6.    % inequality constraints
7.    % Authors: Ploskas, N., & Samaras, N.
8.    %
9.    % Syntax: [A, c, b, Eqin, infeasible] = ...
10.   %   eliminateDualSingletonInequalityConstraints(A, c, b, Eqin)
11.   %
12.   % Input:
13.   % -- A: matrix of coefficients of the constraints
14.   %    (size m x n)
15.   % -- c: vector of coefficients of the objective function
16.   %    (size n x 1)
17.   % -- b: vector of the right-hand side of the constraints
18.   %    (size m x 1)
19.   % -- Eqin: vector of the type of the constraints
20.   %    (size m x 1)
21.   %
22.   % Output:
```

```
23.   % -- A: presolved matrix of coefficients of the constraints
24.   %    (size m x n)
25.   % -- c: presolved vector of coefficients of the objective
26.   %    function (size n x 1)
27.   % -- b: presolved vector of the right-hand side of the
28.   %    constraints (size m x 1)
29.   % -- Eqin: presolved vector of the type of the constraints
30.   %    (size m x 1)
31.   % -- infeasible: flag variable showing if the LP is
32.   %    infeasible or not
33.
34.   infeasible = 0;
35.   % compute the number of nonzero elements in each column
36.   delcols = sum(spones(A));
37.   % find the columns that have only one nonzero element and set
38.   % all the other columns equal to zero
39.   singleton = (delcols == 1);
40.   % find the number of the columns that have only one nonzero
41.   % element
42.   cardcols = nnz(singleton);
43.   countercols = 0;
44.   counterrows = 0;
45.   if cardcols > 0 % if dual singleton constraints exist
46.       % get the indices of the dual singleton constraints
47.       idscols = int16(find(singleton));
48.       % compute the number of the dual singleton constraints
49.       cardcols = length(idscols);
50.       % for each dual singleton inequality constraint
51.       for j = cardcols:-1:1
52.           jj = idscols(j); % get the index of the constraint
53.           % find the nonzero elements of this column
54.           [row, ~] = find(A(:, jj));
55.           if ~isempty(row) % if there are nonzero elements
56.               if Eqin(row) == -1 % constraint type <=
57.                   % eliminate the dual singleton inequality
58.                   % constraints (Case 1)
59.                   if (A(row, jj) > 0) && (c(jj) > 0)
60.                       A(:, jj) = [];
61.                       c(jj) = [];
62.                       countercols = countercols + 1;
63.                   % check if the LP problem is infeasible
64.                   % (Case 2)
65.                   elseif (A(row, jj) < 0) && (c(jj) < 0)
66.                       infeasible = 1;
67.                       return;
68.                   % eliminate the dual singleton inequality
69.                   % constraints (Case 3)
70.                   elseif (A(row,jj) > 0) && (c(jj) == 0)
71.                       A(:, jj) = [];
72.                       c(jj) = [];
73.                       countercols = countercols + 1;
74.                   % eliminate the dual singleton inequality
75.                   % constraints (Case 4)
76.                   elseif (A(row, jj) < 0) && (c(jj) == 0)
```

```
77.                        A(:, jj) = [];
78.                        c(jj) = [];
79.                        A(row, :) = [];
80.                        b(row) = [];
81.                        Eqin(row) = [];
82.                        countercols = countercols + 1;
83.                        counterrows = counterrows + 1;
84.                    end
85.                elseif Eqin(row) == 1 % constraint type >=
86.                    % check if the LP problem is infeasible
87.                    % (Case 5)
88.                    if (A(row, jj) > 0) && (c(jj) < 0)
89.                        infeasible = 1;
90.                        return;
91.                    % eliminate the dual singleton inequality
92.                    % constraints (Case 6)
93.                    elseif (A(row, jj) < 0) && (c(jj) > 0)
94.                        A(:, jj) = [];
95.                        c(jj) = [];
96.                        countercols = countercols + 1;
97.                    % eliminate the dual singleton inequality
98.                    % constraints (Case 7)
99.                    elseif (A(row,jj) > 0) && (c(jj) == 0)
100.                       A(:, jj) = [];
101.                       c(jj) = [];
102.                       A(row, :) = [];
103.                       b(row) = [];
104.                       Eqin(row) = [];
105.                       countercols = countercols + 1;
106.                       counterrows = counterrows + 1;
107.                   % eliminate the dual singleton inequality
108.                   % constraints (Case 8)
109.                   elseif (A(row, jj) < 0) && (c(jj) == 0)
110.                       A(:, jj) = [];
111.                       c(jj) = [];
112.                       countercols = countercols + 1;
113.                   end
114.               end
115.           end
116.   end
117.   if (countercols > 0) || (counterrows > 0)
118.       fprintf(['ELIMINATE DUAL SINGLETON INEQUALITY ' ...
119.           'CONSTRAINTS: %i constraints and %i columns ' ...
120.           'were eliminated\n'], counterrows, ...
121.           countercols);
122.   end
123. end
124. end
```

4.10   Eliminate Implied Free Singleton Columns

4.10.1   Mathematical Formulation

A constraint that implies a free singleton column can be formulated as:

$$A_{i.}x + A_{is}x_s = b \tag{4.15}$$

where $i = 1, 2, \cdots, m$, and the singleton column inside it ($A_{is} \neq 0$) is redundant iff:

$$A_{is} > 0 \wedge A_{ij} \leq 0, j \neq s, j = 1, 2, \cdots, n \tag{4.16}$$

or:

$$A_{is} < 0 \wedge A_{ij} \geq 0, j \neq s, j = 1, 2, \cdots, n \tag{4.17}$$

In that case, we can delete variable x_s from the LP problem. In addition, we can delete constraint i. If $c_s = 0$, then we delete only the constraint i. If $c_s \neq 0$, then we update vector c and the constant term of the objective function (c_0):

$$c = c - \frac{c_s}{A_{is}}A_{i.}^T \tag{4.18}$$

$$c_0 = c_0 + \frac{c_s}{A_{is}}b_i \tag{4.19}$$

4.10.2   Numerical Example

This subsection demonstrates the presolve method that eliminates implied free singleton columns through two illustrative examples.

1st example

The LP problem that will be presolved is the following:

$$
\begin{aligned}
\min z = \quad & x_1 + 2x_2 - 4x_3 - 3x_4 \\
\text{s.t.} \quad & 3x_1 \qquad\quad + 5x_3 + 2x_4 \leq 5 \quad (1) \\
& -x_1 + x_2 \qquad\quad - 3x_4 = 8 \quad (2) \\
& -2x_1 \qquad\quad - 2x_3 + x_4 \geq 6 \quad (3) \\
& x_j \geq 0, \quad (j = 1, 2, 3, 4)
\end{aligned}
$$

In matrix notation the above LP problem is written as follows:

$$A = \begin{bmatrix} 3 & 0 & 5 & 2 \\ -1 & 1 & 0 & -3 \\ -2 & 0 & -2 & 1 \end{bmatrix}, c = \begin{bmatrix} 1 \\ 2 \\ -4 \\ -3 \end{bmatrix}, b = \begin{bmatrix} 5 \\ 8 \\ 6 \end{bmatrix}, Eqin = \begin{bmatrix} -1 \\ 0 \\ 1 \end{bmatrix}$$

Initially, we start searching for columns that have only one nonzero element. We observe that all elements of the second column are equal to zero except the second element. According to the first case of the previous subsection ($A_{is} > 0 \wedge A_{ij} \leq 0, j \neq s$), the second column is redundant and can be deleted. First, we update vector c and calculate the constant term of the objective function (c_0), assuming that its initial value is zero:

$$c_0 = c_0 + \frac{c_2}{A_{22}} b_2 = 0 + \frac{2}{1} 8 = 16 \quad c = c - \frac{c_2}{A_{22}} A_{2.}^T = \begin{bmatrix} 1 \\ 2 \\ -4 \\ -3 \end{bmatrix} - \frac{2}{1} \begin{bmatrix} -1 \\ 1 \\ 0 \\ -3 \end{bmatrix} = \begin{bmatrix} 3 \\ 0 \\ -4 \\ 3 \end{bmatrix}$$

So, we can delete the second row and the second column from matrix A and the second element from vectors c, b, and $Eqin$:

$$A = \begin{bmatrix} 3 & 5 & 2 \\ -2 & -2 & 1 \end{bmatrix}, c = \begin{bmatrix} 3 \\ -4 \\ -3 \end{bmatrix}, b = \begin{bmatrix} 5 \\ 6 \end{bmatrix}, Eqin = \begin{bmatrix} -1 \\ 1 \end{bmatrix}, c_0 = 16$$

Finally, the equivalent LP problem after presolve is the following:

$$\begin{aligned}
\min z = \quad & 3x_1 - 4x_3 - 3x_4 + 16 \\
\text{s.t.} \quad & 3x_1 + 5x_3 + 2x_4 \leq 5 \\
& -2x_1 - 2x_3 + x_4 \geq 6 \\
x_j \geq 0, \quad & (j = 1, 3, 4)
\end{aligned}$$

2nd example

The LP problem that will be presolved is the following:

$$\begin{aligned}
\min z = \quad & -2x_1 - x_2 - 3x_3 - x_4 \\
\text{s.t.} \quad & x_1 + 3x_2 \qquad\quad + x_4 \leq 10 \quad (1) \\
& 2x_1 + x_2 - 2x_3 + 3x_4 = -5 \quad (2) \\
& -x_1 + 2x_2 \qquad\quad + x_4 \leq 9 \quad (3) \\
x_j \geq 0, \quad & (j = 1, 2, 3, 4)
\end{aligned}$$

In matrix notation the above LP problem is written as follows:

$$A = \begin{bmatrix} 1 & 3 & 0 & 1 \\ 2 & 1 & -2 & 3 \\ -1 & 2 & 0 & 1 \end{bmatrix}, c = \begin{bmatrix} -2 \\ -1 \\ -3 \\ -1 \end{bmatrix}, b = \begin{bmatrix} 10 \\ -5 \\ 9 \end{bmatrix}, Eqin = \begin{bmatrix} -1 \\ 0 \\ -1 \end{bmatrix}$$

Initially, we start searching for columns that have only one nonzero element. We observe that all elements of the third column are equal to zero except the second element. According to the second case of the previous subsection ($A_{is} < 0 \wedge A_{ij} \geq 0, j \neq s$), the third column is redundant and can be deleted. First, we update vector c and calculate the constant term of the objective function (c_0), assuming that its initial value is zero:

$$c_0 = c_0 + \frac{c_3}{A_{23}} b_2 = 0 + \frac{-3}{-2}(-5) = -15/2c = c - \frac{c_3}{A_{23}} A_{2}^T.$$

$$= \begin{bmatrix} -2 \\ -1 \\ -3 \\ -1 \end{bmatrix} - \frac{-3}{-2} \begin{bmatrix} 2 \\ 1 \\ -2 \\ 3 \end{bmatrix} = \begin{bmatrix} -5 \\ -5/2 \\ 0 \\ -11/2 \end{bmatrix} \quad (4.20)$$

So, we can delete the second row and third column from matrix A, the third element from vector c, and the second element from vectors b and $Eqin$:

$$A = \begin{bmatrix} 1 & 3 & 1 \\ -1 & 2 & 1 \end{bmatrix}, c = \begin{bmatrix} -2 \\ -1 \\ -1 \end{bmatrix}, b = \begin{bmatrix} 10 \\ 9 \end{bmatrix}, Eqin = \begin{bmatrix} -1 \\ -1 \end{bmatrix}$$

Finally, the equivalent LP problem after presolve is the following:

$$\begin{aligned} \min z = -2x_1 &- x_2 - x_4 \\ \text{s.t.} \quad x_1 + 3x_2 &+ x_4 \leq 10 \\ -x_1 + 2x_2 &+ x_4 \leq 9 \\ x_j \geq 0, \quad (j &= 1, 2, 4) \end{aligned}$$

4.10.3   Implementation in MATLAB

This subsection presents the implementation in MATLAB of the presolve method that eliminates implied free singleton constraints (filename: eliminateImplied FreeSingletonColumns.m). Some necessary notations should be introduced before the presentation of the implementation of the presolve method that eliminates implied free singleton constraints. Let A be a $m \times n$ matrix with the coefficients of the constraints, c be a $n \times 1$ vector with the coefficients of the objective function,

b be a $m \times 1$ vector with the right-hand side of the constraints, *Eqin* be a $m \times 1$ vector with the type of the constraints, and $c0$ be the constant term of the objective function.

The function for the presolve method that eliminates implied free singleton constraints takes as input the matrix of coefficients of the constraints (matrix A), the vector of the coefficients of the objective function (vector c), the vector of the right-hand side of the constraints (vector b), the vector of the type of the constraints (vector *Eqin*), and the constant term of the objective function (variable $c0$), and returns as output the presolved matrix of coefficients of the constraints (matrix A), the presolved vector of the coefficients of the objective function (vector c), the presolved vector of the right-hand side of the constraints (vector b), the presolved vector of the type of the constraints (vector *Eqin*), and the updated constant term of the objective function (variable $c0$).

In lines 35–38, we find the columns that have only one nonzero element and set all the other columns equal to zero. Moreover, we find the number of the columns that have only one nonzero element (line 41). If any implied free singleton column exists, then we eliminate it (lines 44–154). For each implied free singleton column, we examine the cases that were presented in subsection 4.10.1 in order to eliminate the implied free singleton columns. Finally, the number of the eliminated implied free singleton columns is printed (lines 155–159).

```
1.    function [A, c, b, Eqin, c0] = ...
2.       eliminateImpliedFreeSingletonColumns(A, c, b, Eqin, c0)
3.    % Filename: eliminateImpliedFreeSingletonColumns.m
4.    % Description: the function is an implementation of the
5.    % presolve technique that eliminates implied free
6.    % singleton constraints
7.    % Authors: Ploskas, N., & Samaras, N.
8.    %
9.    % Syntax: [A, c, b, Eqin, c0] = ...
10.   %    eliminateImpliedFreeSingletonColumns(A, c, b, Eqin, c0)
11.   %
12.   % Input:
13.   % -- A: matrix of coefficients of the constraints
14.   %      (size m x n)
15.   % -- c: vector of coefficients of the objective function
16.   %      (size n x 1)
17.   % -- b: vector of the right-hand side of the constraints
18.   %      (size m x 1)
19.   % -- Eqin: vector of the type of the constraints
20.   %      (size m x 1)
21.   % -- c0: constant term of the objective function
22.   %
23.   % Output:
24.   % -- A: presolved matrix of coefficients of the constraints
25.   %      (size m x n)
26.   % -- c: presolved vector of coefficients of the objective
27.   %      function (size n x 1)
28.   % -- b: presolved vector of the right-hand side of the
29.   %      constraints (size m x 1)
```

```
30.    % -- Eqin: presolved vector of the type of the constraints
31.    %      (size m x 1)
32.    % -- c0: updated constant term of the objective function
33.
34.    % compute the number of nonzero elements in each column
35.    delcols = sum(spones(A));
36.    % find the columns that have only one nonzero element and
37.    % set all the other columns equal to zero
38.    singleton = (delcols == 1);
39.    % find the number of the columns that have only one nonzero
40.    % element
41.    cardcols = nnz(singleton);
42.    idscols = find(singleton);
43.    countercols = 0;
44.    for j = cardcols:-1:1 % for each dual singleton constraint
45.        jj = idscols(j); % get the index of the constraint
46.        % find the nonzero elements of this column
47.        [row, ~] = find(A(:, jj));
48.        if ~isempty(row) % if there are nonzero elements
49.            if Eqin(row) == 0 % constraint type =
50.                % if the element is greater than zero
51.                if A(row, jj) > 0
52.                    % find the number of columns
53.                    n0 = size(A, 2);
54.                    % compute the indices of all other columns
55.                    set = setdiff(1:n0, jj);
56.                    % if all the elements of the row except
57.                    % the one in the specific column are less
58.                    % than or equal to zero and the right-hand
59.                    % side greater than or equal to zero,
60.                    % then update matrix A, vectors c, b, and
61.                    % Eqin, and variable c0
62.                    if all(A(row, set) <= 0) && b(row) >= 0
63.                        if c(jj) ~= 0
64.                            c0 = c0 + c(jj) * (b(row) / A(row, jj));
65.                            c = c - (c(jj) / A(row, jj)) * A(row, :)';
66.                        end
67.                        c(jj) = [];
68.                        A(:, jj) = [];
69.                        A(row, :) = [];
70.                        b(row) = [];
71.                        Eqin(row) = [];
72.                        countercols = countercols + 1;
73.                    end
74.                elseif A(row, jj) < 0 % if the element is less
75.                    % than zero
76.                    n0 = size(A, 2); % find the number of columns
77.                    % compute the indices of all other columns
78.                    set = setdiff(1:n0, jj);
79.                    % if all the elements of the row except the
80.                    % one in the specific column are greater than
81.                    % or equal to zero and the right-hand side
82.                    % less than or equal to zero, then update
83.                    % matrix A, vectors c, b, and Eqin, and
```

```
84.              % variable c0
85.              if all(A(row, set) >= 0) && b(row) <= 0
86.                  if c(jj) ~= 0
87.                      c0 = c0 + c(jj) * (b(row) / A(row, jj));
88.                      c = c - (c(jj) / A(row,jj)) * A(row, :)';
89.                  end
90.                  c(jj) = [];
91.                  A(:, jj) = [];
92.                  A(row, :) = [];
93.                  b(row) = [];
94.                  Eqin(row) = [];
95.                  countercols = countercols + 1;
96.              end
97.          end
98.      elseif Eqin(row) == 1 % constraint type <=
99.          if A(row, jj) > 0 % if the element is greater
100.             % than zero
101.             n0 = size(A, 2); % find the number of columns
102.             % compute the indices of all other columns
103.             set = setdiff(1:n0, jj);
104.             % if all the elements of the row except
105.             % the one in the specific column are less
106.             % than or equal to zero and the right-hand
107.             % side greater than or equal to -1, then
108.             % update matrix A, vectors c, b, and Eqin, and
109.             % variable c0
110.             if all(A(row, set) <= 0) && b(row) >= -1
111.                 if c(jj) ~= 0
112.                     c0 = c0 + c(jj) * (b(row) ...
113.                         / A(row, jj));
114.                     c = c - (c(jj) / A(row, jj)) ...
115.                         * A(row, :)';
116.                 end
117.                 c(jj) = [];
118.                 A(:, jj) = [];
119.                 A(row, :) = [];
120.                 b(row) = [];
121.                 Eqin(row) = [];
122.                 countercols = countercols + 1;
123.             end
124.         end
125.     elseif Eqin(row) == -1 % constraint type >=
126.         if A(row, jj) < 0 % if the element is less
127.             % than zero
128.             n0 = size(A, 2); % find the number of columns
129.             % compute the indices of all other columns
130.             set = setdiff(1:n0, jj);
131.             % if all the elements of the row except
132.             % the one in the specific column are greater
133.             % than or equal to zero and the right-hand
134.             % side less than or equal to 1, then update
135.             % matrix A, vectors c, b, and Eqin, and
136.             % variable c0
137.             if all(A(row, set) >= 0) && b(row) <= 1
```

```
138.                    if c(jj) ~= 0
139.                        c0 = c0 + c(jj) * (b(row) ...
140.                            / A(row, jj));
141.                        c = c - (c(jj) / A(row, jj)) ...
142.                            * A(row, :)';
143.                    end
144.                    c(jj) = [];
145.                    A(:, jj) = [];
146.                    A(row, :) = [];
147.                    b(row) = [];
148.                    Eqin(row) = [];
149.                    countercols = countercols + 1;
150.                end
151.            end
152.        end
153.    end
154. end
155. if countercols > 0
156.     fprintf(['ELIMINATE IMPLIED FREE SINGLETON COLUMNS: ' ...
157.         '%i rows and %i columns were eliminated\n'], ...
158.         countercols, countercols);
159. end
160. end
```

4.11 Eliminate Redundant Columns

4.11.1 Mathematical Formulation

Theorem 4.5 *A linear constraint of the form:*

$$A_{i1}x_1 + A_{i2}x_2 + \cdots + A_{ik}x_k = 0 \qquad (4.21)$$

with all $A_{ij} > 0$, $i = 1, 2, \cdots, m$, $j = 1, 2, \cdots, k$, $1 \leq k \leq n$, implies that $x_j = 0$.

Proof It is very easy to check the correctness of the above statement. Assume that there is a feasible solution to the LP problem. As a consequence $x_j \geq 0$, $j = 1, 2, \cdots, n$. According to Equation (4.21) and with all $A_i \geq 0$, $i = 1, 2, \cdots, k$, we can conclude that either $x_j < 0$ with $j \neq i$ or $x_j = 0$, $j = 1, 2, \cdots, k$. The first case ($x_j < 0$ with $j \neq i$) is not possible because it is against the natural constraints $x_j \geq 0$. Consequently, $x_j = 0$, $j = 1, 2, \cdots, k$. □

Theorem 4.6 *A linear constraint of the form:*

$$A_{i1}x_1 + A_{i2}x_2 + \cdots + A_{ik}x_k = 0 \qquad (4.22)$$

with all $A_{ij} < 0$, $i = 1, 2, \cdots, m$, $j = 1, 2, \cdots, k$, $1 \leq k \leq n$, implies that $x_j = 0$.

Proof Similarly, there are two possible cases. The first claims that all $x_j = 0, j = 1, 2, \cdots, k$ and the second that $x_j < 0$ with $j \neq i$, which is not possible because this case is excluded by the natural constraints $x_j \geq 0$ that stand for all variables in the LP problem.   □

In both cases, all variables $x_j = 0, j = 1, 2, \cdots, k, 1 \leq k \leq n$, are redundant and can be deleted. Consequently, the constraints (4.21) and (4.22) are also linearly dependent and they can be deleted from the LP problem.

4.11.2 Numerical Example

This subsection demonstrates the presolve method that eliminates redundant columns through an illustrative example. The LP problem that will be presolved is the following:

$$
\begin{aligned}
\min z = -6x_1 + \ & 4x_2 - \ 2x_3 - 8x_4 \\
\text{s.t.} \quad 2x_1 - \ & 4x_2 + 2x_3 + 2x_4 \qquad\qquad\ = \ 20 \quad (1) \\
- \ & 6x_2 - \ 2x_3 + 2x_4 + x_5 \qquad = \ 26 \quad (2) \\
& 2x_2 + 8x_3 \qquad\qquad\quad + x_6 = \ \ 0 \quad (3) \\
& 16x_2 - 12x_3 - 8x_4 \qquad\qquad = -84 \quad (4)
\end{aligned}
$$
$$
x_j \geq 0, \quad (j = 1, 2, 3, 4, 5, 6)
$$

In matrix notation the above LP problem is written as follows:

$$
A = \begin{bmatrix} 2 & -4 & 2 & 2 & 0 & 0 \\ 0 & -6 & -2 & 2 & 1 & 0 \\ 0 & 2 & 8 & 0 & 0 & 1 \\ 0 & 16 & -12 & -8 & 0 & 0 \end{bmatrix}, c = \begin{bmatrix} -6 \\ 4 \\ -2 \\ -8 \\ 0 \\ 0 \end{bmatrix}, b = \begin{bmatrix} 20 \\ 26 \\ 0 \\ -84 \end{bmatrix}, Eqin = \begin{bmatrix} 0 \\ 0 \\ 0 \\ 0 \end{bmatrix}
$$

Initially, we start searching for equality constraints with a zero right-hand side. We observe that the third constraint has a zero right-hand side. All elements in the third row are greater than or equal to zero.

$$
2x_2 + 8x_3 + x_6 = 0
$$

Variables x_2, x_3, and x_6 are redundant and can be deleted:

$$
A = \begin{bmatrix} 2 & 2 & 0 \\ 0 & 2 & 1 \\ 0 & 0 & 0 \\ 0 & -8 & 0 \end{bmatrix}, c = \begin{bmatrix} -6 \\ -8 \\ 0 \end{bmatrix}, b = \begin{bmatrix} 20 \\ 26 \\ 0 \\ -84 \end{bmatrix}, Eqin = \begin{bmatrix} 0 \\ 0 \\ 0 \\ 0 \end{bmatrix}
$$

Note that the third constraint $0x_1 + 0x_4 + 0x_5 = 0$ is a zero constraint and can be eliminated. The equivalent LP problem after presolve is the following:

$$\begin{aligned}
\min z = {-6x_1} &- 8x_4 \\
\text{s.t.} \quad 2x_1 &+ 2x_4 && = 20 \\
&+ 2x_4 + x_5 &&= 26 \\
&- 8x_4 && = -84 \\
x_j \geq 0, \quad (j = 1, 4, 5)
\end{aligned}$$

4.11.3   Implementation in MATLAB

This subsection presents the implementation in MATLAB of the presolve method that eliminates redundant columns (filename: `eliminateRedundant Columns.m`). Some necessary notations should be introduced before the presentation of the implementation of the presolve method that eliminates redundant columns. Let A be a $m \times n$ matrix with the coefficients of the constraints, c be a $n \times 1$ vector with the coefficients of the objective function, b be a $m \times 1$ vector with the right-hand side of the constraints, and $Eqin$ be a $m \times 1$ vector with the type of the constraints.

The function for the presolve method that eliminates redundant columns takes as input the matrix of coefficients of the constraints (matrix A), the vector of the coefficients of the objective function (vector c), the vector of the right-hand side of the constraints (vector b), and the vector of the type of the constraints (vector $Eqin$), and returns as output the presolved matrix of coefficients of the constraints (matrix A), and the presolved vector of the coefficients of the objective function (vector c).

In lines 25–30, we find the equality constraints with a zero right-hand side and if such constraints exist and all elements of these constraints are greater than zero or less than zero, then we eliminate them (lines 34–61). For each constraint, we examine the cases that were presented in subsection 4.11.1 in order to eliminate the redundant columns. Finally, the number of the eliminated redundant columns is printed (lines 56–60).

```
1.    function [A, c] = eliminateRedundantColumns(A, c, b, Eqin)
2.    % Filename: eliminateRedundantColumns.m
3.    % Description: the function is an implementation of the
4.    % presolve technique that eliminates redundant columns
5.    % Authors: Ploskas, N., & Samaras, N.
6.    %
7.    % Syntax: [A, c] = eliminateRedundantColumns(A, c, b, Eqin)
8.    %
9.    % Input:
10.   % -- A: matrix of coefficients of the constraints
11.   %    (size m x n)
12.   % -- c: vector of coefficients of the objective function
13.   %    (size n x 1)
```

```
14.  % -- b: vector of the right-hand side of the constraints
15.  %      (size m x 1)
16.  % -- Eqin: vector of the type of the constraints
17.  %      (size m x 1)
18.  %
19.  % Output:
20.  % -- A: presolved matrix of coefficients of the constraints
21.  %      (size m x n)
22.  % -- c: presolved vector of coefficients of the objective
23.  %      function (size n x 1)
24.
25.  [row, ~] = find(Eqin == 0); % find the equality constraints
26.  % find the constraints with a zero right-hand side
27.  [row1, ~] = find(b == 0);
28.  % get the intersection of these sets, i.e., the equality
29.  % constraints with a zero right-hand side
30.  idsrows = intersect(row, row1);
31.  % find the number of these constraints
32.  cardrows = length(idsrows);
33.  countercols = 0;
34.  if cardrows > 0 % if such constraints exist
35.      for i = cardrows:-1:1 % for each of these constraints
36.          ii = idsrows(i); % get the index of the constraint
37.          % find the nonzero elements of this row
38.          [row, j] = find(A(ii, :));
39.          if ~isempty(row) % if there are nonzero elements
40.              % if all elements are greater than zero,
41.              % eliminate the columns and update matrix A
42.              % and vector c
43.              if all(A(ii, j) > 0)
44.                  c(j) = [];
45.                  A(:, j) = [];
46.                  countercols = countercols + length(j);
47.              % if all elements are less than zero, eliminate
48.              % the columns and update matrix A and vector c
49.              elseif all(A(ii, j) < 0)
50.                  c(j) = [];
51.                  A(:, j) = [];
52.                  countercols = countercols + length(j);
53.              end
54.          end
55.      end
56.      if countercols > 0
57.          fprintf(['ELIMINATE REDUNDANT COLUMNS: %i ' ...
58.              'redundant columns were eliminated' ...
59.              '\n'], countercols);
60.      end
61.  end
62.  end
```

4.12 Eliminate Implied Bounds on Rows

4.12.1 Mathematical Formulation

A constraint that implies new bounds for the constraints can be formulated as:

$$\underline{b}_i \leq A_i.x \leq \overline{b}_i \tag{4.23}$$

and

$$\underline{x} \leq x \leq \overline{x} \tag{4.24}$$

where $\underline{x} = 0$ and $\overline{x} = +\infty$. These new bounds can be computed by:

$$\underline{b}_i' = \inf_{\underline{x} \leq x \leq \overline{x}} A_i.x = \sum_{A_{ij} \geq 0} A_{ij}\underline{x}_j + \sum_{A_{ij} \leq 0} A_{ij}\overline{x}_j \tag{4.25}$$

$$\overline{b}_i' = \sup_{\underline{x} \leq x \leq \overline{x}} A_i.x = \sum_{A_{ij} \geq 0} A_{ij}\overline{x}_j + \sum_{A_{ij} \leq 0} A_{ij}\underline{x}_j \tag{4.26}$$

The function *inf* (inferior) calculates the greatest from the inferior bounds and the function *sup* (superior) calculates the smallest from the superior bounds. If $\left[\underline{b}_i', \overline{b}_i'\right] \cap \left[\underline{b}_i, \overline{b}_i\right] = \varnothing$, then the LP problem is infeasible. If $\left[\underline{b}_i', \overline{b}_i'\right] \subseteq \left[\underline{b}_i, \overline{b}_i\right]$, then constraint i is redundant and can be deleted. All possible cases are distinguished in the following theorem.

Theorem 4.7 *For each constraint that implies new bounds, we distinguish the following cases:*

1. *Constraint type $\leq$, $A_i. \leq 0$, and $\overline{b}_i \geq 0$: Row i is redundant and can be deleted.*
2. *Constraint type $\geq$, $A_i. \geq 0$, and $\underline{b}_i \leq 0$: Row i is redundant and can be deleted.*

Proof

1. If the constraint is an inequality of the form $\leq$, $A_i. \leq 0$, and $\overline{b}_i \geq 0$, then the new bounds of the constraint i can be calculated according to Equations (4.25) and (4.26):

$$\underline{b}_i' = \sum_{A_{ij} \geq 0} A_{ij}\underline{x}_j + \sum_{A_{ij} \leq 0} A_{ij}\overline{x}_j = \sum_{A_{ij} \leq 0} A_{ij}\overline{x}_j = -\infty$$

and

$$\overline{b}_i' = \sum_{A_{ij}\geq 0} A_{ij}\overline{x}_j + \sum_{A_{ij}\leq 0} A_{ij}\underline{x}_j = \sum_{A_{ij}\leq 0} A_{ij}\underline{x}_j = 0$$

The initial bounds of the constraint i, as shown in Equation (4.23), are $[\underline{b}_i, \overline{b}_i]$. If $[\underline{b}_i', \overline{b}_i'] \subseteq [\underline{b}_i, \overline{b}_i]$, then constraint i is redundant and can be deleted. Since $[\underline{b}_i', \overline{b}_i'] \subseteq [\underline{b}_i, \overline{b}_i] \Rightarrow (\infty, 0] \subseteq (\infty, \overline{b}_i]$ holds, constraint i is redundant and can be deleted.

2. If the constraint is an inequality of the form $\geq$, $A_{i.} \geq 0$, and $\underline{b}_i \leq 0$, then the new bounds of the constraint i can be calculated according to Equations (4.25) and (4.26):

$$\underline{b}_i' = \sum_{A_{ij}\geq 0} A_{ij}\underline{x}_j + \sum_{A_{ij}\leq 0} A_{ij}\overline{x}_j = \sum_{A_{ij}\geq 0} A_{ij}\underline{x}_j = 0$$

and

$$\overline{b}_i' = \sum_{A_{ij}\geq 0} A_{ij}\overline{x}_j + \sum_{A_{ij}\leq 0} A_{ij}\underline{x}_j = \sum_{A_{ij}\geq 0} A_{ij}\overline{x}_j = \infty$$

The initial bounds of the constraint i, as shown in Equation (4.23), are $[\underline{b}_i, \overline{b}_i]$. If $[\underline{b}_i', \overline{b}_i'] \subseteq [\underline{b}_i, \overline{b}_i]$, then constraint i is redundant and can be deleted. Since $[\underline{b}_i', \overline{b}_i'] \subseteq [\underline{b}_i, \overline{b}_i] \Rightarrow [0, \infty) \subseteq [\underline{b}_i, \infty]$ holds, constraint i is redundant and can be deleted.

□

4.12.2 Numerical Example

This subsection demonstrates the presolve method that eliminates implied bounds on rows through an illustrative example. The LP problem that will be presolved is the following:

$$
\begin{aligned}
\min z = \quad & 3x_1 - 4x_2 + 5x_3 - 2x_4 \\
\text{s.t.} \quad & -2x_1 - x_2 - 4x_3 - 2x_4 \leq 4 \quad (1) \\
& 5x_1 + 3x_2 + x_3 + 2x_4 \leq 18 \quad (2) \\
& 5x_1 + 3x_2 + x_4 \geq -13 \quad (3) \\
& 4x_1 + 6x_2 + 2x_3 + 5x_4 \geq -10 \quad (4) \\
& x_j \geq 0, \quad (j = 1, 2, 3, 4)
\end{aligned}
$$

In matrix notation, the above LP problem is written as follows:

$$
A = \begin{bmatrix} -2 & -1 & -4 & -2 \\ 5 & 3 & 1 & 2 \\ 5 & 3 & 0 & 1 \\ 4 & 6 & 2 & 5 \end{bmatrix}, c = \begin{bmatrix} 3 \\ -4 \\ 5 \\ -2 \end{bmatrix}, b = \begin{bmatrix} 4 \\ 18 \\ -13 \\ -10 \end{bmatrix}, Eqin = \begin{bmatrix} -1 \\ -1 \\ 1 \\ 1 \end{bmatrix}
$$

Initially, we start searching for inequality constraints. The first constraint is an inequality constraint ($\leq$). According to the first case of the previous subsection ($A_{i.} \leq 0$ and $\bar{b}_i \geq 0$), the first row is redundant and can be deleted. The presolved LP problem is now:

$$
A = \begin{bmatrix} 5 & 3 & 1 & 2 \\ 5 & 3 & 0 & 1 \\ 4 & 6 & 2 & 5 \end{bmatrix}, c = \begin{bmatrix} 3 \\ -4 \\ 5 \\ -2 \end{bmatrix}, b = \begin{bmatrix} 18 \\ -13 \\ -10 \end{bmatrix}, Eqin = \begin{bmatrix} -1 \\ 1 \\ 1 \end{bmatrix}
$$

Similarly, the new first constraint (initially second) is an inequality constraint ($\leq$). According to the first case of the previous subsection ($A_{i.} \leq 0$ and $\bar{b}_i \geq 0$), the first row is not redundant. Moreover, the new second constraint (initially third) is an inequality constraint ($\geq$). According to the second case of the previous subsection ($A_{i.} \geq 0$ and $\underline{b}_i \leq 0$), the second row is redundant and can be deleted. The equivalent presolved LP problem is now:

$$
A = \begin{bmatrix} 5 & 3 & 1 & 2 \\ 4 & 6 & 2 & 5 \end{bmatrix}, c = \begin{bmatrix} 3 \\ -4 \\ 5 \\ -2 \end{bmatrix}, b = \begin{bmatrix} 18 \\ -10 \end{bmatrix}, Eqin = \begin{bmatrix} -1 \\ 1 \end{bmatrix}
$$

Similarly, the second constraint (initially fourth) is an inequality constraint ($\geq$). According to the second case of the previous subsection ($A_{i.} \geq 0$ and $\underline{b}_i \leq 0$), the second row is redundant and can be deleted:

$$
A = \begin{bmatrix} 5 & 3 & 1 & 2 \end{bmatrix}, c = \begin{bmatrix} 3 \\ -4 \\ 5 \\ -2 \end{bmatrix}, b = \begin{bmatrix} 18 \end{bmatrix}, Eqin = \begin{bmatrix} -1 \end{bmatrix}
$$

Finally, the equivalent LP problem after presolve is the following:

$$
\begin{aligned}
\min z = {} & 3x_1 - 4x_2 + 5x_3 - 2x_4 \\
\text{s.t.} \quad & 5x_1 + 3x_2 + x_3 + 2x_4 \leq 18 \\
& x_j \geq 0, \quad (j = 1, 2, 3, 4)
\end{aligned}
$$

4.12.3   *Implementation in MATLAB*

This subsection presents the implementation in MATLAB of the presolve method that eliminates implied bounds on rows (filename: eliminateImpliedBounds onRows.m). Some necessary notations should be introduced before the presentation of the implementation of the presolve method that eliminates implied bounds on rows. Let A be a $m \times n$ matrix with the coefficients of the constraints, b be a $m \times 1$ vector with the right-hand side of the constraints, and $Eqin$ be a $m \times 1$ vector with the type of the constraints.

The function for the presolve method that eliminates implied bounds on rows takes as input the matrix of coefficients of the constraints (matrix A), the vector of the right-hand side of the constraints (vector b), and the vector of the type of the constraints (vector $Eqin$), and returns as output the presolved matrix of coefficients of the constraints (matrix A), the presolved vector of the right-hand side of the constraints (vector b), and the presolved vector of the type of the constraints (vector $Eqin$).

In lines 28–30, we find the indices and the number of the inequality constraints and if implied bounds in these constraints exist, then we eliminate them (lines 32–59). For each constraint, we examine the cases that were presented in subsection 4.12.1 in order to eliminate the implied bounds on rows. Finally, the number of the eliminated implied bounds on rows is printed (lines 60–63).

```
1.    function [A, b, Eqin] = eliminateImpliedBoundsonRows(A, ...
2.        b, Eqin)
3.    % Filename: eliminateImpliedBoundsonRows.m
4.    % Description: the function is an implementation of the
5.    % presolve technique that eliminates implied bounds on
6.    % rows
7.    % Authors: Ploskas, N., & Samaras, N.
8.    %
9.    % Syntax: [A, b, Eqin] = eliminateImpliedBoundsonRows(A, ...
10.   %     b, Eqin)
11.   %
12.   % Input:
13.   % -- A: matrix of coefficients of the constraints
14.   %      (size m x n)
15.   % -- b: vector of the right-hand side of the constraints
16.   %      (size m x 1)
17.   % -- Eqin: vector of the type of the constraints
18.   %      (size m x 1)
19.   %
20.   % Output:
21.   % -- A: presolved matrix of coefficients of the constraints
22.   %      (size m x n)
23.   % -- b: presolved vector of the right-hand side of the
24.   %      constraints (size m x 1)
25.   % -- Eqin: presolved vector of the type of the constraints
26.   %      (size m x 1)
27.
```

```
28.   [row, ~] = find(Eqin ~= 0); % find the inequality constraints
29.   % find the number of these constraints
30.   cardrows = length(row);
31.   counterrows = 0;
32.   for i = cardrows:-1:1 % for each of these constraints
33.       ii = row(i); % get the index of the constraint
34.       % if all elements in that constraint are less than or
35.       % equal to zero, the constraint has a right-hand side
36.       % greater than or equal to zero, and the constraint
37.       % type is <=, then eliminate the constraint and update
38.       % matrix A and vectors b and Eqin (Case 1)
39.       if all((A(ii, :) <= 0)) && (b(ii) >= 0)
40.           if Eqin(ii) == -1
41.               A(ii, :) = [];
42.               Eqin(ii) = [];
43.               b(ii) = [];
44.               counterrows = counterrows + 1;
45.           end
46.       % if all elements in that constraint are greater than
47.       % or equal to zero, the constraint has a right-hand
48.       % side less than or equal to zero, and the constraint
49.       % type is >=, then eliminate the constraint and update
50.       % matrix A and vectors b and Eqin (Case 1)
51.       elseif all((A(ii, :) >= 0)) && (b(ii) <= 0)
52.           if Eqin(ii) == 1
53.               A(ii, :) = [];
54.               Eqin(ii) = [];
55.               b(ii) = [];
56.               counterrows = counterrows + 1;
57.           end
58.       end
59.   end
60.   if counterrows > 0
61.       fprintf(['ELIMINATE IMPLIED BOUNDS ON ROWS: %i ' ...
62.           'constraints were eliminated\n'], counterrows);
63.   end
64.   end
```

4.13 Eliminate Redundant Rows

4.13.1 Mathematical Formulation

Two constraints i and k, with $i \neq k$, are linearly dependent iff $A_{i.} = \lambda A_{k.}$ with $\lambda \in \mathbb{R}$, $i, k \in \{1, 2, , \cdots, m\}$. A plethora of methods have been proposed to identify linearly dependent rows [6, 7, 20]. The Gauss-Jordan elimination with partial pivoting is used to spot all linearly dependent constraints.

Linearly dependent constraints can be identified while calculating the rank of coefficient matrix A using the augmented matrix $[A|b]$ and performing adequate elementary row operations until the identity matrix is derived. Hence, redundant constraints that can be deleted have the following form:

$$A_{i1}x_1 + A_{i2}x_2 + \cdots + A_{in}x_n = 0 \qquad (4.27)$$

where $A_{ij} = 0$, $i = 1, 2, \cdots, m$, and $j = 1, 2, \cdots, n$. The LP problem is infeasible if a constraint of the following form appears when performing elementary row operations:

$$A_{i1}x_1 + A_{i2}x_2 + \cdots + A_{in}x_n = b_i \qquad (4.28)$$

where $A_{ij} = 0$, $b_i \neq 0$, $i = 1, 2, \cdots, m$, and $j = 1, 2, \cdots, n$.

The LP problem must be in its standard form in order to perform this presolve method.

4.13.2   Numerical Example

This subsection demonstrates the presolve method that eliminates redundant rows through two illustrative examples.

1st example

The LP problem that will be presolved is the following:

$$
\begin{aligned}
\min z = \quad & x_1 + x_2 - 2x_3 - 3x_4 \\
\text{s.t.} \quad & x_1 + 2x_2 + 4x_3 + 5x_4 = 10 \quad (1) \\
& 3x_1 + 5x_2 + 8x_3 + 4x_4 = 2 \quad (2) \\
& 0.5x_1 + x_2 + 2x_3 + 2.5x_4 = 5 \quad (3) \\
& x_j \geq 0, \quad (j = 1, 2, 3, 4)
\end{aligned}
$$

In matrix notation the above LP problem is written as follows:

$$
A = \begin{bmatrix} 1 & 2 & 4 & 5 \\ 3 & 5 & 8 & 4 \\ 0.5 & 1 & 2 & 2.5 \end{bmatrix}, c = \begin{bmatrix} 1 \\ 1 \\ -2 \\ -3 \end{bmatrix}, b = \begin{bmatrix} 10 \\ 2 \\ 5 \end{bmatrix}, Eqin = \begin{bmatrix} 0 \\ 0 \\ 0 \end{bmatrix}
$$

Initially, we formulate the augmented matrix $[A|b]$:

$$
Augmented = \begin{bmatrix} 1 & 2 & 4 & 5 & | & 10 \\ 3 & 5 & 8 & 4 & | & 2 \\ 0.5 & 1 & 2 & 2.5 & | & 5 \end{bmatrix}
$$

We can observe that the first and the third constraints are linearly dependent:

$$Augmented_{3.} = \frac{1}{2}Augmented_1$$

Hence, we can delete either the first or the third constraint:

$$A = \begin{bmatrix} 1\ 2\ 4\ 5 \\ 3\ 5\ 8\ 4 \end{bmatrix}, c = \begin{bmatrix} 1 \\ 1 \\ -2 \\ -3 \end{bmatrix}, b = \begin{bmatrix} 10 \\ 2 \end{bmatrix}, Eqin = \begin{bmatrix} 0 \\ 0 \end{bmatrix}$$

Finally, the equivalent LP problem after presolve is the following:

$$
\begin{aligned}
\min z = \quad & x_1 + x_2 - 2x_3 - 3x_4 \\
\text{s.t.} \quad & x_1 + 2x_2 + 4x_3 + 5x_4 = 10 \\
& 3x_1 + 5x_2 + 8x_3 + 4x_4 = 2 \\
& x_j \geq 0, \quad (j = 1, 2, 3, 4)
\end{aligned}
$$

2nd example

The LP problem that will be presolved is the following:

$$
\begin{aligned}
\min z = \quad & -x_1 + 3x_2 + x_3 - 2x_4 - x_5 \\
\text{s.t.} \quad & x_1 - 2x_2 + x_3 + 3x_4 - 2x_5 && = 3 \quad (1)\\
& -x_1 + x_2 + 2x_3 - x_4 && - 3x_6 = 7 \quad (2)\\
& 2x_1 - x_2 - x_3 - 2x_4 && = 10 \quad (3)\\
& -2/3x_1 + 2/3x_2 + 4/3x_3 - 2/3x_4 && - 2x_6 = 14/3 \quad (4)\\
& x_j \geq 0, \quad (j = 1, 2, 3, 4, 5, 6)
\end{aligned}
$$

In matrix notation the above LP problem is written as follows:

$$A = \begin{bmatrix} 1 & -2 & 1 & 3 & -2 & 0 \\ -1 & 1 & 2 & -1 & 0 & -3 \\ 2 & -1 & -1 & -2 & 0 & 0 \\ -2/3 & 2/3 & 4/3 & -2/3 & 0 & -2 \end{bmatrix}, c = \begin{bmatrix} -1 \\ 3 \\ 1 \\ -2 \\ -1 \\ 0 \end{bmatrix}, b = \begin{bmatrix} 3 \\ 7 \\ 10 \\ 14/3 \end{bmatrix}, Eqin = \begin{bmatrix} 0 \\ 0 \\ 0 \\ 0 \end{bmatrix}$$

Initially, we formulate the augmented matrix $[A|b]$:

$$Augmented = \begin{bmatrix} 1 & -2 & 1 & 3 & -2 & 0 & | & 3 \\ -1 & 1 & 2 & -1 & 0 & -3 & | & 7 \\ 2 & -1 & -1 & -2 & 0 & 0 & | & 10 \\ -2/3 & 2/3 & 4/3 & -2/3 & 0 & -2 & | & 14/3 \end{bmatrix}$$

We apply row operations in the above augmented matrix to form an identity matrix, if possible. Since we use the Gaussian elimination with partial pivoting, we select the first pivot element to be $A_{31} = 2$. Therefore, we interchange the first and third rows in the augmented matrix:

$$Augmented = \begin{bmatrix} 2 & -1 & -1 & -2 & 0 & 0 & | & 10 \\ -1 & 1 & 2 & -1 & 0 & -3 & | & 7 \\ 1 & -2 & 1 & 3 & -2 & 0 & | & 3 \\ -2/3 & 2/3 & 4/3 & -2/3 & 0 & -2 & | & 14/3 \end{bmatrix}$$

Before applying any row operations, multiply the first row by $1/2$ to make the pivot element equal to 1:

$$Augmented = \begin{bmatrix} 1 & -1/2 & -1/2 & -1 & 0 & 0 & | & 5 \\ -1 & 1 & 2 & -1 & 0 & -3 & | & 7 \\ 1 & -2 & 1 & 3 & -2 & 0 & | & 3 \\ -2/3 & 2/3 & 4/3 & -2/3 & 0 & -2 & | & 14/3 \end{bmatrix}$$

Next, we apply the following row operations in order to set the values of the first column equal to zero:

- Add row 1 to row 2 ($r_2 = r_2 + r_1$).
- Add -1 times row 1 to row 3 ($r_3 = r_3 - r_1$).
- Add $2/3$ times row 1 to row 4 ($r_4 = r_4 + 2/3 r_1$).

The new augmented matrix is the following:

$$Augmented = \begin{bmatrix} 1 & -1/2 & -1/2 & -1 & 0 & 0 & | & 5 \\ 0 & 1/2 & 3/2 & -2 & 0 & -3 & | & 12 \\ 0 & -3/2 & 3/2 & 4 & -2 & 0 & | & -2 \\ 0 & 1/3 & 1 & -4/3 & 0 & -2 & | & 24/3 \end{bmatrix}$$

Now, use as pivot the element $A_{32} = -3/2 \neq 0$. Therefore, we interchange the second and third rows in the augmented matrix:

$$Augmented = \begin{bmatrix} 1 & -1/2 & -1/2 & -1 & 0 & 0 & | & 5 \\ 0 & -3/2 & 3/2 & 4 & -2 & 0 & | & -2 \\ 0 & 1/2 & 3/2 & -2 & 0 & -3 & | & 12 \\ 0 & 1/3 & 1 & -4/3 & 0 & -2 & | & 24/3 \end{bmatrix}$$

Before applying any row operations, multiply the second row by $-2/3$ to make the pivot element equal to 1.

$$Augmented = \begin{bmatrix} 1 & -1/2 & -1/2 & -1 & 0 & 0 & | & 5 \\ 0 & 1 & -1 & -8/3 & 4/3 & 0 & | & 4/3 \\ 0 & 1/2 & 3/2 & -2 & 0 & -3 & | & 12 \\ 0 & 1/3 & 1 & -4/3 & 0 & -2 & | & 24/3 \end{bmatrix}$$

Apply the following row operations in order to set the values of the second column equal to zero:

- Add $1/2$ times row 2 to row 1 ($r_1 = r_1 + 1/2 r_2$).
- Add $-1/2$ times row 2 to row 3 ($r_3 = r_3 - 1/2 r_2$).
- Add $-1/3$ times row 2 to row 4 ($r_4 = r_4 - 1/3 r_2$).

The new augmented matrix is the following:

$$Augmented = \begin{bmatrix} 1 & 0 & -1 & -7/3 & 2/3 & 0 & | & 17/3 \\ 0 & 1 & -1 & -8/3 & 4/3 & 0 & | & 4/3 \\ 0 & 0 & 2 & -2/3 & -2/3 & -3 & | & 34/3 \\ 0 & 0 & 4/3 & -4/9 & -4/9 & -2 & | & 68/9 \end{bmatrix}$$

Now, use as pivot the element $A_{33} = 2 \neq 0$. Before applying any row operations, multiply the third row by $1/2$ to make the pivot element equal to 1.

$$Augmented = \begin{bmatrix} 1 & 0 & -1 & -7/3 & 2/3 & 0 & | & 17/3 \\ 0 & 1 & -1 & -8/3 & 4/3 & 0 & | & 4/3 \\ 0 & 0 & 1 & -1/3 & -1/3 & -3/2 & | & 17/3 \\ 0 & 0 & 4/3 & -4/9 & -4/9 & -2 & | & 68/9 \end{bmatrix}$$

Apply the following row operations in order to set the values of the third column equal to zero:

- Add row 3 to row 1 ($r_1 = r_1 + r_3$).
- Add row 3 to row 2 ($r_2 = r_2 + r_3$).
- Add $-4/3$ times row 3 to row 4 ($r_4 = r_4 - 4/3 r_3$).

The new augmented matrix is the following:

$$Augmented = \begin{bmatrix} 1 & 0 & 0 & -8/3 & 1/3 & -3/2 & | & 34/3 \\ 0 & 1 & 0 & -3 & 1 & -3/2 & | & 7 \\ 0 & 0 & 1 & -1/3 & -1/3 & -3/2 & | & 17/3 \\ 0 & 0 & 0 & 0 & 0 & 0 & | & 0 \end{bmatrix}$$

The fourth constraint is a zero row. That means that the third (initially second) and the fourth constraints are linearly dependent:

$$Augmented_{2.} = \frac{2}{3} Augmented_4$$

Hence, we can delete either the second or the fourth constraint.

$$
A = \begin{bmatrix} 1 & -2 & 1 & 3 & -2 & 0 \\ -1 & 1 & 2 & -1 & 0 & -3 \\ 2 & -1 & -1 & -2 & 0 & 0 \end{bmatrix}, c = \begin{bmatrix} -1 \\ 3 \\ 1 \\ -2 \\ -1 \\ 0 \end{bmatrix}, b = \begin{bmatrix} 3 \\ 7 \\ 10 \end{bmatrix}, Eqin = \begin{bmatrix} 0 \\ 0 \\ 0 \end{bmatrix}
$$

Finally, the equivalent LP problem after presolve is the following:

$$
\begin{aligned}
\min z = -x_1 + 3x_2 + & x_3 - 2x_4 - x_5 \\
\text{s.t.} \quad x_1 - 2x_2 + & x_3 + 3x_4 - 2x_5 && = 3 \quad (1) \\
-x_1 + & x_2 + 2x_3 - x_4 && - 3x_6 = 7 \quad (2) \\
2x_1 - & x_2 - x_3 - 2x_4 && = 10 \quad (3) \\
x_j \geq 0, \quad (j = 1,2,3,4,5,6) &
\end{aligned}
$$

4.13.3 Implementation in MATLAB

This subsection presents the implementation in MATLAB of the presolve method that eliminates redundant rows (filename: eliminateRedundantRows.m). Some necessary notations should be introduced before the presentation of the implementation of the presolve method that eliminates redundant rows. Let A be a $m \times n$ matrix with the coefficients of the constraints, b be a $m \times 1$ vector with the right-hand side of the constraints, $Eqin$ be a $m \times 1$ vector with the type of the constraints, and *infeasible* be a flag variable showing if the LP problem is infeasible or not.

The function for the presolve method that eliminates redundant rows takes as input the matrix of coefficients of the constraints (matrix A), the vector of the right-hand side of the constraints (vector b), and the vector of the type of the constraints (vector $Eqin$), and returns as output the presolved matrix of coefficients of the constraints (matrix A), the presolved vector of the right-hand side of the constraints (vector b), the presolved vector of the type of the constraints (vector $Eqin$), and a flag variable showing if the LP problem is infeasible or not (variable *infeasible*).

In lines 30–36, we find the equality constraints and create matrix $Aeqin$ and vector $beqin$ that store the rows of matrix A and the elements of vector b, respectively, that correspond to equality constraints. Then, we perform row operations using Gauss–Jordan elimination with partial pivoting (lines 50–85). If all elements of row h of matrix A and the corresponding right-hand side are equal to 0, then row h is redundant and can be deleted (lines 91–93). If all elements of row h of matrix A are equal to 0 and the corresponding right-hand side is not equal to 0, then the LP problem is infeasible (lines 97–100). Then, we find the rows and the number of rows that have been marked with 1 in the *rowindex* vector (lines 103–110) and if such rows exist, then we eliminate them and update matrix *A* and vectors *b* and *Eqin* (lines 111–119). Finally, the number of the redundant rows is printed (lines 120–121).

```
1.  function [A, b, Eqin, infeasible] = ...
2.      eliminateRedundantRows(A, b, Eqin)
3.  % Filename: eliminateRedundantRows.m
4.  % Description: the function is an implementation of the
5.  % presolve technique that eliminates redundant rows
6.  % Authors: Ploskas, N., & Samaras, N.
7.  %
8.  % Syntax: [A, b, Eqin, infeasible] = ...
9.  %    eliminateRedundantRows(A, b, Eqin)
10. %
11. % Input:
12. % -- A: matrix of coefficients of the constraints
13. %    (size m x n)
14. % -- b: vector of the right-hand side of the constraints
15. %    (size m x 1)
16. % -- Eqin: vector of the type of the constraints
17. %    (size m x 1)
18. %
19. % Output:
20. % -- A: presolved matrix of coefficients of the
21. %    constraints (size m x n)
22. % -- b: presolved vector of the right-hand side of the
23. %    constraints (size m x 1)
24. % -- Eqin: presolved vector of the type of the
25. %    constraints (size m x 1)
26. % -- infeasible: flag variable showing if the LP problem
27. %    is infeasible or not
28.
29. infeasible = 0;
30. row = find(Eqin == 0); % find the equality constraints
31. % get the rows of A that correspond to equality
32. % constraints
33. Aeqin = A(row, :);
34. % get the elements of b that correspond to equality
35. % constraints
36. beqin = b(row, :);
37. % find the number of the equality constraints
38. [m1, n] = size(Aeqin);
39. % create a 2 x m1 zero matrix
40. rowindex = zeros(2, m1);
41. % store the indices of the equality constraints in the
42. % first row
43. rowindex(1, 1:m1) = row';
44. % compute the tolerance
45. tol = max(m1, n) * eps * norm(A, 'inf');
46. i = 1;
47. j = 1;
48. % perform row operations using Gauss - Jordan elimination
49. % with partial pivoting
50. while (i <= m1) && (j <= n)
```

```
51.      % find the maximum element of jth column of Aeqin in
52.      % absolute value
53.      [p, k] = max(abs(Aeqin(i:m1, j)));
54.      % if the maximum element is less than tol, then set
55.      % all the elements of jth column equal to zero
56.      if p < tol
57.          Aeqin(i:m1, j) = zeros(m1 - i + 1, 1);
58.          j = j + 1;
59.      % perform the pivoting operation and update matrix
60.      % Aeqin and vector beqin
61.      else
62.          if p ~= 0
63.              k = k + i - 1;
64.              rowindex(:, [i k]) = rowindex(:, [k i]);
65.              beqin(i, :) = beqin(i, :) / Aeqin(i, j);
66.              Aeqin(i, j:n) = Aeqin(i, j:n) / Aeqin(i, j);
67.              i_nz = find(Aeqin(:, j));
68.              i_nz = setdiff(i_nz, i);
69.              for t = i_nz
70.                  if beqin(i) ~= 0
71.                      beqin(t) = beqin(t) - ...
72.                          Aeqin(t, j) * beqin(i);
73.                      toler = abs(beqin) <= tol;
74.                      beqin(toler == 1) = 0;
75.                  end
76.                  Aeqin(t, j:n) = Aeqin(t, j:n) ...
77.                      - Aeqin(t, j) * Aeqin(i, j:n);
78.                  toler = abs(Aeqin) <= tol;
79.                  Aeqin(toler == 1) = 0;
80.              end
81.              i = i + 1;
82.              j = j + 1;
83.          end
84.      end
85.  end
86.  i = 1;
87.  for h = [1:i - 1 i + 1:m1]
88.      % if all elements of row h are equal to zero and the
89.      % corresponding right-hand side is equal to 0, then
90.      % row h is redundant and can be deleted
91.      if (Aeqin(h, :) == 0) & (beqin(h) == 0)
92.          rowindex(2, h) = 1;
93.      end
94.      % if all elements of row h are equal to zero and the
95.      % corresponding right-hand side is not equal to 0,
96.      % then the LP problem is infeasible
97.      if (Aeqin(h, :) == 0) & (beqin(h) ~= 0)
98.          infeasible = 1;
99.          return;
100.     end
101. end
102. % find the rows that have been marked with 1
103. if any(rowindex(2, :) == 1)
104.     row = find(rowindex(2, :) == 1);
```

```
105. end
106. row = (rowindex(2, :) == 1);
107. row = rowindex(1, row);
108. row = sort(row);
109. % find the number of the rows that have been marked with 1
110. cardrows = length(row);
111. if cardrows > 0 % if such rows exist
112.     % eliminate these rows and update matrix A and vectors
113.     % b and Eqin
114.     for i = cardrows:-1:1
115.         ii = row(i);
116.         A(ii, :) = [];
117.         b(ii) = [];
118.         Eqin(ii) = [];
119.     end
120.     fprintf(['ELIMINATE REDUNDANT ROWS: %i redundant ' ...
121.         'rows were eliminated\n'], cardrows);
122. end
123. end
```

4.14 Make Coefficient Matrix Structurally Full Rank

4.14.1 Mathematical Formulation

The rank of a matrix is the number of linearly independent rows and/or columns. A matrix $A \in \mathbb{R}^{m \times n}$ is said to have full rank, if:

$$rank(A) = \begin{Bmatrix} m, \text{ if } m < n \\ m, \text{ if } m = n \\ n, \text{ if } m > n \end{Bmatrix} \qquad (4.29)$$

The aim of this presolve method is to identify the rank of the coefficient matrix A and make it structurally full rank (if it does not already have full rank).

4.14.2 Numerical Example

This subsection demonstrates the presolve method that makes the coefficient matrix A structurally full rank through an illustrative example. The LP problem that will be presolved is the following:

$$\begin{aligned}
\min z = \ & x_1 + 2x_2 - x_3 - 5x_4 \\
\text{s.t.} \quad & 2x_1 + 6x_2 + 10x_3 + 18x_4 = 16 \quad (1) \\
& 3x_1 + 3x_2 - 8x_3 - 4x_4 = 2 \quad (2) \\
& x_1 + 3x_2 + 5x_3 + 9x_4 = 8 \quad (3) \\
x_j \geq 0, \quad & (j = 1, 2, 3, 4)
\end{aligned}$$

In matrix notation the above LP problem is written as follows:

$$A = \begin{bmatrix} 2 & 6 & 10 & 18 \\ 3 & 3 & -8 & -4 \\ 1 & 3 & 5 & 9 \end{bmatrix}, c = \begin{bmatrix} 1 \\ 2 \\ -1 \\ -5 \end{bmatrix}, b = \begin{bmatrix} 16 \\ 2 \\ 8 \end{bmatrix}, Eqin = \begin{bmatrix} 0 \\ 0 \\ 0 \end{bmatrix}$$

Initially, we calculate the rank of the matrix (see function *rank* or *sprank* of MATLAB). The matrix has fewer rows than columns; so its maximum rank is equal to the maximum number of linearly independent rows. Since the first and the third rows are linearly dependent ($A_1. = 2A_3.$), the matrix has two linearly independent rows; so its rank is two ($rank(A) = 2$). Hence, we can delete either the first or the third constraint in order to formulate a full rank matrix (with rank 2):

$$A = \begin{bmatrix} 2 & 6 & 10 & 18 \\ 3 & 3 & -8 & -4 \end{bmatrix}, c = \begin{bmatrix} 1 \\ 2 \\ -1 \\ -5 \end{bmatrix}, b = \begin{bmatrix} 16 \\ 2 \end{bmatrix}, Eqin = \begin{bmatrix} 0 \\ 0 \end{bmatrix}$$

Finally, the equivalent LP problem after presolve is the following:

$$\begin{aligned}
\min z = \ & x_1 + 2x_2 - x_3 - 5x_4 \\
\text{s.t.} \quad & 2x_1 + 6x_2 + 10x_3 + 18x_4 = 16 \\
& 3x_1 + 3x_2 - 8x_3 - 4x_4 = 2 \\
x_j \geq 0, \quad & (j = 1, 2, 3, 4)
\end{aligned}$$

4.14.3   Implementation in MATLAB

This subsection presents the implementation in MATLAB of the presolve method that makes the coefficient matrix structurally full rank (filename: fullRank.m). Some necessary notations should be introduced before the presentation of the implementation of the presolve method that makes the coefficient matrix structurally full rank. Let A be a $m \times n$ matrix with the coefficients of the constraints, *Atemp* be a matrix with the coefficients of the constraints after adding slack variables, b be a $m \times 1$ vector with the right-hand side of the constraints, and *Eqin* be a $m \times 1$ vector with the type of the constraints.

The function for the presolve method that makes the coefficient matrix structurally full rank takes as input the matrix of coefficients of the constraints (matrix A), the matrix of the coefficients of the constraints after adding slack variables (matrix *Atemp*), the vector of the right-hand side of the constraints (vector b), and the vector of the type of the constraints (vector *Eqin*), and returns as output the presolved matrix of coefficients of the constraints (matrix A), the presolved vector of the right-hand side of the constraints (vector b), and the presolved vector of the type of the constraints (vector *Eqin*).

In line 32, we find the rank of matrix *Atemp* (using MATLAB's built-in function *sprank*) and if the matrix is not of full rank, we transform it to be of full rank using MATLAB's built-in function *dmperm* (lines 35–43). Finally, the number of the eliminated rows is printed (lines 41–42).

```
1.    function [A, b, Eqin] = fullRank(A, Atemp, b, Eqin)
2.    % Filename: fullRank.m
3.    % Description: the function is an implementation of the
4.    % presolve technique that makes the coefficient matrix
5.    % structurally full rank
6.    % Authors: Ploskas, N., & Samaras, N.
7.    %
8.    % Syntax: [A, b, Eqin] = fullRank(A, Atemp, b, Eqin)
9.    %
10.   % Input:
11.   % -- A: matrix of coefficients of the constraints before
12.   %    adding slack variables (size m x n)
13.   % -- Atemp: matrix of coefficients of the constraints after
14.   %    adding slack variables (depending on the type of the
15.   %    constraints: if all constraints are equalities then
16.   %    Atemp = A and the size of Atemp = m x n, if all
17.   %    constraints are inequalities then the size of Atemp
18.   %    = m x (m + n))
19.   % -- b: vector of the right-hand side of the constraints
20.   %    (size m x 1)
21.   % -- Eqin: vector of the type of the constraints
22.   %    (size m x 1)
23.   %
24.   % Output:
25.   % -- A: presolved matrix of coefficients of the constraints
26.   %    (size m x n)
27.   % -- b: presolved vector of the right-hand side of the
28.   %    constraints (size m x 1)
29.   % -- Eqin: presolved vector of the type of the constraints
30.   %    (size m x 1)
31.
32.   sp_rank = sprank(Atemp'); % find the rank of matrix Atemp
33.   [m, n] = size(Atemp); % size of Atemp
34.   t = min(m, n);
35.   if sp_rank < t % if the matrix is not of full rank
```

```
36.      [rmp, ~] = dmperm(Atemp);
37.      rows = rmp(1:sp_rank);
38.      A = A(rows, :);
39.      b = b(rows);
40.      Eqin = Eqin(rows);
41.      fprintf(['MAKE COEFFICIENT MATRIX STRUCTURALLY FULL ' ...
42.          'RANK: %i rows were eliminated\n'], t - sp_rank);
43.   end
44.   end
```

4.15 MATLAB Code for the Presolve Analysis

This section presents a code that calls all presolve methods presented in this chapter (filename: `presolve.m`). The order that the presolve methods are called is not a de facto one, but we have come with this order after some computational experiments. However, a thorough computational study should be performed on many benchmark LPs to decide the order of the presolve methods and when the presolve analysis should be terminated. Some necessary notations should be introduced before the presentation of this code. Let A be a $m \times n$ matrix with the coefficients of the constraints, c be a $n \times 1$ vector with the coefficients of the objective function, b be a $m \times 1$ vector with the right-hand side of the constraints, $Eqin$ be a $m \times 1$ vector with the type of the constraints, $c0$ be the constant term of the objective function, *infeasible* be a flag variable showing if the LP problem is infeasible or not, and *unbounded* be a flag variable showing if the LP problem is unbounded or not.

The function takes as input the matrix of coefficients of the constraints (matrix A), the vector of the coefficients of the objective function (vector c), the vector of the right-hand side of the constraints (vector b), the vector of the type of the constraints (vector $Eqin$), and the constant term of the objective function (variable $c0$), and returns as output the presolved matrix of coefficients of the constraints (matrix A), the presolved vector of the coefficients of the objective function (vector c), the presolved vector of the right-hand side of the constraints (vector b), the presolved vector of the type of the constraints (vector $Eqin$), the updated constant term of the objective function (variable $c0$), a flag variable showing if the LP problem is infeasible or not (variable *infeasible*), and a flag variable showing if the LP problem is unbounded or not (variable *unbounded*).

In lines 37–41, we compute the initial size of matrix A and store it in order to find out the difference after the presolve analysis. If the constant term does not exist, we set it to zero (lines 45–47). Then, we call the presolve methods until no change in matrix A is found (lines 51–107). Moreover, we add slack variables (lines 108–122) and call the presolve method that makes the coefficient matrix structurally full rank (line 125) and the presolve method that eliminates redundant rows (lines 127–128). Finally, we output the statistics after presolve (lines 132–143).

```
1.    function [A, c, b, Eqin, c0, infeasible, unbounded] = ...
2.        presolve(A, c, b, Eqin, c0)
```

```
3.    % Filename: presolve.m
4.    % Description: the function is an implementation of the
5.    % main file that calls all the presolve techniques
6.    % Authors: Ploskas, N., & Samaras, N.
7.    %
8.    % Syntax: [A, c, b, Eqin, c0, infeasible, unbounded] = ...
9.    %   presolve(A, c, b, Eqin, c0)
10.   %
11.   % Input:
12.   % -- A: matrix of coefficients of the constraints
13.   %     (size m x n)
14.   % -- c: vector of coefficients of the objective function
15.   %     (size n x 1)
16.   % -- b: vector of the right-hand side of the constraints
17.   %     (size m x 1)
18.   % -- Eqin: vector of the type of the constraints
19.   %     (size m x 1)
20.   % -- c0: constant term of the objective function
21.   %
22.   % Output:
23.   % -- A: presolved matrix of coefficients of the constraints
24.   %     (size m x n)
25.   % -- c: presolved vector of coefficients of the objective
26.   %     function (size n x 1)
27.   % -- b: presolved vector of the right-hand side of the
28.   %     constraints (size m x 1)
29.   % -- Eqin: presolved vector of the type of the constraints
30.   %     (size m x 1)
31.   % -- c0: updated constant term of the objective function
32.   % -- infeasible: flag variable showing if the LP is
33.   %     infeasible or not
34.   % -- unbounded: flag variable showing if the LP is unbounded
35.   %     or not
36.
37.   [m, n] = size(A); % find size of A
38.   % store the initial size of A to calculate the difference
39.   % after the presolve analysis
40.   minit = m;
41.   ninit = n;
42.   infeasible = 0;
43.   unbounded = 0;
44.   % if the constant term does not exist, set it to zero
45.   if(~exist('c0'))
46.       c0 = 0;
47.   end
48.   mprevious = 0;
49.   nprevious = 0;
50.   % call presolve methods until no change in matrix A
51.   while(mprevious ~= m || nprevious ~= n)
52.       % call presolve technique to eliminate zero rows
53.       [A, b, Eqin, infeasible] = eliminateZeroRows(A, b, Eqin);
54.       if(infeasible == 1)
55.           return;
56.       end
```

```
57.     % call presolve technique to eliminate zero columns
58.     [A, c, unbounded] = eliminateZeroColumns(A, c);
59.     if(unbounded == 1)
60.         return;
61.     end
62.     % call presolve technique to eliminate kton equality
63.     % constraints (as an example, we set kton = 2, i.e.,
64.     % we eliminate doubleton and singleton constraints)
65.     [A, c, b, Eqin, c0, infeasible] = ...
66.         eliminateKtonEqualityConstraints(A, c, b, ...
67.             Eqin, c0, 2);
68.     if(infeasible == 1)
69.         return;
70.     end
71.     % call presolve technique to eliminate singleton
72.     % inequality constraints
73.     [A, c, b, Eqin, infeasible] = ...
74.         eliminateSingletonInequalityConstraints(A, c, ...
75.             b, Eqin);
76.     if(infeasible == 1)
77.         return;
78.     end
79.     % call presolve technique to eliminate dual singleton
80.     % equality constraints
81.     [A, c, b, Eqin, infeasible] = ...
82.         eliminateDualSingletonInequalityConstraints(A, ...
83.             c, b, Eqin);
84.     if(infeasible == 1)
85.         return;
86.     end
87.     % call presolve technique to eliminate implied free
88.     % singleton columns
89.     [A, c, b, Eqin, c0] = ...
90.         eliminateImpliedFreeSingletonColumns(A, c, b, ...
91.          Eqin, c0);
92.     % call presolve technique to eliminate redundant
93.     % columns
94.     [A, c] = eliminateRedundantColumns(A, c, b, Eqin);
95.     % call presolve technique to eliminate implied bounds
96.     % on rows
97.     [A, b, Eqin] = eliminateImpliedBoundsonRows(A, ...
98.          b, Eqin);
99.     % call presolve technique to eliminate zero columns
100.    [A, c, unbounded] = eliminateZeroColumns(A, c);
101.    if(unbounded == 1)
102.        return;
103.    end
104.    mprevious = m;
105.    nprevious = n;
106.    [m, n] = size(A);
107. end
108. % add slack variables
109. Atemp = A;
110. id = find(Eqin ~= 0);
```

```
111. [m, ~] = size(A);
112. for i = 1:length(id)
113.     if(Eqin(id(i)) == -1) % <= constraints
114.         ei = zeros(m, 1);
115.         ei(id(i)) = 1;
116.         Atemp = [Atemp ei];
117.     else % >= constraints
118.         ei = zeros(m, 1);
119.         ei(id(i)) = -1;
120.         Atemp = [Atemp ei];
121.     end
122. end
123. % call presolve technique to make the coefficient matrix
124. % structurally full rank
125. [A, b, Eqin] = fullRank(A, Atemp, b, Eqin);
126. % call presolve technique to eliminate redundant rows
127. [A, b, Eqin, infeasible] = eliminateRedundantRows(A, ...
128.     b, Eqin);
129. if(infeasible == 1)
130.     return;
131. end
132. % output statistics after the presolve analysis
133. [m, n] = size(A);
134. fprintf('==========================================\n')
135. fprintf(' (%i Constraints eliminated) \n', minit - m);
136. fprintf(' (%i Variables eliminated) \n', ninit - n);
137. fprintf(' ======================================= \n')
138. fprintf(' ========FINAL REDUCTION (RESULTS)======= \n')
139. fprintf(' ======================================= \n')
140. fprintf(' %i Constraints \n', m);
141. fprintf(' %i Variables \n', n);
142. fprintf(' %i NonZeros \n', nnz(A));
143. fprintf(' ======================================= \n')
144. end
```

4.16 Postsolve

A complete preprocessing procedure includes presolve (Chapter 4), scaling (Chapter 5), and the problem reformulation (Chapter 2) of the original LP problem. After solving the transformed LP problem with an LP algorithm, we get a solution in terms of the variables and constraints of the transformed LP problem. Of course, the transformed LP problem that was solved is equivalent to the original LP problem. However, it is not identical to the original LP problem. Moreover, we want to get the solution in terms of the variables and constraints of the original LP problem. Hence, a procedure called postsolve needs to be applied in order to get the solution of the original LP problem. This essentially means that we need to undo the changes performed during presolve, scaling, and problem reformulation.

In order to undo changes, we should keep track on the changes after every operation of presolve, scaling, and problem reformulation. These changes are usually stored in a stack type data structure since we need to undo the changes in the reverse order. The most recently made change is on the top of the stack. It is easier to keep separate stack data structures for the three different preprocessing procedures, i.e., presolve, scaling, and problem reformulation.

The postsolve routine works as follows:

1. Undo presolve: This task is the most complicated since each presolve method performs different operations on the original problem. Some methods do not require any action, e.g., the elimination of empty rows, while some other need complicate postsolve procedures. Interested readers are referred to [2, 9, 18] for details.
2. Undo scaling: At this stage, we want to express the solution of the transformed (scaled) LP problem in terms of the original (unscaled) LP problem. As will be discussed in Chapter 5, each scaling technique involves multiplications of matrix A, vector b, and vector c with the row and column scaling factors. Let r_i be the row scaling factor for row i and s_j be the column scaling factor for column j. The scaled matrix is expressed as $X = RAS$, where $R = diag(r_1 \ldots r_m)$ and $S = diag(s_1 \ldots s_n)$. Hence, the unscaled values of the original LP problem are $S\bar{x} = x$, where $\bar{x}$ is the solution of the transformed (scaled) LP problem. Note that we should also divide the right-hand side values and the objective function coefficients with the row and column scaling factors, respectively.
3. Undo problem reformulation: The original problem was transformed to its standard form. The changes should be recorded and should be undone in the reverse order.

4.17 Computational Study

This section presents a computational study. The aim of this computational study is twofold: (i) compare the execution time and the reduction to the problem size of the aforementioned presolve methods, and (ii) investigate the impact of preprocessing prior to the application of LP algorithms. The execution time and the number of iterations with and without preprocessing are presented. The test set used in this computational study is 19 LPs from the Netlib (optimal, Kennington, and infeasible LPs) and Mészáros problem sets that were presented in Section 3.6. Tables 4.2 and 4.3 present the results from the execution of the implementations of the presolve methods. The following abbreviations are used: (i) P1—presolve method that eliminates zero rows, (ii) P2—presolve method that eliminates zero columns, (iii) P3—presolve method that eliminates singleton and doubleton equality constraints ($k = 2$), (iv) P4—presolve method that eliminates singleton inequality constraints, (v) P5—presolve method that eliminates dual singleton inequality constraints, (vi) P6—presolve method that eliminates implied free singleton columns, (vii) P7—

Table 4.2 Total number of eliminations (constraints and/or variables) that each presolve method performed over the benchmark set

Name	Before presolve			After presolve			P1	P2	P3	P4	P5	P6	P7	P8	P9	P10	Total size reduction (%)	Nonzeros reduction (%)
	Constraints	Variables	Nonzeros	Constraints	Variables	Nonzeros												
beaconfd	173	262	3,375	85	141	1,422	6	5	85	37	4	46	18	8	0	0	48.05%	57.87%
cari	400	1,200	152,800	400	1,200	152,800	0	0	0	0	0	0	0	0	0	0	0.00%	0.00%
farm	7	12	36	6	10	32	0	0	2	0	1	0	0	0	0	0	15.79%	11.11%
itest6	11	8	20	7	1	7	0	0	2	0	7	0	0	2	0	0	57.89%	65.00%
klein2	477	54	4,585	477	54	4,585	0	0	0	0	0	0	0	0	0	0	0.00%	0.00%
nsic1	451	463	2,853	451	463	2,853	0	0	0	0	0	0	0	0	0	0	0.00%	0.00%
nsic2	465	463	3,015	459	463	3,015	6	0	0	0	0	0	0	0	0	0	0.65%	0.00%
osa-07	1,118	23,949	143,694	1,081	23,949	88,235	0	0	0	0	0	0	0	37	0	0	0.15%	38.60%
osa-14	2,337	52,460	314,760	2,300	52,460	194,416	0	0	0	0	0	0	0	37	0	0	0.07%	38.23%
osa-30	4,350	100,024	600,138	4,313	100,024	373,091	0	0	0	0	0	0	0	37	0	0	0.04%	37.83%
rosen1	520	1,024	23,274	520	1,024	23,274	0	0	0	0	0	0	0	0	0	0	0.00%	0.00%
rosen10	2,056	4,096	62,136	2,056	4,096	62,136	0	0	0	0	0	0	0	0	0	0	0.00%	0.00%
rosen2	1,032	2,048	46,504	1,032	2,048	46,504	0	0	0	0	0	0	0	0	0	0	0.00%	0.00%
rosen7	264	512	7,770	264	512	7,770	0	0	0	0	0	0	0	0	0	0	0.00%	0.00%
rosen8	520	1,024	15,538	520	1,024	15,538	0	0	0	0	0	0	0	0	0	0	0.00%	0.00%
sc205	205	203	551	164	163	455	1	0	41	39	0	0	0	0	0	0	19.85%	17.42%
scfxm1	330	457	2,589	296	420	2,295	3	0	27	12	0	16	9	4	0	0	9.02%	11.36%
sctap2	1,090	1,880	6,714	1,033	1,880	6,489	0	0	0	0	0	0	0	57	0	0	1.92%	3.35%
sctap3	1,480	2,480	8,874	1,408	2,480	8,595	0	0	0	0	0	0	0	72	0	0	1.82%	3.14%
Shifted geometric mean							1.37	1.10	2.06	1.68	1.26	1.42	1.32	3.57	1.00	1.00	14.29%	18.54%

Table 4.3   Total execution time of each presolve method over the benchmark set

Name	Constraints	Variables	Nonzeros	P1	P2	P3	P4	P5	P6	P7	P8	P9	P10	Total time
beaconfd	173	262	3,375	0.0006	0.0006	0.0212	0.0025	0.0031	0.0238	0.0013	0.0020	0.0004	0.0210	0.0765
cari	400	1,200	152,800	0.0061	0.0018	0.0122	0.0061	0.0144	0.1906	0.0002	0.0001	0.0019	6.8189	7.0524
farm	7	12	36	0.0001	0.0001	0.0014	0.0001	0.0002	0.0006	0.0004	0.0002	0.0001	0.0003	0.0034
itest6	11	8	20	0.0001	0.0001	0.0020	0.0009	0.0004	0.0001	0.0006	0.0005	0.0001	0.0001	0.0049
klein2	477	54	4,585	0.0002	0.0001	0.0008	0.0004	0.0001	0.0001	0.0002	0.0071	0.0001	0.0001	0.0093
nsic1	451	463	2,853	0.0001	0.0001	0.0008	0.0072	0.0001	0.0001	0.0002	0.0116	0.0001	0.0285	0.0489
nsic2	465	463	3,015	0.0005	0.0002	0.0016	0.0159	0.0002	0.0002	0.0005	0.0231	0.0001	0.0282	0.0704
osa-07	1,118	23,949	143,694	0.0083	0.0029	0.0158	0.0077	0.0068	0.0066	0.0006	1.5845	0.0008	0.0008	1.6347
osa-14	2,337	52,460	314,760	0.0196	0.0088	0.0385	0.0190	0.0165	0.0163	0.0007	7.2238	0.0020	0.0020	7.3471
osa-30	4,350	100,024	600,138	0.0375	0.0185	0.0750	0.0369	0.0328	0.0324	0.0007	28.1522	0.0041	0.0042	28.3944
rosen1	520	1,024	23,274	0.0006	0.0002	0.0011	0.0005	0.0004	0.0004	0.0003	0.0275	0.0002	0.0002	0.0314
rosen10	2,056	4,096	62,136	0.0018	0.0006	0.0035	0.0017	0.0015	0.0015	0.0003	0.3653	0.0005	0.0004	0.3770
rosen2	1,032	2,048	46,504	0.0014	0.0004	0.0024	0.0012	0.0010	0.0011	0.0003	0.0990	0.0003	0.0003	0.1073
rosen7	264	512	7,770	0.0003	0.0001	0.0004	0.0002	0.0002	0.0002	0.0003	0.0084	0.0001	0.0001	0.0103
rosen8	520	1,024	15,538	0.0004	0.0002	0.0008	0.0004	0.0003	0.0003	0.0003	0.0274	0.0001	0.0001	0.0303
sc205	205	203	551	0.0013	0.0011	0.0194	0.0066	0.0010	0.0008	0.0241	0.0341	0.0000	0.0262	0.1145
scfxm1	330	457	2,589	0.0005	0.0002	0.0082	0.0044	0.0010	0.0094	0.0066	0.0104	0.0001	0.3567	0.3975
sctap2	1,090	1,880	6,714	0.0005	0.0003	0.0008	0.0004	0.0117	0.2284	0.0005	0.0746	0.0001	0.1484	0.4657
sctap3	1,480	2,480	8,874	0.0006	0.0003	0.0011	0.0005	0.0154	0.3409	0.0005	0.1379	0.0001	0.2103	0.7076
Geometric mean				0.0009	0.0004	0.0034	0.0018	0.0013	0.0026	0.0006	0.0296	0.0002	0.0039	0.1580

presolve method that eliminates redundant columns, (viii) P8—presolve method that eliminates implied bounds on rows, (ix) P9—presolve method that eliminates redundant rows, and (x) P10—presolve method that makes the coefficient matrix structurally full rank. In this computational study, we use the function *presolve* that was presented in Section 4.15 and report the number of the eliminations that it performed (Table 4.2) and the execution time of each presolve method (Table 4.3). All times are measured in seconds with MATLAB's built-in functions *tic* and *toc*. For each instance, we averaged times over 10 runs. The average value of multiple instances is computed using the shifted geometric means of the individual instances. We use a shift of 0.1 for size and nonzeros reductions and 1 for the eliminations of the presolve methods. Let m and n be the number of constraints and variables, respectively, before presolve, and m_{new} and n_{new} be the number of constraints and variables, respectively, after presolve. Let nnz be the number of nonzeros in the constraint matrix A before presolve and nnz_{new} the number of nonzeros in the constraint matrix A after presolve. The column "Total size reduction" in Table 4.2 is calculated as follows: $-(m_{new} + n_{new} - m - n)/(m + n)$. In addition, the column "Nonzeros reduction" is calculated as: $-(nnz_{new} - nnz)/nnz$.

Table 4.2 reveals that the preprocessing analysis can significantly reduce the size of LPs (rows and columns of the coefficient matrix of the constraints). The size of the LPs is reduced by 14.29% on average, while the nonzeros are reduced by 18.54% on average. Moreover, we can also observe that most eliminations are made by the following presolve methods: (i) the presolve method that eliminates singleton and doubleton equality constraints (P3), (ii) the presolve method that eliminates singleton inequality constraints (P4), and (iii) the presolve method that eliminates implied bounds on rows (P8). The first two methods need a relatively small amount of time to perform these eliminations, while the elimination of implied bounds on rows is more time-consuming (Table 4.3). On the other hand, the presolve method that makes the coefficient matrix structurally full rank is time-consuming and does not offer any eliminations on the specific benchmark set. Hence, we can exclude it when using the proposed preprocessing procedure.

Then, this computational study is extended in order to study the impact of preprocessing prior to the application of an interior point method, the revised primal simplex algorithm, the revised dual simplex algorithm, and the exterior point simplex algorithm. These algorithms are executed with and without preprocessing. The implementations used in this computational study are the implementation of the revised primal simplex algorithm that will be presented in Section 8.5, the implementation of the revised dual simplex algorithm that will be presented in Section 9.5, the implementation of the exterior point simplex algorithm that will be presented in Section 10.5, and the implementation of Mehrotra's Predictor-Corrector interior point method that will be presented in Section 11.6. Tables 4.4, 4.5, 4.6, and 4.7 present the results from the execution of Mehrotra's Predictor-Corrector interior point method, the exterior point simplex algorithm, the revised primal simplex algorithm, and the revised dual simplex algorithm, respectively, over the same benchmark set. Tables 4.4, 4.5, 4.6, and 4.7 report the execution time and

Table 4.4 Results of interior point method with and without preprocessing

Name	Problem			Time without preprocessing	Time with preprocessing	Iterations without preprocessing	Iterations with preprocessing
	Constraints	Variables	Nonzeros				
beaconfd	173	262	3,375	0.0305	0.0095	12	11
cari	400	1,200	152,800	1.5290	1.5230	20	20
farm	7	12	36	0.0058	0.0037	8	8
itest6	11	8	20	0.0621	0.0030	35	7
klein2	477	54	4,585	0.0439	0.0445	1	1
nsic1	451	463	2,853	0.0559	0.0514	9	9
nsic2	465	463	3,015	0.4309	0.4329	69	72
osa-07	1,118	23,949	143,694	0.7817	0.3918	21	19
osa-14	2,337	52,460	314,760	1.6751	0.8349	19	16
osa-30	4,350	100,024	600,138	4.0450	1.8915	22	18
rosen1	520	1,024	23,274	0.1869	0.1654	14	14
rosen10	2,056	4,096	62,136	3.7851	3.7347	16	16
rosen2	1,032	2,048	46,504	0.8306	0.8128	15	15
rosen7	264	512	7,770	0.0447	0.0426	12	12
rosen8	520	1,024	15,538	0.1728	0.1712	15	15
sc205	205	203	551	0.0140	0.0097	11	11
scfxm1	330	457	2,589	0.0767	0.0531	21	21
sctap2	1,090	1,880	6,714	0.1298	0.1227	13	13
sctap3	1,480	2,480	8,874	0.2162	0.1888	14	13
Geometric mean				0.1858	0.1218	14.33	12.76

Table 4.5 Results of exterior point simplex algorithm with and without preprocessing

| Name | Problem | | | Time without preprocessing | Time with preprocessing | Iterations without preprocessing | Iterations with preprocessing |
	Constraints	Variables	Nonzeros				
beaconf'd	173	262	3,375	0.2023	0.0476	56	41
cari	400	1,200	152,800	9.9945	9.8213	914	914
farm	7	12	36	0.0041	0.0022	5	5
itest6	11	8	20	0.0032	0.0015	4	2
klein2	477	54	4,585	0.4062	0.4044	500	500
nsic1	451	463	2,853	0.3557	0.3683	384	384
nsic2	465	463	3,015	0.4861	0.5032	535	535
osa-07	1,118	23,949	143,694	72.7601	68.0223	8,559	8,720
osa-14	2,337	52,460	314,760	447.3447	376.5277	20,284	19,720
osa-30	4,350	100,024	600,138	1,912.5663	1,637.4515	41,039	39,315
rosen1	520	1,024	23,274	3.0047	3.0665	1,661	1,661
rosen10	2,056	4,096	62,136	24.4213	24.2797	3,893	3,893
rosen2	1,032	2,048	46,504	15.6035	15.6566	3,581	3,581
rosen7	264	512	7,770	0.3722	0.3708	505	505
rosen8	520	1,024	15,538	1.6341	1.6568	1,171	1,171
sc205	205	203	551	0.2828	0.1787	248	246
scfxm1	330	457	2,589	1.1521	0.8491	426	440
sctap2	1,090	1,880	6,714	1.8277	1.6929	788	782
sctap3	1,480	2,480	8,874	3.5044	3.4535	1,147	1,157
Geometric mean				1.8847	1.5258	694.44	657.72

Table 4.6 Results of revised primal simplex algorithm with and without preprocessing

Problem				Time without preprocessing	Time with preprocessing	Iterations without preprocessing	Iterations with preprocessing
Name	Constraints	Variables	Nonzeros				
beaconfd	173	262	3,375	0.1620	0.0311	38	26
cari	400	1,200	152,800	8.9107	9.0362	963	963
farm	7	12	36	0.0023	0.0019	5	5
itest6	11	8	20	0.0022	0.0014	4	2
klein2	477	54	4,585	0.4066	0.4154	500	500
nsic1	451	463	2,853	0.2066	0.2050	369	369
nsic2	465	463	3,015	0.2400	0.2382	473	473
osa-07	1,118	23,949	143,694	6.4731	5.4660	1,043	997
osa-14	2,337	52,460	314,760	49.4608	45.6242	2,457	2,454
osa-30	4,350	100,024	600,138	275.1153	243.0262	4,874	4,681
rosen1	520	1,024	23,274	3.2865	3.2685	1,893	1,893
rosen10	2,056	4,096	62,136	40.5450	40.4708	5,116	5,116
rosen2	1,032	2,048	46,504	19.8664	19.8402	4,242	4,242
rosen7	264	512	7,770	0.2345	0.2328	420	420
rosen8	520	1,024	15,538	1.3890	1.3722	1,045	1,045
sc205	205	203	551	0.1818	0.1117	143	165
scfxm1	330	457	2,589	1.0581	0.7474	301	339
sctap2	1,090	1,880	6,714	1.6809	1.5009	1,061	1,001
sctap3	1,480	2,480	8,874	4.2757	3.8493	1,844	1,809
Geometric mean				1.1546	0.9490	488.19	463.78

Table 4.7 Results of revised dual simplex algorithm with and without preprocessing

| Name | Problem | | | Time without preprocessing | Time with preprocessing | Iterations without preprocessing | Iterations with preprocessing |
	Constraints	Variables	Nonzeros				
beaconfd	173	262	3,375	0.1789	0.0318	48	25
cari	400	1,200	152,800	9.2769	9.1942	745	745
farm	7	12	36	0.0018	0.0013	2	2
itest6	11	8	20	0.0049	0.0013	2	1
klein2	477	54	4,585	1.3467	1.2896	1,520	1,520
nsic1	451	463	2,853	0.1893	0.1931	335	335
nsic2	465	463	3,015	0.3166	0.3112	571	571
osa-07	1,118	23,949	143,694	1.9390	1.4524	431	427
osa-14	2,337	52,460	314,760	9.6657	8.5234	964	963
osa-30	4,350	100,024	600,138	36.8929	34.5432	1,942	1,941
rosen1	520	1,024	23,274	0.7131	0.7054	621	621
rosen10	2,056	4,096	62,136	11.5846	11.0089	2,281	2,281
rosen2	1,032	2,048	46,504	2.8738	2.7837	1,197	1,197
rosen7	264	512	7,770	0.0818	0.0824	183	183
rosen8	520	1,024	15,538	0.4569	0.4501	492	492
sc205	205	203	551	0.1762	0.2216	122	354
scfxm1	330	457	2,589	0.9745	0.6798	354	303
sctap2	1,090	1,880	6,714	0.7698	0.7242	570	570
sctap3	1,480	2,480	8,874	1.3851	1.2546	740	740
Geometric mean				0.6252	0.4969	296.55	289.66

the number of iterations for each algorithm with and without presolve. Execution times do not include the time of the presolve and scaling analysis.

Tables 4.4–4.7 reveal that all algorithms can solve the LPs faster when they have been presolved. The preprocessing analysis affects more the execution time rather than the number of iterations. On the other hand, if we also add to the execution time the time to preprocess the LPs, then we observe that all algorithms can solve the LPs faster when they have not been presolved. That is due to the relative large execution time to presolve the LPs. However, taking into account that some LPs cannot be solved without preprocessing and most presolve methods perform many eliminations in a relatively small amount of time, the preprocessing analysis should be always performed prior to the application of an LP algorithm.

Note that these findings cannot be generalized, because only a sample of benchmark LPs is included in this computational study. Interested readers are referred to the computational studies performed in [2, 8, 10, 17].

4.18 Chapter Review

Presolve methods are used prior to the application of an LP algorithm in order to: (i) eliminate redundant constraints, (ii) fix variables, (iii) transform bounds of single structural variables, and (iv) reduce the number of variables and constraints by eliminations. In this chapter, we presented eleven presolve methods used prior to the execution of an LP algorithm: (i) eliminate zero rows, (ii) eliminate zero columns, (iii) eliminate singleton equality constraints, (iv) eliminate $k$ton equality constraints, (v) eliminate singleton inequality constraints, (vi) eliminate dual singleton inequality constraints, (vii) eliminate implied free singleton columns, (viii) eliminate redundant columns, (ix) eliminate implied bounds on rows, (x) eliminate redundant rows, and (xi) make coefficient matrix structurally full rank. Each method was presented with: (i) its mathematical formulation, (ii) a thorough numerical example, and (iii) its implementation in MATLAB. Finally, we performed a computational study in order to compare the execution time and the reduction to the problem size of the presolve methods and to investigate the impact of preprocessing prior to the application of LP algorithms. Computational results showed that the preprocessing analysis can significantly reduce the size and the nonzeros of LPs. Moreover, the LP algorithms can solve faster LPs when preprocessing analysis has already been performed. Hence, taking also into account that some LPs cannot be solved without preprocessing and most presolve methods perform many eliminations in a relatively small amount of time, the preprocessing analysis should be always performed prior to the application of an LP algorithm. On the other hand, the presolve methods and their order of appearance in the preprocessing analysis should be carefully selected after many experiments.

References

1. Adler, I., Resende, M. G., Veiga, G., & Karmarkar, N. (1989). An implementation of Karmarkar's algorithm for linear programming. *Mathematical Programming, 44*(1–3), 297–335.
2. Andersen, E. D., & Andersen, K. D. (1995). Presolving in linear programming. *Mathematical Programming, 71*(2), 221–245.
3. Baricelli, P., Mitra, G., & Nygreen, B. (1998). Modelling of augmented makespan assignment problems (AMAPs): Computational experience of applying integer presolve at the modelling stage. *Annals of Operations Research, 82*, 269–288.
4. Brearley, A. L., Mitra, G., & Williams, H. P. (1975). Analysis of mathematical programming problems prior to applying the simplex algorithm. *Mathematical Programming, 8*(1), 54–83.
5. Chang, S. F., & McCormick, S. T. (1992). A hierarchical algorithm for making sparse matrices sparser. *Mathematical Programming, 56*(1–3), 1–30.
6. Cosnard, M., Marrakchi, M., Robert, Y., & Trystram, D. (1986). Gauss elimination algorithms for MIMD computers. In *Proceeding of CONPAR 86* (pp. 247–254). Berlin/Heidelberg: Springer.
7. Douglas, A. (1971). Examples concerning efficient strategies for gaussian elimination. *Computing, 8*(3–4), 382–394.
8. Elble, J. M. (2010). *Computational experience with linear optimization and related problems.* Doctoral dissertation, University of Illinois at Urbana-Champaign, Chicago, USA.
9. Gondzio, J. (1997). Presolve analysis of linear programs prior to applying an interior point method. *INFORMS Journal on Computing, 9*(1), 73–91.
10. Gould, N., & Toint, P. L. (2004). Preprocessing for quadratic programming. *Mathematical Programming, 100*(1), 95–132.
11. Hall, J. A., & McKinnon, K. I. (2005). Hyper-sparsity in the revised simplex method and how to exploit it. *Computational Optimization and Applications, 32*(3), 259–283.
12. Ioslovich, I. (2001). Robust reduction of a class of large-scale linear programs. *SIAM Journal on Optimization, 12*(1), 262–282.
13. Lustig, I. J. (1989). An analysis of an available set of linear programming test problems. *Computers & Operations Research, 16*(2), 173–184.
14. Lustig, I. J., Marsten, R. E., & Shanno, D. F. (1994). Interior point methods for linear programming: Computational state of the art. *ORSA Journal on Computing, 6*(1), 1–14.
15. McCormick, S. T. (1990). Making sparse matrices sparser: Computational results. *Mathematical Programming, 49*(1–3), 91–111.
16. Mészáros, C., & Suhl, U. H. (2003). Advanced preprocessing techniques for linear and quadratic programming. *OR Spectrum, 25*(4), 575–595.
17. Swietanowski, A. (1995). *A modular presolve procedure for large scale linear programming.* Working paper WP 95–113, International Institute for Applied Systems Analysis, Laxenburg, Austria.
18. Tomlin, J. A., & Welch, J. S. (1983). Formal optimization of some reduced linear programming problems. *Mathematical Programming, 27*(2), 232–240.
19. Urdaneta, A. J., Perez, L. G., Gomez, J. F., Feijoo, B., & Gonzalez, M. (2001). Presolve analysis and interior point solutions of the linear programming coordination problem of directional overcurrent relays. *International Journal of Electrical Power & Energy Systems, 23*(8), 819–825.
20. Weispfenning, V. (2004). Solving constraints by elimination methods. In *Proceedings of the International Joint Conference on Automated Reasoning* (pp. 336–341). Berlin/Heidelberg: Springer.
21. Ye, Y. (1989). Eliminating columns in the simplex method for linear programming. *Journal of Optimization Theory and Applications, 63*(1), 69–77.

Chapter 5
Scaling Techniques

Abstract Preconditioning techniques are important in solving LPs, as they improve their computational properties. One of the most widely used preconditioning technique in LP solvers is scaling. Scaling is used prior to the application of an LP algorithm in order to: (i) produce a compact representation of the variable bounds, (ii) reduce the condition number of the constraint matrix, (iii) improve the numerical behavior of the algorithms, (iv) reduce the number of iterations required to solve LPs, and (v) simplify the setup of the tolerances. This chapter presents eleven scaling techniques used prior to the execution of an LP algorithm: (i) arithmetic mean, (ii) de Buchet for the case $p = 1$, (iii) de Buchet for the case $p = 2$, (iv) de Buchet for the case $p = \infty$, (v) entropy, (vi) equilibration, (vii) geometric mean, (viii) IBM MPSX, (ix) L_p-norm for the case $p = 1$, (x) L_p-norm for the case $p = 2$, and (xi) L_p-norm for the case $p = \infty$. Each technique is presented with: (i) its mathematical formulation, (ii) a thorough illustrative numerical example, and (iii) its implementation in MATLAB. Finally, a computational study is performed. The aim of the computational study is twofold: (i) compare the execution time of the scaling techniques, and (ii) investigate the impact of scaling prior to the application of LP algorithms. The execution time and the number of iterations with and without scaling are presented.

5.1 Chapter Objectives

- Discuss the importance of scaling prior to the execution of an LP algorithm.
- Provide the mathematical formulation of the scaling techniques.
- Understand the scaling techniques through a thorough numerical example.
- Implement the scaling techniques using MATLAB.
- Compare the computational performance of the scaling techniques.

© Springer International Publishing AG 2017   219
N. Ploskas, N. Samaras, *Linear Programming Using MATLAB®*,
Springer Optimization and Its Applications 127, DOI 10.1007/978-3-319-65919-0_5

5.2 Introduction

As in the solution of any large-scale mathematical system, the computational time
and the numerical accuracy for large LPs are of major concern. Preconditioning
techniques can be applied to LPs prior to the application of a solver in order to
improve their computational properties. Scaling is the most widely used preconditioning technique in linear optimization solvers. A matrix is badly scaled if its
nonzero elements are of different magnitudes. Scaling is an operation in which
the rows and columns of a matrix are multiplied by positive scalars and these
operations lead to nonzero numerical values of similar magnitude. Scaling is used
prior to the application of an LP algorithm for five reasons [11]: (i) produce a
compact representation of the variable bounds, (ii) reduce the condition number
of the constraint matrix, (iii) improve the numerical behavior of the algorithms, (iv)
reduce the number of the iterations required to solve LPs, and (v) simplify the setup
of the tolerances.

Tomlin [11] presented a thesis on scaling LPs and a computational study comparing arithmetic mean, geometric mean, equilibration, Curtis and Reid [2] scaling
technique, Fulkerson and Wolfe [5] scaling technique, and various combinations on
six test problems of varying size. Tomlin concluded that the geometric mean scaling
technique, optionally followed by the equilibration or the Curtis - Reid method, are
the best combined scaling techniques. Larsson [7] expanded on Tomlin's study by
presenting and comparing entropy, L_p-norm [6], and de Buchet [3] scaling models
over 135 randomly generated problems of varying size. Larsson remarked that the
entropy model is often able to improve the conditioning of the randomly generated
LPs. Elble and Sahinidis [4] performed a computational study comparing arithmetic
mean, de Buchet, entropy, equilibration, geometric mean, IBM MPSX method, L_p-
norm, binormalization, and various combinations of these scaling techniques over
Netlib and Kennington set. Elble and Sahinidis used four measures to evaluate the
quality of each scaling technique: (a) scaling time, (b) solution time, (c) solution
iterations, and (d) maximum condition number. Elble and Sahinidis concluded that
on average no scaling technique outperforms the equilibration scaling technique
despite the added complexity and the computational cost. Ploskas and Samaras [9]
presented a computational study of the impact of scaling prior to the application
of MATLAB's interior point method, exterior point simplex algorithm, and revised
simplex algorithm. The scaling techniques that were included in their computational
study are arithmetic mean, equilibration, and geometric mean. Computational
results showed that equilibration is the best scaling technique and that the effect
of scaling is significant to MATLAB's interior point method and revised simplex
algorithm, while the exterior point simplex algorithm is scaling invariant [12].
Ploskas and Samaras [10] performed a computational study comparing arithmetic
mean, de Buchet for the case $p = 1$, de Buchet for the case $p = 2$, entropy,
equilibration, geometric mean, IBM MPSX, L_p-norm for the case $p = 1$, L_p-norm
for the case $p = 2$, L_p-norm for the case $p = \infty$, and de Buchet for the case
$p = \infty$. Ploskas and Samaras [10] concluded that arithmetic mean, equilibration,
and geometric mean are the best scaling techniques in terms of execution time.

This chapter presents eleven scaling techniques used prior to the execution of an LP algorithm: (i) arithmetic mean, (ii) de Buchet for the case $p = 1$, (iii) de Buchet for the case $p = 2$, (iv) de Buchet for the case $p = \infty$, (v) entropy, (vi) equilibration, (vii) geometric mean, (viii) IBM MPSX, (ix) L_p-norm for the case $p = 1$, (x) L_p-norm for the case $p = 2$, and (xi) L_p-norm for the case $p = \infty$. Each technique is presented with: (i) its mathematical formulation, (ii) a thorough numerical example, and (iii) its implementation in MATLAB. Finally, a computational study is performed. The aim of the computational study is twofold: (i) compare the execution time of the scaling techniques, and (ii) investigate the impact of scaling prior to the application of LP algorithms. The execution time and the number of iterations with and without scaling are presented.

The structure of this chapter is as follows. In Section 5.3, some necessary mathematical preliminaries are introduced. In Sections 5.4–5.10, eleven scaling techniques are presented. More specifically, Section 5.4 presents the arithmetic mean scaling method, Section 5.5 presents the de Buchet for the case $p = 1$, 2, and ∞ scaling methods, Section 5.6 presents the entropy scaling method, Section 5.7 presents the equilibration scaling method, Section 5.8 presents the geometric mean scaling method, Section 5.9 presents the IBM MPSX scaling method, while Section 5.10 presents the L_p-norm for the case $p = 1$, 2, and ∞ scaling methods. Section 5.11 presents a computational study that compares the execution time of the above scaling techniques and investigates the impact of scaling prior to the application of LP algorithms. Finally, conclusions and further discussion are presented in Section 5.12.

5.3 Mathematical Preliminaries

Some necessary mathematical preliminaries should be introduced, before the presentation of the scaling techniques. Let A be a m x n matrix. Let r_i be the row scaling factor for row i and s_j be the column scaling factor for column j. Let $N_i = \{j | A_{ij} \neq 0\}$, where $i = 1, \ldots, m$, and $M_j = \{i | A_{ij} \neq 0\}$, where $j = 1, \ldots, n$. Let n_i and m_j be the cardinality numbers of the sets N_i and M_j, respectively ($n_i = |N_i|$ and $m_j = |M_j|$). Furthermore, we assume that there are no zero rows or columns in matrix A (or if exist, we remove them with the application of a presolve technique). The scaled matrix is expressed as $X = RAS$, where $R = diag(r_1 \ldots r_m)$ and $S = diag(s_1 \ldots s_n)$. All scaling methods presented in this chapter perform first a scaling of the rows and then a scaling of the columns.

Row and column scales are applied iteratively. Let $X^0 = A$. Let k be the index of the scaling iterations, X^k be the scaled matrix after k iterations, $X^{k+1/2}$ be the scaled matrix only by row scaling factors in iteration $k + 1$, X^{k+1} be the scaled matrix both by row and column scaling factors in iteration $k + 1$, r^{k+1} be the row scaling factors in iteration $k + 1$, and s^{k+1} be the column scaling factors in iteration $k + 1$. Then, each iteration is given by:

$$X^{k+1/2} = R^{k+1}X^k,$$
$$X^{k+1} = X^{k+1/2}S^{k+1}$$

(5.1)

where:

$$R = \prod_{k=1}^{t} R^k,$$

(5.2)

$$S = \prod_{k=1}^{t} S^k$$

where t is the number of iterations for a scaling method to converge to the scaled matrix.

5.4 Arithmetic Mean

5.4.1 Mathematical Formulation

Arithmetic mean aims to decrease the variance between the nonzero elements in the coefficient matrix A. In this method, each row is divided by the arithmetic mean of the absolute value of the elements in that row. The row scaling factors are presented in Equation (5.3):

$$r_i^{k+1} = n_i / \sum_{j \in N_i} |X_{ij}^k|$$

(5.3)

Similarly, each column is divided by the arithmetic mean of the absolute value of the elements in that column. The column scaling factors are presented in Equation (5.4):

$$s_j^{k+1} = m_j / \sum_{i \in M_j} |X_{ij}^{k+1/2}|$$

(5.4)

5.4.2 Numerical Example

This subsection demonstrates the arithmetic mean scaling technique through an illustrative example. The LP problem that will be scaled is the following:

$$\begin{aligned}
\min z = {}& 7/2x_1 & & + 453x_3 + & 6x_4 \\
\text{s.t.} \quad & 956x_1 & & + x_3 + 258/5x_4 = & 4 \\
& 5/2x_1 + & 4x_2 + 13/2x_3 + & 149/5x_4 = & 7/2 \\
& x_1 + 3/2x_2 & & + 67/10x_4 = & 55
\end{aligned}$$

$$x_j \geq 0, \quad (j = 1, 2, 3, 4)$$

In matrix notation the above LP problem is written as follows:

$$A = \begin{bmatrix} 956 & 0 & 1 & 258/5 \\ 5/2 & 4 & 13/2 & 149/5 \\ 1 & 3/2 & 0 & 67/10 \end{bmatrix}, c = \begin{bmatrix} 7/2 \\ 0 \\ 453 \\ 6 \end{bmatrix}, b = \begin{bmatrix} 4 \\ 7/2 \\ 55 \end{bmatrix}, Eqin = \begin{bmatrix} 0 \\ 0 \\ 0 \end{bmatrix}$$

Initially, the row scaling factors are calculated according to Equation (5.3):

$$r = \begin{bmatrix} 3/(956 + 1 + 258/5) \\ 4/(5/2 + 4 + 13/2 + 149/5) \\ 3/(1 + 3/2 + 67/10) \end{bmatrix} = \begin{bmatrix} 5/1,681 \\ 10/107 \\ 15/46 \end{bmatrix}$$

Then, matrix A and vector b are scaled by multiplying the ith row of matrix A and the ith element of vector b by the ith element of vector r:

$$A = \begin{bmatrix} 956 & 0 & 1 & 258/5 \\ 5/2 & 4 & 13/2 & 149/5 \\ 1 & 3/2 & 0 & 67/10 \end{bmatrix} \begin{matrix} \times 5/1,681 \\ \times 10/107 \\ \times 15/46 \end{matrix} =$$

$$\begin{bmatrix} 2,599/914 & 0 & 5/1,681 & 258/1,681 \\ 25/107 & 40/107 & 65/107 & 298/107 \\ 15/46 & 45/92 & 0 & 201/92 \end{bmatrix},$$

$$b = \begin{bmatrix} 4 \\ 7/2 \\ 55 \end{bmatrix} \begin{matrix} \times 5/1,681 \\ \times 10/107 \\ \times 15/46 \end{matrix} = \begin{bmatrix} 20/1,681 \\ 35/107 \\ 825/46 \end{bmatrix}$$

Next, the column scaling factors are calculated according to Equation (5.4):

$$s = \begin{bmatrix} 3/(2,599/914 + 25/107 + 15/46) \\ 2/(40/107 + 45/92) \\ 2/(5/1,681 + 65/107) \\ 3/(258/1,681 + 298/107 + 201/92) \end{bmatrix} = \begin{bmatrix} 305/346 \\ 540/233 \\ 842/257 \\ 592/1,011 \end{bmatrix}$$

Then, matrix A and vector c are scaled by multiplying the jth column of matrix A and the jth element of vector c by the jth element of vector s:

$$305/346 \quad 540/233 \quad 842/257 \quad 592/1,011$$
$$\times \qquad \times \qquad \times \qquad \times$$

$$A = \begin{bmatrix} 2,599/914 & 0 & 5/1,681 & 258/1,681 \\ 25/107 & 40/107 & 65/107 & 298/107 \\ 15/46 & 45/92 & 0 & 201/92 \end{bmatrix}$$

$$= \begin{bmatrix} 950/379 & 0 & 107/10,980 & 63/701 \\ 159/772 & 843/973 & 1,634/821 & 1,895/1,162 \\ 403/1,402 & 823/726 & 0 & 2,029/1,586 \end{bmatrix},$$

$$c = \begin{bmatrix} 7/2 \\ 0 \\ 453 \\ 6 \end{bmatrix} \begin{matrix} \times 305/346 \\ \times 540/233 \\ \times 842/257 \\ \times 592/1,011 \end{matrix} = \begin{bmatrix} 2,533/821 \\ 0 \\ 40,072/27 \\ 1,184/337 \end{bmatrix}$$

Finally, the row and column scaling factors, the scaled matrix A, and vectors b and c are the following:

$$r = \begin{bmatrix} 5/1,681 \\ 10/107 \\ 15/46 \end{bmatrix}, s = \begin{bmatrix} 305/346 \\ 540/233 \\ 842/257 \\ 592/1,011 \end{bmatrix}$$

$$A = \begin{bmatrix} 950/379 & 0 & 107/10,980 & 63/701 \\ 159/772 & 843/973 & 1,634/821 & 1,895/1,162 \\ 403/1,402 & 823/726 & 0 & 2,029/1,586 \end{bmatrix}, c = \begin{bmatrix} 2,533/821 \\ 0 \\ 40,072/27 \\ 1,184/337 \end{bmatrix},$$

$$b = \begin{bmatrix} 20/1,681 \\ 35/107 \\ 825/46 \end{bmatrix}$$

The final scaled LP problem is the following:

$$\begin{aligned}
\min z = \; & 2,533/821x_1 & & & + \; 40,072/27x_3 + & 1,184/337x_4 \\
\text{s.t.} \quad & 950/379x_1 & & & + \; 107/10,980x_3 + & 63/701x_4 = 20/1,681 \\
& 159/772x_1 + & 843/973x_2 + & 1,634/821x_3 + & 1,895/1,162x_4 = & 35/107 \\
& 403/1,402x_1 + & 823/726x_2 & & + \; 2,029/1,586x_4 = & 825/46 \\
& x_j \geq 0, \quad (j = 1,2,3,4)
\end{aligned}$$

5.4.3 Implementation in MATLAB

This subsection presents the implementation in MATLAB of the arithmetic mean scaling technique (filename: `arithmeticMean.m`). Some necessary notations should be introduced before the presentation of the implementation of the arithmetic mean scaling technique. Let *row_multi* be a $m \times 1$ vector with row scaling factors and *col_multi* be a $n \times 1$ vector with column scaling factors. Let *row_sum* be a $m \times 1$ vector with the sum of each row's elements in absolute value and *col_sum* be a $n \times 1$ vector with the sum of each column's elements in absolute value.

The function for the arithmetic mean scaling technique takes as input the matrix of coefficients of the constraints (matrix A), the vector of coefficients of the objective function (vector c), and the vector of the right-hand side of the constraints (vector b), and returns as output the scaled matrix of coefficients of the constraints (matrix A), the scaled vector of coefficients of the objective function (vector c), the scaled vector of the right-hand side of the constraints (vector b), and the row and column scaling factors (vectors *row_multi* and *col_multi*).

Lines 33–34 include the pre-allocation of vectors *row_sum* and *row_multi*. In the first for-loop (lines 35–52), the row scaling factors are calculated as the number of the nonzero elements of each row to the sum in absolute value of the same row (line 46). Then, matrix A and vector b are updated (lines 48–50). Next, lines 54–55 include the pre-allocation of vectors *col_sum* and *col_multi*. Finally, in the second for-loop (lines 56–74), the column scaling factors are calculated as the number of the nonzero elements of each column to the sum in absolute value of the same column (line 67). Then, matrix A and vector c are updated (lines 70–72).

```
1.   function [A, c, b, row_multi, col_multi] = ...
2.       arithmeticMean(A, c, b)
3.   % Filename: arithmeticMean.m
4.   % Description: the function is an implementation of the
5.   % arithmetic mean scaling technique
6.   % Authors: Ploskas, N., & Samaras, N.
7.   %
8.   % Syntax: [A, c, b, row_multi, col_multi] = ...
9.   %     arithmeticMean(A, c, b)
10.  %
11.  % Input:
12.  % -- A: matrix of coefficients of the constraints
13.  %    (size m x n)
14.  % -- c: vector of coefficients of the objective function
15.  %    (size n x 1)
16.  % -- b: vector of the right-hand side of the constraints
17.  %    (size m x 1)
18.  %
19.  % Output:
20.  % -- A: scaled matrix of coefficients of the constraints
21.  %    (size m x n)
22.  % -- c: scaled vector of coefficients of the objective
23.  %    function (size n x 1)
24.  % -- b: scaled vector of the right-hand side of the
```

```
25.  %      constraints (size m x 1)
26.  % -- row_multi: vector of the row scaling factors
27.  %      (size m x 1)
28.  % -- col_multi: vector of the column scaling factors
29.  %      (size n x 1)
30.
31.  [m, n] = size(A); % size of matrix A
32.  % first apply row scaling
33.  row_sum = zeros(m, 1);
34.  row_multi = zeros(m, 1);
35.  for i = 1:m
36.      % find the indices of the nonzero elements of the
37.      % specific row
38.      ind = find(A(i, :));
39.      % if the specific row contains at least one nonzero
40.      % element
41.      if ~isempty(ind)
42.          % sum of the absolute value of the nonzero elements of
43.          % the specific row
44.          row_sum(i) = sum(sum(abs(A(i, ind))));
45.          % calculate the specific row scaling factor
46.          row_multi(i) = length(ind) / row_sum(i);
47.          % scale the elements of the specific row of matrix A
48.          A(i, :) = A(i, :) * row_multi(i);
49.          % scale the elements of vector b
50.          b(i) = b(i) * row_multi(i);
51.      end
52.  end
53.  % then apply column scaling
54.  col_sum = zeros(n, 1);
55.  col_multi = zeros(n, 1);
56.  for j = 1:n
57.      % find the indices of the nonzero elements of the
58.      % specific column
59.      ind = find(A(:, j));
60.      % if the specific column contains at least one nonzero
61.      % element
62.      if ~isempty(ind)
63.          % sum of the absolute value of the nonzero elements of
64.          % the specific column
65.          col_sum(j) = sum(sum(abs(A(ind, j))));
66.          % calculate the specific column scaling factor
67.          col_multi(j) = length(ind) / col_sum(j);
68.          % scale the elements of the specific column of
69.          % matrix A
70.          A(:, j) = A(:, j) * col_multi(j);
71.          % scale the elements of vector c
72.          c(j) = c(j) * col_multi(j);
73.      end
74.  end
75.  end
```

5.5 de Buchet

5.5.1 Mathematical Formulation

The de Buchet scaling model is based on the relative divergence and is formulated as shown in Equation (5.5):

$$\min_{(r,s>0)} \left(\sum_{(i,j)\in\overline{Z}} \{A_{ij}r_i s_j + 1/\left(A_{ij}r_i s_j\right)\}^p \right)^{1/p} \tag{5.5}$$

where p is a positive integer and $\overline{Z}$ is the number of the nonzero elements of A.

We focus now our attention on the cases $p = 1, 2$, and ∞. For the case $p = 1$, Equation (5.5) is formulated as shown in Equation (5.6):

$$\min_{(r,s>0)} \sum_{(i,j)\in\overline{Z}} A_{ij}r_i s_j + 1/\left(A_{ij}r_i s_j\right) \tag{5.6}$$

The row scaling factors are presented in Equation (5.7):

$$r_i^{k+1} = \left\{ \left(\sum_{j\in N_i} 1/\left|X_{ij}^k\right| \right) \left(\sum_{j\in N_i} \left|X_{ij}^k\right| \right) \right\}^{1/2} \tag{5.7}$$

Similarly, the column scaling factors are presented in Equation (5.8):

$$s_j^{k+1} = \left\{ \left(\sum_{i\in M_j} 1/\left|X_{ij}^{k+1/2}\right| \right) \left(\sum_{i\in M_j} \left|X_{ij}^{k+1/2}\right| \right) \right\}^{1/2} \tag{5.8}$$

For the case $p = 2$, Equation (5.5) is stated as shown in Equation (5.9):

$$\min_{(r,s>0)} \left(\sum_{(i,j)\in\overline{Z}} \{A_{ij}r_i s_j + 1/\left(A_{ij}r_i s_j\right)\}^2 \right)^{1/2} \tag{5.9}$$

The row scaling factors are presented in Equation (5.10):

$$r_i^{k+1} = \left\{ \left(\sum_{j\in N_i} \left(1/\left|X_{ij}^k\right|\right)^2 \right) \left(\sum_{j\in N_i} \left(\left|X_{ij}^k\right|\right)^2 \right) \right\}^{1/4} \tag{5.10}$$

Similarly, the column scaling factors are presented in Equation (5.11):

$$s_j^{k+1} = \left\{ \left(\sum_{i \in M_j} \left(1/\left|X_{ij}^{k+1/2}\right| \right)^2 \right) \left(\sum_{i \in M_j} \left(\left|X_{ij}^{k+1/2}\right| \right)^2 \right) \right\}^{1/4} \qquad (5.11)$$

Finally, for the case $p = \infty$, Equation (5.5) is formulated as shown in Equation (5.12):

$$\min_{(r,s>0)} \max_{(i,j) \in \overline{Z}} \left| \log \left(A_{ij} r_i s_j \right) \right| \qquad (5.12)$$

The row scaling factors are described in Equation (5.13):

$$r_i^{k+1} = 1 / \left\{ \left(\max_{j \in N_i} \left|X_{ij}^k\right| \right) \left(\min_{j \in N_i} \left|X_{ij}^k\right| \right) \right\}^{1/2} \qquad (5.13)$$

Similarly, the column scaling factors are presented in Equation (5.14):

$$s_j^{k+1} = 1 / \left\{ \left(\max_{i \in M_j} \left|X_{ij}^{k+1/2}\right| \right) \left(\min_{i \in M_j} \left|X_{ij}^{k+1/2}\right| \right) \right\}^{1/2} \qquad (5.14)$$

The de Buchet for the case $p = \infty$ scaling technique corresponds to the geometric mean scaling technique that will be described in Section 5.8.

5.5.2 Numerical Example

This subsection demonstrates the de Buchet scaling technique through an illustrative example. The LP problem that will be scaled is the following:

$$
\begin{aligned}
\min z = 7/2x_1 & & & + 453x_3 + & 6x_4 & \\
\text{s.t.} \quad 956x_1 & & & + & x_3 + 258/5x_4 & = 4 \\
5/2x_1 & + & 4x_2 & + 13/2x_3 & + 149/5x_4 & = 7/2 \\
x_1 & + & 3/2x_2 & & + 67/10x_4 & = 55 \\
x_j \geq 0, & \quad (j = 1,2,3,4) & & & &
\end{aligned}
$$

In matrix notation the above LP problem is written as follows:

$$
A = \begin{bmatrix} 956 & 0 & 1 & 258/5 \\ 5/2 & 4 & 13/2 & 149/5 \\ 1 & 3/2 & 0 & 67/10 \end{bmatrix}, c = \begin{bmatrix} 7/2 \\ 0 \\ 453 \\ 6 \end{bmatrix}, b = \begin{bmatrix} 4 \\ 7/2 \\ 55 \end{bmatrix}, Eqin = \begin{bmatrix} 0 \\ 0 \\ 0 \end{bmatrix}
$$

First of all, the de Buchet scaling technique for the case $p = 1$ will be performed. Initially, the row scaling factors are calculated according to Equation (5.7):

$$r = \begin{bmatrix} \left(\dfrac{1/956+1/1+1/(258/5)}{956+1+258/5} \right)^{1/2} \\ \left(\dfrac{1/(5/2)+1/4+1/(13/2)+1/(149/5)}{5/2+4+13/2+149/5} \right)^{1/2} \\ \left(\dfrac{1/1+1/(3/2)+1/(67/10)}{1+3/2+67/10} \right)^{1/2} \end{bmatrix} = \begin{bmatrix} 41/1289 \\ 295/2109 \\ 1184/2665 \end{bmatrix}$$

Then, matrix A and vector b are scaled:

$$A = \begin{bmatrix} 8,271/272 & 0 & 41/1,289 & 1,185/722 \\ 1,475/4,218 & 1,180/2,109 & 771/848 & 4,606/1,105 \\ 1,184/2,665 & 889/1,334 & 0 & 1,658/557 \end{bmatrix}, b = \begin{bmatrix} 164/1,289 \\ 352/719 \\ 2,077/85 \end{bmatrix}$$

Next, the column scaling factors are calculated according to Equation (5.8):

$$s = \begin{bmatrix} \left(\dfrac{1/(8,271/272)+1/(1,475/4,218)+1/(1,184/2,665)}{8,271/272+1,475/4,218+1,184/2,665} \right)^{1/2} \\ \left(\dfrac{1/(1,180/2,109)+1/(889/1,334)}{1,180/2,109+889/1,334} \right)^{1/2} \\ \left(\dfrac{1/(41/1,289)+1/(771/848)}{41/1,289+771/848} \right)^{1/2} \\ \left(\dfrac{1/(1,185/722)+1/(4,606/1,105)+1/(1,658/557)}{1,185/722+4,606/1,105+1,658/557} \right)^{1/2} \end{bmatrix} = \begin{bmatrix} 365/899 \\ 1,261/770 \\ 3,146/535 \\ 469/1,277 \end{bmatrix}$$

Then, matrix A and vector c are scaled:

$$A = \begin{bmatrix} 3,605/292 & 0 & 280/1,497 & 1,082/1,795 \\ 461/3,247 & 996/1,087 & 1,497/280 & 793/518 \\ 57/316 & 1,087/996 & 0 & 1,114/1,019 \end{bmatrix}, c = \begin{bmatrix} 2,555/1,798 \\ 0 \\ 55,940/21 \\ 1,591/722 \end{bmatrix}$$

Finally, the row and column scaling factors, the scaled matrix A, and vectors b and c are the following:

$$r = \begin{bmatrix} 41/1,289 \\ 295/2,109 \\ 1,184/2,665 \end{bmatrix}, s = \begin{bmatrix} 365/899 \\ 1,261/770 \\ 3,146/535 \\ 469/1,277 \end{bmatrix}$$

$$A = \begin{bmatrix} 3,605/292 & 0 & 280/1,497 & 1,082/1,795 \\ 461/3,247 & 996/1,087 & 1,497/280 & 793/518 \\ 57/316 & 1,087/996 & 0 & 1,114/1,019 \end{bmatrix}, c = \begin{bmatrix} 2,555/1,798 \\ 0 \\ 55,940/21 \\ 1,591/722 \end{bmatrix},$$

$$b = \begin{bmatrix} 164/1,289 \\ 352/719 \\ 2,077/85 \end{bmatrix}$$

The final scaled LP problem is the following:

$$
\begin{array}{llll}
\min z = 2,555/1,798x_1 & + 55,940/21x_3 + & 1,591/722x_4 \\
\text{s.t.} \quad 3,605/292x_1 & + 280/1,497x_3 + 1,082/1,795x_4 = 164/1,289 \\
461/3,247x_1 + 996/1,087x_2 + 1,497/280x_3 + & 793/518x_4 = & 352/719 \\
57/316x_1 + 1,087/996x_2 & + 1,114/1,019x_4 = & 2,077/85 \\
\end{array}
$$
$$x_j \geq 0, \quad (j = 1,2,3,4)$$

Next, the de Buchet scaling technique for the case $p = 2$ is applied to the same example. Initially, the row scaling factors are calculated according to Equation (5.10):

$$
r = \begin{bmatrix}
\left(\frac{956^{-2}+1^{-2}+(258/5)^{-2}}{956^2+1^2+(258/5)^2} \right)^{1/4} \\
\left(\frac{(5/2)^{-2}+4^{-2}+(13/2)^{-2}+(149/5)^{-2}}{(5/2)^2+4^2+(13/2)^2+(149/5)^2} \right)^{1/4} \\
\left(\frac{1^{-2}+(3/2)^{-2}+(67/10)^{-2}}{1^2+(3/2)^2+(67/10)^2} \right)^{1/4}
\end{bmatrix}
= \begin{bmatrix} 327/10,117 \\ 869/6,846 \\ 526/1,259 \end{bmatrix}
$$

Then, matrix A and vector b are scaled:

$$
A = \begin{bmatrix}
9,548/309 & 0 & 327/10,117 & 487/292 \\
496/1,563 & 787/1,550 & 1,283/1,555 & 2,402/635 \\
526/1,259 & 1,160/1,851 & 0 & 2,119/757
\end{bmatrix}, b = \begin{bmatrix} 1,259/9,738 \\ 869/1,956 \\ 3,217/140 \end{bmatrix}
$$

Next, the column scaling factors are calculated according to Equation (5.11):

$$
s = \begin{bmatrix}
3/(9,548/309 + 496/1,563 + 526/1,259) \\
2/(787/1,550 + 1,160/1,851) \\
2/(327/10,117 + 1,283/1,555) \\
3/(487/292 + 2,402/635 + 2,119/757)
\end{bmatrix}
= \begin{bmatrix} 1,134/3,169 \\ 1,810/1,021 \\ 2,131/348 \\ 735/1,901 \end{bmatrix}
$$

Then, matrix A and vector c are scaled:

$$
A = \begin{bmatrix}
6,767/612 & 0 & 515/2,602 & 581/901 \\
325/2,862 & 820/911 & 2,602/515 & 1,249/854 \\
286/1,913 & 911/820 & 0 & 1,802/1,665
\end{bmatrix}, c = \begin{bmatrix} 640/511 \\ 0 \\ 99,863/36 \\ 1,661/716 \end{bmatrix}
$$

Finally, the row and column scaling factors, the scaled matrix A, and vectors b and c are the following:

$$r = \begin{bmatrix} 327/10, 117 \\ 869/6, 846 \\ 526/1, 259 \end{bmatrix}, s = \begin{bmatrix} 1, 134/3, 169 \\ 1, 810/1, 021 \\ 2, 131/348 \\ 735/1, 901 \end{bmatrix}$$

$$A = \begin{bmatrix} 6, 767/612 & 0 & 515/2, 602 & 581/901 \\ 325/2, 862 & 820/911 & 2, 602/515 & 1, 249/854 \\ 286/1, 913 & 911/820 & 0 & 1, 802/1, 665 \end{bmatrix}, c = \begin{bmatrix} 640/511 \\ 0 \\ 99, 863/36 \\ 1, 661/716 \end{bmatrix},$$

$$b = \begin{bmatrix} 1, 259/9, 738 \\ 869/1, 956 \\ 3, 217/140 \end{bmatrix}$$

The final scaled LP problem is the following:

$$
\begin{array}{lllll}
\min z = & 640/511x_1 & & + 99, 863/36x_3 + & 1, 661/716x_4 \\
\text{s.t.} & 6, 767/612x_1 & & + 515/2, 602x_3 + & 581/901x_4 = 1, 259/9, 738 \\
& 325/2, 862x_1 & + 820/911x_2 & + 2, 602/515x_3 + & 1, 249/854x_4 = 869/1, 956 \\
& 286/1, 913x_1 & + 911/820x_2 & & + 1, 802/1, 665x_4 = 3, 217/140 \\
x_j \geq 0, & (j = 1, 2, 3, 4) & & &
\end{array}
$$

Finally, the de Buchet scaling technique for the case $p = \infty$ is applied to the same example. Initially, the row scaling factors are calculated according to Equation (5.13):

$$r = \begin{bmatrix} (956 \times 1)^{-1/2} \\ (149/5 \times 5/2)^{-1/2} \\ (67/10 \times 1)^{-1/2} \end{bmatrix} = \begin{bmatrix} 161/4, 978 \\ 217/1, 873 \\ 294/761 \end{bmatrix}$$

Then, matrix A and vector b are scaled:

$$A = \begin{bmatrix} 4, 978/161 & 0 & 161/4, 978 & 1, 013/607 \\ 769/2, 655 & 868/1, 873 & 552/733 & 2, 655/769 \\ 294/761 & 441/761 & 0 & 761/294 \end{bmatrix}, b = \begin{bmatrix} 322/2, 489 \\ 1, 637/4, 037 \\ 3, 251/153 \end{bmatrix}$$

Next, the column scaling factors are calculated according to Equation (5.14):

$$s = \begin{bmatrix} (4, 978/161 \times 769/2, 655)^{-1/2} \\ (441/761 \times 868/1, 873)^{-1/2} \\ (552/733 \times 161/4, 978)^{-1/2} \\ (2, 655/769 \times 1, 013/607)^{-1/2} \end{bmatrix} = \begin{bmatrix} 539/1, 613 \\ 2, 771/1, 436 \\ 2, 185/341 \\ 532/1, 277 \end{bmatrix}$$

Then, matrix A and vector c are scaled:

$$A = \begin{bmatrix} 2,552/247 & 0 & 1,489/7,185 & 527/758 \\ 247/2,552 & 296/331 & 7,185/1,489 & 758/527 \\ 256/1,983 & 331/296 & 0 & 991/919 \end{bmatrix}, c = \begin{bmatrix} 1,145/979 \\ 0 \\ 66,761/23 \\ 3,182/1,273 \end{bmatrix}$$

Finally, the row and column scaling factors, the scaled matrix A, and vectors b and c are the following:

$$r = \begin{bmatrix} 161/4,978 \\ 217/1,873 \\ 294/761 \end{bmatrix}, s = \begin{bmatrix} 539/1,613 \\ 2,771/1,436 \\ 2,185/341 \\ 532/1,277 \end{bmatrix}$$

$$A = \begin{bmatrix} 2,552/247 & 0 & 1,489/7,185 & 527/758 \\ 247/2,552 & 296/331 & 7,185/1,489 & 758/527 \\ 256/1,983 & 331/296 & 0 & 991/919 \end{bmatrix}, c = \begin{bmatrix} 1,145/979 \\ 0 \\ 66,761/23 \\ 3,182/1,273 \end{bmatrix},$$

$$b = \begin{bmatrix} 322/2,489 \\ 1,637/4,037 \\ 3,251/153 \end{bmatrix}$$

The final scaled LP problem is the following:

$$
\begin{aligned}
\min z = {}& 1,145/979 x_1 & & & + {}& 66,761/23 x_3 & + {}& 3,182/1,273 x_4 \\
\text{s.t.} \quad & 2,552/247 x_1 & & & + {}& 1,489/7,185 x_3 + {}& 527/758 x_4 = {}& 322/2,489 \\
& 247/2,552 x_1 & + {}& 296/331 x_2 & + {}& 7,185/1,489 x_3 + {}& 758/527 x_4 = {}& 1,637/4,037 \\
& 256/1,983 x_1 & + {}& 331/296 x_2 & & & + {}& 991/919 x_4 = {}& 3,251/153 \\
& x_j \geq 0, \quad (j = 1,2,3,4)
\end{aligned}
$$

5.5.3 Implementation in MATLAB

This subsection presents the implementation in MATLAB of the de Buchet scaling technique (filenames: `debuchet1.m`, `debuchet2.m`, and `debuchetinf.m`). Some necessary notations should be introduced before the presentation of the implementation of the de Buchet scaling technique. Let *row_multi* be a $m \times 1$ vector with row scaling factors and *col_multi* be a $n \times 1$ vector with column scaling factors. Let *row_sum* be a $m \times 1$ vector with the sum of each row's elements and *col_sum* be a $n \times 1$ vector with the sum of each column's elements, while *row_inv_sum* is a $m \times 1$ vector with the sum of the inverse of each row's elements in absolute value and *col_inv_sum* is a $n \times 1$ vector with the sum of the inverse of each column's elements in absolute value. Let *row_pow* be a $m \times 1$ vector with the sum of each row's elements to the power of 2 and *col_pow* be a $n \times 1$ vector with the sum of each column's elements to the power of 2, while *row_inv_pow* is a $m \times 1$ vector with the sum of the inverse of each row's elements to the power of 2 and *col_inv_sum* is a $n \times 1$ vector with the sum of the inverse of each column's elements to the power of 2. Let *row_max* be a $m \times 1$ vector with the maximum in absolute value of each row's elements and *col_max* be a $n \times 1$ vector with the maximum in absolute value of each column's elements, while *row_min* is a $m \times 1$ vector with the minimum in absolute value of each row's elements and *col_min* is a $n \times 1$ vector with the minimum in absolute value of each column's elements.

The function for the de Buchet scaling technique for the case $p = 1$ takes as input the matrix of coefficients of the constraints (matrix A), the vector of coefficients of the objective function (vector c), and the vector of the right-hand side of the constraints (vector b), and returns as output the scaled matrix of coefficients of the constraints (matrix A), the scaled vector of coefficients of the objective function (vector c), the scaled vector of the right-hand side of the constraints (vector b), and the row and column scaling factors (vectors *row_multi* and *col_multi*).

Lines 32–34 include the pre-allocation of vectors *row_sum*, *row_inv_sum*, and *row_multi*. In the first for-loop (lines 35–55), the row scaling factors are calculated as the square root of the division of the inverse sum of each row to the sum of the same row (line 49). Then, matrix A and vector b are updated (lines 51–53). Next, lines 57–59 include the pre-allocation of vectors *col_sum*, *col_inv_sum*, and *col_multi*. Finally, in the second for-loop (lines 60–81), the column scaling factors are calculated as the square root of the division of the inverse sum of each column to the sum of the same column (line 74). Then, matrix A and vector c are updated (lines 77–79).

```
1.    function [A, c, b, row_multi, col_multi] = debuchet1(A, c, b)
2.    % Filename: debuchet1.m
3.    % Description: the function is an implementation of the de
4.    % Buchet for the case p = 1 scaling technique
5.    % Authors: Ploskas, N., & Samaras, N.
6.    %
7.    % Syntax: [A, c, b, row_multi, col_multi] = ...
8.    %     debuchet1(A, c, b)
9.    %
10.   % Input:
11.   % -- A: matrix of coefficients of the constraints
12.   %    (size m x n)
13.   % -- c: vector of coefficients of the objective function
14.   %    (size n x 1)
15.   % -- b: vector of the right-hand side of the constraints
16.   %    (size m x 1)
17.   %
18.   % Output:
19.   % -- A: scaled matrix of coefficients of the constraints
20.   %    (size m x n)
21.   % -- c: scaled vector of coefficients of the objective
22.   %    function (size n x 1)
```

```
23.   % -- b: scaled vector of the right-hand side of the
24.   %      constraints (size m x 1)
25.   % -- row_multi: vector of the row scaling factors
26.   %      (size m x 1)
27.   % -- col_multi: vector of the column scaling factors
28.   %      (size n x 1)
29.   %
30.   [m, n] = size(A); % size of matrix A
31.   % first apply row scaling
32.   row_sum = zeros(m, 1);
33.   row_inv_sum = zeros(m, 1);
34.   row_multi = zeros(m, 1);
35.   for i = 1:m
36.       % find the indices of the nonzero elements of the
37.       % specific row
38.       ind = find(A(i, :));
39.       % if the specific row contains at least one nonzero
40.       % element
41.       if ~isempty(ind)
42.           % sum of the absolute value of the nonzero elements of
43.           % the specific row
44.           row_sum(i) = sum(sum(abs(A(i, ind))));
45.           % sum of the inverse of the absolute value of the
46.           % nonzero elements of the specific row
47.           row_inv_sum(i) = sum(sum(abs(1 ./ A(i, ind))));
48.           % calculate the specific row scaling factor
49.           row_multi(i) = (row_inv_sum(i) / row_sum(i))^(1 / 2);
50.           % scale the elements of the specific row of matrix A
51.           A(i, :) = A(i, :) * row_multi(i);
52.           % scale the elements of vector b
53.           b(i) = b(i) * row_multi(i);
54.       end
55.   end
56.   % then apply column scaling
57.   col_sum = zeros(n, 1);
58.   col_inv_sum = zeros(n, 1);
59.   col_multi = zeros(n, 1);
60.   for j = 1:n
61.       % find the indices of the nonzero elements of the
62.       % specific column
63.       ind = find(A(:, j));
64.       % if the specific column contains at least one
65.       % nonzero element
66.       if ~isempty(ind)
67.           % sum of the absolute value of the nonzero elements of
68.           % the specific column
69.           col_sum(j) = sum(sum(abs(A(ind, j))));
70.           % sum of the inverse of the absolute value of the
71.           % nonzero elements of the specific column
72.           col_inv_sum(j) = sum(sum(abs(1 ./ A(ind, j))));
73.           % calculate the specific column scaling factor
74.           col_multi(j) = (col_inv_sum(j) / col_sum(j)) ^ (1/2);
75.           % scale the elements of the specific column of
76.           % matrix A
```

```
77.            A(:, j) = A(:, j) * col_multi(j);
78.            % scale the elements of vector c
79.            c(j) = c(j) * col_multi(j);
80.        end
81.    end
82.    end
```

The function for the de Buchet scaling technique for the case $p = 2$ takes as input the matrix of coefficients of the constraints (matrix A), the vector of coefficients of the objective function (vector c), and the vector of the right-hand side of the constraints (vector b), and returns as output the scaled matrix of coefficients of the constraints (matrix A), the scaled vector of coefficients of the objective function (vector c), the scaled vector of the right-hand side of the constraints (vector b), and the row and column scaling factors (vectors *row_multi* and *col_multi*).

Lines 32–34 include the pre-allocation of vectors *row_pow*, *row_inv_pow*, and *row_multi*. In the first for-loop (lines 35–56), the row scaling factors are calculated as the division of the sum of the inverse of the value to the power of 2 of the nonzero elements of each row to the sum of the value to the power of 2 of the nonzero elements of the same row to the power of $1/4$ (line 50). Then, matrix A and vector b are updated (lines 52–54). Next, lines 58–60 include the pre-allocation of vectors *col_pow*, *col_inv_pow*, and *col_multi*. Finally, in the second for-loop (lines 61–83), the column scaling factors are calculated as the division of the sum of the inverse of the value to the power of 2 of the nonzero elements of each column to the sum of the value to the power of 2 of the nonzero elements of the same column to the power of $1/4$ (line 76). Then, matrix A and vector c are updated (lines 79–81).

```
1.    function [A, c, b, row_multi, col_multi] = debuchet2(A, c, b)
2.    % Filename: debuchet2.m
3.    % Description: the function is an implementation of the de
4.    % Buchet for the case p = 2 scaling technique
5.    % Authors: Ploskas, N., & Samaras, N.
6.    %
7.    % Syntax: [A, c, b, row_multi, col_multi] = ...
8.    %     debuchet2(A, c, b)
9.    %
10.   % Input:
11.   % -- A: matrix of coefficients of the constraints
12.   %     (size m x n)
13.   % -- c: vector of coefficients of the objective function
14.   %     (size n x 1)
15.   % -- b: vector of the right-hand side of the constraints
16.   %     (size m x 1)
17.   %
18.   % Output:
19.   % -- A: scaled matrix of coefficients of the constraints
20.   %     (size m x n)
21.   % -- c: scaled vector of coefficients of the objective
22.   %     function (size n x 1)
23.   % -- b: scaled vector of the right-hand side of the
24.   %     constraints (size m x 1)
25.   % -- row_multi: vector of the row scaling factors
```

```
26.  %      (size m x 1)
27.  % -- col_multi: vector of the column scaling factors
28.  %      (size n x 1)
29.  %
30.  [m, n] = size(A); % size of matrix A
31.  % first apply row scaling
32.  row_pow = zeros(m, 1);
33.  row_inv_pow = zeros(m, 1);
34.  row_multi = zeros(m, 1);
35.  for i = 1:m
36.      % find the indices of the nonzero elements of the
37.      % specific row
38.      ind = find(A(i, :));
39.      % if the specific row contains at least one nonzero
40.      % element
41.      if ~isempty(ind)
42.          % sum of the value to the power of 2 of the nonzero
43.          % elements of the specific row
44.          row_pow(i) = sum(sum(A(i,ind) .* A(i, ind)));
45.          % sum of the inverse of the value to the power of 2 of
46.          % the nonzero elements of the specific row
47.          row_inv_pow(i) = sum(sum(1 ./ ...
48.              (abs(A(i, ind) .* A(i, ind)))));
49.          % calculate the specific row scaling factor
50.          row_multi(i) = (row_inv_pow(i) / row_pow(i))^(1 / 4);
51.          % scale the elements of the specific row of matrix A
52.          A(i, :) = A(i, :) * row_multi(i);
53.          % scale the elements of vector b
54.          b(i) = b(i) * row_multi(i);
55.      end
56.  end
57.  % then apply column scaling
58.  col_pow = zeros(n, 1);
59.  col_inv_pow = zeros(n, 1);
60.  col_multi = zeros(n, 1);
61.  for j = 1:n
62.      % find the indices of the nonzero elements of the
63.      % specific column
64.      ind = find(A(:, j));
65.      % if the specific column contains at least one nonzero
66.      % element
67.      if ~isempty(ind)
68.          % sum of the value to the power of 2 of the nonzero
69.          % elements of the specific column
70.          col_pow(j) = sum(sum(A(ind, j) .* A(ind, j)));
71.          % sum of the inverse of the value to the power of 2 of
72.          % the nonzero elements of the specific column
73.          col_inv_pow(j) = sum(sum(1 ./  ...
74.              (abs(A(ind,j) .* A(ind, j)))));
75.          % calculate the specific column scaling factor
76.          col_multi(j) = (col_inv_pow(j) / col_pow(j))^(1 / 4);
77.          % scale the elements of the specific column of
78.          % matrix A
79.          A(:, j) = A(:, j) * col_multi(j);
```

```
80.              % scale the elements of vector c
81.              c(j) = c(j) * col_multi(j);
82.       end
83.   end
84.   end
```

The function for the de Buchet scaling technique for the case $p = \infty$ takes as input the matrix of coefficients of the constraints (matrix A), the vector of coefficients of the objective function (vector c), and the vector of the right-hand side of the constraints (vector b), and returns as output the scaled matrix of coefficients of the constraints (matrix A), the scaled vector of coefficients of the objective function (vector c), the scaled vector of the right-hand side of the constraints (vector b), and the row and column scaling factors (vectors *row_multi* and *col_multi*).

Lines 33–35 include the pre-allocation of vectors *row_max*, *row_min*, and *row_multi*. In the first for-loop (lines 36–56), the row scaling factors are calculated as the inverse of the square root of the product of the maximum and the minimum element of the same row (line 50). Then, matrix A and vector b are updated (lines 52–54). Next, lines 58–60 include the pre-allocation of vectors *col_max*, *col_min*, and *col_multi*. Finally, in the second for-loop (lines 61–82), the column scaling factors are calculated as the inverse of the square root of the product of the maximum and the minimum element of the same column (line 75). Then, matrix A and vector c are updated (lines 78–80).

```
1.    function [A, c, b, row_multi, col_multi] = ...
2.         debuchetinf(A, c, b)
3.    % Filename: debuchetinf.m
4.    % Description: the function is an implementation of the de
5.    % Buchet for the case p = inf scaling technique
6.    % Authors: Ploskas, N., & Samaras, N.
7.    %
8.    % Syntax: [A, c, b, row_multi, col_multi] = ...
9.    %      debuchetinf(A, c, b)
10.   %
11.   % Input:
12.   % -- A: matrix of coefficients of the constraints
13.   %      (size m x n)
14.   % -- c: vector of coefficients of the objective function
15.   %      (size n x 1)
16.   % -- b: vector of the right-hand side of the constraints
17.   %      (size m x 1)
18.   %
19.   % Output:
20.   % -- A: scaled matrix of coefficients of the constraints
21.   %      (size m x n)
22.   % -- c: scaled vector of coefficients of the objective
23.   %      function (size n x 1)
24.   % -- b: scaled vector of the right-hand side of the
25.   %      constraints (size m x 1)
26.   % -- row_multi: vector of the row scaling factors
27.   %      (size m x 1)
28.   % -- col_multi: vector of the column scaling factors
29.   %      (size n x 1)
30.
```

```
31.   [m, n] = size(A); % size of matrix A
32.   % first apply row scaling
33.   row_max = zeros(m, 1);
34.   row_min = zeros(m, 1);
35.   row_multi = zeros(m, 1);
36.   for i = 1:m
37.      % find the indices of the nonzero elements of the
38.      % specific row
39.      ind = find(A(i, :));
40.      % if the specific row contains at least one nonzero
41.      % element
42.      if ~isempty(ind)
43.            % find the maximum in absolute value of the nonzero
44.            % elements of the specific row
45.            row_max(i) = max(max(abs(A(i, ind))));
46.            % find the minimum in absolute value of the nonzero
47.            % elements of the specific row
48.            row_min(i) = min(min(abs(A(i, ind))));
49.            % calculate the specific row scaling factor
50.            row_multi(i) = 1 / (sqrt(row_max(i) * row_min(i)));
51.            % scale the elements of the specific row of matrix A
52.            A(i, :) = A(i, :) * row_multi(i);
53.            % scale the elements of vector b
54.            b(i) = b(i) * row_multi(i);
55.      end
56.   end
57.   % then apply column scaling
58.   col_max = zeros(n, 1);
59.   col_min = zeros(n, 1);
60.   col_multi = zeros(n, 1);
61.   for j = 1:n
62.      % find the indices of the nonzero elements of the
63.      % specific column
64.      ind = find(A(:, j));
65.      % if the specific column contains at least one nonzero
66.      % element
67.      if ~isempty(ind)
68.            % find the maximum in absolute value of the nonzero
69.            % elements of the specific column
70.            col_max(j) = max(max(abs(A(ind, j))));
71.            % find the minimum in absolute value of the nonzero
72.            % elements of the specific column
73.            col_min(j) = min(min(abs(A(ind, j))));
74.            % calculate the specific column scaling factor
75.            col_multi(j) = 1 / (sqrt(col_max(j) * col_min(j)));
76.            % scale the elements of the specific column of
77.            % matrix A
78.            A(:, j) = A(:, j) * col_multi(j);
79.            % scale the elements of vector c
80.            c(j) = c(j) * col_multi(j);
81.      end
82.   end
83.   end
```

5.6 Entropy

5.6.1 Mathematical Formulation

The entropy model was first presented by Larsson [7] and is attributed to Dantzig and Erlander. This technique solves the model presented in Equation (5.15), in order to identify a scaling X, with all $x_{ij} \neq 0$ of magnitude one:

$$\min \sum_{(i,j) \in \overline{Z}} X_{ij} \left(\log \left(X_{ij}/A_{ij} \right) - 1 \right)$$

$$\text{s.t.} \sum_{j \in N_i)} X_{ij} = n_i \quad i = 1, \ldots, m$$

$$\sum_{i \in M_j} X_{ij} = m_j \quad j = 1, \ldots, n$$

$$X_{ij} \geq 0 \quad \forall \, (i,j) \in \overline{Z} \tag{5.15}$$

According to Larsson [7], any algorithm for solving entropy programs can be used for scaling matrices, but it is recommended to apply the row and column scaling factors presented in Equations (5.16) and (5.17):

$$r_i^{k+1} = n_i / \sum_{j \in N_i} \left| X_{ij}^k \right| \tag{5.16}$$

$$s_j^{k+1} = m_j / \sum_{i \in M_j} \left| X_{ij}^{k+1/2} \right| \tag{5.17}$$

The entropy scaling technique corresponds to the arithmetic mean scaling technique as described in Section 5.4.

5.6.2 Numerical Example

This subsection demonstrates the entropy scaling technique through an illustrative example. The LP problem that will be scaled is the following:

$$\begin{aligned}
\min z = {}& 7/2x_1 & + \ 453x_3 + & \ 6x_4 \\
\text{s.t.} \quad & 956x_1 & + \ x_3 + 258/5x_4 = & \ 4 \\
& 5/2x_1 + \ 4x_2 + 13/2x_3 + 149/5x_4 = & \ 7/2 \\
& x_1 + 3/2x_2 & + \ 67/10x_4 = & \ 55 \\
& x_j \geq 0, \quad (j = 1, 2, 3, 4)
\end{aligned}$$

In matrix notation the above LP problem is written as follows:

$$A = \begin{bmatrix} 956 & 0 & 1 & 258/5 \\ 5/2 & 4 & 13/2 & 149/5 \\ 1 & 3/2 & 0 & 67/10 \end{bmatrix}, c = \begin{bmatrix} 7/2 \\ 0 \\ 453 \\ 6 \end{bmatrix}, b = \begin{bmatrix} 4 \\ 7/2 \\ 55 \end{bmatrix}, Eqin = \begin{bmatrix} 0 \\ 0 \\ 0 \end{bmatrix}$$

Initially, the row scaling factors are calculated according to Equation (5.16):

$$r = \begin{bmatrix} 3/(956 + 1 + 258/5) \\ 4/(5/2 + 4 + 13/2 + 149/5) \\ 3/(1 + 3/2 + 67/10) \end{bmatrix} = \begin{bmatrix} 5/1,681 \\ 10/107 \\ 15/46 \end{bmatrix}$$

Then, matrix A and vector b are scaled:

$$A = \begin{bmatrix} 2,599/914 & 0 & 5/1,681 & 258/1,681 \\ 25/107 & 40/107 & 65/107 & 298/107 \\ 15/46 & 45/92 & 0 & 201/92 \end{bmatrix}, b = \begin{bmatrix} 20/1,681 \\ 35/107 \\ 825/46 \end{bmatrix}$$

Next, the column scaling factors are calculated according to Equation (5.17):

$$s = \begin{bmatrix} 3/(2,599/914 + 25/107 + 15/46) \\ 2/(40/107 + 45/92) \\ 2/(5/1,681 + 65/107) \\ 3/(258/1,681 + 298/107 + 201/92) \end{bmatrix} = \begin{bmatrix} 305/346 \\ 540/233 \\ 842/257 \\ 592/1,011 \end{bmatrix}$$

Then, matrix A and vector c are scaled:

$$A = \begin{bmatrix} 950/379 & 0 & 107/10,980 & 63/701 \\ 159/772 & 843/973 & 1,634/821 & 1,895/1,162 \\ 403/1,402 & 823/726 & 0 & 2,029/1,586 \end{bmatrix}, c = \begin{bmatrix} 2,533/821 \\ 0 \\ 40,072/27 \\ 1,184/337 \end{bmatrix}$$

Finally, the row and column scaling factors, the scaled matrix A, and vectors b and c are the following:

$$r = \begin{bmatrix} 5/1,681 \\ 10/107 \\ 15/46 \end{bmatrix}, s = \begin{bmatrix} 305/346 \\ 540/233 \\ 842/257 \\ 592/1,011 \end{bmatrix}$$

$$A = \begin{bmatrix} 950/379 & 0 & 107/10,980 & 63/701 \\ 159/772 & 843/973 & 1,634/821 & 1,895/1,162 \\ 403/1,402 & 823/726 & 0 & 2,029/1,586 \end{bmatrix}, c = \begin{bmatrix} 2,533/821 \\ 0 \\ 40,072/27 \\ 1,184/337 \end{bmatrix},$$

$$b = \begin{bmatrix} 20/1,681 \\ 35/107 \\ 825/46 \end{bmatrix}$$

The final scaled LP problem is the following:

$$
\begin{aligned}
\min z = {} & 2,533/821x_1 & & + 40,072/27x_3 + & 1,184/337x_4 & \\
\text{s.t.} \quad & 950/379x_1 & & + 107/10,980x_3 + & 63/701x_4 & = 20/1,681 \\
& 159/772x_1 + 843/973x_2 & & + 1,634/821x_3 + 1,895/1,162x_4 & = & 35/107 \\
& 403/1,402x_1 + 823/726x_2 & & + 2,029/1,586x_4 & = & 825/46 \\
x_j \geq 0, \quad & (j = 1,2,3,4) & & &
\end{aligned}
$$

5.6.3 Implementation in MATLAB

This subsection presents the implementation in MATLAB of the entropy scaling technique (filename: `entropy.m`). Some necessary notations should be introduced before the presentation of the implementation of the entropy scaling technique. Let *row_multi* be a $m \times 1$ vector with row scaling factors and *col_multi* be a $n \times 1$ vector with column scaling factors. Let *row_sum* be a $m \times 1$ vector with the sum of each row's elements in absolute value and *col_sum* be a $n \times 1$ vector with the sum of each column's elements in absolute value.

The function for the entropy scaling technique takes as input the matrix of coefficients of the constraints (matrix A), the vector of coefficients of the objective function (vector c), and the vector of the right-hand side of the constraints (vector b), and returns as output the scaled matrix of coefficients of the constraints (matrix A), the scaled vector of coefficients of the objective function (vector c), the scaled vector of the right-hand side of the constraints (vector b), and the row and column scaling factors (vectors *row_multi* and *col_multi*).

Lines 31–32 include the pre-allocation of vectors *row_sum* and *row_multi*. In the first for-loop (lines 33–50), the row scaling factors are calculated as the number of the nonzero elements of each row to the sum in absolute value of the same row (line 44). Then, matrix A and vector b are updated (lines 46–48). Next, lines 52–53 include the pre-allocation of vectors *col_sum* and *col_multi*. Finally, in the second for-loop (lines 54–72), the column scaling factors are calculated as the number of the nonzero elements of each column to the sum in absolute value of the same column (line 65). Then, matrix A and vector c are updated (lines 68–70).

```
1.    function [A, c, b, row_multi, col_multi] = entropy(A, c, b)
2.    % Filename: entropy.m
3.    % Description: the function is an implementation of the
4.    % entropy scaling technique
5.    % Authors: Ploskas, N., & Samaras, N.
6.    %
7.    % Syntax: [A, c, b, row_multi, col_multi] = entropy(A, c, b)
8.    %
9.    % Input:
10.   % -- A: matrix of coefficients of the constraints
11.   %      (size m x n)
12.   % -- c: vector of coefficients of the objective function
13.   %      (size n x 1)
14.   % -- b: vector of the right-hand side of the constraints
15.   %      (size m x 1)
16.   %
17.   % Output:
18.   % -- A: scaled matrix of coefficients of the constraints
19.   %      (size m x n)
20.   % -- c: scaled vector of coefficients of the objective
21.   %      function (size n x 1)
22.   % -- b: scaled vector of the right-hand side of the
23.   %      constraints (size m x 1)
24.   % -- row_multi: vector of the row scaling factors
25.   %      (size m x 1)
26.   % -- col_multi: vector of the column scaling factors
27.   %      (size n x 1)
28.
29.   [m, n] = size(A); % size of matrix A
30.   % first apply row scaling
31.   row_sum = zeros(m, 1);
32.   row_multi = zeros(m, 1);
33.   for i = 1:m
34.       % find the indices of the nonzero elements of the
35.       % specific row
36.       ind = find(A(i, :));
37.       % if the specific row contains at least one nonzero
38.       % element
39.       if ~isempty(ind)
40.           % sum of the absolute value of the nonzero elements of
41.           % the specific row
42.           row_sum(i) = sum(sum(abs(A(i, ind))));
43.           % calculate the specific row scaling factor
44.           row_multi(i) = length(ind) / row_sum(i);
45.           % scale the elements of the specific row of matrix A
46.           A(i, :) = A(i, :) * row_multi(i);
47.           % scale the elements of vector b
48.           b(i) = b(i) * row_multi(i);
49.       end
50.   end
51.   % then apply column scaling
52.   col_sum = zeros(n, 1);
53.   col_multi = zeros(n, 1);
54.   for j = 1:n
```

```
55.      % find the indices of the nonzero elements of the
56.      % specific column
57.      ind = find(A(:, j));
58.      % if the specific column contains at least one nonzero
59.      % element
60.      if ~isempty(ind)
61.          % sum of the absolute value of the nonzero elements of
62.          % the specific column
63.          col_sum(j) = sum(sum(abs(A(ind, j))));
64.          % calculate the specific column scaling factor
65.          col_multi(j) = length(ind) / col_sum(j);
66.          % scale the elements of the specific column of
67.          % matrix A
68.          A(:, j) = A(:, j) * col_multi(j);
69.          % scale the elements of vector c
70.          c(j) = c(j) * col_multi(j);
71.      end
72.  end
73.  end
```

5.7 Equilibration

5.7.1 Mathematical Formulation

For each row of the coefficient matrix A the largest element in absolute value is found, and the specified row of matrix A and the corresponding element of vector b are multiplied by the inverse of the largest element. Then, for each column of the coefficient matrix A that does not include 1 as the largest element in absolute value, the largest element in absolute value is found, and the specified column of matrix A and the corresponding element of vector c are multiplied by the inverse of the largest element. Consequently, all the elements of matrix A will have values between -1 and 1.

5.7.2 Numerical Example

This subsection demonstrates the equilibration scaling technique through an illustrative example. The LP problem that will be scaled is the following:

$$
\begin{aligned}
\min z = 7/2x_1 && + 453x_3 + && 6x_4 \\
\text{s.t.} \quad 956x_1 && + x_3 + 258/5x_4 = && 4 \\
5/2x_1 + && 4x_2 + 13/2x_3 + 149/5x_4 = && 7/2 \\
x_1 + 3/2x_2 && + 67/10x_4 = && 55 \\
x_j \geq 0, \quad (j = 1, 2, 3, 4) &&&&
\end{aligned}
$$

In matrix notation the above LP problem is written as follows:

$$A = \begin{bmatrix} 956 & 0 & 1 & 258/5 \\ 5/2 & 4 & 13/2 & 149/5 \\ 1 & 3/2 & 0 & 67/10 \end{bmatrix}, c = \begin{bmatrix} 7/2 \\ 0 \\ 453 \\ 6 \end{bmatrix}, b = \begin{bmatrix} 4 \\ 7/2 \\ 55 \end{bmatrix}, Eqin = \begin{bmatrix} 0 \\ 0 \\ 0 \end{bmatrix}$$

Initially, the row scaling factors are calculated:

$$r = \begin{bmatrix} 1/956 \\ 1/(149/5) \\ 1/(67/10) \end{bmatrix} = \begin{bmatrix} 1/956 \\ 5/149 \\ 10/67 \end{bmatrix}$$

Then, matrix A and vector b are scaled:

$$A = \begin{bmatrix} 1 & 0 & 1/956 & 129/2,390 \\ 25/298 & 20/149 & 65/298 & 1 \\ 10/67 & 15/67 & 0 & 1 \end{bmatrix}, b = \begin{bmatrix} 1/239 \\ 35/298 \\ 550/67 \end{bmatrix}$$

Next, the column scaling factors are calculated:

$$s = \begin{bmatrix} 1/1 \\ 1/(15/67) \\ 1/(65/298) \\ 1/1 \end{bmatrix} = \begin{bmatrix} 1 \\ 67/15 \\ 298/65 \\ 1 \end{bmatrix}$$

Then, matrix A and vector c are scaled:

$$A = \begin{bmatrix} 1 & 0 & 149/31,070 & 129/2,390 \\ 25/298 & 268/447 & 1 & 1 \\ 10/67 & 1 & 0 & 1 \end{bmatrix}, c = \begin{bmatrix} 7/2 \\ 0 \\ 134,994/65 \\ 6 \end{bmatrix}$$

Finally, the row and column scaling factors, the scaled matrix A, and vectors b and c are the following:

$$r = \begin{bmatrix} 1/956 \\ 5/149 \\ 10/67 \end{bmatrix}, s = \begin{bmatrix} 1 \\ 67/15 \\ 298/65 \\ 1 \end{bmatrix}$$

$$A = \begin{bmatrix} 1 & 0 & 149/31,070 & 129/2,390 \\ 25/298 & 268/447 & 1 & 1 \\ 10/67 & 1 & 0 & 1 \end{bmatrix}, c = \begin{bmatrix} 7/2 \\ 0 \\ 134,994/65 \\ 6 \end{bmatrix},$$

$$b = \begin{bmatrix} 1/239 \\ 35/298 \\ 550/67 \end{bmatrix}$$

The final scaled LP problem is the following:

$$
\begin{aligned}
\min z = \quad & 7/2x_1 && + 134,994/65x_3 + && 6x_4 \\
\text{s.t.} \quad & x_1 && + 149/31,070x_3 + 129/2,390x_4 = && 1/239 \\
& 25/298x_1 + 268/447x_2 + && x_3 + && x_4 = 35/298 \\
& 10/67x_1 + \quad x_2 && + && x_4 = 550/67 \\
& x_j \geq 0, \quad (j = 1,2,3,4)
\end{aligned}
$$

5.7.3 Implementation in MATLAB

This subsection presents the implementation in MATLAB of the equilibration scaling technique (filename: equilibration.m). Some necessary notations should be introduced before the presentation of the implementation of the equilibration scaling technique. Let *row_multi* be a $m \times 1$ vector with row scaling factors and *col_multi* be a $n \times 1$ vector with column scaling factors. Let *row_max* be a $m \times 1$ vector with the maximum in absolute value of each row's elements and *col_max* be a $n \times 1$ vector with the maximum in absolute value of each column's elements.

The function for the equilibration scaling technique takes as input the matrix of coefficients of the constraints (matrix A), the vector of coefficients of the objective function (vector c), and the vector of the right-hand side of the constraints (vector b), and returns as output the scaled matrix of coefficients of the constraints (matrix A), the scaled vector of coefficients of the objective function (vector c), the scaled vector of the right-hand side of the constraints (vector b), and the row and column scaling factors (vectors *row_multi* and *col_multi*).

Lines 33–34 include the pre-allocation of vectors *row_max* and *row_multi*. In the first for-loop (lines 35–52), the row scaling factors are calculated as the inverse of the maximum element of each row (line 46). Then, matrix A and vector b are updated (lines 48–50). Next, lines 54–55 include the pre-allocation of vectors *col_max* and *col_multi*. Finally, in the second for-loop (lines 56–74), the column scaling factors are calculated as the inverse of the maximum element of each column (line 67). Then, matrix A and vector c are updated (lines 70–72).

```
1.     function [A, c, b, row_multi, col_multi] = ...
2.          equilibration(A, c, b)
3.     % Filename: equilibration.m
4.     % Description: the function is an implementation of the
5.     % equilibration scaling technique
6.     % Authors: Ploskas, N., & Samaras, N.
7.     %
8.     % Syntax: [A, c, b, row_multi, col_multi] = ...
9.     %    equilibration(A, c, b)
10.    %
11.    % Input:
12.    % -- A: matrix of coefficients of the constraints
13.    %     (size m x n)
14.    % -- c: vector of coefficients of the objective function
15.    %     (size n x 1)
16.    % -- b: vector of the right-hand side of the constraints
17.    %     (size m x 1)
18.    %
19.    % Output:
20.    % -- A: scaled matrix of coefficients of the constraints
21.    %     (size m x n)
22.    % -- c: scaled vector of coefficients of the objective
23.    %     function (size n x 1)
24.    % -- b: scaled vector of the right-hand side of the
25.    %     constraints (size m x 1)
26.    % -- row_multi: vector of the row scaling factors
27.    %     (size m x 1)
28.    % -- col_multi: vector of the column scaling factors
29.    %     (size n x 1)
30.
31.    [m, n] = size(A); % size of matrix A
32.    % first apply row scaling
33.    row_max = zeros(m, 1);
34.    row_multi = zeros(m, 1);
35.    for i = 1:m
36.        % find the indices of the nonzero elements of the
37.        % specific row
38.        ind = find(A(i, :));
39.        % if the specific row contains at least one nonzero
40.        % element
41.        if ~isempty(ind)
42.            % find the maximum in absolute value of the nonzero
43.            % elements of the specific row
44.            row_max(i) = max(max(abs(A(i, ind))));
45.            % calculate the specific row scaling factor
46.            row_multi(i) = 1 / row_max(i);
47.            % scale the elements of the specific row of matrix A
48.            A(i, :) = A(i, :) * row_multi(i);
49.            % scale the elements of vector b
50.            b(i) = b(i) * row_multi(i);
51.        end
52.    end
53.    % then apply column scaling
54.    col_max = zeros(n, 1);
```

```
55.  col_multi = zeros(n, 1);
56.  for j = 1:n
57.       % find the indices of the nonzero elements of the
58.       % specific column
59.       ind = find(A(:, j));
60.       % if the specific column contains at least one nonzero
61.       % element
62.       if ~isempty(ind)
63.            % find the maximum in absolute value of the nonzero
64.            % elements of the specific column
65.            col_max(j) = max(max(abs(A(ind, j))));
66.            % calculate the specific column scaling factor
67.            col_multi(j) = 1 / col_max(j);
68.            % scale the elements of the specific column of
69.            % matrix A
70.            A(:, j) = A(:, j) * col_multi(j);
71.            % scale the elements of vector c
72.            c(j) = c(j) * col_multi(j);
73.       end
74.  end
75.  end
```

5.8 Geometric Mean

5.8.1 Mathematical Formulation

Like arithmetic mean, geometric mean also aims to decrease the variance between the nonzero elements in a matrix. In this method, the row scaling factors are calculated as shown in Equation (5.18):

$$r_i^{k+1} = \left(\max_{j \in N_i} X_{ij}^k \min_{j \in N_i} X_{ij}^k \right)^{-1/2} \tag{5.18}$$

Similarly, the column scaling factors are presented in Equation (5.19):

$$s_j^{k+1} = \left(\max_{j \in M_j} X_{ij}^{k+1/2} \min_{j \in M_j} X_{ij}^{k+1/2} \right)^{-1/2} \tag{5.19}$$

5.8.2 Numerical Example

This subsection demonstrates the geometric mean scaling technique through an illustrative example. The LP problem that will be scaled is the following:

$$\min z = 7/2x_1 \qquad\qquad + \ 453x_3 + \quad 6x_4$$
$$\text{s.t.} \qquad 956x_1 \qquad\qquad + \qquad x_3 + 258/5x_4 = \quad 4$$
$$5/2x_1 + \quad 4x_2 + 13/2x_3 + 149/5x_4 = 7/2$$
$$x_1 + 3/2x_2 \qquad\qquad\quad + 67/10x_4 = \quad 55$$
$$x_j \geq 0, \quad (j = 1, 2, 3, 4)$$

In matrix notation the above LP problem is written as follows:

$$A = \begin{bmatrix} 956 & 0 & 1 & 258/5 \\ 5/2 & 4 & 13/2 & 149/5 \\ 1 & 3/2 & 0 & 67/10 \end{bmatrix}, c = \begin{bmatrix} 7/2 \\ 0 \\ 453 \\ 6 \end{bmatrix}, b = \begin{bmatrix} 4 \\ 7/2 \\ 55 \end{bmatrix}, Eqin = \begin{bmatrix} 0 \\ 0 \\ 0 \end{bmatrix}$$

Initially, the row scaling factors are calculated according to Equation (5.18):

$$r = \begin{bmatrix} (956 \times 1)^{-1/2} \\ (149/5 \times 5/2)^{-1/2} \\ (67/10 \times 1)^{-1/2} \end{bmatrix} = \begin{bmatrix} 161/4,978 \\ 217/1,873 \\ 294/761 \end{bmatrix}$$

Then, matrix A and vector b are scaled:

$$A = \begin{bmatrix} 4,978/161 & 0 & 161/4,978 & 1,013/607 \\ 769/2,655 & 868/1,873 & 552/733 & 2,655/769 \\ 294/761 & 441/761 & 0 & 761/294 \end{bmatrix}, b = \begin{bmatrix} 322/2,489 \\ 1,637/4,037 \\ 3,251/153 \end{bmatrix}$$

Next, the column scaling factors are calculated according to Equation (5.19):

$$s = \begin{bmatrix} (4,978/161 \times 769/2,655)^{-1/2} \\ (441/761 \times 868/1,873)^{-1/2} \\ (552/733 \times 161/4,978)^{-1/2} \\ (2,655/769 \times 1,013/607)^{-1/2} \end{bmatrix} = \begin{bmatrix} 539/1,613 \\ 2,771/1,436 \\ 2,185/341 \\ 532/1,277 \end{bmatrix}$$

Then, matrix A and vector c are scaled:

$$A = \begin{bmatrix} 2,552/247 & 0 & 1,489/7,185 & 527/758 \\ 247/2,552 & 296/331 & 7,185/1,489 & 758/527 \\ 256/1,983 & 331/296 & 0 & 991/919 \end{bmatrix}, c = \begin{bmatrix} 1,145/979 \\ 0 \\ 66,761/23 \\ 3,182/1,273 \end{bmatrix}$$

Finally, the row and column scaling factors, the scaled matrix A, and vectors b and c are the following:

$$r = \begin{bmatrix} 161/4,978 \\ 217/1,873 \\ 294/761 \end{bmatrix}, s = \begin{bmatrix} 539/1,613 \\ 2,771/1,436 \\ 2,185/341 \\ 532/1,277 \end{bmatrix}$$

$$A = \begin{bmatrix} 2,552/247 & 0 & 1,489/7,185 & 527/758 \\ 247/2,552 & 296/331 & 7,185/1,489 & 758/527 \\ 256/1,983 & 331/296 & 0 & 991/919 \end{bmatrix}, c = \begin{bmatrix} 1,145/979 \\ 0 \\ 66,761/23 \\ 3,182/1,273 \end{bmatrix},$$

$$b = \begin{bmatrix} 322/2,489 \\ 1,637/4,037 \\ 3,251/153 \end{bmatrix}$$

The final scaled LP problem is the following:

$$
\begin{array}{llll}
\min z = 1,145/979x_1 & & + \quad 66,761/23x_3 + 3,182/1,273x_4 & \\
\text{s.t.} \quad 2,552/247x_1 & & + 1,489/7,185x_3 + & 527/758x_4 = 322/2,489 \\
247/2,552x_1 & + 296/331x_2 & + 7,185/1,489x_3 + & 758/527x_4 = 1,637/4,037 \\
256/1,983x_1 & + 331/296x_2 & + & 991/919x_4 = 3,251/153 \\
x_j \geq 0, \quad (j = 1,2,3,4) & & &
\end{array}
$$

5.8.3 Implementation in MATLAB

This subsection presents the implementation in MATLAB of the geometric mean scaling technique (filename: geometricMean.m). Some necessary notations should be introduced before the presentation of the implementation of the geometric mean scaling technique. Let *row_multi* be a $m \times 1$ vector with row scaling factors and *col_multi* be a $n \times 1$ vector with column scaling factors. Let *row_max* be a $m \times 1$ vector with the maximum in absolute value of each row's elements and *col_max* be a $n \times 1$ vector with the maximum in absolute value of each column's elements, while *row_min* is a $m \times 1$ vector with the minimum in absolute value of each row's elements and *col_min* is a $n \times 1$ vector with the minimum in absolute value of each column's elements.

The function for the geometric mean scaling technique takes as input the matrix of coefficients of the constraints (matrix A), the vector of coefficients of the objective function (vector c), and the vector of the right-hand side of the constraints (vector b), and returns as output the scaled matrix of coefficients of the constraints (matrix A), the scaled vector of coefficients of the objective function (vector c), the scaled vector of the right-hand side of the constraints (vector b), and the row and column scaling factors (vectors *row_multi* and *col_multi*).

Lines 33–35 include the pre-allocation of vectors *row_max*, *row_min*, and *row_multi*. In the first for-loop (lines 36–56), the row scaling factors are calculated as the inverse of the square root of the product of the maximum and the minimum element of the same row (line 50). Then, matrix *A* and vector *b* are updated (lines 52–54). Next, lines 58–60 include the pre-allocation of vectors *col_max*, *col_min*, and *col_multi*. Finally, in the second for-loop (lines 61–82), the column scaling factors are calculated as the inverse of the square root of the product of the maximum and the minimum element of the same column (line 75). Then, matrix *A* and vector *c* are updated (lines 78–80).

```
1.    function [A, c, b, row_multi, col_multi] = ...
2.        geometricMean(A, c, b)
3.    % Filename: geometricMean.m
4.    % Description: the function is an implementation of the
5.    % geometric mean scaling technique
6.    % Authors: Ploskas, N., & Samaras, N.
7.    %
8.    % Syntax: [A, c, b, row_multi, col_multi] = ...
9.    %   geometricMean(A, c, b)
10.   %
11.   % Input:
12.   % -- A: matrix of coefficients of the constraints
13.   %    (size m x n)
14.   % -- c: vector of coefficients of the objective function
15.   %    (size n x 1)
16.   % -- b: vector of the right-hand side of the constraints
17.   %    (size m x 1)
18.   %
19.   % Output:
20.   % -- A: scaled matrix of coefficients of the constraints
21.   %    (size m x n)
22.   % -- c: scaled vector of coefficients of the objective
23.   %    function (size n x 1)
24.   % -- b: scaled vector of the right-hand side of the
25.   %    constraints (size m x 1)
26.   % -- row_multi: vector of the row scaling factors
27.   %    (size m x 1)
28.   % -- col_multi: vector of the column scaling factors
29.   %    (size n x 1)
30.
31.   [m, n] = size(A); % size of matrix A
32.   % first apply row scaling
33.   row_max = zeros(m, 1);
34.   row_min = zeros(m, 1);
35.   row_multi = zeros(m, 1);
36.   for i = 1:m
37.       % find the indices of the nonzero elements of the
38.       % specific row
39.       ind = find(A(i, :));
40.       % if the specific row contains at least one nonzero
41.       % element
42.       if ~isempty(ind)
```

```
43.              % find the maximum in absolute value of the nonzero
44.              % elements of the specific row
45.              row_max(i) = max(max(abs(A(i, ind))));
46.              % find the minimum in absolute value of the nonzero
47.              % elements of the specific row
48.              row_min(i) = min(min(abs(A(i, ind))));
49.              % calculate the specific row scaling factor
50.              row_multi(i) = 1 / (sqrt(row_max(i) * row_min(i)));
51.              % scale the elements of the specific row of matrix A
52.              A(i, :) = A(i, :) * row_multi(i);
53.              % scale the elements of vector b
54.              b(i) = b(i) * row_multi(i);
55.      end
56.  end
57.  % then apply column scaling
58.  col_max = zeros(n, 1);
59.  col_min = zeros(n, 1);
60.  col_multi = zeros(n, 1);
61.  for j = 1:n
62.      % find the indices of the nonzero elements of the
63.      % specific column
64.      ind = find(A(:, j));
65.      % if the specific column contains at least one nonzero
66.      % element
67.      if ~isempty(ind)
68.              % find the maximum in absolute value of the nonzero
69.              % elements of the specific column
70.              col_max(j) = max(max(abs(A(ind, j))));
71.              % find the minimum in absolute value of the nonzero
72.              % elements of the specific column
73.              col_min(j) = min(min(abs(A(ind, j))));
74.              % calculate the specific column scaling factor
75.              col_multi(j) = 1 / (sqrt(col_max(j) * col_min(j)));
76.              % scale the elements of the specific column of
77.              % matrix A
78.              A(:, j) = A(:, j) * col_multi(j);
79.              % scale the elements of vector c
80.              c(j) = c(j) * col_multi(j);
81.      end
82.  end
83.  end
```

5.9 IBM MPSX

5.9.1 Mathematical Formulation

IBM MPSX scaling method was proposed by Benichou et al. [1] and was later adopted by IBM, which used this method in IBM's MPSX linear optimization solver. This method combines the geometric mean and the equilibration scaling technique. Initially, geometric mean is performed four times or until the relation (5.20) holds true.

$$\frac{1}{|\overline{Z}|} \left(\sum_{(i,j)\in\overline{Z}} (X_{ij}^k)^2 - \left(\sum_{(i,j)\in\overline{Z}} (X_{ij}^k)^2 \right)^2 / |\overline{Z}| \right) < \varepsilon \qquad (5.20)$$

where $|\overline{Z}|$ is the cardinality number of the nonzero elements of X^k and ε is a tolerance (a small positive number), which often is set below ten. Then, the equilibration scaling technique is applied.

5.9.2 Numerical Example

This subsection demonstrates the IBM MPSX scaling technique through an illustrative example. The LP problem that will be scaled is the following:

$$\begin{aligned}
\min z = 7/2x_1 & & + 453x_3 + & 6x_4 \\
\text{s.t.} \quad 956x_1 & & + x_3 + 258/5x_4 = & 4 \\
5/2x_1 + & 4x_2 + 13/2x_3 + 149/5x_4 = & 7/2 \\
x_1 + 3/2x_2 & & + 67/10x_4 = & 55 \\
x_j \geq 0, \quad (j = 1, 2, 3, 4)
\end{aligned}$$

In matrix notation the above LP problem is written as follows:

$$A = \begin{bmatrix} 956 & 0 & 1 & 258/5 \\ 5/2 & 4 & 13/2 & 149/5 \\ 1 & 3/2 & 0 & 67/10 \end{bmatrix}, c = \begin{bmatrix} 7/2 \\ 0 \\ 453 \\ 6 \end{bmatrix}, b = \begin{bmatrix} 4 \\ 7/2 \\ 55 \end{bmatrix}, Eqin = \begin{bmatrix} 0 \\ 0 \\ 0 \end{bmatrix}$$

According to IBM MPSX scaling technique, the geometric mean scaling technique is performed four times or until the relation (5.20) holds true. So, let's start with applying the geometric mean scaling technique for the first time. Initially, the row scaling factors are calculated according to Equation (5.18):

$$r = \begin{bmatrix} (956 \times 1)^{-1/2} \\ (149/5 \times 5/2)^{-1/2} \\ (67/10 \times 1)^{-1/2} \end{bmatrix} = \begin{bmatrix} 161/4,978 \\ 217/1,873 \\ 294/761 \end{bmatrix}$$

Then, matrix A and vector b are scaled:

$$A = \begin{bmatrix} 4,978/161 & 0 & 161/4,978 & 1,013/607 \\ 769/2,655 & 868/1,873 & 552/733 & 2,655/769 \\ 294/761 & 441/761 & 0 & 761/294 \end{bmatrix}, b = \begin{bmatrix} 322/2,489 \\ 1,637/4,037 \\ 3,251/153 \end{bmatrix}$$

Next, the column scaling factors are calculated according to Equation (5.19):

$$s = \begin{bmatrix} (4,978/161 \times 769/2,655)^{-1/2} \\ (441/761 \times 868/1,873)^{-1/2} \\ (552/733 \times 161/4,978)^{-1/2} \\ (2,655/769 \times 1,013/607)^{-1/2} \end{bmatrix} = \begin{bmatrix} 539/1,613 \\ 2,771/1,436 \\ 2,185/341 \\ 532/1,277 \end{bmatrix}$$

Then, matrix A and vector c are scaled:

$$A = \begin{bmatrix} 2.552/247 & 0 & 1,489/7,185 & 527/758 \\ 247/2,552 & 296/331 & 7,185/1,489 & 758/527 \\ 256/1,983 & 331/296 & 0 & 991/919 \end{bmatrix}, c = \begin{bmatrix} 1,145/979 \\ 0 \\ 66,761/23 \\ 3,182/1,273 \end{bmatrix}$$

The row and column scaling factors, the scaled matrix A, and vectors b and c are the following:

$$r = \begin{bmatrix} 161/4,978 \\ 217/1,873 \\ 294/761 \end{bmatrix}, s = \begin{bmatrix} 539/1,613 \\ 2,771/1,436 \\ 2,185/341 \\ 532/1,277 \end{bmatrix}$$

$$A = \begin{bmatrix} 2.552/247 & 0 & 1,489/7,185 & 527/758 \\ 247/2,552 & 296/331 & 7,185/1,489 & 758/527 \\ 256/1,983 & 331/296 & 0 & 991/919 \end{bmatrix}, c = \begin{bmatrix} 1,145/979 \\ 0 \\ 66,761/23 \\ 3,182/1,273 \end{bmatrix},$$

$$b = \begin{bmatrix} 322/2,489 \\ 1,637/4,037 \\ 3,251/153 \end{bmatrix}$$

Then, we check if relation (5.20) holds true.

$$\frac{1}{|\bar{Z}|} \left(\sum_{(i,j)\in\bar{Z}} \left(X_{ij}^k\right)^2 - \left(\sum_{(i,j)\in\bar{Z}} \left(X_{ij}^k\right)^2 \right)^2 / |\bar{Z}| \right) =$$

$$\frac{1}{10} (5,163/38 - (9,965/23/10)) = 10,411/1,125 = 9.2542 < 10$$

The relation (5.20) holds true (9.2542 < 10), so the equilibration scaling technique should be applied. Initially, the row scaling factors are calculated:

$$r = \begin{bmatrix} 1/(2,552/247) \\ 1/(7,185/1,489) \\ 1/(331/296) \end{bmatrix} = \begin{bmatrix} 247/2,552 \\ 1,489/7,185 \\ 296/331 \end{bmatrix}$$

Then, matrix A and vector b are scaled:

$$A = \begin{bmatrix} 1 & 0 & 104/5,185 & 273/4,057 \\ 104/5,185 & 346/1,867 & 1 & 2,370/7,951 \\ 358/3,101 & 1 & 0 & 946/981 \end{bmatrix}, b = \begin{bmatrix} 81/6,469 \\ 759/9,032 \\ 12,161/640 \end{bmatrix}$$

Next, the column scaling factors are calculated:

$$s = \begin{bmatrix} 1/1 \\ 1/1 \\ 1/1 \\ 1/(946/981) \end{bmatrix} = \begin{bmatrix} 1 \\ 1 \\ 1 \\ 981/946 \end{bmatrix}$$

Then, matrix A and vector c are scaled:

$$A = \begin{bmatrix} 1 & 0 & 104/5,185 & 487/6,979 \\ 104/5,185 & 346/1,867 & 1 & 421/1,362 \\ 358/3,101 & 1 & 0 & 1 \end{bmatrix}, c = \begin{bmatrix} 1,145/979 \\ 0 \\ 66,761/23 \\ 2,097/809 \end{bmatrix}$$

Finally, the row and column scaling factors, the scaled matrix A, and vectors b and c are the following:

$$r = \begin{bmatrix} 247/2,552 \\ 1,489/7,185 \\ 296/331 \end{bmatrix}, s = \begin{bmatrix} 1 \\ 1 \\ 1 \\ 981/946 \end{bmatrix}$$

$$A = \begin{bmatrix} 1 & 0 & 104/5,185 & 487/6,979 \\ 104/5,185 & 346/1,867 & 1 & 421/1,362 \\ 358/3,101 & 1 & 0 & 1 \end{bmatrix}, c = \begin{bmatrix} 1,145/979 \\ 0 \\ 66,761/23 \\ 2,097/809 \end{bmatrix},$$

$$b = \begin{bmatrix} 81/6,469 \\ 759/9,032 \\ 12,161/640 \end{bmatrix}$$

The final scaled LP problem is the following:

$$\min z = 1,145/979x_1 \qquad\qquad + 66,761/23x_3 + 2,097/809x_4$$

$$\text{s.t.} \qquad\qquad x_1 \qquad\quad + 104/5,185x_3 + 487/6,979x_4 = \quad 81/6,469$$
$$104/5,185x_1 + 346/1,867x_2 + \qquad x_3 + 421/1,362x_4 = \quad 759/9,032$$
$$358/3,101x_1 + \qquad x_2 \qquad\qquad + \qquad x_4 = 12,161/640$$
$$x_j \geq 0, \quad (j = 1,2,3,4)$$

5.9.3   Implementation in MATLAB

This subsection presents the implementation in MATLAB of the IBM MPSX scaling technique (filename: ibmmpsx.m). Some necessary notations should be introduced before the presentation of the implementation of the geometric mean scaling technique. Let *row_multigm* be a $m \times 1$ vector with row scaling factors after applying geometric mean scaling technique and *col_multigm* be a $n \times 1$ vector with column scaling factors after applying geometric mean scaling technique, while *row_multieq* is a $m \times 1$ vector with row scaling factors after applying equilibration scaling technique and *col_multieq* is a $n \times 1$ vector with column scaling factors after applying equilibration scaling technique. Let *max_A* be the maximum element of matrix A and *min_A* be the minimum element of matrix A. Let *tole* be the tolerance ε in Equation (5.20) and *tol* a very small positive number.

The function for the IBM MPSX scaling technique takes as input the matrix of coefficients of the constraints (matrix A), the vector of coefficients of the objective function (vector c), and the vector of the right-hand side of the constraints (vector b), and returns as output the scaled matrix of coefficients of the constraints (matrix A), the scaled vector of coefficients of the objective function (vector c), the scaled vector of the right-hand side of the constraints (vector b), the row and column scaling factors after applying the geometric mean scaling technique (vectors *row_multigm* and *col_multigm*), and the row and column scaling factors after applying the equilibration scaling technique (vectors *row_multieq* and *col_multieq*).

Lines 36–39 include the pre-allocation of vectors *row_multigm*, *col_multigm*, *row_multieq*, and *col_multieq*. If the difference in absolute value of the maximum to the minimum element of matrix A is less than a very small positive number (*tol*), then there is no need for scaling (line 47). Otherwise, the geometric mean (lines 51–52) is performed four times (line 50) or until the relation (5.20) holds true (lines 54–55). Finally, the equilibration scaling technique is performed (lines 61–62).

```
1.    function [A, c, b, row_multigm, col_multigm, ...
2.       row_multieq, col_multieq] = ibmmpsx(A, c, b)
3.    % Filename: ibmmpsx.m
4.    % Description: the function is an implementation of the IBM
5.    % MPSX scaling technique
6.    % Authors: Ploskas, N., & Samaras, N.
```

```
7.    %
8.    % Syntax: [A, c, b, row_multigm, col_multigm, ...
9.    %   row_multieq, col_multieq] = ibmmpsx(A, c, b)
10.   %
11.   % Input:
12.   % -- A: matrix of coefficients of the constraints
13.   %    (size m x n)
14.   % -- c: vector of coefficients of the objective function
15.   %    (size n x 1)
16.   % -- b: vector of the right-hand side of the constraints
17.   %    (size m x 1)
18.   %
19.   % Output:
20.   % -- A: scaled matrix of coefficients of the constraints
21.   %    (size m x n)
22.   % -- c: scaled vector of coefficients of the objective
23.   %    function (size n x 1)
24.   % -- b: scaled vector of the right-hand side of the
25.   %    constraints (size m x 1)
26.   % -- row_multigm: vector of the row scaling factors from
27.   %    geometric mean scaling technique (size m x 1)
28.   % -- col_multigm: vector of the column scaling factors from
29.   %    geometric mean scaling technique (size n x 1)
30.   % -- row_multieq: vector of the row scaling factors from
31.   %    equilibration scaling technique (size m x 1)
32.   % -- col_multieq: vector of the column scaling factors from
33.   %    equilibration scaling technique (size n x 1)
34.
35.   [m, n] = size(A); % size of matrix A
36.   row_multigm = zeros(m, 1);
37.   col_multigm = zeros(n, 1);
38.   row_multieq = zeros(m, 1);
39.   col_multieq = zeros(n, 1);
40.   max_A = max(max(A)); % find the maximum element of matrix A
41.   min_A = min(min(A)); % find the minimum element of matrix A
42.   tol = max(m, n) * eps * norm(A, 'inf');
43.   tole = 10; % initialize tolerance
44.   % if the absolute value of the difference of the maximum
45.   % element and the minimum element are below tol, then do not
46.   % scale
47.   if abs(max_A - min_A) > tol
48.       counter = 0;
49.       % geometric mean is performed four times
50.       while(counter < 4)
51.           [A, c, b, row_multigm, col_multigm] = ...
52.               geometricMean(A, c, b);
53.           % or until the following relation holds true
54.           if((1 / nnz(A)) * (sum(sum(power(A, 2))) - ...
55.                   power(sum(sum(abs(A))), 2) / nnz(A)) < tole)
56.               break;
57.           end
58.           counter = counter + 1;
59.       end
60.       % then equilibration scaling technique is performed
```

```
61.        [A, c, b, row_multieq, col_multieq] = ...
62.            equilibration(A, c, b);
63.    end
64.    end
```

5.10 L_p-Norm

5.10.1   Mathematical Formulation

The L_p-norm scaling model is formulated as shown in Equation (5.21):

$$\min_{(r,s>0)} \left(\sum_{(i,j) \in \overline{Z}} \left| \log \left(A_{ij} r_i s_j \right) \right|^p \right)^{1/p} \tag{5.21}$$

where p is a positive integer and $|\overline{Z}|$ is the cardinality number of the nonzero elements of the constraint matrix A.

We focus now our attention on the cases $p = 1, 2$, and ∞. For the case $p = 1$, the model is formulated as shown in Equation (5.22):

$$\min_{(r,s>0)} \sum_{(i,j) \in \overline{Z}} \left| \log \left(A_{ij} r_i s_j \right) \right| \tag{5.22}$$

We divide each row and column by the median of the absolute value of the nonzero elements. The row scaling factors are presented in Equation (5.23):

$$r_i^{k+1} = 1/\text{median} \left\{ X_{ij}^k \middle| j \in N_i \right\} \tag{5.23}$$

Similarly, the column scaling factors are presented in Equation (5.24):

$$s_j^{k+1} = 1/\text{median} \left\{ X_{ij}^{k+1/2} \middle| i \in M_j \right\} \tag{5.24}$$

For the case $p = 2$, the model is stated as shown in Equation (5.25):

$$\min_{(r,s>0)} \left(\sum_{(i,j) \in \overline{Z}} \left| \log \left(A_{ij} r_i s_j \right) \right|^2 \right)^{1/2} \tag{5.25}$$

The row scaling factors are calculated as shown in Equation (5.26):

$$r_i^{k+1} = 1/\prod_{j \in N_i} \left(X_{ij}^k \right)^{1/n_i} \tag{5.26}$$

Similarly, the column scaling factors are presented in Equation (5.27):

$$s_j^{k+1} = 1/ \prod_{i \in M_j} \left(x_{ij}^{k+1/2} \right)^{1/m_j} \tag{5.27}$$

Finally, for the case $p = \infty$, the model is equivalent to the de Buchet method (case $p = \infty$) and geometric mean scaling techniques.

5.10.2 Numerical Example

This subsection demonstrates the L_p-norm scaling technique through an illustrative example. The LP problem that will be scaled is the following:

$$
\begin{aligned}
\min z = {}& 7/2x_1 && + 453x_3 + && 6x_4 \\
\text{s.t.} \quad & 956x_1 && + \quad x_3 + 258/5x_4 = && 4 \\
& 5/2x_1 + \quad 4x_2 + {}& 13/2x_3 + 149/5x_4 = && 7/2 \\
& x_1 + 3/2x_2 && + 67/10x_4 = && 55 \\
& x_j \geq 0, \quad (j = 1, 2, 3, 4)
\end{aligned}
$$

In matrix notation the above LP problem is written as follows:

$$
A = \begin{bmatrix} 956 & 0 & 1 & 258/5 \\ 5/2 & 4 & 13/2 & 149/5 \\ 1 & 3/2 & 0 & 67/10 \end{bmatrix}, c = \begin{bmatrix} 7/2 \\ 0 \\ 453 \\ 6 \end{bmatrix}, b = \begin{bmatrix} 4 \\ 7/2 \\ 55 \end{bmatrix}, Eqin = \begin{bmatrix} 0 \\ 0 \\ 0 \end{bmatrix}
$$

First of all, the L_p-norm scaling technique for the case $p = 1$ will be performed. Initially, the row scaling factors are calculated according to Equation (5.23):

$$
r = \begin{bmatrix} 1/(258/5) \\ 1/((4 + 13/2)/2) \\ 1/(3/2) \end{bmatrix} = \begin{bmatrix} 5/258 \\ 4/21 \\ 2/3 \end{bmatrix}
$$

Then, matrix A and vector b are scaled:

$$
A = \begin{bmatrix} 2,390/129 & 0 & 5/258 & 1 \\ 10/21 & 16/21 & 26/21 & 596/105 \\ 2/3 & 1 & 0 & 67/15 \end{bmatrix}, b = \begin{bmatrix} 10/129 \\ 2/3 \\ 110/3 \end{bmatrix}
$$

Next, the column scaling factors are calculated according to Equation (5.24):

$$s = \begin{bmatrix} 1/(2/3) \\ 1/((16/21+1)/2) \\ 1/((5/258+26/21)/2) \\ 1/(67/15) \end{bmatrix} = \begin{bmatrix} 3/2 \\ 42/37 \\ 1,204/757 \\ 15/67 \end{bmatrix}$$

Then, matrix A and vector c are scaled:

$$A = \begin{bmatrix} 1,195/43 & 0 & 70/2,271 & 15/67 \\ 5/7 & 32/37 & 575/292 & 596/469 \\ 1 & 42/37 & 0 & 1 \end{bmatrix}, c = \begin{bmatrix} 21/4 \\ 0 \\ 42.509/59 \\ 90/67 \end{bmatrix}$$

Finally, the row and column scaling factors, the scaled matrix A, and vectors b and c are the following:

$$r = \begin{bmatrix} 5/258 \\ 4/21 \\ 2/3 \end{bmatrix}, s = \begin{bmatrix} 3/2 \\ 42/37 \\ 1,204/757 \\ 15/67 \end{bmatrix}$$

$$A = \begin{bmatrix} 1,195/43 & 0 & 70/2,271 & 15/67 \\ 5/7 & 32/37 & 575/292 & 596/469 \\ 1 & 42/37 & 0 & 1 \end{bmatrix}, c = \begin{bmatrix} 21/4 \\ 0 \\ 42,509/59 \\ 90/67 \end{bmatrix}, b = \begin{bmatrix} 10/129 \\ 2/3 \\ 110/3 \end{bmatrix}$$

The final scaled LP problem is the following:

$$
\begin{aligned}
\min z = \quad & 21/4x_1 & & + 42,509/59x_3 + & 90/67x_4 \\
\text{s.t.} \quad & 1,195/43x_1 & & + 70/2,271x_3 + & 15/67x_4 = 10/129 \\
& 5/7x_1 + 32/37x_2 + & & 575/292x_3 + 596/469x_4 = & 2/3 \\
& x_1 + 42/37x_2 & & + & x_4 = 110/3 \\
x_j \geq 0, \quad & (j = 1,2,3,4)
\end{aligned}
$$

Next, the L_p-norm scaling technique for the case $p = 2$ is applied to the same example. Initially, the row scaling factors are calculated according to Equation (5.26):

$$r = \begin{bmatrix} (956 \times 1 \times 258/5)^{-1/3} \\ (5/2 \times 4 \times 13/2 \times 149/5)^{-1/4} \\ (1 \times 3/2 \times 67/10)^{-1/3} \end{bmatrix} = \begin{bmatrix} 323/11,846 \\ 604/4,007 \\ 424/915 \end{bmatrix}$$

Then, matrix A and vector b are scaled:

$$A = \begin{bmatrix} 8,967/344 & 0 & 323/11,846 & 809/575 \\ 563/1,494 & 2,416/4,007 & 824/841 & 557/124 \\ 424/915 & 212/305 & 0 & 2,313/745 \end{bmatrix}, b = \begin{bmatrix} 243/2,228 \\ 947/1,795 \\ 4,664/183 \end{bmatrix}$$

Next, the column scaling factors are calculated according to Equation (5.27):

$$s = \begin{bmatrix} (8,967/344 \times 563/1,494 \times 424/915)^{-1/3} \\ (2,416/4,007 \times 212/305)^{-1/2} \\ (323/11,846 \times 824/841)^{-1/2} \\ (809/575 \times 557/124 \times 2,313/745)^{-1/3} \end{bmatrix} = \begin{bmatrix} 604/1,001 \\ 2,039/1,320 \\ 2,227/364 \\ 1,387/3,741 \end{bmatrix}$$

Then, matrix A and vector c are scaled:

$$A = \begin{bmatrix} 5,159/328 & 0 & 361/2,164 & 675/1,294 \\ 1,111/4,886 & 1,642/1,763 & 2,164/361 & 3,106/1,865 \\ 654/2,339 & 1,763/1,642 & 0 & 1,059/920 \end{bmatrix}, c = \begin{bmatrix} 302/143 \\ 0 \\ 91,460/33 \\ 961/432 \end{bmatrix}$$

Finally, the row and column scaling factors, the scaled matrix A, and vectors b and c are the following:

$$r = \begin{bmatrix} 323/11,846 \\ 604/4,007 \\ 424/915 \end{bmatrix}, s = \begin{bmatrix} 604/1,001 \\ 2,039/1,320 \\ 2,227/364 \\ 1,387/3,741 \end{bmatrix}$$

$$A = \begin{bmatrix} 5,159/328 & 0 & 361/2,164 & 675/1,294 \\ 1,111/4,886 & 1,642/1,763 & 2,164/361 & 3,106/1,865 \\ 654/2,339 & 1,763/1,642 & 0 & 1,059/920 \end{bmatrix}, c = \begin{bmatrix} 302/143 \\ 0 \\ 91,460/33 \\ 961/432 \end{bmatrix},$$

$$b = \begin{bmatrix} 243/2,228 \\ 947/1,795 \\ 4,664/183 \end{bmatrix}$$

The final scaled LP problem is the following:

$$
\begin{aligned}
\min z = \quad & 302/143x_1 & & + 91,460/33x_3 + & 961/432x_4 \\
\text{s.t.} \quad & 5,159/328x_1 & & + 361/2,164x_3 + & 675/1,294x_4 = 243/2,228 \\
& 1,111/4,886x_1 + 1,642/1,763x_2 & & + 2,164/361x_3 + 3,106/1,865x_4 = 947/1,795 \\
& 654/2,339x_1 + 1,763/1,642x_2 & & + & 1,059/920x_4 = 4,664/183 \\
& x_j \geq 0, \quad (j = 1,2,3,4)
\end{aligned}
$$

Finally, the L_p-norm scaling technique for the case $p = \infty$ is applied to the same example. Initially, the row scaling factors are calculated according to Equation (5.18):

$$r = \begin{bmatrix} (956 \times 1)^{-1/2} \\ (149/5 \times 5/2)^{-1/2} \\ (67/10 \times 1)^{-1/2} \end{bmatrix} = \begin{bmatrix} 161/4,978 \\ 217/1,873 \\ 294/761 \end{bmatrix}$$

Then, matrix A and vector b are scaled:

$$A = \begin{bmatrix} 4,978/161 & 0 & 161/4,978 & 1,013/607 \\ 769/2,655 & 868/1,873 & 552/733 & 2,655/769 \\ 294/761 & 441/761 & 0 & 761/294 \end{bmatrix}, b = \begin{bmatrix} 322/2,489 \\ 1,637/4,037 \\ 3,251/153 \end{bmatrix}$$

Next, the column scaling factors are calculated according to Equation (5.19):

$$s = \begin{bmatrix} (4,978/161 \times 769/2,655)^{-1/2} \\ (441/761 \times 868/1,873)^{-1/2} \\ (552/733 \times 161/4,978)^{-1/2} \\ (2,655/769 \times 1,013/607)^{-1/2} \end{bmatrix} = \begin{bmatrix} 539/1,613 \\ 2,771/1,436 \\ 2,185/341 \\ 532/1,277 \end{bmatrix}$$

Then, matrix A and vector c are scaled:

$$A = \begin{bmatrix} 2,552/247 & 0 & 1,489/7,185 & 527/758 \\ 247/2,552 & 296/331 & 7,185/1,489 & 758/527 \\ 256/1,983 & 331/296 & 0 & 991/919 \end{bmatrix}, c = \begin{bmatrix} 1,145/979 \\ 0 \\ 6,6761/23 \\ 3,182/1,273 \end{bmatrix}$$

Finally, the row and column scaling factors, the scaled matrix A, and vectors b and c are the following:

$$r = \begin{bmatrix} 161/4,978 \\ 217/1,873 \\ 294/761 \end{bmatrix}, s = \begin{bmatrix} 539/1,613 \\ 2,771/1,436 \\ 2,185/341 \\ 532/1,277 \end{bmatrix}$$

$$A = \begin{bmatrix} 2,552/247 & 0 & 1,489/7,185 & 527/758 \\ 247/2,552 & 296/331 & 7,185/1,489 & 758/527 \\ 256/1,983 & 331/296 & 0 & 991/919 \end{bmatrix}, c = \begin{bmatrix} 1,145/979 \\ 0 \\ 6,6761/23 \\ 3,182/1,273 \end{bmatrix},$$

$$b = \begin{bmatrix} 322/2,489 \\ 1,637/4,037 \\ 3,251/153 \end{bmatrix}$$

The final scaled LP problem is the following:

$$
\begin{aligned}
\min z = \; & 1,145/979x_1 & & + & 66,761/23x_3 & + 3,182/1,273x_4 & \\
\text{s.t.} \quad & 2,552/247x_1 & & + 1,489/7,185x_3 \; + & 527/758x_4 & = & 322/2,489 \\
& 247/2,552x_1 + 296/331x_2 & + 7,185/1,489x_3 \; + & & 758/527x_4 & = & 1,637/4,037 \\
& 256/1,983x_1 + 331/296x_2 & & + & 991/919x_4 & = & 3,251/153 \\
x_j \geq 0, \quad & (j = 1,2,3,4) & & & & &
\end{aligned}
$$

5.10.3 Implementation in MATLAB

This subsection presents the implementation in MATLAB of the L_p-norm scaling technique (filenames: lpnorm1.m, lpnorm2.m and lpnorminf.m). Some necessary notations should be introduced before the presentation of the implementation of the L_p-norm scaling technique. Let *row_multi* be a $m \times 1$ vector with row scaling factors and *col_multi* be a $n \times 1$ vector with column scaling factors. Let *row_med* be a $m \times 1$ vector with the median in absolute value of each row's elements and *col_med* be a $n \times 1$ vector with the median in absolute value of each column's elements, while *row_prod* is a $m \times 1$ vector with the inverse of the product of each row's elements is absolute value to the power of the negative inverse of the number of each row's nonzero elements and *col_pow* is a $n \times 1$ vector with the inverse of the product of each column's elements to the power of the negative inverse of the number of each column's nonzero elements. Let *row_max* be a $m \times 1$ vector with the maximum in absolute value of each row's elements and *col_max* be a $n \times 1$ vector with the maximum in absolute value of each column's elements, while *row_min* is a $m \times 1$ vector with the minimum in absolute value of each row's elements and *col_min* is a $n \times 1$ vector with the minimum in absolute value of each column's elements.

The function for the L_p-norm scaling technique for the case $p = 1$ takes as input the matrix of coefficients of the constraints (matrix A), the vector of coefficients of the objective function (vector c), and the vector of the right-hand side of the constraints (vector b), and returns as output the scaled matrix of coefficients of the constraints (matrix A), the scaled vector of coefficients of the objective function (vector c), the scaled vector of the right-hand side of the constraints (vector b), and the row and column scaling factors (vectors *row_multi* and *col_multi*).

Lines 31–32 include the pre-allocation of vectors *row_med* and *row_multi*. In the first for-loop (lines 33–51), the row scaling factors are calculated as the inverse of the median of the nonzero elements in absolute value of each row (line 44). Then, matrix A and vector b are updated (lines 47–49). Next, lines 53–54 include the pre-allocation of vectors *col_med* and *col_multi*. Finally, in the second for-loop (lines 55–73), the column scaling factors are calculated as the inverse of the median of the nonzero elements in absolute value of each column (line 66). Then, matrix A and vector c are updated (lines 69–71).

```
1.    function [A, c, b, row_multi, col_multi] = lpnorm1(A, c, b)
2.    % Filename: lpnorm1.m
3.    % Description: the function is an implementation of the
4.    % Lp-norm for the case p = 1 scaling technique
5.    % Authors: Ploskas, N., & Samaras, N.
6.    %
7.    % Syntax: [A, c, b, row_multi, col_multi] = lpnorm1(A, c, b)
8.    %
9.    % Input:
10.   % -- A: matrix of coefficients of the constraints
11.   %    (size m x n)
12.   % -- c: vector of coefficients of the objective function
13.   %    (size n x 1)
14.   % -- b: vector of the right-hand side of the constraints
15.   %    (size m x 1)
16.   %
17.   % Output:
18.   % -- A: scaled matrix of coefficients of the constraints
19.   %    (size m x n)
20.   % -- c: scaled vector of coefficients of the objective
21.   %    function (size n x 1)
22.   % -- b: scaled vector of the right-hand side of the
23.   %    constraints (size m x 1)
24.   % -- row_multi: vector of the row scaling factors
25.   %    (size m x 1)
26.   % -- col_multi: vector of the column scaling factors
27.   %    (size n x 1)
28.
29.   [m, n] = size(A); % size of matrix A
30.   % first apply row scaling
31.   row_med = zeros(m, 1);
32.   row_multi = zeros(m, 1);
33.   for i = 1:m
34.       % find the indices of the nonzero elements of the
35.       % specific row
36.       ind = find(A(i, :));
37.       % if the specific row contains at least one nonzero
38.       % element
39.       if ~isempty(ind)
40.           % find the median in absolute value of the nonzero
41.           % elements of the specific row
42.           row_med(i) = median(abs(A(i, ind)));
43.           % calculate the specific row scaling factor
44.           row_multi(i) = 1 / row_med(i);
45.           % scale the elements of the specific row of
46.           % matrix A
47.           A(i, :) = A(i, :) * row_multi(i);
48.           % scale the elements of vector b
49.           b(i) = b(i) * row_multi(i);
50.       end
51.   end
52.   % then apply column scaling
53.   col_med = zeros(n, 1);
54.   col_multi = zeros(n, 1);
```

```
55.   for j = 1:n
56.       % find the indices of the nonzero elements of the
57.       % specific column
58.       ind = find(A(:, j));
59.       % if the specific column contains at least one
60.       % nonzero element
61.       if ~isempty(ind)
62.           % find the median in absolute value of the nonzero
63.           % elements of the specific column
64.           col_med(j) = median(abs(A(ind, j)));
65.           % calculate the specific column scaling factor
66.           col_multi(j) = 1 / col_med(j);
67.           % scale the elements of the specific column of
68.           % matrix A
69.           A(:, j) = A(:, j) * col_multi(j);
70.           % scale the elemens of vector c
71.           c(j) = c(j) * col_multi(j);
72.       end
73.   end
74.   end
```

The function for the L_p-norm scaling technique for the case $p = 2$ takes as input the matrix of coefficients of the constraints (matrix A), the vector of coefficients of the objective function (vector c), and the vector of the right-hand side of the constraints (vector b), and returns as output the scaled matrix of coefficients of the constraints (matrix A), the scaled vector of coefficients of the objective function (vector c), the scaled vector of the right-hand side of the constraints (vector b), and the row and column scaling factors (vectors *row_multi* and *col_multi*).

Lines 31–32 include the pre-allocation of vectors *row_prod* and *row_multi*. In the first for-loop (lines 33–52), the row scaling factors are calculated as the inverse of the product of each row's elements is absolute value to the power of the negative inverse of the number of each row's nonzero elements (line 46). Then, matrix A and vector b are updated (lines 48–50). Next, lines 54–55 include the pre-allocation of vectors *col_prod* and *col_multi*. Finally, in the second for-loop (lines 56–75), the column scaling factors are calculated as the inverse of the product of each column's elements is absolute value to the power of the negative inverse of the number of each column's nonzero elements (line 68). Then, matrix A and vector c are updated (lines 71–73).

```
1.    function [A, c, b, row_multi, col_multi] = lpnorm2(A, c, b)
2.    % Filename: lpnorm2.m
3.    % Description: the function is an implementation of the
4.    % Lp-norm for the case p = 2 scaling technique
5.    % Authors: Ploskas, N., & Samaras, N.
6.    %
7.    % Syntax: [A, c, b, row_multi, col_multi] = lpnorm2(A, c, b)
8.    %
9.    % Input:
10.   % -- A: matrix of coefficients of the constraints
11.   %    (size m x n)
12.   % -- c: vector of coefficients of the objective function
```

```
13.  %     (size n x 1)
14.  % -- b: vector of the right-hand side of the constraints
15.  %     (size m x 1)
16.  %
17.  % Output:
18.  % -- A: scaled matrix of coefficients of the constraints
19.  %     (size m x n)
20.  % -- c: scaled vector of coefficients of the objective
21.  %     function (size n x 1)
22.  % -- b: scaled vector of the right-hand side of the
23.  %     constraints (size m x 1)
24.  % -- row_multi: vector of the row scaling factors
25.  %     (size m x 1)
26.  % -- col_multi: vector of the column scaling factors
27.  %     (size n x 1)
28.
29.  [m, n] = size(A); % size of matrix A
30.  % first apply row scaling
31.  row_prod = zeros(m, 1);
32.  row_multi = zeros(m, 1);
33.  for i = 1:m
34.      % find the indices of the nonzero elements of the
35.      % specific row
36.      ind = find(A(i, :));
37.      % if the specific row contains at least one nonzero
38.      % element
39.      if ~isempty(ind)
40.          % find the inverse of the product in absolute
41.          % value of the nonzero elements of the specific
42.          % row
43.          row_prod(i) = 1 / prod(abs(A(i, ind)))^ ...
44.              (-1 / length(ind));
45.          % calculate the specific row scaling factor
46.          row_multi(i) = 1 / row_prod(i);
47.          % scale the elements of the specific row of matrix A
48.          A(i, :) = A(i, :) * row_multi(i);
49.          % scale the elements of vector b
50.          b(i) = b(i) * row_multi(i);
51.      end
52.  end
53.  % then apply column scaling
54.  col_prod = zeros(n, 1);
55.  col_multi = zeros(n, 1);
56.  for j = 1:n
57.      % find the indices of the nonzero elements of the
58.      % specific row
59.      ind = find(A(:, j));
60.      % if the specific column contains at least one nonzero
61.      % element
62.      if ~isempty(ind)
63.          % find the inverse of the product in absolute value
64.          % of the nonzero elements of the specific column
65.          col_prod(j) = 1 / prod(abs(A(ind, j)))^ ...
66.              (-1 / length(ind));
```

```
67.            % calculate the specific column scaling factor
68.            col_multi(j) = 1 / col_prod(j);
69.            % scale the elements of the specific column of
70.            % matrix A
71.            A(:, j) = A(:, j) * col_multi(j);
72.            % scale the elements of vector c
73.            c(j) = c(j) * col_multi(j);
74.      end
75.  end
76.  end
```

The function for the L_p-norm scaling technique for the case $p = \infty$ takes as input the matrix of coefficients of the constraints (matrix A), the vector of coefficients of the objective function (vector c), and the vector of the right-hand side of the constraints (vector b), and returns as output the scaled matrix of coefficients of the constraints (matrix A), the scaled vector of coefficients of the objective function (vector c), the scaled vector of the right-hand side of the constraints (vector b), and the row and column scaling factors (vectors *row_multi* and *col_multi*).

Lines 33–35 include the pre-allocation of vectors *row_max*, *row_min*, and *row_multi*. In the first for-loop (lines 36–56), the row scaling factors are calculated as the inverse of the square root of the product of the maximum and the minimum element of the same row (line 50). Then, matrix A and vector b are updated (lines 52–54). Next, lines 58–60 include the pre-allocation of vectors *col_max*, *col_min*, and *col_multi*. Finally, in the second for-loop (lines 61–82), the column scaling factors are calculated as the inverse of the square root of the product of the maximum and the minimum element of the same column (line 75). Then, matrix A and vector c are updated (lines 78–80).

```
1.   function [A, c, b, row_multi, col_multi] = ...
2.        lpnorminf(A, c, b)
3.   % Filename: lpnorminf.m
4.   % Description: the function is an implementation of the
5.   % Lp-norm for the case p = inf scaling technique
6.   % Authors: Ploskas, N., & Samaras, N.
7.   %
8.   % Syntax: [A, c, b, row_multi, col_multi] = ...
9.   %    lpnorminf(A, c, b)
10.  %
11.  % Input:
12.  % -- A: matrix of coefficients of the constraints
13.  %    (size m x n)
14.  % -- c: vector of coefficients of the objective function
15.  %    (size n x 1)
16.  % -- b: vector of the right-hand side of the constraints
17.  %    (size m x 1)
18.  %
19.  % Output:
20.  % -- A: scaled matrix of coefficients of the constraints
21.  %    (size m x n)
22.  % -- c: scaled vector of coefficients of the objective
23.  %    function (size n x 1)
```

```
24.   % -- b: scaled vector of the right-hand side of the
25.   %      constraints (size m x 1)
26.   % -- row_multi: vector of the row scaling factors
27.   %      (size m x 1)
28.   % -- col_multi: vector of the column scaling factors
29.   %      (size n x 1)
30.
31.   [m, n] = size(A); % size of matrix A
32.   % first apply row scaling
33.   row_max = zeros(m, 1);
34.   row_min = zeros(m, 1);
35.   row_multi = zeros(m, 1);
36.   for i = 1:m
37.       % find the indices of the nonzero elements of the
38.       % specific row
39.       ind = find(A(i, :));
40.       % if the specific row contains at least one nonzero
41.       % element
42.       if ~isempty(ind)
43.           % find the maximum in absolute value of the nonzero
44.           % elements of the specific row
45.           row_max(i) = max(max(abs(A(i, ind))));
46.           % find the minimum in absolute value of the nonzero
47.           % elements of the specific row
48.           row_min(i) = min(min(abs(A(i, ind))));
49.           % calculate the specific row scaling factor
50.           row_multi(i) = 1 / (sqrt(row_max(i) * row_min(i)));
51.           % scale the elements of the specific row of matrix A
52.           A(i, :) = A(i, :) * row_multi(i);
53.           % scale the elements of vector b
54.           b(i) = b(i) * row_multi(i);
55.       end
56.   end
57.   % then apply column scaling
58.   col_max = zeros(n, 1);
59.   col_min = zeros(n, 1);
60.   col_multi = zeros(n, 1);
61.   for j = 1:n
62.       % find the indices of the nonzero elements of the
63.       % specific column
64.       ind = find(A(:, j));
65.       % if the specific column contains at least one nonzero
66.       % element
67.       if ~isempty(ind)
68.           % find the maximum in absolute value of the nonzero
69.           % elements of the specific column
70.           col_max(j) = max(max(abs(A(ind, j))));
71.           % find the minimum in absolute value of the nonzero
72.           % elements of the specific column
73.           col_min(j) = min(min(abs(A(ind, j))));
74.           % calculate the specific column scaling factor
75.           col_multi(j) = 1 / (sqrt(col_max(j) * col_min(j)));
76.           % scale the elements of the specific column of
77.           % matrix A
```

```
78.                A(:, j) = A(:, j) * col_multi(j);
79.                % scale the elements of vector c
80.                c(j) = c(j) * col_multi(j);
81.        end
82.    end
83.    end
```

5.11 Computational Study

This section presents a computational study. The aim of this computational study
is twofold: (i) compare the execution time of the previously presented scaling
techniques, and (ii) investigate the impact of scaling prior to the application of LP
algorithms. The test set used in this computational study is 19 LPs from the Netlib
(optimal, Kennington, and infeasible LPs) and Mészáros problem sets that were
presented in Section 3.6. Table 5.1 presents the results from the execution of the
implementations of the scaling methods. The following abbreviations are used: (i)
S1—arithmetic mean, (ii) S2—de Buchet for the case $p = 1$, (iii) S3—de Buchet for
the case $p = 2$, (iv) S4—de Buchet for the case $p = A\infty$, (v) S5—entropy, (vi) S6—
equilibration, (vii) S7—geometric mean, (viii) S8—IBM MPSX, (ix) S9—L_p-norm
for the case $p = 1$, (x) S10—L_p-norm for the case $p = 2$, and (xi) S11—L_p-norm
for the case $p = A\infty$. The execution times for de Buchet for the case $p = A\infty$
(S4), geometric mean (S7), and Lp-norm for the case $p = A\infty$ (S11) are grouped
together (S4—S7—S11) since these methods have the exact same implementation,
as already stated in previous sections of this chapter. Moreover, the execution times
for arithmetic mean (S1) and entropy (S5) are also grouped together (S1—S5) for
the same reason. For each instance, we averaged times over 10 runs. All times are
measured in seconds.

Table 5.1 shows that the arithmetic mean, equilibration, and geometric mean
scaling techniques require less execution time to scale the LPs than the other scaling
techniques.

Then, this computational study is extended in order to study the impact of scaling
prior to the application of Mehrotra's Predictor-Corrector interior point method,
the exterior point simplex algorithm, the revised primal simplex algorithm, and
the revised dual simplex algorithm. So, these algorithms are executed with and
without scaling. The implementations used in this computational study are the
implementation of the revised primal simplex algorithm that will be presented in
Section 8.5, the implementation of the revised dual simplex algorithm that will
be presented in Section 9.5, the implementation of the exterior primal simplex
algorithm that will be presented in Section 10.5, and Mehrotra's Predictor-Corrector
interior point method that will be presented in Section 11.6. Three different scaling
techniques have been used: (i) arithmetic mean, (ii) equilibration, and (iii) geometric
mean. These scaling techniques were selected as the best in terms of execution time
(Table 5.1). Tables 5.2, 5.3, 5.4, and 5.5 present the results from the execution

Table 5.1 Total execution time of each scaling technique over the benchmark set

Name	Problem			S1—S5	S2	S3	S4_S7—S11	S6	S8	S9	S10
	Constraints	Variables	Nonzeros								
beaconfd	173	262	3,375	0.0167	0.0199	0.0218	0.0191	0.0167	0.0570	0.0340	0.0220
cari	400	1,200	152,800	1.2757	1.2961	1.3273	1.2842	1.2759	6.5538	1.3688	3.8640
farm	7	12	36	0.0005	0.0006	0.0007	0.0006	0.0005	0.0011	0.0013	0.0006
itest6	11	8	20	0.0005	0.0006	0.0007	0.0006	0.0005	0.0011	0.0012	0.0006
klein2	477	54	4,585	0.0332	0.0373	0.0398	0.0374	0.0332	0.0709	0.0558	0.0367
nsic1	451	463	2,853	0.0388	0.0455	0.0491	0.0429	0.0392	0.1314	0.0746	0.0450
nsic2	465	463	3,015	0.0399	0.0479	0.0523	0.0444	0.0407	0.1362	0.0767	0.0464
osa-07	1,118	23,949	143,694	4.2315	4.4437	4.5572	4.3339	4.2272	8.8815	5.6977	4.5623
osa-14	2,337	52,460	314,760	19.8264	20.3499	21.1999	20.0586	19.7962	40.4866	23.2962	21.0539
osa-30	4,350	100,024	600,138	80.9368	81.2840	83.4521	81.0922	81.0302	163.3031	86.4114	82.6966
rosen1	520	1,024	23,274	0.1724	0.1838	0.1902	0.1740	0.1715	0.3553	0.2350	0.1844
rosen10	2,056	4,096	62,136	1.7965	1.8568	1.8891	1.8558	1.8095	3.7266	2.0905	1.1238
rosen2	1,032	2,048	46,504	0.6270	0.6609	0.6695	0.6620	0.6278	1.2827	0.7632	0.5103
rosen7	264	512	7,770	0.0427	0.0490	0.0530	0.0491	0.0431	0.0922	0.0761	0.0480
rosen8	520	1,024	15,538	0.1362	0.1496	0.1572	0.1488	0.1367	0.2862	0.2038	0.0945
sc205	205	203	551	0.0110	0.0145	0.0159	0.0145	0.0113	0.0261	0.0271	0.0138
scfxm1	330	457	2,589	0.0298	0.0360	0.0391	0.0360	0.0300	0.0671	0.0607	0.0353
sctap2	1,090	1,880	6,714	0.1712	0.1970	0.2091	0.1876	0.1726	0.3730	0.2872	0.1920
sctap3	1,480	2,480	8,874	0.2923	0.3247	0.3425	0.3058	0.2943	0.6113	0.4458	0.3209
Geometric mean				0.1358	0.1529	0.1620	0.1499	0.1358	0.3283	0.2200	0.1513

Table 5.2 Results of interior point method with and without scaling

Problem				IPM without scaling		IPM with AM scaling		IPM with EQ scaling		IPM with GM scaling	
Name	Constraints	Variables	Nonzeros	Time	Iterations	Time	Iterations	Time	Iterations	Time	Iterations
beaconfd	173	262	3,375	0.01	9	0.01	12	0.01	11	0.01	12
cari	400	1,200	152,800	1.16	13	2.42	29	1.62	20	3.93	45
farm	7	12	36	0.01	16	0.01	10	0.01	8	0.01	11
itest6	11	8	20	0.01	9	0.01	7	0.01	7	0.01	7
klein2	477	54	4,585	0.03	1	0.03	1	0.02	1	0.02	1
nsic1	451	463	2,853	0.09	13	0.07	12	0.07	11	0.05	9
nsic2	465	463	3,015	0.49	86	0.51	89	0.42	72	0.37	63
osa-07	1,118	23,949	143,694	0.65	32	0.32	17	0.38	19	0.43	22
osa-14	2,337	52,460	314,760	1.55	32	0.85	17	0.78	16	0.98	20
osa-30	4,350	100,024	600,138	3.75	37	1.79	17	1.88	18	2.84	28
rosen1	520	1,024	23,274	0.23	18	0.20	15	0.18	14	0.20	15
rosen10	2,056	4,096	62,136	4.61	20	3.67	16	3.77	16	4.38	18
rosen2	1,032	2,048	46,504	1.13	21	0.86	16	0.86	16	0.92	17
rosen7	264	512	7,770	0.06	15	0.05	13	0.05	12	0.05	13
rosen8	520	1,024	15,538	0.22	20	0.22	18	0.20	17	0.19	17
sc205	205	203	551	0.01	12	0.02	11	0.01	11	0.01	11
scfxm1	330	457	2,589	0.04	18	0.05	19	0.08	21	0.08	25
sctap2	1,090	1,880	6,714	0.16	16	0.15	16	0.11	13	0.15	16
sctap3	1,480	2,480	8,874	0.29	17	0.30	18	0.20	13	0.27	16
Geometric mean				0.16	16.19	0.14	14.11	0.14	13.03	0.15	14.95

Table 5.3 Results of exterior point simplex algorithm with and without scaling

Problem				EPSA without scaling		EPSA with AM scaling		EPSA with EQ scaling		EPSA with GM scaling	
Name	Constraints	Variables	Nonzeros	Time	Iterations	Time	Iterations	Time	Iterations	Time	Iterations
beaconfd	173	262	3,375	0.05	41	0.05	41	0.05	41	0.05	60
cari	400	1,200	152,800	10.34	957	10.21	867	10.63	914	10.36	915
farm	7	12	36	0.00	6	0.00	5	0.00	5	0.00	5
itest6	11	8	20	0.00	4	0.00	2	0.00	2	0.00	2
klein2	477	54	4,585	0.23	2,275	0.38	499	0.49	500	0.21	283
nsic1	451	463	2,853	0.30	332	0.35	388	0.34	384	0.33	371
nsic2	465	463	3,015	0.53	525	0.56	538	0.50	535	0.53	519
osa-07	1,118	23,949	143,694	116.96	13,063	25.58	2,921	69.75	8,720	13.72	1,454
osa-14	2,337	52,460	314,760	971.09	42,159	50.51	2,057	378.20	19,720	68.37	2,329
osa-30	4,350	100,024	600,138	1,612.15	35,057	284.77	4,310	1,668.47	39,315	1,326.20	27,179
rosen1	520	1,024	23,274	3.18	1,661	2.80	1,540	3.16	1,661	3.40	1,756
rosen10	2,056	4,096	62,136	23.41	3,785	33.16	4,602	24.26	3,893	24.24	3,850
rosen2	1,032	2,048	46,504	15.18	3,580	11.31	3,025	15.29	3,581	13.07	3,253
rosen7	264	512	7,770	0.39	505	0.55	636	0.41	505	0.38	460
rosen8	520	1,024	15,538	1.70	1,169	1.62	1,119	1.74	1,171	1.77	1,175
sc205	205	203	551	0.16	235	0.17	244	0.16	246	0.19	269
scfxm1	330	457	2,589	0.86	472	0.89	515	0.85	440	0.96	547
sctap2	1,090	1,880	6,714	1.81	794	1.33	609	1.70	782	1.68	761
sctap3	1,480	2,480	8,874	4.60	1,176	3.28	856	3.93	1,157	4.01	1,062
Geometric mean				1.70	784.43	1.18	482.21	1.54	657.72	1.23	519.13

Table 5.4 Results of revised primal simplex algorithm with and without scaling

Problem				RSA without scaling		RSA with AM scaling		RSA with EQ scaling		RSA with GM scaling	
Name	Constraints	Variables	Nonzeros	Time	Iterations	Time	Iterations	Time	Iterations	Time	Iterations
beaconfd	173	262	3,375	0.03	26	0.03	26	0.04	26	0.04	49
cari	400	1,200	152,800	8.93	774	9.55	1,113	9.28	963	9.63	1,198
farm	7	12	36	0.00	7	0.00	5	0.00	5	0.00	6
itest6	11	8	20	0.00	4	0.00	2	0.00	2	0.00	2
klein2	477	54	4,585	0.22	275	0.38	499	0.43	500	0.21	283
nsic1	451	463	2,853	0.25	347	0.25	431	0.23	369	0.21	373
nsic2	465	463	3,015	0.28	498	0.25	382	0.29	473	0.20	341
osa-07	1,118	23,949	143,694	7.06	1,047	4.55	647	5.82	997	6.73	819
osa-14	2,337	52,460	314,760	54.87	2,441	34.75	1,439	48.15	2,454	49.34	1,774
osa-30	4,350	100,024	600,138	296.41	4,793	206.71	2,951	249.08	4,681	311.48	3,678
rosen1	520	1,024	23,274	7.07	3,445	4.02	2,090	3.21	1,893	5.21	2,591
rosen10	2,056	4,096	62,136	79.24	8,307	47.62	5,733	40.22	5,116	69.76	7,732
rosen2	1,032	2,048	46,504	38.07	6,878	23.91	4,741	19.88	4,242	29.10	5,398
rosen7	264	512	7,770	0.45	659	0.47	738	0.25	420	0.37	562
rosen8	520	1,024	15,538	1.71	1,300	1.98	1,443	1.32	1,045	2.03	1,488
sc205	205	203	551	0.11	152	0.10	144	0.11	165	0.10	145
scfxm1	330	457	2,589	0.71	372	0.80	369	0.69	339	0.79	370
sctap2	1,090	1,880	6,714	2.37	1,532	1.18	671	1.48	1,001	1.29	805
sctap3	1,480	2,480	8,874	5.63	2,315	3.75	1,631	3.86	1,809	3.64	1,514
Geometric mean				1.19	548.54	0.98	447.30	0.97	463.78	1.04	477.69

Table 5.5 Results of revised dual simplex algorithm with and without scaling

Name	Problem Constraints	Variables	Nonzeros	RDSA without scaling Time	Iterations	RDSA with AM Scaling Time	Iterations	RDSA with EQ Scaling Time	Iterations	RDSA with GM Scaling Time	Iterations
beaconfd	173	262	3,375	0.03	26	0.04	55	0.03	25	0.04	57
cari	400	1,200	152,800	9.96	1,019	8.66	507	9.30	745	8.66	617
farm	7	12	36	0.00	3	0.00	2	0.00	2	0.00	3
itest6	11	8	20	0.00	2	0.00	1	0.00	1	0.00	1
klein2	477	54	4,585	0.52	720	0.62	882	1.00	1,520	1.01	1,040
nsic1	451	463	2,853	0.21	379	0.22	336	0.18	335	0.22	356
nsic2	465	463	3,015	0.29	459	0.32	526	0.33	571	0.31	477
osa-07	1,118	23,949	143,694	3.92	1,025	1.69	430	1.62	427	3.58	848
osa-14	2,337	52,460	314,760	23.37	2,296	8.71	919	8.69	963	19.40	1,859
osa-30	4,350	100,024	600,138	91.11	4,017	34.65	1,874	32.06	1,941	90.03	3,760
rosen1	520	1,024	23,274	2.77	1,410	1.58	902	0.70	621	1.95	1,068
rosen10	2,056	4,096	62,136	28.63	4,092	21.01	2,965	11.17	2,281	27.93	3,529
rosen2	1,032	2,048	46,504	12.84	2,794	5.04	1,472	2.95	1,197	8.13	1,858
rosen7	264	512	7,770	0.46	629	0.18	287	0.09	183	0.21	333
rosen8	520	1,024	15,538	1.67	1,263	0.84	691	0.46	492	1.11	867
sc205	205	203	551	0.09	125	0.16	260	0.19	354	0.12	173
scfxm1	330	457	2,589	0.80	377	0.74	287	0.74	303	0.81	323
sctap2	1,090	1,880	6,714	1.06	794	0.58	411	0.76	570	0.75	542
sctap3	1,480	2,480	8,874	1.96	1,103	1.19	661	1.34	740	1.19	629
Geometric mean				0.80	428.02	0.56	297.93	0.49	289.66	0.71	362.58

of Mehrotra's Predictor-Corrector interior point method (IPM), the exterior point simplex algorithm (EPSA), the revised primal simplex algorithm (RSA), and the revised dual simplex algorithm (RDSA), respectively, over the same benchmark set. The following abbreviations are used: (i) AM—arithmetic mean, (ii) EQ—equilibration, and (iii) GM—geometric mean. Execution times do not include the time of the presolve and scaling analysis.

Table 5.2 reveals that scaling has a small impact on Mehrotra's Predictor-Corrector interior point method. Both the number of iterations and the execution time are slightly reduced. Tables 5.3, 5.4, and 5.5 show that the exterior point simplex algorithm, the revised primal simplex algorithm, and the revised dual simplex algorithm are faster when scaling has been applied. Moreover, equilibration is the best scaling technique for the revised primal and dual simplex algorithm, while arithmetic mean is the best scaling technique for the exterior point simplex algorithm. However, if we also add to the execution time the time to scale the LPs, then we observe that Mehrotra's Predictor-Corrector interior point method can solve the LPs faster when they have not been scaled. That is due to the relative large execution time to scale the LPs. However, taking into account that some LPs cannot be solved without scaling, the scaling methods should be always performed prior to the application of an LP algorithm.

Note that these findings cannot be generalized, because only a sample of benchmark LPs is included in this computational study. Interested readers are referred to the computational studies performed in [4, 7–11].

5.12   Chapter Review

Scaling is a prerequisite step that is applied prior to the application of LP algorithms in order to: (i) produce a compact representation of the variable bounds, (ii) reduce the condition number of the constraint matrix, (iii) improve the numerical behavior of the algorithms, (iv) reduce the number of iterations required to solve LPs, and (v) simplify the setup of the tolerances. In this chapter, we presented eleven scaling techniques used prior to the execution of an LP algorithm: (i) arithmetic mean, (ii) de Buchet for the case $p = 1$, (iii) de Buchet for the case $p = 2$, (iv) de Buchet for the case $p = \infty$, (v) entropy, (vi) equilibration, (vii) geometric mean, (viii) IBM MPSX, (ix) L_p-norm for the case $p = 1$, (x) L_p-norm for the case $p = 2$, and (xi) L_p-norm for the case $p = \infty$. Each technique was presented with: (i) its mathematical formulation, (ii) a thorough numerical example, and (iii) its implementation in MATLAB. Finally, we performed a computational study in order to compare the execution time of the aforementioned scaling techniques and to investigate the impact of scaling prior to the application of LP algorithms. Computational results showed that all algorithms can solve faster LPs that have been scaled. Equilibration was the best scaling technique for the interior point method, the revised primal simplex algorithm, and the revised dual simplex algorithm, while arithmetic mean was the best scaling technique for the exterior point simplex algorithm.

References

1. Benichou, M., Gautier, J., Hentges, G., & Ribiere, G. (1977). The efficient solution of large-scale linear programming problems. *Mathematical Programming, 13*, 280–322.
2. Curtis, A. R., & Reid, J. K. (1972). On the automatic scaling of matrices for Gaussian elimination. *IMA Journal of Applied Mathematics, 10*, 118–124.
3. de Buchet, J. (1966). Experiments and statistical data on the solving of large–scale linear programs. In *Proceedings of the Fourth International Conference on Operational Research* (pp. 3–13). New York: Wiley-Interscience.
4. Elble, J. M., & Sahinidis, N. V. (2012). Scaling linear optimization problems prior to application of the simplex method. *Computational Optimization and Applications, 52*(2), 345–371 .
5. Fulkerson, D. R., & Wolfe, P. (1962). An algorithm for scaling matrices. *SIAM Review, 4*, 142–146.
6. Hamming, R. W. (1971). *Introduction to applied numerical analysis*. New York: McGraw-Hill.
7. Larsson, T. (1993). On scaling linear programs – Some experimental results. *Optimization, 27*, 335–373.
8. Ploskas, N. (2014). *Hybrid optimization algorithms: implementation on GPU*. Ph.D. thesis, Department of Applied Informatics, University of Macedonia.
9. Ploskas, N., & Samaras, N. (2013). The impact of scaling on simplex type algorithms. In *Proceedings of the 6th Balkan Conference in Informatics* (pp. 17–22), 19–21 September (ACM, Thessaloniki, Greece).
10. Ploskas, N., & Samaras, N. (2015). A computational comparison of scaling techniques for linear optimization problems on a graphical processing unit. *International Journal of Computer Mathematics, 92*(2), 319–336.
11. Tomlin, J. A. (1975). On scaling linear programming problems. *Mathematical Programming Studies, 4*, 146–166.
12. Triantafyllidis, C., & Samaras, N. (2014). Three nearly scaling-invariant versions of an exterior point algorithm for linear programming. *Optimization, 64*(10), 2163–2181.

Chapter 6
Pivoting Rules

Abstract Simplex-type algorithms perform successive pivoting operations (or iterations) in order to reach the optimal solution. The choice of the pivot element at each iteration is one of the most critical steps in simplex-type algorithms. The flexibility of the entering and leaving variable selection allows to develop various pivoting rules. This chapter presents six pivoting rules used in each iteration of the simplex algorithm to determine the entering variable: (i) Bland's rule, (ii) Dantzig's rule, (iii) Greatest Increment Method, (iv) Least Recently Considered Method, (v) Partial Pricing rule, and (vi) Steepest Edge rule. Each technique is presented with: (i) its mathematical formulation, (ii) a thorough numerical example, and (iii) its implementation in MATLAB. Finally, a computational study is performed. The aim of the computational study is twofold: (i) compare the execution time of the presented pivoting rules, and (ii) highlight the impact of the choice of the pivoting rule in the number of iterations and the execution time of the revised simplex algorithm.

6.1 Chapter Objectives

- Discuss the importance of the choice of the pivoting rule in the number of iterations and the execution time of the revised simplex algorithm.
- Provide the mathematical formulation of the pivoting rules.
- Understand the pivoting rules through a thorough numerical example.
- Implement the pivoting rules using MATLAB.
- Compare the computational performance of the pivoting rules.

6.2 Introduction

A critical step in solving an LP problem is the selection of the entering variable in each iteration. It is obvious that simplex-type algorithms behavior can be improved by modifying: (i) the initial solution and (ii) the pivoting rule. Good choices of the entering variable can lead to fast convergence to the optimal solution, while poor

© Springer International Publishing AG 2017   277
N. Ploskas, N. Samaras, *Linear Programming Using MATLAB®*,
Springer Optimization and Its Applications 127, DOI 10.1007/978-3-319-65919-0_6

choices lead to more iterations and worst execution times or even no solutions of the LPs. The pivoting rule applied to a simplex-type algorithm is the key factor that will determine the number of iterations that the algorithm performs [13]. Different pivoting rules yield different basis sequences in the simplex algorithm. None of the existing pivoting rules admits polynomial complexity. Fukuda and Terlaky [8] proved that starting from any basis, there exists a short admissible pivot sequence containing at most n-pivot steps leading to an optimal basis. This result is very promising because it indicates that polynomial pivot algorithms might exist. It remains an open question whether there exists a strongly polynomial algorithm for LP whose running time depends only on the number of constraints and variables. Many pivoting rules have been proposed in the literature. A complete presentation of them can be found in [20]. Six of them are implemented and compared in this chapter, namely: (i) Bland's rule, (ii) Dantzig's rule, (iii) Greatest Increment Method, (iv) Least Recently Considered Method, (v) Partial Pricing rule, and (vi) Steepest Edge rule. Other well-known pivoting rules include Devex [11], Modified Devex [2], Steepest Edge approximation scheme [19], Murty's Bard type scheme [14], Edmonds-Fukuda rule [7], and its variants [4, 23, 25, 26].

Forrest and Goldfarb [6] presented several new implementations of the steepest edge pivoting rule and compared them with Devex variants and Dantzig's rule over large LPs. They concluded that steepest edge variants are clearly superior to Devex variants and Dantzig's rule for solving difficult large-scale LPs. Thomadakis [21] implemented and compared the following five pivoting rules: (i) Dantzig's rule, (ii) Bland's rule, (iii) Least-Recently Considered Method, (iv) Greatest-Increment Method, and (v) Steepest Edge rule. Thomadakis examined the trade-off between the number of iterations and the execution time per iteration and concluded that: (i) Bland's rule has the shortest execution time per iteration, but it usually needs many more iterations than the other pivoting rules to converge to the optimal solution of an LP problem, (ii) Dantzig's rule and Least Recently Considered Method perform comparably, but the latter requires fewer iterations when degenerate pivots exist, (iii) Greatest Increment Method has the worst execution time per iteration, but it usually needs fewer iterations to converge to the optimal solution, and (iv) Steepest Edge rule requires fewer iterations than all the other pivoting rules and its execution time per iteration is lower than Greatest Increment Method but higher than the other three pivoting rules. Ploskas and Samaras [17] implemented and compared the implementations of eight pivoting rules over small- and medium-sized Netlib LPs: (i) Bland's rule, (ii) Dantzig's rule, (iii) Greatest Increment Method, (iv) Least Recently Considered Method, (v) Partial Pricing rule, (vi) Queue rule, (vii) Stack rule, and (viii) Steepest Edge rule. In contrast to Thomadakis, Ploskas, and Samaras also examined the total execution time of the revised simplex algorithm relating to the pivoting rule that is used and concluded that: (i) with a limit of 70, 000 iterations, only Dantzig's rule has solved all instances of the test set, while Bland's rule solved 45 out of 48 instances, Greatest Increment method 46 out of 48, Least Recently

Considered method 45 out of 48, Partial Pricing rule 45 out of 48, Queue's rule 41 out of 48, Stack's rule 43 out of 48, and Steepest Edge rule 46 out of 48, (ii) Dantzig's rule requires the shortest execution time both on average and on almost all instances, while Steepest Edge rule has the worst execution time both on average and on almost all instances, and (iii) despite its computational cost, Steepest Edge rule needs the fewest number of iterations than all the other pivoting rules, while Bland's rule is by far the worst pivoting rule in terms of the number of iterations. Ploskas and Samaras [18] proposed GPU-based implementations of some of the most well-known pivoting rules for the revised simplex algorithm. They performed a computational study on large-scale randomly generated optimal dense LPs, the Netlib (optimal, Kennington, and infeasible LPs), and the Mészáros set and found that only Steepest Edge rule is suitable for GPUs. The maximum speedup gained from the pivoting operation of Steepest Edge rule was 16.72.

In this chapter, we present six pivoting rules used in each iteration of the revised simplex algorithm to determine the entering variable. Each technique is presented with: (i) its mathematical formulation, (ii) a thorough numerical example, and (iii) its implementation in MATLAB. Finally, a computational study is performed. The aim of the computational study is twofold: (i) compare the execution time of the pivoting rules, and (ii) highlight the impact of the choice of a pivoting rule in the number of iterations and the execution time of the revised simplex algorithm.

The structure of this chapter is as follows. In Sections 6.3–6.8, six pivoting rules are presented. More specifically, Section 6.3 presents Bland's rule, Section 6.4 presents Dantzig's rule, Section 6.5 presents Greatest Increment method, Section 6.6 presents Least Recently Considered Method, Section 6.7 presents partial pricing rule, while Section 6.8 presents Steepest Edge rule. Section 6.9 presents a computational study that compares the execution time of the pivoting rules and highlights the impact of the choice of a pivoting rule in the number of iterations and the execution time of the revised simplex algorithm. Finally, conclusions and further discussion are presented in Section 6.10.

6.3 Bland's Rule

6.3.1 Mathematical Formulation

Bland's rule [3] selects as entering variable the first among the eligible ones, i.e., the leftmost among columns with negative relative cost coefficient $(j|s_j < 0 \wedge j \in N)$. If the simplex algorithm uses Bland's rule, it terminates in finite time with an optimal solution. Whereas Bland's rule avoids cycling, it has been observed in practice that this pivoting rule can lead to stalling, a phenomenon where long degenerate paths are produced.

6.3.2 Numerical Example

This subsection demonstrates Bland's pivoting rule through an illustrative example.
Bland's pivoting rule will be used to select the entering variable in the first iteration
of the LP problem shown below:

$$\min z = -2x_1 + x_2 - 4x_3$$
$$\text{s.t.} \quad 2x_1 + 4x_2 + 2x_3 \le 6$$
$$x_1 - 2x_2 + 6x_3 \le 5$$
$$-x_1 + 4x_2 + 3x_3 \le 8$$
$$x_j \ge 0, \quad (j = 1, 2, 3)$$

We transform the above LP problem to its standard form by introducing the slack
variables x_4, x_5, and x_6:

$$\min z = -2x_1 + x_2 - 4x_3$$
$$\text{s.t.} \quad 2x_1 + 4x_2 + 2x_3 + x_4 \qquad\qquad = 6$$
$$x_1 - 2x_2 + 6x_3 \qquad + x_5 \qquad = 5$$
$$-x_1 + 4x_2 + 3x_3 \qquad\qquad + x_6 = 8$$
$$x_j \ge 0, \quad (j = 1, 2, 3, 4, 5, 6)$$

The above LP problem can be rewritten in matrix notation as follows:

$$c = \begin{bmatrix} -2 \\ 1 \\ -4 \\ 0 \\ 0 \\ 0 \end{bmatrix} \quad A = \begin{bmatrix} 2 & 4 & 2 & 1 & 0 & 0 \\ 1 & -2 & 6 & 0 & 1 & 0 \\ -1 & 4 & 3 & 0 & 0 & 1 \end{bmatrix} \quad b = \begin{bmatrix} 6 \\ 5 \\ 8 \end{bmatrix}$$

We determine the partition (B, N), where $B = \begin{bmatrix} 4 & 5 & 6 \end{bmatrix}$ and $N = \begin{bmatrix} 1 & 2 & 3 \end{bmatrix}$. Hence:

$$A_B = \begin{bmatrix} 1 & 0 & 0 \\ 0 & 1 & 0 \\ 0 & 0 & 1 \end{bmatrix}$$

$$A_N = \begin{bmatrix} 2 & 4 & 2 \\ 1 & -2 & 6 \\ -1 & 4 & 3 \end{bmatrix}$$

$$A_B^{-1} = A_B = I$$

$$c_B = \begin{bmatrix} 0 \\ 0 \\ 0 \end{bmatrix}$$

$$c_N = \begin{bmatrix} -2 \\ 1 \\ -4 \end{bmatrix}$$

$$x_B = A_B^{-1}b = \begin{bmatrix} 6 \\ 5 \\ 8 \end{bmatrix}$$

$$w = c_B^T A_B^{-1} = \begin{bmatrix} 0 & 0 & 0 \end{bmatrix}$$

$$s_N = c_N^T - wA_N = \begin{bmatrix} -2 & 1 & -4 \end{bmatrix}$$

It is easily seen that the partition (B, N) is feasible because $x_B \geq 0$. So, we do not have to apply Phase I (more details in Chapter 8). Then, we proceed with the first iteration of the algorithm. We check if $s_N \geq 0$; $s_N \not\geq 0$, so we proceed with the selection of the entering variable. The eligible variables that can be selected as the entering one are x_1 and x_3. According to Bland's rule, the entering variable is x_1, because it is the first among the eligible ones.

6.3.3 Implementation in MATLAB

This subsection presents the implementation in MATLAB of Bland's pivoting rule (filename: bland.m). Some necessary notations should be introduced before the presentation of the implementation of Bland's pivoting rule. Let Sn be a $1 \times (n - m)$ vector with the reduced costs, $NonBasicList$ be a $1 \times (n - m)$ vector with the indices of the nonbasic variables, and $index$ be the index of the entering variable.

The function for Bland's pivoting rule takes as input the vector of the reduced costs (vector Sn) and the vector of indices of the nonbasic variables (vector $NonBasicList$), and returns as output the index of the entering variable (variable $index$).

In line 17, the indices of the eligible variables that can enter the basis are found. In line 19, the leftmost among the eligible variables is found. Finally, the index of the entering variable is calculated (line 22).

```
1.    function [index] = bland(Sn, NonBasicList)
2.    % Filename: bland.m
3.    % Description: the function is an implementation of the
4.    % Bland's pivoting rule
5.    % Authors: Ploskas, N., & Samaras, N.
```

```
6.   %
7.   % Syntax: [index] = bland(Sn, NonBasicList)
8.   %
9.   % Input:
10.  % -- Sn: vector of reduced costs (size 1 x (n - m))
11.  % -- NonBasicList: vector of indices of the nonbasic
12.  %    variables (size 1 x (n - m))
13.  %
14.  % Output:
15.  % -- index: the index of the entering variable
16.
17.  temp = find(Sn < 0); % find the indices of the eligible
18.  % variables that can enter the basis
19.  [~, b] = min(NonBasicList(temp)); % find the leftmost among
20.  % the eligible variables
21.  % calculate the index of the entering variable
22.  index = temp(b);
23.  end
```

6.4 Dantzig's Rule

6.4.1 Mathematical Formulation

Dantzig's rule or largest coefficient rule [5] is the first pivoting rule that was proposed for the simplex algorithm. It has been widely used in simplex implementations [1, 15]. This pivoting rule selects the column A_l with the most negative $\bar{s}_l$ ($s_l = \min \{s_j : s_j < 0 \land j \in N\}$). Dantzig's rule ensures the largest reduction in the objective value per unit of nonbasic variable $\bar{s}_l$ increase. Although its worst-case complexity is exponential [12], Dantzig's rule is considered as simple but powerful enough to guide the simplex algorithm into short paths [21].

6.4.2 Numerical Example

The LP problem that we will demonstrate Dantzig's rule is the same as in the previous section and has the following variables:

$$c = \begin{bmatrix} -2 \\ 1 \\ -4 \\ 0 \\ 0 \\ 0 \end{bmatrix}$$

$$A = \begin{bmatrix} 2 & 4 & 2 & 1 & 0 & 0 \\ 1 & -2 & 6 & 0 & 1 & 0 \\ -1 & 4 & 3 & 0 & 0 & 1 \end{bmatrix}$$

$$b = \begin{bmatrix} 6 \\ 5 \\ 8 \end{bmatrix}$$

$$A_B = \begin{bmatrix} 1 & 0 & 0 \\ 0 & 1 & 0 \\ 0 & 0 & 1 \end{bmatrix}$$

$$A_N = \begin{bmatrix} 2 & 4 & 2 \\ 1 & -2 & 6 \\ -1 & 4 & 3 \end{bmatrix}$$

$$A_B^{-1} = A_B = I$$

$$c_B = \begin{bmatrix} 0 \\ 0 \\ 0 \end{bmatrix}$$

$$c_N = \begin{bmatrix} -2 \\ 1 \\ -4 \end{bmatrix}$$

$$x_B = A_B^{-1} b = \begin{bmatrix} 6 \\ 5 \\ 8 \end{bmatrix}$$

$$w = c_B^T A_B^{-1} = \begin{bmatrix} 0 & 0 & 0 \end{bmatrix}$$

$$s_N = c_N^T - w A_N = \begin{bmatrix} -2 & 1 & -4 \end{bmatrix}$$

According to Dantzig's rule, the entering variable is x_3, because it is the one with the most negative $\bar{s}_l = \bar{s}_3 = -4$.

6.4.3 Implementation in MATLAB

This subsection presents the implementation in MATLAB of Dantzig's pivoting rule (filename: dantzig.m). Some necessary notations should be introduced before the presentation of the implementation of Dantzig's pivoting rule. Let Sn be a $1 \times (n-m)$ vector with the reduced costs and *index* be the index of the entering variable.

The function for Dantzig's pivoting rule takes as input the vector of the reduced costs (vector *Sn*) and returns as output the index of the entering variable (variable *index*).

In line 15, the index of the variable with the most negative *Sn* is calculated.

```
1.    function [index] = dantzig(Sn)
2.    % Filename: dantzig.m
3.    % Description: the function is an implementation of the
4.    % Dantzig's pivoting rule
5.    % Authors: Ploskas, N., & Samaras, N.
6.    %
7.    % Syntax: [index] = dantzig(Sn)
8.    %
9.    % Input:
10.   % -- Sn: vector of reduced costs (size 1 x (n - m))
11.   %
12.   % Output:
13.   % -- index: the index of the entering variable
14.
15.   [~, index] = min(Sn); % find the index of the variable
16.   % with the most negative Sn
17.   end
```

6.5   Greatest Increment Method

6.5.1   *Mathematical Formulation*

The entering variable in Greatest Increment Method [12] is the one with the largest total objective value improvement. The improvement of the objective value is computed for each nonbasic variable. The variable that offers the largest improvement in the objective value over all possibilities is selected as the entering variable. This pivoting rule can lead to fast convergence to the optimal solution. However, this advantage is eliminated by the additional computational cost per iteration. This pivoting rule may lead to cycling. Finally, Gärtner [9] constructed LPs where Greatest Increment Method showed exponential complexity.

6.5.2   *Numerical Example*

The LP problem that we will demonstrate Greatest Increment Method is the same as in previous sections and has the following variables:

$$c = \begin{bmatrix} -2 \\ 1 \\ -4 \\ 0 \\ 0 \\ 0 \end{bmatrix}$$

$$A = \begin{bmatrix} 2 & 4 & 2 & 1 & 0 & 0 \\ 1 & -2 & 6 & 0 & 1 & 0 \\ -1 & 4 & 3 & 0 & 0 & 1 \end{bmatrix}$$

$$b = \begin{bmatrix} 6 \\ 5 \\ 8 \end{bmatrix}$$

$$A_B = \begin{bmatrix} 1 & 0 & 0 \\ 0 & 1 & 0 \\ 0 & 0 & 1 \end{bmatrix}$$

$$A_N = \begin{bmatrix} 2 & 4 & 2 \\ 1 & -2 & 6 \\ -1 & 4 & 3 \end{bmatrix}$$

$$A_B^{-1} = A_B = I$$

$$c_B = \begin{bmatrix} 0 \\ 0 \\ 0 \end{bmatrix}$$

$$c_N = \begin{bmatrix} -2 \\ 1 \\ -4 \end{bmatrix}$$

$$x_B = A_B^{-1}b = \begin{bmatrix} 6 \\ 5 \\ 8 \end{bmatrix}$$

$$w = c_B^T A_B^{-1} = \begin{bmatrix} 0 & 0 & 0 \end{bmatrix}$$

$$s_N = c_N^T - wA_N = \begin{bmatrix} -2 & 1 & -4 \end{bmatrix}$$

According to Greatest Increment Method, the entering variable will be the one with the largest total objective value improvement. The eligible variables are x_1 and x_3. If variable x_1 enters the basis, then the pivot column will be:

$$h_1 = A_B^{-1} A_1 = \begin{bmatrix} 2 \\ 1 \\ -1 \end{bmatrix}$$

Hence, the improvement of the objective value will be:

$$\theta_0 = \min\left(\frac{x_{B[1]}}{h_{11}}, \frac{x_{B[2]}}{h_{12}}, x\right) = \min\left(\frac{6}{2}, \frac{5}{1}, x\right) = 3$$

$$Improvement_1 = \theta_0 s_N(1) = 3 \times -2 = -6$$

If variable x_3 enters the basis, then the pivot column will be:

$$h_3 = A_B^{-1} A_3 = \begin{bmatrix} 2 \\ 6 \\ 3 \end{bmatrix}$$

Hence, the improvement of the objective value will be:

$$\theta_0 = \min\left(\frac{x_{B[1]}}{h_{31}}, \frac{x_{B[2]}}{h_{32}}, \frac{x_{B[3]}}{h_{33}}\right) = \min\left(\frac{6}{2}, \frac{5}{6}, \frac{8}{3}\right) = \frac{5}{6}$$

$$Improvement_3 = \theta_0 s_N(3) = \frac{5}{6} \times -4 = -\frac{20}{6}$$

The largest total objective value improvement will occur if variable x_1 enters the basis. Therefore, the entering variable is x_1.

6.5.3 Implementation in MATLAB

This subsection presents the implementation in MATLAB of Greatest Increment Method (filename: gim.m). Some necessary notations should be introduced before the presentation of the implementation of Greatest Increment Method. Let Sn be a $1 \times (n - m)$ vector with the reduced costs, $NonBasicList$ be a $1 \times (n - m)$ vector of indices of the nonbasic variables, A be a $m \times n$ matrix of coefficients of the constraints, $BasisInv$ be a $m \times m$ matrix with the basis inverse, xb be a $m \times 1$ vector with the values of the basic variables, $tole$ be the value of the tolerance for the pivoting step, and $index$ be the index of the entering variable.

The function for Greatest Increment Method takes as input the vector of the reduced costs (vector Sn), the vector of indices of the nonbasic variables (vector $NonBasicList$), the matrix of coefficients of the constraints (matrix A), the matrix with the basis inverse (matrix $BasisInv$), the vector of the values of the basic variables (vector *xb*), and the value of the tolerance for the pivoting step (variable *tole*), and returns as output the index of the entering variable (variable *index*).

In lines 29–57, the improvement of the objective value is computed for each nonbasic variable and we find the variable with the largest total objective value improvement (lines 53–56). The if-statement in lines 30–32 makes sure that we only consider the eligible variables. The index of the candidate variable is stored in variable *l* (line 34) and the pivot column is computed (line 35). Then, we set equal to zero the values of the pivot column that are less than or equal to the given tolerance (lines 38–39). Moreover, we find the number of the variables in the pivot column that are greater than zero (line 40). If there is not any variable in the pivot column greater than zero, then the problem is unbounded and the function is terminated (lines 43–46). Otherwise, the total objective value improvement is calculated if variable *l* enters the basis (lines 48–50) and if this improvement is greater than the current maximum improvement, then this value is stored to the variable *maxDecrease* and its index to the variable *index* (lines 53–56) Finally, if the entering variable is still empty, then we select as entering variable the variable with the most negative Sn (lines 60–62).

```
1.    function [index] = gim(Sn, NonBasicList, A, ...
2.        BasisInv, xb, tole)
3.    % Filename: gim.m
4.    % Description: the function is an implementation of the
5.    % Greatest Increment Method pivoting rule
6.    % Authors: Ploskas, N., & Samaras, N.
7.    %
8.    % Syntax: [index] = gim(Sn, NonBasicList, A, ...
9.    %     BasisInv, xb, tole)
10.   %
11.   % Input:
12.   % -- Sn: vector of reduced costs (size 1 x (n - m))
13.   % -- NonBasicList: vector of indices of the nonbasic
14.   %     variables (size 1 x (n - m))
15.   % -- A: matrix of coefficients of the constraints
16.   %     (size m x n)
17.   % -- BasisInv: matrix with the basis inverse (size m x m)
18.   % -- xb: vector with the values of the basic variables
19.   %     (size m x 1)
20.   % -- tole: the value of the tolerance for the pivoting step
21.   %
22.   % Output:
23.   % -- index: the index of the entering variable
24.
25.   index = -1;
26.   maxDecrease = 1;
27.   % calculate the improvement of the objective value for each
28.   % nonbasic variable
29.   for i = 1:length(NonBasicList)
30.       if Sn(i) >= 0 % only consider the eligible variables
31.           continue;
32.       end
33.       % store index of the candidate variable
```

```
34.      l = NonBasicList(i);
35.      hl = BasisInv * A(:, l); % compute the pivot column
36.      % set equal to zero the values of the pivot column that
37.      % are less than or equal to the given tolerance
38.      toler = abs(hl) <= tole;
39.      hl(toler == 1) = 0;
40.      mrt = find(hl > 0);
41.      % if there is not any variable in the pivot column
42.      % greater than zero, then the problem is unbounded
43.      if numel(mrt) < 1
44.           index = -1;
45.           return;
46.      end
47.      % calculate the total objective value improvement
48.      xbDIVhl = xb(mrt) ./ hl(mrt);
49.      theta0 = min(xbDIVhl(xbDIVhl > 0));
50.      currDecrease = theta0 * Sn(i);
51.      % if this improvement is greater than the current
52.      % maximum improvement, store the value and the index
53.      if currDecrease < maxDecrease
54.           maxDecrease = currDecrease;
55.           index = i;
56.      end
57. end
58. % if the entering variable is empty, then find the index
59. % of the variable with the most negative Sn
60. if index < 0
61.      [~, index] = min(Sn);
62. end
63. end
```

6.6   Least Recently Considered Method

6.6.1   *Mathematical Formulation*

In the first iteration of the Least Recently Considered Method [24], the entering variable l is selected according to Dantzig's rule, i.e., the variable with the most negative $\bar{s}_l$. In the next iterations, it starts searching for the first eligible variable with index greater than l. If $l = n$ then the Least Recently Considered Method starts searching from the first column again. Least Recently Considered Method prevents stalling and has been shown that it performs fairly well in practice [21]. However, its worst-case complexity has not been proven yet.

6.6.2 Numerical Example

The LP problem that we will demonstrate Least Recently Considered Method is the same as in previous sections and has the following variables:

$$c = \begin{bmatrix} -2 \\ 1 \\ -4 \\ 0 \\ 0 \\ 0 \end{bmatrix}$$

$$A = \begin{bmatrix} 2 & 4 & 2 & 1 & 0 & 0 \\ 1 & -2 & 6 & 0 & 1 & 0 \\ -1 & 4 & 3 & 0 & 0 & 1 \end{bmatrix}$$

$$b = \begin{bmatrix} 6 \\ 5 \\ 8 \end{bmatrix}$$

$$A_B = \begin{bmatrix} 1 & 0 & 0 \\ 0 & 1 & 0 \\ 0 & 0 & 1 \end{bmatrix}$$

$$A_N = \begin{bmatrix} 2 & 4 & 2 \\ 1 & -2 & 6 \\ -1 & 4 & 3 \end{bmatrix}$$

$$A_B^{-1} = A_B = I$$

$$c_B = \begin{bmatrix} 0 \\ 0 \\ 0 \end{bmatrix}$$

$$c_N = \begin{bmatrix} -2 \\ 1 \\ -4 \end{bmatrix}$$

$$x_B = A_B^{-1}b = \begin{bmatrix} 6 \\ 5 \\ 8 \end{bmatrix}$$

$$w = c_B^T A_B^{-1} = \begin{bmatrix} 0 & 0 & 0 \end{bmatrix}$$

$$s_N = c_N^T - wA_N = \begin{bmatrix} -2 & 1 & -4 \end{bmatrix}$$

In the first iteration, the entering variable is selected according to Dantzig's rule. Therefore, the entering variable is x_3, because it is the one with the most negative $\bar{s}_l$ = $\bar{s}_3$ = −4. In the following iterations, the entering variable will be the first eligible one with index greater than 3. If there is not such a variable, the entering variable will be the first eligible one (like Bland's rule).

6.6.3 Implementation in MATLAB

This subsection presents the implementation in MATLAB of Least Recently Considered Method (filename: lrcm.m). Some necessary notations should be introduced before the presentation of the implementation of Least Recently Considered Method. Let Sn be a $1 \times (n - m)$ vector with the reduced costs, *NonBasicList* be a $1 \times (n - m)$ vector of indices of the nonbasic variables, *lrcmlast* be the index of the last selected entering variable, *iteration* be the current iteration of the simplex algorithm, and *index* be the index of the entering variable.

The function for Least Recently Considered Method takes as input the vector of the reduced costs (vector Sn), the vector of indices of the nonbasic variables (vector *NonBasicList*), the index of the last selected entering variable (variable *lrcmLast*), and the current iteration of the simplex algorithm (variable *iteration*), and returns as output the index of the entering variable (variable *index*).

In the first iteration of Least Recently Considered Method, the entering variable is selected according to Dantzig's rule (lines 24–25). In the following iterations, it starts searching for the first eligible variable with index greater than *lrcmLast* (lines 29–31). If the Least Recently Considered Method cannot find any eligible variable with index greater than *lrcmLast* (lines 34), then it starts searching from the first column again (lines 35–36).

```
1.     function [index] = lrcm(Sn, NonBasicList, lrcmLast, ...
2.         iteration)
3.     % Filename: lrcm.m
4.     % Description: the function is an implementation of the
5.     % Least Recently Considered Method pivoting rule
6.     % Authors: Ploskas, N., & Samaras, N.
7.     %
8.     % Syntax: [index] = lrcm(Sn, NonBasicList, lrcmLast, ...
9.     %     iteration)
10.    %
11.    % Input:
12.    % -- Sn: vector of reduced costs (size 1 x (n - m))
13.    % -- NonBasicList: vector of indices of the nonbasic
14.    %     variables (size 1 x (n - m))
15.    % -- lrcmLast: the index of the last selected entering
16.    %     variable
17.    % -- iteration: the current iteration of the simplex
18.    %     algorithm
19.
20.    % Output:
```

```
21.  % -- index: the index of the entering variable
22.
23.  % use Dantzig's rule in the first iteration
24.  if iteration == 1
25.      [~, index] = min(Sn);
26.  else
27.      % searching for the first eligible variable with index
28.      % greater than lrcmLast
29.      temp1 = find(Sn < 0);
30.      temp2 = NonBasicList(temp1);
31.      temp3 = find(temp2 > lrcmLast);
32.      % if not any eligible variable with index greater than
33.      % lrcmLast, start searching from the first column again
34.      if isempty(temp3)
35.          [~, b] = min(temp2);
36.          index = temp1(b);
37.       else
38.          index = temp1(min(temp3));
39.      end
40.  end
41.  end
```

6.7 Partial Pricing Rule

6.7.1 Mathematical Formulation

Full pricing considers the selection of the entering variable from all eligible variables. Partial pricing methods are variants of the standard rules that take only a part of the nonbasic variables into account. Here, we use the partial pricing rule as a variant of Dantzig's rule. In static partial pricing, nonbasic variables are divided into equal segments with predefined size. Then, the pricing operation is carried out segment by segment until a reduced cost is found. In dynamical partial pricing, segment size is determined dynamically during the execution of the algorithm.

6.7.2 Numerical Example

Partial pricing rule can be used when the number of the nonbasic variables is large. However, we demonstrate it to a simple example for the sake of completeness. The LP problem that we will demonstrate the partial pricing rule is the same as in previous sections and has the following variables:

$$c = \begin{bmatrix} -2 \\ 1 \\ -4 \\ 0 \\ 0 \\ 0 \end{bmatrix}$$

$$A = \begin{bmatrix} 2 & 4 & 2 & 1 & 0 & 0 \\ 1 & -2 & 6 & 0 & 1 & 0 \\ -1 & 4 & 3 & 0 & 0 & 1 \end{bmatrix}$$

$$b = \begin{bmatrix} 6 \\ 5 \\ 8 \end{bmatrix}$$

$$A_B = \begin{bmatrix} 1 & 0 & 0 \\ 0 & 1 & 0 \\ 0 & 0 & 1 \end{bmatrix}$$

$$A_N = \begin{bmatrix} 2 & 4 & 2 \\ 1 & -2 & 6 \\ -1 & 4 & 3 \end{bmatrix}$$

$$A_B^{-1} = A_B = I$$

$$c_B = \begin{bmatrix} 0 \\ 0 \\ 0 \end{bmatrix}$$

$$c_N = \begin{bmatrix} -2 \\ 1 \\ -4 \end{bmatrix}$$

$$x_B = A_B^{-1} b = \begin{bmatrix} 6 \\ 5 \\ 8 \end{bmatrix}$$

$$w = c_B^T A_B^{-1} = \begin{bmatrix} 0 & 0 & 0 \end{bmatrix}$$

Let's also assume that the segment size is equal to 2. So, the partial pricing technique will initially calculate only the first two elements of the vector s_N:

$$s_N = c_N^T(1:2) - w A_N(1:2) = \begin{bmatrix} -2 & 1 \end{bmatrix}$$

According to Dantzig's rule, the entering variable is x_1, because it is the one with the most negative $\bar{s}_l = \bar{s}_1 = -2$ in the current segment. Note that if a negative $\bar{s}_l$ value does not exist in the current segment, then we proceed to calculate the vector s_N for the next segment.

6.7.3 Implementation in MATLAB

This subsection presents the implementation in MATLAB of the static partial pricing method using Dantzig's rule, i.e., nonbasic variables are divided into equal segments with predefined size and apply Dantzig's rule to find the entering variable in a segment. Some necessary notations should be introduced before the presentation of the implementation of the Partial Pricing pivoting rule. Let Sn be a $1 \times (n - m)$ vector with the reduced costs, A be a $m \times n$ matrix of coefficients of the constraints, *BasicList* be a $1 \times m$ vector of indices of the basic variables, *NonBasicList* be a $1 \times (n - m)$ vector of indices of the nonbasic variables, *BasisInv* be a $m \times m$ matrix with the basis inverse, *tole* be the value of the tolerance for the pivoting step, *segmentSize* be the size of each segment, and *index* be the index of the entering variable.

The partial pricing method is implemented with two functions: (i) one function that finds the entering variable in the selected segment and it is similar to the function implemented previously for Dantzig's pivoting rule (filename: partialPricingInitial.m), and (ii) one function that updates vector Sn (filename: partialPricingUpdate.m).

The function for the selection of the entering variable of the partial pricing method takes as input the vector of the reduced costs (vector Sn) and returns as output the index of the entering variable (variable *index*).

In line 18, the index of the variable with the most negative Sn is calculated. Unlike Dantzig's rule, partial pricing method does not search all candidate variables, but only the variables in a specific segment.

```
1.    function [index] = partialPricingInitial(Sn)
2.    % Filename: partialPricing.m
3.    % Description: the function is an implementation of the
4.    % static partial pricing method using Dantzig's pivoting
5.    % rule
6.    % Authors: Ploskas, N., & Samaras, N.
7.    %
8.    % Syntax: [index] = partialPricing(Sn)
9.    %
10.   % Input:
11.   % -- Sn: vector of reduced costs (size 1 x (n - m))
12.   %
13.   % Output:
14.   % -- index: the index of the entering variable
```

```
15.
16.    % find the index of the variable with the most
17.    % negative Sn
18.    [~, index] = min(Sn);
19.    end
```

The function of the partial pricing method that updates vector *Sn* takes as input the matrix of coefficients of the constraints (matrix *A*), the vector of coefficients of the objective function (vector *c*), the vector of indices of the basic variables (vector *BasicList*), the vector of indices of the nonbasic variables (vector *NonBasicList*), the matrix with the basis inverse (matrix *BasisInv*), the value of the tolerance for the pivoting step (variable *tole*), and the size of each segment (variable *segmentSize*), and returns as output the vector of the reduced costs (vector *Sn*).

In lines 29–35, vectors *cb*, *cn*, *w*, and *Sn*, and variables *currentSegment*, *negativeFound*, and *Nend* are initialized. Starting from the first segment, we search each segment in order to find a negative value in vector *Sn*. The while loop stops when a negative value is found on a segment or when we finish searching all segments (lines 38–60).

```
1.     function [Sn] = partialPricingUpdate(A, c, BasicList, ...
2.         NonBasicList, BasisInv, tole, segmentSize)
3.     % Filename: partialPricingUpdate.m
4.     % Description: the function is an implementation of the
5.     % update routine for the static partial pricing method
6.     % using Dantzig's pivoting rule
7.     % Authors: Ploskas, N., & Samaras, N.
8.     %
9.     % Syntax: [index] = partialpricingupdate(Sn, A, BasicList, ...
10.    %   NonBasicList, BasisInv, tole, segmentSize)
11.    %
12.    % Input:
13.    % -- A: matrix of coefficients of the constraints
14.    %    (size m x n)
15.    % -- c: vector of coefficients of the objective function
16.    %    (size n x 1)
17.    % -- BasicList: vector of indices of the basic variables
18.    %    (size 1 x m)
19.    % -- NonBasicList: vector of indices of the nonbasic
20.    %    variables (size 1 x (n - m))
21.    % -- BasisInv: matrix with the basis inverse (size m x m)
22.    % -- tole: the value of the tolerance for the pivoting step
23.    % -- segmentSize: the size of the segment
24.
25.    % Output:
26.    % -- Sn: vector of reduced costs (size 1 x (n - m))
27.    %
28.    % initialize vectors cb, cn, w and Sn
29.    cb = c(BasicList);
30.    cn = c(NonBasicList);
31.    w = cb * BasisInv;
32.    Sn = zeros(size(cn), 1);
33.    currentSegment = 1; % initialize current segment
```

```
34.   negativeFound = false;
35.   Nend = 0;
36.   % stop when a negative value is found in Sn or when we
37.   % finished searching all segments
38.   while negativeFound == false && Nend < length(cn)
39.       % the start of the segment
40.       Nstart = (currentSegment - 1) * segmentSize + 1;
41.       % the end of the segment
42.       Nend = Nstart + segmentSize - 1;
43.       % if the end is greater than the length of Sn, set
44.       % Nend = length(Sn)
45.       if Nend > length(cn)
46.           Nend = length(cn);
47.       end
48.       % calculate Sn
49.       N = A(:, NonBasicList);
50.       HRN = w * N(:, Nstart:Nend);
51.       Sn(Nstart:Nend) = cn(Nstart:Nend) - HRN;
52.       % set to zero, the values of Sn that are less than or
53.       % equal to tole
54.       toler = abs(Sn) <= tole;
55.       Sn(toler == 1) = 0;
56.       % check if a negative value exists in Sn or otherwise
57.       % continue with the next segment
58.       negativeFound = ~isempty(Sn < 0);
59.       currentSegment = currentSegment + 1;
60.   end
61.   end
```

A more simplified (and more time-consuming) version of the partial pricing is presented in file partialPricing.m. We will use this version in Chapter 8 in order to avoid adding further complexity to the simplex algorithm when the partial pricing rule is used.

6.8 Steepest Edge Rule

6.8.1 *Mathematical Formulation*

Steepest Edge rule or All-Variable Gradient Method [10] selects as entering variable the variable with the most objective value reduction per unit distance, as shown in Equation (6.1):

$$d_j = \min \left\{ \frac{s_l}{\sqrt{1 + \sum_{i=1}^{m} x_{il}^2}} : l = 1, 2, \ldots, n \right\} \tag{6.1}$$

This pivoting rule can lead to fast convergence to the optimal solution. However, this advantage is debatable due to the additional computational cost. Approximate methods have been proposed in order to improve the computational efficiency of this method [19, 22].

6.8.2 Numerical Example

The LP problem that we will demonstrate Steepest Edge rule is the same as in previous sections and has the following variables:

$$c = \begin{bmatrix} -2 \\ 1 \\ -4 \\ 0 \\ 0 \\ 0 \end{bmatrix}$$

$$A = \begin{bmatrix} 2 & 4 & 2 & 1 & 0 & 0 \\ 1 & -2 & 6 & 0 & 1 & 0 \\ -1 & 4 & 3 & 0 & 0 & 1 \end{bmatrix}$$

$$b = \begin{bmatrix} 6 \\ 5 \\ 8 \end{bmatrix}$$

$$A_B = \begin{bmatrix} 1 & 0 & 0 \\ 0 & 1 & 0 \\ 0 & 0 & 1 \end{bmatrix}$$

$$A_N = \begin{bmatrix} 2 & 4 & 2 \\ 1 & -2 & 6 \\ -1 & 4 & 3 \end{bmatrix}$$

$$A_B^{-1} = A_B = I$$

$$c_B = \begin{bmatrix} 0 \\ 0 \\ 0 \end{bmatrix}$$

$$c_N = \begin{bmatrix} -2 \\ 1 \\ -4 \end{bmatrix}$$

$$x_B = A_B^{-1}b = \begin{bmatrix} 6 \\ 5 \\ 8 \end{bmatrix}$$

$$w = c_B^T A_B^{-1} = \begin{bmatrix} 0 & 0 & 0 \end{bmatrix}$$

$$s_N = c_N^T - wA_N = \begin{bmatrix} -2 & 1 & -4 \end{bmatrix}$$

According to Steepest Edge rule, the entering variable is x_1, because it is the one with the most objective value reduction per unit distance, as shown below:

$$d_j = \min \left\{ \frac{s_l}{\sqrt{1 + \sum_{i=1}^m x_{il}^2}} : l = 1, 2, \ldots, n \right\} = \min \{-0.7559, 0.1644, -0.5657\} =$$

$$-0.7559$$

6.8.3 *Implementation in MATLAB*

This subsection presents the implementation in MATLAB of Steepest Edge pivoting rule (filename: `steepestEdge.m`). Some necessary notations should be introduced before the presentation of the implementation of Steepest Edge pivoting rule. Let Sn be a $1 \times (n-m)$ vector with the reduced costs, *NonBasicList* be a $1 \times (n-m)$ vector of indices of the nonbasic variables, A be a $m \times n$ matrix of coefficients of the constraints, *BasisInv* be a $m \times m$ matrix with the basis inverse, and *index* be the index of the entering variable.

The function for Steepest Edge pivoting rule takes as input the vector of the reduced costs (vector Sn), the vector of indices of the nonbasic variables (vector *NonBasicList*), the matrix of coefficients of the constraints (matrix A), and the matrix with the basis inverse (matrix *BasisInv*), and returns as output the index of the entering variable (variable *index*).

In lines 23–24, the denominator of Equation (6.1) is computed (vector nd). Finally, the index of the entering variable is calculated by finding the index of the minimum value of the division of vector Sn to vector nd (lines 27–28).

```
1.    function [index] = steepestEdge(Sn, NonBasicList, ...
2.       A, BasisInv)
3.    % Filename: steepestEdge.m
4.    % Description: the function is an implementation of the
5.    % Steepest Edge pivoting rule
6.    % Authors: Ploskas, N., & Samaras, N.
7.    %
8.    % Syntax: [index] = steepestEdge(Sn, NonBasicList, ...
```

```
9.    %    A, BasisInv)
10.   %
11.   % Input:
12.   % -- Sn: vector of reduced costs (size 1 x (n - m))
13.   % -- NonBasicList: vector of indices of the nonbasic
14.   %      variables (size 1 x (n - m))
15.   % -- A: matrix of coefficients of the constraints
16.   %      (size m x n)
17.   % -- BasisInv: matrix with the basis inverse (size m x m)
18.   %
19.   % Output:
20.   % -- index: the index of the entering variable
21.
22.   % calculate the denominator of the equation
23.   Y = BasisInv * A(:, NonBasicList);
24.   nd = sqrt(1 + diag(Y' * Y));
25.   % calculate the index of the minimum value of the division
26.   % of the vector Sn to vector nd
27.   [~, j] = min(Sn'./ nd);
28.   index = j(1);
29.   end
```

6.9   Computational Study

In this section, we present a computational study. The aim of this computational study is to highlight the impact of the choice of a pivoting rule in the number of iterations and the execution time of the simplex algorithm. The test set used in the computational study is 19 LPs from the Netlib (optimal, Kennington, and infeasible LPs) and Mészáros problem sets that were presented in Section 3.6. Table 6.1 presents the results from the execution of the revised primal simplex algorithm, which will be presented in Chapter 8, using each pivoting rule, while Table 6.2 presents the number of iterations of the revised primal simplex algorithm using each pivoting rule. The following abbreviations are used: (i) BLAND—Bland's rule, (ii) DANTZIG—Dantzig's rule, (iii) GIM—Greatest Increment Method, (iv) LRCM—Least Recently Considered Method, (v) PP—Partial Pricing, and (vi) SE—Steepest Edge. For each instance, we averaged times over 10 runs. All times are measured in seconds. A time limit of 1 h was set. Execution times do not include the time of the presolve and scaling analysis.

Steepest Edge cannot solve four LPs (osa-07, osa-14, osa-30, and rosen10), Bland's rule cannot solve one LP problem (scfxm1), while Greatest Increment Method cannot solve two LPs (osa-14 and osa-30). Moreover, Steepest Edge has less number of iterations but also higher execution time almost on all LPs compared to the other pivoting rules. Finally, Dantzig's rule is the best pivoting rule in terms of execution time.

Table 6.1 Total execution time of the revised simplex algorithm using each pivoting rule over the benchmark set

Name	Problem			BLAND	DANTZIG	GIM	LRCM	PP	SE
	Constraints	Variables	Nonzeros						
beaconfd	173	262	3,375	0.0467	0.0316	0.0539	0.0467	0.0346	0.0382
cari	400	1,200	152,800	9.9669	8.4223	12.1699	9.5898	10.8548	38.0474
farm	7	12	36	0.0022	0.0019	0.0024	0.0032	0.0020	0.0018
itest6	11	8	20	0.0019	0.0010	0.0011	0.0014	0.0011	0.0010
klein2	477	54	4,585	1.4145	0.4387	0.5791	1.6363	1.2011	0.9924
nsic1	451	463	2,853	0.2402	0.2039	2.3609	0.3191	0.2673	1.2925
nsic2	465	463	3,015	0.2709	0.2372	2.4084	0.4258	0.3444	1.1487
osa-07	1,118	23,949	143,694	12.1577	5.8196	431.0771	27.3094	23.9310	–
osa-14	2,337	52,460	314,760	82.4874	46.3920	–	230.8852	192.6567	–
osa-30	4,350	100,024	600,138	384.9059	245.6443	–	1,339.6000	1,182.3412	–
rosen1	520	1,024	23,274	58.1259	3.0349	37.2976	22.3079	9.5456	150.7600
rosen10	2,056	4,096	62,136	979.4402	37.2525	672.7435	477.6082	234.8841	–
rosen2	1,032	2,048	46,504	458.0748	19.9285	235.1951	142.7747	65.0874	1,404.5467
rosen7	264	512	7,770	4.4192	0.2427	4.1507	2.3669	0.7965	9.1547
rosen8	520	1,024	15,538	22.9467	1.3889	25.5244	9.2055	3.7609	79.9199
sc205	205	203	551	0.0972	0.1119	0.2188	0.2132	0.1198	0.2675
scfxm1	330	457	2,589	–	0.1021	0.1907	0.1921	0.1098	0.2572
sctap2	1,090	1,880	6,714	5.9337	1.3781	84.7093	8.0476	10.6484	161.4832
sctap3	1,480	2,480	8,874	10.5489	3.5675	162.6104	12.3484	14.3926	453.1744
Geometric mean				3.1821	0.8324	3.3506	2.8695	1.9777	2.7111

Table 6.2 Number of iterations of the revised simplex algorithm using each pivoting rule over the benchmark set

Name	Problem Constraints	Variables	Nonzeros	BLAND	DANTZIG	GIM	LRCM	PP	SE
beaconfd	173	262	3,375	47	26	26	41	26	26
cari	400	1,200	152,800	1,925	963	494	1,369	2,174	293
farm	7	12	36	6	5	5	6	5	4
itest6	11	8	20	2	2	2	2	2	2
klein2	477	54	4,585	1,610	500	237	1,560	1,323	350
nsic1	451	463	2,853	485	369	315	433	448	357
nsic2	465	463	3,015	553	473	328	584	574	290
osa-07	1,118	23,949	143,694	2,849	997	898	6,194	4,805	–
osa-14	2,337	52,460	314,760	6,483	2,454	–	23,004	16,921	–
osa-30	4,350	100,024	600,138	14,148	4,681	–	65,995	50,116	–
rosen1	520	1,024	23,274	36,673	1,893	985	17,448	8,479	540
rosen10	2,056	4,096	62,136	131,329	5,116	2,656	90,810	44,731	1,611
rosen2	1,032	2,048	46,504	97,307	4,242	2,077	48,382	22,635	897
rosen7	264	512	7,770	7,127	420	318	3,202	1,420	239
rosen8	520	1,024	15,538	18,823	1,045	729	8,198	3,868	423
sc205	205	203	551	151	165	116	238	162	117
scfxm1	330	457	2,589	–	165	116	238	162	117
sctap2	1,090	1,880	6,714	2,035	1,001	1,377	2,496	3,376	390
sctap3	1,480	2,480	8,874	2,061	1,809	1,754	2,328	2,800	782
Geometric mean				1,602.88	446.53	260.30	1,477.09	1,123.68	174.32

Note that these findings cannot be generalized, because only a sample of benchmark LPs is included in this computational study. Interested readers are referred to the computational studies performed in [6, 16–18, 21].

6.10 Chapter Review

The choice of the pivot element at each iteration is one of the most critical steps in simplex-type algorithms. In this chapter, we presented six pivoting rules used in each iteration of the simplex algorithm to determine the entering variable: (i) Bland's rule, (ii) Dantzig's rule, (iii) Greatest Increment Method, (iv) Least Recently Considered Method, (v) Partial Pricing rule, and (vi) Steepest Edge rule. Each technique was presented with: (i) its mathematical formulation, (ii) a thorough numerical example, and (iii) its implementation in MATLAB. Finally, we performed a computational study in order to compare the execution time of the aforementioned pivoting rules and to highlight the impact of the choice of a pivoting rule in the number of iterations and the execution time of the simplex algorithm. Computational results showed that Dantzig's rule is the best pivoting rule in terms of execution time, while Steepest Edge is the best pivoting rule in terms of iterations. Interested readers are referred to articles that present more advanced pivoting rules [2, 11, 19].

References

1. Bazaraa, M. S., Jarvis, J. J., & Sherali, H. D. (1990). *Linear programming and network flows*. New York: John Wiley & Sons, Inc.
2. Benichou, M., Gautier, J., Hentges, G., & Ribiere, G. (1977). The efficient solution of large–scale linear programming problems. *Mathematical Programming, 13*, 280–322.
3. Bland, R. G. (1977). New finite pivoting rules for the simplex method. *Mathematics of Operations Research, 2*(2), 103–107.
4. Clausen, J. (1987). A note on Edmonds-Fukuda's pivoting rule for the simplex method. *European Journal of Operations Research, 29*, 378–383.
5. Dantzig, G. B. (1963). *Linear programming and extensions*. Princeton, NJ: Princeton University Press.
6. Forrest, J. J., & Goldfarb, D. (1992). Steepest-edge simplex algorithms for linear programming. *Mathematical Programming, 57*(1–3), 341–374.
7. Fukuda, K. (1982). *Oriented matroid programming*. Ph.D. Thesis, Waterloo University, Waterloo, Ontario, Canada.
8. Fukuda, K., & Terlaky, T. (1999). On the existence of a short admissible pivot sequence for feasibility and linear optimization problems. *Pure Mathematics and Applications, 10*(4), 431–447
9. Gärtner, B. (1995). *Randomized optimization by simplex-type methods*. Ph.D. thesis, Freien Universität, Berlin.
10. Goldfarb, D., & Reid, J. K. (1977). A practicable steepest–edge simplex algorithm. *Mathematical Programming, 12*(3), 361–371.
11. Harris, P. M. J. (1973). Pivot selection methods for the Devex LP code. *Mathematical Programming, 5*, 1–28.

12. Klee, V., & Minty, G. J. (1972). How good is the simplex algorithm. In O. Shisha (ed.) *Inequalities – III*. New York and London: Academic Press Inc.
13. Maros, I., & Khaliq, M. H. (1999). Advances in design and implementation of optimization software. *European Journal of Operational Research, 140*(2), 322–337.
14. Murty, K. G. (1974). A note on a Bard type scheme for solving the complementarity problem. *Opsearch, 11*, 123–130.
15. Papadimitriou, C. H., & Steiglitz, K. (1982). *Combinatorial optimization: algorithms and complexity*. Englewood Cliffs, NJ: Prentice-Hall, Inc.
16. Ploskas, N. (2014). *Hybrid optimization algorithms: implementation on GPU*. Ph.D. thesis, Department of Applied Informatics, University of Macedonia.
17. Ploskas, N., & Samaras, N. (2014). Pivoting rules for the revised simplex algorithm. *Yugoslav Journal of Operations Research, 24*(3), 321–332.
18. Ploskas, N., & Samaras, N. (2014). GPU accelerated pivoting rules for the simplex algorithm. *Journal of Systems and Software, 96*, 1–9.
19. Świetanowski, A. (1998). A new steepest edge approximation for the simplex method for linear programming. *Computational Optimization and and Applications, 10*(3), 271–281.
20. Terlaky, T., & Zhang, S. (1993). Pivot rules for linear programming: a survey on recent theoretical developments. *Annals of Operations Research, 46*(1), 203–233.
21. Thomadakis, M. E. (1994). *Implementation and evaluation of primal and dual simplex methods with different pivot-selection techniques in the LPBench environment*. A research report. Texas A&M University, Department of Computer Science.
22. Van Vuuren, J. H., & Grundlingh, W. R. (2001). An active decision support system for optimality in open air reservoir release strategies. *International Transactions in Operational Research, 8*(4), 439–464.
23. Wang, Z. (1989). A modified version of the Edmonds-Fukuda algorithm for LP problems in the general form. *Asia-Pacific Journal of Operational Research, 8*(1), 55–61.
24. Zadeh, N. (1980). *What is the worst case behavior of the simplex algorithm*. Technical report, Department of Operations Research, Stanford University.
25. Zhang, S. (1991). On anti-cycling pivoting rules for the simplex method. *Operations Research Letters, 10*, 189–192.
26. Ziegler, G. M. (1990). *Linear programming in oriented matroids*. Technical Report No. 195, Institut für Mathematik, Universität Augsburg, Germany.

Chapter 7
Basis Inverse and Update Methods

Abstract The computation of the basis inverse is the most time-consuming step in simplex-type algorithms. The basis inverse does not have to be computed from scratch at each iteration, but updating schemes can be applied to accelerate this calculation. This chapter presents two basis inverse and two basis update methods used in simplex-type algorithms: (i) Gauss-Jordan elimination basis inverse method, (ii) LU Decomposition basis inverse method, (iii) Product Form of the Inverse basis update method, and (iv) Modification of the Product Form of the Inverse basis update method. Each technique is presented with: (i) its mathematical formulation, (ii) a thorough numerical example, and (iii) its implementation in MATLAB. Finally, a computational study is performed. The aim of the computational study is to compare the execution time of the basis inverse and update methods and highlight the significance of the choice of the basis update method on simplex-type algorithms and the reduction that it can offer to the solution time.

7.1 Chapter Objectives

- Discuss the importance of the choice of the basis update method in the execution time of simplex-type algorithms.
- Provide the mathematical formulation of the basis inverse and update methods.
- Understand the basis inverse and update methods through a thorough numerical example.
- Implement the basis inverse and update methods using MATLAB.
- Compare the computational performance of the basis inverse and update methods.

7.2 Introduction

As in the solution of any large-scale mathematical system, the computational time for large LPs is a major concern. The basis inverse dictates the total computational effort of an iteration of simplex-type algorithms. This inverse does not have to be

© Springer International Publishing AG 2017   303
N. Ploskas, N. Samaras, *Linear Programming Using MATLAB®*,
Springer Optimization and Its Applications 127, DOI 10.1007/978-3-319-65919-0_7

computed from scratch at each iteration, but can be updated through a number of updating schemes. All efficient versions of the simplex algorithm work with some factorization of the basic matrix A_B or its inverse A_B^{-1}.

Dantzig and Orchard-Hays [6] proposed the Product Form of the Inverse (PFI), which maintains the basis inverse using a set of eta vectors. Benhamadou [3] proposed a Modification of the Product Form of the Inverse (MPFI). The key idea is that the current basis inverse $(A_{\overline{B}})^{-1}$ can be computed from the previous basis inverse $(A_B)^{-1}$ using a simple outer product of two vectors and one matrix addition.

LU Decomposition produces generally sparser factorizations that PFI [4]. A complete LU factorization requires $O(n^3)$ operations. The LU factorization for the basis inverse has been proposed by Markowitz [11]. Markowitz used LU Decomposition to fully invert a matrix, but used the PFI scheme to update the basis inverse during simplex iterations. Bartels and Golub [2] have later proposed a scheme to update a sparse factorization, which was more stable than PFI. Their computational experiments, however, proved that it was more computationally expensive. Forrest and Tomlin [7] created a variant of the Bartels-Golub method by sacrificing some stability characteristics causing the algorithm to have a smaller growth rate in the number of nonzero elements relative to PFI scheme. Reid [21] proposed two variants of the Bartels-Golub updating scheme that aim to balance the sparsity and the numerical stability of the factorization. A variant of the Forrest-Tomlin update was proposed by Suhl and Suhl [23]. Other important updating techniques can be found in Saunders [22] and Goldfarb [8]. A full description of most of these updating methods can be found in Nazareth [13] and Chvatal [5].

There have been many reviews and variants of these methods individually, but only a few comparisons between them. McCoy and Tomlin [12] reported the results of some experiments on measuring the accuracy of the PFI scheme, Bartels-Golub method, and Forrest-Tomlin scheme. Lim et al. [10] provided a comparative study between Bartels-Golub method, Forrest-Tomlin method, and Reid method. Badr et al. [1] performed a computational evaluation of PFI and MPFI updating schemes. Ploskas et al. [20] studied the impact of basis update on exterior point simplex algorithms and proposed a parallel implementation of the PFI scheme. Ploskas et al. [18] performed a computational study to compare four basis update methods: (i) LU Decomposition, (ii) Bartels–Golub, (iii) Forrest–Tomlin, and (iv) Sherman–Morrison–Woodbury. Ploskas and Samaras [15] performed a computational study among three basis inverse and update methods: (i) MATLAB's built-in method *inv*, (ii) Product Form of the Inverse, and (iii) Modification of the Product Form of the Inverse. They incorporated these methods on the exterior point simplex algorithm and the revised simplex algorithm in order to study the significance of the choice of the basis update method in simplex-type algorithms. Finally, Ploskas and Samaras [16] compared PFI and MPFI updating schemes both on their serial and their parallel implementation. The results of the computational study showed that MPFI updating scheme is the fastest when solving large dense LPs. Interested readers are referred to [9] for matrix computations.

This chapter presents two basis inverse and two basis update methods used in simplex algorithm. Each technique is presented with: (i) its mathematical formulation, (ii) a thorough numerical example, and (iii) its implementation in MATLAB. Finally, a computational study is performed. The aim of the computational study is to compare the execution time of the basis inverse and update methods and highlight the significance of the choice of the basis update method on simplex-type algorithms and the reduction that it can offer to the solution time.

The structure of this chapter is as follows. In Sections 7.3–7.6, two basis inverse and two basis update methods are presented. More specifically, Section 7.3 presents the Gauss-Jordan elimination method, Section 7.4 presents the LU Decomposition method, Section 7.5 presents the Product Form of the Inverse update method, while Section 7.6 presents the Modification of the Product Form of the Inverse update method. Section 7.7 presents a computational study that compares the execution time of the basis inverse and update methods and highlights the significance of the choice of the basis update method on simplex-type algorithms and the reduction that it can offer to the solution time. Finally, conclusions and further discussion are presented in Section 7.8.

7.3 Gauss–Jordan Elimination

7.3.1 Mathematical Formulation

Gauss-Jordan elimination is a method for solving systems of linear equations, which can be used to compute the inverse of a matrix. Gauss-Jordan elimination performs the following two steps: (i) Forward elimination: reduces the given matrix to a triangular or echelon form, and (ii) Back Substitution: finds the solution of the given system. Gauss-Jordan elimination with partial pivoting requires $O(n^3)$ time complexity.

We implemented Gauss-Jordan elimination in MATLAB using the *mldivide* operator. In order to find the basis inverse using Gauss-Jordan elimination, one can use Equation (7.1):

$$(A_B)^{-1} = A_B \setminus I \tag{7.1}$$

Applying a sequence of row operations to matrix A_B, we reduce that matrix to the identity matrix I. Then, the basis inverse $(A_B)^{-1}$ can be computed by applying the same sequence of operations to the identity matrix I. If matrix A_B cannot be reduced to the identity matrix I, then $(A_B)^{-1}$ does not exist (a row of zeros will appear in matrix A_B in that situation). The elementary row operations that are used to reduce matrix A_B to the identity matrix I are:

- interchanging two rows ($R_i \leftrightarrow R_j$),
- multiplying a row by a constant factor ($R_i \leftarrow kR_i, k \neq 0$), and
- adding a multiple of one row to another row ($R_i \leftarrow R_i + kR_j, k \neq 0$).

7.3.2 Numerical Example

This subsection demonstrates Gauss-Jordan elimination basis inverse method through an illustrative example. The Gauss-Jordan elimination basis inverse method will be used to invert the following matrix:

$$A_B = \begin{bmatrix} 1 & 3 & 3 \\ 1 & 4 & 3 \\ 2 & 7 & 7 \end{bmatrix}$$

We place A_B and I adjacent to each other.

$$\begin{bmatrix} 1 & 3 & 3 \\ 1 & 4 & 3 \\ 2 & 7 & 7 \end{bmatrix} \begin{bmatrix} 1 & 0 & 0 \\ 0 & 1 & 0 \\ 0 & 0 & 1 \end{bmatrix}$$

Initially, we proceed by reducing A_B to an upper triangular form, i.e., reducing A_B to the form

$$\begin{bmatrix} 1 & * & * \\ 0 & 1 & * \\ 0 & 0 & 1 \end{bmatrix}$$

where $*$ can be any number. This can be achieved by performing the elementary row operations presented in the previous subsection.

Subtract row 1 from row 2 and twice row 1 from row 3 on both matrices:

$$\begin{bmatrix} 1 & 3 & 3 \\ 0 & 1 & 0 \\ 0 & 1 & 1 \end{bmatrix} \begin{bmatrix} 1 & 0 & 0 \\ -1 & 1 & 0 \\ -2 & 0 & 1 \end{bmatrix}$$

Then, subtract row 2 from row 3:

$$\begin{bmatrix} 1 & 3 & 3 \\ 0 & 1 & 0 \\ 0 & 0 & 1 \end{bmatrix} \begin{bmatrix} 1 & 0 & 0 \\ -1 & 1 & 0 \\ -1 & -1 & 1 \end{bmatrix}$$

At this point, matrix A_B has been reduced to an upper triangular form. We continue the row operations in order to reduce the elements above the leading diagonal to zero.

Subtract 3 times row 3 from row 1:

$$\begin{bmatrix} 1 & 3 & 0 \\ 0 & 1 & 0 \\ 0 & 0 & 1 \end{bmatrix} \begin{bmatrix} 4 & 3 & -3 \\ -1 & 1 & 0 \\ -1 & -1 & 1 \end{bmatrix}$$

Then, subtract 3 times row 2 from row 1:

$$\begin{bmatrix} 1 & 0 & 0 \\ 0 & 1 & 0 \\ 0 & 0 & 1 \end{bmatrix} \begin{bmatrix} 7 & 0 & -3 \\ -1 & 1 & 0 \\ -1 & -1 & 1 \end{bmatrix}$$

The left matrix is the identity matrix, so the right matrix is the inverse of the initial matrix A_B.

$$(A_B)^{-1} = \begin{bmatrix} 7 & 0 & -3 \\ -1 & 1 & 0 \\ -1 & -1 & 1 \end{bmatrix}$$

Hence, $A_B A_B^{-1} = A_B^{-1} A_B = I$.

7.3.3 Implementation in MATLAB

This subsection presents the implementation in MATLAB of the Gauss-Jordan elimination basis inverse method (filename: gaussjordanElimination.m). Some necessary notations should be introduced before the presentation of the implementation of the Gauss-Jordan elimination. Let A be a $m \times n$ matrix of coefficients of the constraints, *BasicList* be a $1 \times m$ vector of indices of the basic variables, and *BasisInv* be a $m \times m$ matrix with the basis inverse.

The function for the Gauss-Jordan elimination takes as input the matrix of coefficients of the constraints (matrix A) and the vector of indices of the basic variables (vector *BasicList*), and returns as output the new basis inverse (matrix *BasisInv*).

Gauss-Jordan elimination has been implemented in MATLAB using the *mldivide* operator (line 18). Even though the *mldivide* operator includes a variety of algorithms, we decided to use this operator for the Gauss-Jordan elimination implementation since MATLAB's built-in operators are optimized.

```
1.    function [BasisInv] = gaussjordanElimination(A, BasicList)
2.    % Filename: gaussjordanElimination.m
3.    % Description: the function is an implementation of the
4.    % Gauss-Jordan elimination basis inverse method
5.    % Authors: Ploskas, N., & Samaras, N.
6.    %
7.    % Syntax: [BasisInv] = gaussjordanElimination(A, BasicList)
```

```
8.   %
9.   % Input:
10.  % -- A: matrix of coefficients of the constraints
11.  %      (size m x n)
12.  % -- BasicList: vector of indices of the basic variables
13.  %      (size 1 x m)
14.  %
15.  % Output:
16.  % -- BasisInv: matrix with the new basis inverse (size m x m)
17.
18.  BasisInv = A(:, BasicList) \ speye(length(BasicList));
19.  end
```

7.4 LU Decomposition

7.4.1 Mathematical Formulation

LU Decomposition method factorizes a matrix as the product of an upper (U) and a lower (L) triangular factor, which can be used to compute the inverse of a matrix. LU Decomposition can be computed in $O(n^3)$ time complexity.

Initially, matrix A_B should be decomposed to L and U:

$$A_B = L \times U = \begin{bmatrix} 1 & 0 & 0 \\ l_{21} & 1 & 0 \\ l_{31} & l_{32} & 1 \end{bmatrix} \times \begin{bmatrix} u_{11} & u_{12} & u_{13} \\ 0 & u_{22} & u_{23} \\ 0 & 0 & u_{33} \end{bmatrix}$$

Matrix U is the same as the coefficient matrix at the end of the forward elimination in the Gauss-Jordan elimination procedure. Matrix L has 1 in its diagonal entries. The elements below the main diagonal are the negatives of the multiples that were found during the elimination procedure.

The LU Decomposition can be applied to find the inverse matrix $(A_B)^{-1}$ according to the following two steps:

- the matrix equation $L \times Z = C$, where C is an identity vector, is solved for Z using forward substitution.
- the matrix equation $U \times X = Z$ is solved for X using backward substitution.

These two steps are performed for each column of the inverse matrix.

7.4.2 Numerical Example

This subsection demonstrates LU Decomposition basis inverse method through an illustrative example. The LU Decomposition basis inverse method will be used to invert the following matrix:

$$A_B = \begin{bmatrix} 1 & 3 & 3 \\ 1 & 4 & 3 \\ 2 & 7 & 7 \end{bmatrix}$$

Applying the Gauss-Jordan elimination procedure, we proceed by reducing A_B to an upper triangular form. Subtract row 1 from row 2 and twice row 1 from row 3 on both matrices:

$$\begin{bmatrix} 1 & 3 & 3 \\ 0 & 1 & 0 \\ 0 & 1 & 1 \end{bmatrix}$$

Then, subtract row 2 from row 3:

$$\begin{bmatrix} 1 & 3 & 3 \\ 0 & 1 & 0 \\ 0 & 0 & 1 \end{bmatrix}$$

At the end of this step, we have found the upper triangular matrix U:

$$U = \begin{bmatrix} 1 & 3 & 3 \\ 0 & 1 & 0 \\ 0 & 0 & 1 \end{bmatrix}$$

Now, we proceed to find matrix L. During the Gauss-Jordan elimination procedure, we subtracted once row 1 from row 2, so $l_{21} = 1$. Then, we subtracted twice row 1 from row 3, so $l_{31} = 2$. Finally, we subtracted once row 2 from row 3, so $l_{32} = 1$. Hence, we have found the lower triangular matrix L:

$$L = \begin{bmatrix} 1 & 0 & 0 \\ 1 & 1 & 0 \\ 2 & 1 & 1 \end{bmatrix}$$

Next, we perform forward and backward substitution for each column of the inverse matrix.

1st column

Solve $L \times Z = C$ for Z:

$$\begin{bmatrix} 1 & 0 & 0 \\ 1 & 1 & 0 \\ 2 & 1 & 1 \end{bmatrix} \times \begin{bmatrix} z_1 \\ z_2 \\ z_3 \end{bmatrix} = \begin{bmatrix} 1 \\ 0 \\ 0 \end{bmatrix}$$

This generates the equations:

$$z_1 = 1$$

$$z_1 + z_2 = 0$$

$$2z_1 + z_2 + z_3 = 0$$

Solve for Z:

$$z_1 = 1$$

$$z_2 = 0 - z_1 = -1$$

$$z_3 = 0 - 2z_1 - z_2 = -1$$

Hence:

$$Z = \begin{bmatrix} 1 \\ -1 \\ -1 \end{bmatrix}$$

Then, solve $U \times X = Z$ for X:

$$\begin{bmatrix} 1 & 3 & 3 \\ 0 & 1 & 0 \\ 0 & 0 & 1 \end{bmatrix} \times \begin{bmatrix} x_1 \\ x_2 \\ x_3 \end{bmatrix} = \begin{bmatrix} 1 \\ -1 \\ -1 \end{bmatrix}$$

This generates the equations:

$$x_1 + 3x_2 + 3x_3 = 1$$

$$x_2 = -1$$

$$x_3 = -1$$

Solve for X:

$$x_3 = -1$$

$$x_2 = -1$$

$$x_1 = 1 - 3x_2 - 3x_3 = 1 + 3 + 3 = 7$$

Hence:

$$X = \begin{bmatrix} 7 \\ -1 \\ -1 \end{bmatrix}$$

X is the first column of the basis inverse of A_B.

2nd column

Solve $L \times Z' = C$ for Z':

$$\begin{bmatrix} 1 & 0 & 0 \\ 1 & 1 & 0 \\ 2 & 1 & 1 \end{bmatrix} \times \begin{bmatrix} z'_1 \\ z'_2 \\ z'_3 \end{bmatrix} = \begin{bmatrix} 0 \\ 1 \\ 0 \end{bmatrix}$$

This generates the equations:

$$z'_1 = 0$$
$$z'_1 + z'_2 = 1$$
$$2z'_1 + z'_2 + z'_3 = 0$$

Solve for Z':

$$z'_1 = 0$$
$$z'_2 = 1 - z'_1 = 1$$
$$z'_3 = 0 - 2z'_1 - z'_2 = -1$$

Hence:

$$Z' = \begin{bmatrix} 0 \\ 1 \\ -1 \end{bmatrix}$$

Then, solve $U \times X' = Z'$ for X':

$$\begin{bmatrix} 1 & 3 & 3 \\ 0 & 1 & 0 \\ 0 & 0 & 1 \end{bmatrix} \times \begin{bmatrix} x'_1 \\ x'_2 \\ x'_3 \end{bmatrix} = \begin{bmatrix} 0 \\ 1 \\ -1 \end{bmatrix}$$

This generates the equations:

$$x'_1 + 3x'_2 + 3x'_3 = 0$$
$$x'_2 = 1$$
$$x'_3 = -1$$

Solve for X':

$$x_3' = -1$$

$$x_2' = 1$$

$$x_1' = 0 - 3x_2' - 3x_3' = 0 + 3 - 3 = 0$$

Hence:

$$X' = \begin{bmatrix} 0 \\ 1 \\ -1 \end{bmatrix}$$

X' is the second column of the basis inverse of A_B.

3rd column

Solve $L \times Z'' = C$ for Z'':

$$\begin{bmatrix} 1 & 0 & 0 \\ 1 & 1 & 0 \\ 2 & 1 & 1 \end{bmatrix} \times \begin{bmatrix} z_1'' \\ z_2'' \\ z_3'' \end{bmatrix} = \begin{bmatrix} 0 \\ 0 \\ 1 \end{bmatrix}$$

This generates the equations:

$$z_1'' = 0$$

$$z_1'' + z_2'' = 0$$

$$2z_1'' + z_2'' + z_3'' = 1$$

Solve for Z'':

$$z_1' = 0$$

$$z_2' = 0 - z_1' = 0$$

$$z_3' = 1 - 2z_1' - z_2' = 1$$

Hence:

$$Z'' = \begin{bmatrix} 0 \\ 0 \\ 1 \end{bmatrix}$$

Then, solve $U \times X'' = Z''$ for X'':

$$\begin{bmatrix} 1 & 3 & 3 \\ 0 & 1 & 0 \\ 0 & 0 & 1 \end{bmatrix} \times \begin{bmatrix} x_1'' \\ x_2'' \\ x_3'' \end{bmatrix} = \begin{bmatrix} 0 \\ 0 \\ 1 \end{bmatrix}$$

This generates the equations:

$$x_1'' + 3x_2'' + 3x_3'' = 0$$
$$x_2'' = 0$$
$$x_3'' = 1$$

Solve for X'':

$$x_3'' = 1$$
$$x_2'' = 0$$
$$x_1'' = 0 - 3x_2'' - 3x_3'' = 0 - 0 - 3 = -3$$

Hence:

$$X'' = \begin{bmatrix} -3 \\ 0 \\ 1 \end{bmatrix}$$

X'' is the third column of the basis inverse of A_B.
To sum up, the basis inverse of A_B is:

$$(A_B)^{-1} = \begin{bmatrix} 7 & 0 & -3 \\ -1 & 1 & 0 \\ -1 & -1 & 1 \end{bmatrix}$$

7.4.3 Implementation in MATLAB

This subsection presents the implementation in MATLAB of the LU Decomposition basis inverse method (filename: luInverse.m). Some necessary notations should be introduced before the presentation of the implementation of the LU Decomposition. Let A be a $m \times n$ matrix of coefficients of the constraints, *BasicList* be a $1 \times m$ vector of indices of the basic variables, and *BasisInv* be a $m \times m$ matrix with the basis inverse.

The function for the calculation of the inverse using LU Decomposition takes as input the matrix of coefficients of the constraints (matrix A) and the vector of indices of the basic variables (vector *BasicList*), and returns as output the new basis inverse (matrix *BasisInv*).

In line 19, L and U matrices are calculated using MATLAB's built-in function *lu*. The basis inverse has been implemented in MATLAB using the *mldivide* operator (line 21). Even though we do not know the exact LU factorization algorithm that the *lu* function implements, we decided to use this function since MATLAB's built-in functions are optimized.

```
1.    function [BasisInv] = luInverse(A, BasicList)
2.    % Filename: luInverse.m
3.    % Description: the function is an implementation of the LU
4.    % Decomposition method to find the inverse of a matrix
5.    % Authors: Ploskas, N., & Samaras, N.
6.    %
7.    % Syntax: [BasisInv] = luInverse(A, BasicList)
8.    %
9.    % Input:
10.   % -- A: matrix of coefficients of the constraints
11.   %    (size m x n)
12.   % -- BasicList: vector of indices of the basic variables
13.   %    (size 1 x m)
14.   %
15.   % Output:
16.   % -- BasisInv: matrix with the new basis inverse (size m x m)
17.
18.   % compute L and U matrices
19.   [L, U] = lu(A(:, BasicList));
20.   % calculate the basis inverse
21.   BasisInv = U \ (L \ (speye(length(BasicList))));
30.   end
```

7.5 Product Form of the Inverse (PFI)

7.5.1 Mathematical Formulation

In order to update the basis inverse (square matrix), the PFI scheme uses only information on the leaving variable along with the current basis inverse $(A_B)^{-1}$. The new basis differs from the previous one only in one column. The new basis inverse can be updated at any iteration using Equation (7.2).

$$(A_{\overline{B}})^{-1} = (A_B E)^{-1} = E^{-1}(A_B)^{-1} \qquad (7.2)$$

where E^{-1} is the inverse of the eta-matrix and can be computed by Equation (7.3):

$$E^{-1} = I - \frac{1}{h_{rl}} (h_l - e_l) e_l^T = \begin{bmatrix} 1 & & -h_{1l} & & \\ & \ddots & \vdots & & \\ & & 1/h_{rl} & & \\ & & \vdots & \ddots & \\ & & -h_{ml}/h_{rl} & & 1 \end{bmatrix} \tag{7.3}$$

If the current basis inverse is computed using regular multiplication, then the complexity of the PFI is $\Theta (m^3)$.

7.5.2   Numerical Example

This subsection demonstrates the PFI basis update method through an illustrative example. The PFI basis update method will be used to update the previous basis inverse matrix:

$$(A_B)^{-1} = \begin{bmatrix} 7 & 0 & -3 \\ -1 & 1 & 0 \\ -1 & -1 & 1 \end{bmatrix}$$

Let's assume that the pivot column and the leaving variable are (more details in Chapter 8):

$$h_l = \begin{bmatrix} -1 \\ 3 \\ -3 \end{bmatrix}$$

$$r = 2$$

Applying the PFI basis update method, we first construct the η column vector:

$$\eta = -h_l/h_{l_r} = \begin{bmatrix} -1 \\ 3 \\ -3 \end{bmatrix} /3 = \begin{bmatrix} 1/3 \\ -1 \\ 1 \end{bmatrix}$$

$$\eta_r = 1/h_{l_r} = 1/3$$

Hence, the η column vector is the following:

$$\eta = \begin{bmatrix} 1/3 \\ 1/3 \\ 1 \end{bmatrix}$$

We proceed by constructing matrix E, which is the identity matrix except its column r $(r = 2)$ is equal to the η column vector:

$$E^{-1} = \begin{bmatrix} 1 & 1/3 & 0 \\ 0 & 1/3 & 0 \\ 0 & 1 & 1 \end{bmatrix}$$

Finally, we multiply matrix E with the previous basis inverse $(A_B)^{-1}$ to form the new basis inverse:

$$(A_{\overline{B}})^{-1} = E^{-1} \times (A_B)^{-1} = \begin{bmatrix} 1 & 1/3 & 0 \\ 0 & 1/3 & 0 \\ 0 & 1 & 1 \end{bmatrix} \times \begin{bmatrix} 7 & 0 & -3 \\ -1 & 1 & 0 \\ -1 & -1 & 1 \end{bmatrix} = \begin{bmatrix} 20/3 & 1/3 & -3 \\ 1/3 & 1/3 & 0 \\ -2 & 0 & 1 \end{bmatrix}$$

7.5.3 Implementation in MATLAB

This subsection presents the implementation in MATLAB of the Product Form of the Inverse basis update method (filename: `pfi.m`). Some necessary notations should be introduced before the presentation of the implementation of the Product Form of the Inverse. Let *BasisInv* be a $m \times m$ matrix with the basis inverse, h_l be a $m \times 1$ vector with the pivot column, and r be the index of the leaving variable.

The function for the Product Form of the Inverse takes as input the matrix with the basis inverse (matrix *BasisInv*), the vector of the pivot column (vector h_l), and the index of the leaving variable (variable r), and returns as output the new basis inverse (matrix *BasisInv*).

In lines 18–19, the *eta* column vector is calculated. The *eta* matrix is computed in lines 20–21. Finally, the basis inverse is updated by multiplying the *eta* matrix and the previous basis inverse (line 22).

```
1.    function [BasisInv] = pfi(BasisInv, h_1, r)
2.    % Filename: pfi.m
3.    % Description: the function is an implementation of the
4.    % Product Form of the Inverse basis update method
5.    % Authors: Ploskas, N., & Samaras, N.
6.    %
7.    % Syntax: [BasisInv] = pfi(BasisInv, h_1, r)
8.    %
9.    % Input:
10.   % -- BasisInv: matrix with the basis inverse (size m x m)
11.   % -- h_1: vector with pivot column (size m x 1)
12.   % -- r: the index of the leaving variable
13.   %
14.   % Output:
15.   % -- BasisInv: matrix with the new basis inverse
16.   %    (size m x m)
17.
```

```
18.   eta = -h_l / h_l(r); % compute the eta column vector
19.   eta(r) = 1 / h_l(r);
20.   EInv = speye(length(BasisInv)); % create eta matrix
21.   EInv(:, r) = eta;
22.   BasisInv = EInv * BasisInv; % calculate new basis inverse
23.   end
```

7.6   Modification of the Product Form of the Inverse (MPFI)

7.6.1   Mathematical Formulation

MPFI updating scheme has been proposed by Benhamadou [3]. The main idea of this method is that the current basis inverse $(A_{\bar{B}})^{-1}$ can be computed from the previous inverse $(A_B)^{-1}$ using a simple outer product of two vectors and one matrix addition, as shown in Equation (7.4):

$$(A_{\bar{B}})^{-1} = (A_{\bar{B}})^{-1}_{r.} + h_l \otimes (A_B)^{-1}_{r.} \tag{7.4}$$

The updating scheme of the inverse is shown in Equation (7.5).

$$(A_{\bar{B}})^{-1} = \begin{vmatrix} b_{11} & \cdots & b_{1m} \\ \vdots & \ddots & \vdots \\ 0 & 0 & 0 \\ \vdots & & \ddots & \vdots \\ b_{m1} & \cdots & b_{mm} \end{vmatrix} + \begin{vmatrix} -\frac{h_{1l}}{h_{rl}} \\ \vdots \\ -\frac{1}{h_{rl}} \\ \vdots \\ -\frac{h_{ml}}{h_{rl}} \end{vmatrix} \quad (A_B)^{-1} : \begin{vmatrix} b_{r1} & \cdots & b_{rr} & \cdots & b_{rm} \end{vmatrix} \tag{7.5}$$

The outer product of Equation (7.4) requires m^2 multiplications and the addition of two matrices requires m^2 additions. Hence, the time complexity is $\Theta(m^2)$.

7.6.2   Numerical Example

This subsection demonstrates the MPFI basis update method through an illustrative example. The MPFI basis update method will be used to update the previous basis inverse matrix:

$$(A_B)^{-1} = \begin{bmatrix} 7 & 0 & -3 \\ -1 & 1 & 0 \\ -1 & -1 & 1 \end{bmatrix}$$

Let's assume that the pivot column and the leaving variable are (more details in Chapter 8):

$$h_l = \begin{bmatrix} -1 \\ 3 \\ -3 \end{bmatrix}$$

$$r = 2$$

Applying the MPFI basis update method, we first construct the η column vector:

$$\eta = -h_l/h_{l_r} = \begin{bmatrix} -1 \\ 3 \\ -3 \end{bmatrix} /3 = \begin{bmatrix} 1/3 \\ -1 \\ 1 \end{bmatrix}$$

$$\eta_r = 1/h_{l_r} = 1/3$$

Hence, the η column vector is the following:

$$\eta = \begin{bmatrix} 1/3 \\ 1/3 \\ 1 \end{bmatrix}$$

We proceed by calculating a simple outer product of two vectors:

$$K = h_l \otimes (A_B)_{r.}^{-1} = \begin{bmatrix} 1/3 \\ 1/3 \\ 1 \end{bmatrix} \otimes \begin{bmatrix} -1 & 1 & 0 \end{bmatrix} = \begin{bmatrix} -1/3 & 1/3 & 0 \\ -1/3 & 1/3 & 0 \\ -1 & 1 & 0 \end{bmatrix}$$

Then, we set equal to zero the elements in the row r of the basis inverse:

$$(A_B)^{-1} = \begin{bmatrix} 7 & 0 & -3 \\ 0 & 0 & 0 \\ -1 & -1 & 1 \end{bmatrix}$$

Finally, we add the above matrices to form the new basis inverse:

$$(A_{\bar{B}})^{-1} = (A_B)^{-1} + K = \begin{bmatrix} 7 & 0 & -3 \\ 0 & 0 & 0 \\ -1 & -1 & 1 \end{bmatrix} + \begin{bmatrix} -1/3 & 1/3 & 0 \\ -1/3 & 1/3 & 0 \\ -1 & 1 & 0 \end{bmatrix} = \begin{bmatrix} 20/3 & 1/3 & -3 \\ 1/3 & 1/3 & 0 \\ -2 & 0 & 1 \end{bmatrix}$$

7.6.3 Implementation in MATLAB

This subsection presents the implementation in MATLAB of the Modification of the Product Form of the Inverse basis update method (filename: mpfi.m). Some necessary notations should be introduced before the presentation of the implementation of the Modification of the Product Form of the Inverse. Let *BasisInv* be a $m \times m$ matrix with the basis inverse, h_l be a $m \times 1$ vector with the pivot column, and r be the index of the leaving variable.

The function for the Modification of the Product Form of the Inverse takes as input the matrix with the basis inverse (matrix *BasisInv*), the vector of the pivot column (vector h_l), and the index of the leaving variable (variable r), and returns as output the new basis inverse (matrix *BasisInv*).

In lines 18–19, the *eta* column vector is calculated. A simple outer product between two vectors is performed in line 20. Finally, we set equal to zero the elements in row r of the basis inverse (line 21) and update it by adding the two matrices (line 23).

```
1.    function [BasisInv] = mpfi(BasisInv, h_l, r)
2.    % Filename: mpfi.m
3.    % Description: the function is an implementation of the
4.    % Modification of the Product Form of the Inverse basis
5.    % update method
6.    % Authors: Ploskas, N., & Samaras, N.
7.    %
8.    % Syntax: [BasisInv] = mpfi(BasisInv, h_l, r)
9.    %
10.   % Input:
11.   % -- BasisInv: matrix with the basis inverse (size m x m)
12.   % -- h_l: vector with pivot column (size m x 1)
13.   % -- r: the index of the leaving variable
14.   %
15.   % Output:
16.   % -- BasisInv: matrix with the new basis inverse (size m x m)
17.
18.   eta = -h_l / h_l(r); % compute the eta column vector
19.   eta(r) = 1 / h_l(r);
20.   K = eta * BasisInv(r, :); % perform a simple outer product
21.   BasisInv(r, :) = 0; % set equal to zero the elements in the
22.   % row r of the basis inverse
23.   BasisInv = BasisInv + K; % add the matrices
24.   end
```

7.7 Computational Study

This section presents a computational study. The aim of the computational study is to highlight the significance of the choice of the basis update method on simplex-type algorithms and the reduction that it can offer to the solution time.

The test set used in the computational study is 19 LPs from the Netlib (optimal, Kennington, and infeasible LPs) and Mészáros problem sets that were presented in Section 3.6. Table 7.1 presents the total execution time of the revised primal simplex algorithm, which will be presented in Chapter 8, using each of the basis inverse or update method, while Table 7.2 presents the execution time of each basis inverse or update method. Table 7.3 presents the total execution time of the revised dual simplex algorithm, which will be presented in Chapter 9, using each of the basis inverse or update method, while Table 7.4 presents the execution time of each basis inverse or update method. In addition, Table 7.5 presents the total execution time of the exterior point simplex algorithm, which will be presented in Chapter 10, using each of the basis inverse or update method, while Table 7.6 presents the execution time of each basis inverse or update method. MATLAB's built-in function inv (used to invert a matrix) has been also included in the computational study. The following abbreviations are used: (i) GE—Gauss-Jordan elimination, (ii) LU—LU Decomposition, (iii) INV—MATLAB's built-in function inv, (iv) MPFI—Modification of the Product Form of the Inverse, and (v) PFI—Product Form of the Inverse. For each instance, we averaged times over 10 runs. All times are measured in seconds. Execution times in Tables 7.1, 7.3, and 7.5 do not include the time needed to presolve and scale the LPs.

Tables 7.1–7.6 show that Product Form of the Inverse is the fastest basis update method both on the primal and dual simplex algorithm and on the exterior point simplex algorithm.

Note that these findings cannot be generalized, because only a sample of benchmark LPs is included in this computational study. Interested readers are referred to the computational studies performed in [1, 10, 12, 14–20].

7.8 Chapter Review

The computation of the basis inverse is the most time-consuming step in simplex-type algorithms. The basis inverse does not have to be computed from scratch at each iteration, but updating schemes can be applied to accelerate this calculation. In this chapter, we presented two basis inverse and two basis update methods used in the simplex algorithm to calculate or update the basis inverse: (i) Gauss-Jordan elimination, (ii) LU Decomposition, (iii) Product Form of the Inverse basis update method, and (iv) Modification of the Product Form of the Inverse basis update method. Each technique was presented with: (i) its mathematical formulation, (ii) a thorough numerical example, and (iii) its implementation in MATLAB. Finally, we performed a computational study in order to compare the execution time of the aforementioned basis inverse and update methods and to highlight the significance of the choice of the basis update method on simplex-type algorithms and the reduction that it can offer to the solution time. Computational results showed that Product Form of the Inverse is the fastest basis update method both on the primal and dual simplex algorithm and on the exterior point simplex algorithm.

Table 7.1 Total execution time of the revised primal simplex algorithm using each basis inverse or update method over the benchmark set

Name	Problem			GE	LU	INV	MPFI	PFI
	Constraints	Variables	Nonzeros					
beaconfd	173	262	3,375	0.0344	0.0338	0.0488	0.0365	0.0319
cari	400	1,200	152,800	16.6087	10.1415	16.7874	8.5551	8.6720
farm	7	12	36	0.0032	0.0019	0.0020	0.0022	0.0022
itest6	11	8	20	0.0011	0.0015	0.0010	0.0010	0.0026
klein2	477	54	4,585	8.3114	3.6184	8.1297	0.5255	0.4337
nsic1	451	463	2,853	3.0588	0.4613	2.9616	0.2341	0.2197
nsic2	465	463	3,015	3.3771	1.4465	3.4713	0.2448	0.2393
osa-07	1,118	23,949	143,694	26.8339	5.9738	26.9484	7.1094	5.4501
osa-14	2,337	52,460	314,760	264.6899	51.3072	261.6627	80.2635	45.6474
osa-30	4,350	100,024	600,138	1,950.4067	282.7747	1,955.1719	529.1137	242.9726
rosen1	520	1,024	23,274	43.8646	76.4467	44.1804	3.2411	3.0490
rosen10	2,056	4,096	62,136	1,415.1950	875.3540	1,329.2462	48.1428	39.6499
rosen2	1,032	2,048	46,504	347.4654	694.4424	345.8603	24.2268	18.3378
rosen7	264	512	7,770	1.9400	1.5618	1.8741	0.2338	0.2508
rosen8	520	1,024	15,538	17.6818	18.7221	17.5344	1.3457	1.3466
sc205	205	203	551	0.3372	0.2311	0.3772	0.1145	0.1108
scfxm1	330	457	2,589	2.0346	2.0758	1.9073	0.7500	0.7640
sctap2	1,090	1,880	6,714	34.7137	2.5135	34.7151	1.4779	1.3675
sctap3	1,480	2,480	8,874	110.4081	7.8032	111.7859	4.1684	3.6009
Geometric mean				6.6650	3.1363	6.5793	1.0734	0.9787

Table 7.2 Execution time of each basis inverse or update method on the revised primal simplex algorithm over the benchmark set

Name	Problem		Nonzeros	GE	LU	INV	MPFI	PFI
	Constraints	Variables						
beaconfd	173	262	3,375	0.0026	0.0031	0.0018	0.0022	0.0012
cari	400	1,200	152,800	8.0150	1.6030	8.0863	0.2092	0.2282
farm	7	12	36	0.0001	0.0001	0.0001	0.0001	0.0001
itest6	11	8	20	0.0001	0.0001	0.0001	0.0001	0.0001
klein2	477	54	4,585	8.1507	3.4696	7.9749	0.3998	0.3149
nsic1	451	463	2,853	2.8899	0.3211	2.8024	0.0762	0.0709
nsic2	465	463	3,015	3.1994	1.2331	3.2820	0.0741	0.0689
osa-07	1,118	23,949	143,694	23.0777	2.4176	23.1982	3.3756	1.7866
osa-14	2,337	52,460	314,760	243.8971	30.9076	241.7203	58.0559	23.4736
osa-30	4,350	100,024	600,138	1,868.7580	206.2005	1,873.2799	444.8742	160.4645
rosen1	520	1,024	23,274	42.9150	75.3475	43.1952	2.4378	2.2755
rosen10	2,056	4,096	62,136	1,407.9921	868.2399	1,322.0202	41.5232	33.6067
rosen2	1,032	2,048	46,504	343.7244	690.5800	342.1048	20.8668	15.2465
rosen7	264	512	7,770	1.8135	1.4431	1.7548	0.1163	0.1263
rosen8	520	1,024	15,538	17.2332	18.2677	17.0908	0.9219	0.9216
sc205	205	203	551	0.2665	0.1678	0.2905	0.0421	0.0383
scfxm1	330	457	2,589	1.4410	1.4916	1.3322	0.1842	0.1586
sctap2	1,090	1,880	6,714	33.8917	1.8280	33.8790	0.8291	0.7246
sctap3	1,480	2,480	8,874	108.8346	6.5149	110.1861	2.8708	2.3088
Geometric mean				3.8441	1.5496	3.7019	0.3668	0.2782

Table 7.3 Total execution time of the revised dual simplex algorithm using each basis inverse or update method over the benchmark set

Name	Problem Constraints	Variables	Nonzeros	GE	LU	INV	MPFI	PFI
beaconfd	173	262	3,375	0.0407	0.0347	0.0403	0.0323	0.0325
cari	400	1,200	152,800	16.0492	22.0346	17.0654	8.6429	9.3946
farm	7	12	36	0.0011	0.0012	0.0012	0.0012	0.0012
itest6	11	8	20	0.0008	0.0007	0.0007	0.0038	0.0008
klein2	477	54	4,585	10.7569	0.3804	2.6020	1.0002	1.0001
nsic1	451	463	2,853	1.0762	0.4386	1.0716	0.1731	0.1706
nsic2	465	463	3,015	2.9153	1.1476	2.9070	0.3028	0.2974
osa-07	1,118	23,949	143,694	1.4871	1.5858	1.4676	1.4305	1.4350
osa-14	2,337	52,460	314,760	16.3658	8.5683	15.7739	7.7850	8.7074
osa-30	4,350	100,024	600,138	51.2721	35.6900	52.0340	31.4383	30.1981
rosen1	520	1,024	23,274	8.1665	6.0426	8.1753	0.6662	0.6595
rosen10	2,056	4,096	62,136	467.0564	122.5778	461.8288	11.4776	10.4739
rosen2	1,032	2,048	46,504	81.0522	80.4951	79.7259	2.8339	2.7021
rosen7	264	512	7,770	0.6566	0.2641	0.5624	0.0816	0.0912
rosen8	520	1,024	15,538	6.3170	3.1808	6.2501	0.4433	0.4978
sc205	205	203	551	0.5625	0.7708	0.5587	0.1992	0.1924
scfxm1	330	457	2,589	1.9375	1.2654	1.9381	0.6483	0.6482
sctap2	1,090	1,880	6,714	17.2722	0.8777	17.2754	0.6989	0.6869
sctap3	1,480	2,480	8,874	44.4986	1.5810	44.4741	1.2476	1.3616
Geometric mean				2.4563	1.0773	2.2561	0.5009	0.4676

Table 7.4 Execution time of each basis inverse or update method on the revised dual simplex algorithm over the benchmark set

Name	Problem Constraints	Variables	Nonzeros	GE	LU	INV	MPFI	PFI
beaconfd	173	262	3,375	0.0094	0.0039	0.0090	0.0017	0.0016
cari	400	1,200	152,800	7.9220	13.4078	8.4607	0.5443	0.5159
farm	7	12	36	0.0001	0.0001	0.0001	0.0001	0.0001
itest6	11	8	20	0.0001	0.0001	0.0001	0.0001	0.0001
klein2	477	54	4,585	10.4863	0.3293	2.5250	0.8365	0.8363
nsic1	451	463	2,853	0.9344	0.2972	0.9296	0.0354	0.0327
nsic2	465	463	3,015	2.6864	0.9190	2.6812	0.0798	0.0757
osa-07	1,118	23,949	143,694	0.0917	0.1892	0.0829	0.0437	0.0386
osa-14	2,337	52,460	314,760	7.9862	0.9277	7.9026	0.1971	0.1872
osa-30	4,350	100,024	600,138	21.0650	3.7840	20.4263	0.7268	0.5372
rosen1	520	1,024	23,274	7.8644	5.7301	7.8692	0.3368	0.3284
rosen10	2,056	4,096	62,136	462.5680	118.3438	457.3906	7.7016	6.9408
rosen2	1,032	2,048	46,504	79.4267	78.9840	78.1214	1.7200	1.6029
rosen7	264	512	7,770	0.5762	0.1908	0.5007	0.0214	0.0226
rosen8	520	1,024	15,538	6.0510	2.9208	5.9898	0.1991	0.2165
sc205	205	203	551	0.4625	0.6678	0.4589	0.0875	0.0804
scfxm1	330	457	2,589	1.3993	0.7297	1.3978	0.1137	0.1115
sctap2	1,090	1,880	6,714	16.7926	0.4627	16.7918	0.2785	0.2698
sctap3	1,480	2,480	8,874	43.7324	0.9586	43.7046	0.6167	0.6584
Geometric mean				1.2997	0.4442	1.1858	0.0850	0.0812

Table 7.5 Total execution time of the exterior point simplex algorithm using each basis inverse or update method over the benchmark set

Name	Problem			GE	LU	INV	MPFI	PFI
	Constraints	Variables	Nonzeros					
beaconfd	173	262	3,375	0.0471	0.0429	0.0472	0.0399	0.0513
cari	400	1,200	152,800	18.0295	12.1792	17.7262	10.7169	9.4911
farm	7	12	36	0.0021	0.0021	0.0021	0.0021	0.0022
itest6	11	8	20	0.0010	0.0010	0.0010	0.0010	0.0024
klein2	477	54	4,585	8.0004	3.3757	7.9929	0.4812	0.4041
nsic1	451	463	2,853	2.9178	0.5854	2.9271	0.3082	0.3045
nsic2	465	463	3,015	3.4620	1.0355	3.6378	0.4487	0.4451
osa-07	1,118	23,949	143,694	86.3060	67.7109	86.9472	65.6599	65.2887
osa-14	2,337	52,460	314,760	571.0387	373.8809	569.8053	385.9542	356.2622
osa-30	4,350	100,024	600,138	3,144.0239	1,751.5993	3,183.9778	1,863.9343	1,543.9909
rosen1	520	1,024	23,274	31.3143	41.6023	30.6840	3.5836	3.0822
rosen10	2,056	4,096	62,136	831.1766	174.6777	820.2575	25.4440	23.9468
rosen2	1,032	2,048	46,504	243.9280	296.2548	240.4429	17.8997	14.8725
rosen7	264	512	7,770	1.9876	1.4657	2.0587	0.3858	0.3755
rosen8	520	1,024	15,538	17.6718	10.6203	16.7293	1.8747	1.7754
sc205	205	203	551	0.4342	0.3761	0.4327	0.1635	0.1590
scfxm1	330	457	2,589	2.6557	2.1441	2.7246	0.8229	0.7852
sctap2	1,090	1,880	6,714	23.0360	1.8424	21.9999	1.7607	1.6440
sctap3	1,480	2,480	8,874	60.9264	5.3182	60.5953	3.7064	3.1557
Geometric mean				6.8823	3.5291	6.8448	1.5251	1.5062

Table 7.6 Execution time of each basis inverse or update method on the exterior point simplex algorithm over the benchmark set

Name	Problem			GE	LU	INV	MPFI	PFI
	Constraints	Variables	Nonzeros					
beaconfd	173	262	3,375	0.0090	0.0048	0.0087	0.0018	0.0023
cari	400	1,200	152,800	8.4798	2.7983	8.4551	0.3007	0.2314
farm	7	12	36	0.0001	0.0001	0.0001	0.0001	0.0001
itest6	11	8	20	0.0001	0.0001	0.0001	0.0001	0.0001
klein2	477	54	4,585	7.8602	3.2487	7.8524	0.3745	0.2969
nsic1	451	463	2,853	2.6620	0.3312	2.6675	0.0615	0.0562
nsic2	465	463	3,015	3.1238	0.6803	3.2706	0.0970	0.0935
osa-07	1,118	23,949	143,694	25.8703	6.6107	25.7109	4.2370	2.6285
osa-14	2,337	52,460	314,760	234.8338	49.3288	234.2490	61.8845	27.0776
osa-30	4,350	100,024	600,138	1,723.5131	329.4812	1,745.0322	462.6788	176.0256
rosen1	520	1,024	23,274	29.3850	39.0397	28.7849	1.5956	1.2150
rosen10	2,056	4,096	62,136	818.7072	160.5298	808.0989	13.7775	12.4611
rosen2	1,032	2,048	46,504	236.7363	286.1417	233.3823	10.5239	7.5154
rosen7	264	512	7,770	1.6773	1.1057	1.7332	0.0822	0.0748
rosen8	520	1,024	15,538	16.4771	9.3449	15.6266	0.6550	0.5612
sc205	205	203	551	0.3022	0.2440	0.3008	0.0356	0.0325
scfxm1	330	457	2,589	2.0143	1.4918	2.0711	0.1894	0.1544
sctap2	1,090	1,880	6,714	21.9486	1.0001	20.9776	0.7254	0.6166
sctap3	1,480	2,480	8,874	59.0996	3.7962	58.7819	1.9974	1.4645
Geometric mean				3.6673	1.3274	3.5027	0.3031	0.2396

References

1. Badr, E. S., Moussa, M., Papparrizos, K., Samaras, N., & Sifaleras, A. (2006). Some computational results on MPI parallel implementations of dense simplex method. In *Proceedings of World Academy of Science, Engineering and Technology, 23, (CISE 2006)*, Cairo, Egypt, 39–32.
2. Bartels, R. H., & Golub, G. H. (1969). The simplex method of linear programming using LU decomposition. *Communications of the ACM, 12*, 266–268.
3. Benhamadou, M. (2002). On the simplex algorithm Šrevised formŠ. *Advances in Engineering Software, 33*(11–12), 769–777.
4. Brayton, R. K., Gustavson, F. G., & Willoughby, R. A. (1970) Some results on sparse matrices. *Mathematics of Computation, 24*, 937–954.
5. Chvatal, V. (1983). *Linear programming*. New York, USA: W.H. Freeman and Company.
6. Dantzig, G. B., & Orchard–Hays, W. (1954). The product form of the inverse in the simplex method. *Mathematical Tables and Other Aids to Computation, 8*, 64–67.
7. Forrest, J. J. H., & Tomlin, J. A. (1972). Updated triangular factors of the basis to maintain sparsity in the product form simplex method. *Mathematical Programming, 2*, 263–278.
8. Goldfarb, D. (1977). On the Bartels-Golub decomposition for linear programming bases. *Mathematical Programming, 13*, 272–279.
9. Golub, G. H., & Van Loan, C. F. (2013). *Matrix computations* (4th ed). JHU Press.
10. Lim, S., Kim, G., & Park, S. (2003). A comparative study between various LU update methods in the simplex method. *Journal of the Military Operations Research Society of Korea, 29*(1), 28–42.
11. Markowitz, H. (1957). The elimination form of the inverse and its applications to linear programming. *Management Science, 3*, 255–269.
12. McCoy, P. F., & Tomlin, J. A. (1974). *Some experiments on the accuracy of three methods of updating the inverse in the simplex method*. Technical Report, Stanford University.
13. Nazareth, J. L. (1987). *Computer solution of linear programs*. Oxford, UK: Oxford University Press.
14. Ploskas, N. (2014). *Hybrid optimization algorithms: implementation on GPU*. Ph.D. thesis, Department of Applied Informatics, University of Macedonia.
15. Ploskas, N., & Samaras, N. (2013). Basis update on simplex type algorithms. In *Book of Abstracts of the EWG-DSS Thessaloniki 2013*, p. 11, Thessaloniki, Greece.
16. Ploskas, N., & Samaras, N. (2013). A computational comparison of basis updating schemes for the simplex algorithm on a CPU-GPU system. *American Journal of Operations Research, 3*, 497–505.
17. Ploskas, N., Samaras, N., & Margaritis, K. (2013). A parallel implementation of the revised simplex algorithm using OpenMP: Some Preliminary Results. In A. Migdalas et al. (Eds.), *Optimization Theory, Decision Making, and Operations Research Applications, Series Title: Springer Proceedings in Mathematics & Statistics 31* (pp. 163–175). New York: Springer.
18. Ploskas, N., Samaras, N., & Papathanasiou, J. (2012). LU decomposition in the revised simplex algorithm. In *Proceedings of the 23th National Conference, Hellenic Operational Research Society* (pp. 77–81), Athens, Greece.
19. Ploskas, N., Samaras, N., & Papathanasiou, J. (2013). A web-based decision support system using basis update on simplex type algorithms. In J. Hernandez et al. (Eds.), *Decision Support Systems II Ű Recent Developments Applied to DSS Network Environments, Lecture Notes in Business Information Processing (LNBIP 164)* (pp. 102–114). New York: Springer.
20. Ploskas, N., Samaras, N., & Sifaleras, A. (2009). A parallel implementation of an exterior point algorithm for linear programming problems. In *Proceedings of the 9th Balcan Conference on Operational Research (BALCOR 2009)*, 2–6 September, Constanta, Romania.

21. Reid, J. (1982). A sparsity-exploiting variant of the Bartels-Golub decomposition for linear programming bases. *Mathematical Programming, 24*, 55–69.
22. Saunders, M. (1976). A fast and stable implementation of the simplex method using Bartels-Golub updating. In J. Bunch, & S. T. Rachev (Eds.), *Sparse Matrix Computation* (pp. 213–226). New York: Academic Press.
23. Suhl, L. M., & Suhl, U. H. (1993). A fast LU update for linear programming. *Annals of Operations Research, 43*(1), 33–47.

Chapter 8
Revised Primal Simplex Algorithm

Abstract The simplex algorithm is one of the top ten algorithms with the greatest influence in the twentieth century and the most widely used method for solving linear programming problems (LPs). Nearly all Fortune 500 companies use the simplex algorithm to optimize several tasks. This chapter presents the revised primal simplex algorithm. Numerical examples are presented in order for the reader to understand better the algorithm. Furthermore, an implementation of the algorithm in MATLAB is presented. The implementation is modular allowing the user to select which scaling technique, pivoting rule, and basis update method will use in order to solve LPs. Finally, a computational study over benchmark LPs and randomly generated sparse LPs is performed in order to compare the efficiency of the proposed implementation with MATLAB's simplex algorithm.

8.1 Chapter Objectives

- Present the revised primal simplex algorithm.
- Understand all different methods that can be used in each step of the revised primal simplex algorithm.
- Solve linear programming problems using the revised primal simplex algorithm.
- Implement the revised primal simplex algorithm using MATLAB.
- Compare the computational performance of the revised primal simplex algorithm with MATLAB's implementation.

8.2 Introduction

The most well-known method for solving LPs is the simplex algorithm developed by George B. Dantzig [4, 5]. It is one of the top ten algorithms with the greatest influence in the twentieth century [6]. The simplex algorithm begins with a primal

---

The original version of this chapter was revised. An erratum to this chapter can be found at
https://doi.org/10.1007/978-3-319-65919-0_13

N. Ploskas, N. Samaras, *Linear Programming Using MATLAB®*,
Springer Optimization and Its Applications 127, DOI 10.1007/978-3-319-65919-0_8

feasible basis and uses pricing operations until an optimum solution is computed. It searches for an optimal solution by moving from one feasible solution to another, along the edges of the feasible region. It also guarantees monotonicity of the objective value. It has been proved that the expected number of iterations in the solution of an LP problem is polynomial under a probabilistic model [2, 10]. Moreover, the worst case complexity has exponential behavior $(O(2^n - 1)[8]$. One of the first computer implementations of the simplex algorithm was a model with 71 decision variables and 48 constraints that took more than 18 hours to solve using SEAC (Standard Eastern Automatic Computer). The computational performance of the revised simplex algorithm on practical problems is usually far better than the theoretical worst-case.

This chapter presents the revised primal simplex algorithm. Initially, the description of the algorithm is given. The various steps of the algorithm are presented. Numerical examples are also presented in order for the reader to understand better the algorithm. Furthermore, an implementation of the algorithm in MATLAB is presented. The implementation is modular allowing the user to select which scaling technique, pivoting rule, and basis update method will use in order to solve LPs. Note that controlling the above parameters may affect the execution time and the number of iterations needed to solve an LP. Tuning the algorithm for a given problem structure can dramatically improve its performance. Finally, a computational study is performed in order to compare the efficiency of the proposed implementation with MATLAB's simplex algorithm.

The structure of this chapter is as follows. In Section 8.3, the revised primal simplex algorithm is described. All different steps of the algorithm are presented in detail. Four examples are solved in Section 8.4 using the revised primal simplex algorithm. Section 8.5 presents the MATLAB code of the revised primal simplex algorithm. Section 8.6 presents a computational study that compares the proposed implementation and MATLAB's simplex algorithm. Finally, conclusions and further discussion are presented in Section 8.7.

8.3   Description of the Algorithm

A formal description of the revised primal simplex algorithm is given in Table 8.1 and a flow diagram of its major steps in Figure 8.1.

Initially, the LP problem is presolved. We use the presolve routine presented in Section 4.15. This routine presolves the original LP problem using the eleven presolve methods presented in Chapter 4. After the presolve analysis, the modified LP problem is scaled. In Chapter 5, we presented eleven scaling techniques. Any of these scaling techniques can be used to scale the LP problem.

Then, we calculate an initial basis to start with. That means that we seek for an invertible basis. This basis includes a subset of the columns. Some columns are obvious choices for this initial basis. Slack variables only have one nonzero value in the corresponding column, so they are good choices to include them in the initial basis. Deciding on the other columns is not as easy, especially on large LPs. A

Table 8.1 Revised primal simplex algorithm

Step 0. *(Initialization).*

Presolve the LP problem.

Scale the LP problem.

Select an initial basic solution (B, N).

if the initial basic solution is feasible then proceed to step 2.

Step 1. *(Phase I).*

Construct an auxiliary problem by adding an artificial variable x_{n+1} with a

coefficient vector equal to $-A_B e$, where $e = (1, 1, \cdots, 1)^T \in \mathbb{R}^m$.

Apply the revised simplex algorithm. If the final basic solution (B, N) is feasible,

then proceed to step 2 in order to solve the initial problem. The LP problem can

be either optimal or infeasible.

Step 2. *(Phase II).*

Step 2.0. *(Initialization).*

Compute $(A_B)^{-1}$ and vectors x_B, w, and s_N.

Step 2.1. *(Test of Optimality).*

if $s_N \geq 0$ then STOP. The LP problem (LP.1) is optimal.

else

    Choose the index l of the entering variable using a pivoting rule.

    Variable x_l enters the basis.

Step 2.2. *(Pivoting).*

Compute the pivot column $h_l = (A_B)^{-1} A_l$.

if $h_l \leq 0$ then STOP. The LP problem (LP.1) is unbounded.

else

    Choose the leaving variable $x_{B[r]} = x_k$ using the following relation:

$x_k = x_{B[r]} = \frac{x_{B[r]}}{h_{il}} = \min \left\{ \frac{x_{B[i]}}{h_{il}} : h_{il} > 0 \right\}$

Step 2.3. *(Update).*

Swap indices k and l. Update the new basis inverse $\left(A_{\bar{B}} \right)^{-1}$, using a basis

update scheme. Update vectors x_B, w, and s_N.

Go to Step 2.1.

---

general procedure to select an initial basis is presented in Section 8.5. Function *lprref* applies Gauss-Jordan elimination with partial pivoting in order to find an initial invertible basis. Many other sophisticated techniques have been proposed for the calculation of the initial basis. Interested readers are referred to [1, 3, 7, 9].'

Consider the following LP problem (LP.1) in the standard form:

$$\min \quad c^T x$$

$$\text{s.t.} \quad Ax = b \qquad (LP.1)$$

$$x \geq 0$$

where $A \in \mathbb{R}^{m \times n}$, $(c, x) \in \mathbb{R}^n$, $b \in \mathbb{R}^m$, and T denotes transposition. We assume that A has full rank, $rank(A) = m$, $m < n$. The dual problem associated with (LP.1) is:

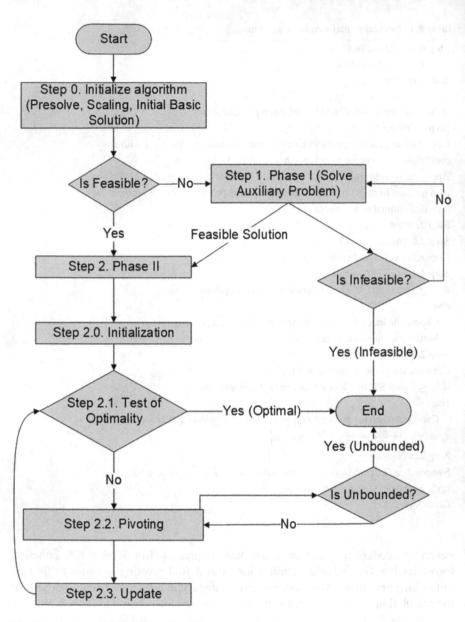

Fig. 8.1 Revised primal simplex algorithm

$$\max \quad b^T w$$

$$\text{s.t.} \quad A^T w + s = c \quad\quad\quad (DP.1)$$

$$s \geq 0$$

where $w \in \mathbb{R}^m$ and $s \in \mathbb{R}^n$.

Let B be a subset of the index set $\{1, 2, \cdots, n\}$. If $x \in \mathbb{R}^n$, we denote by x_B the part of x indexed by B. The set of indices corresponding to nonbasic variables is denoted by $N = \{1, 2, \cdots, n\} \setminus B$. A_B denotes the submatrix of A containing the columns A_j such that $j \in B$. Similarly, A_N denotes the submatrix of A containing the columns A_j such that $j \in N$. Partitioning matrix A as $A = \begin{bmatrix} A_B & A_N \end{bmatrix}$ and with a corresponding partitioning and ordering of $x^T = \begin{bmatrix} x_B & x_N \end{bmatrix}$ and $c^T = \begin{bmatrix} c_B & c_N \end{bmatrix}$, (LP.1) is written as:

$$\min \quad c_B^T x_B + c_N^T x_N$$

$$\text{s.t.} \quad A_B x_B + A_N x_N = b$$

$$x_B, x_N \geq 0$$

In component notation, the ith entry of Ax can be partitioned into a basic and a nonbasic part as follows:

$$\sum_{j=1}^n A_{ij} x_j = \sum_{j \in B} A_{ij} x_j + \sum_{j \in N} A_{ij} x_j \quad\quad\quad (8.1)$$

If $A_B \in \mathbb{R}^{m \times m}$ is non-singular (invertible), then B is called a basis and A_B is a basic matrix. Given a basis B the associated solution $x_B = A_B^{-1} b$, $x_N = 0$ is called a basic solution. This solution is primal feasible iff $x_B \geq 0$ and we proceed to apply the revised simplex algorithm to the original problem (Phase II). Otherwise, it is called infeasible. In this case, we construct an auxiliary problem to find an initial feasible basis (Phase I). In Phase I, it is possible to find out that the LP problem is infeasible. The row vector $w^T = c_B^T A_B^{-1}$ stands for the simplex multipliers. The solution w is feasible to (DP.1) iff $s_N^T = c_N^T - w^T A_N \geq 0$. With s we denote the dual slack variables. The ith row of the coefficient matrix A is denoted by $A_{i.}$ and the jth column by $A_{.j}$.

Definition 8.1 A feasible point in a minimization LP is optimal if for any other feasible point y the relation $c^T x \leq c^T y$ holds true. Similarly, a feasible point in a maximization LP is optimal if for any other feasible point y the relation $c^T x \geq c^T y$ holds true.

Definition 8.2 A feasible minimization (maximization) LP is unbounded if there exists a series of feasible points $(x^1 + x^2 + \cdots)$ such that its objective function value $(c^T x^1 + c^T x^2 + \cdots)$ approaches $+\infty$ $(-\infty)$.

Theorem 8.1 (Optimality Condition) *Let (B, N) be a basic partition of (LP.1). The associated basic solution (x_B, x_N) is optimal if $x_B \geq 0$ and $s_N \geq 0$.*

Proof Since $x_B \geq 0$, the basic solution (x_B, x_N) is feasible. Let y be any other feasible point of (LP.1). Hence, it is sufficient to prove that $c^T x \leq c^T y$ (according to the above definition). Point x is feasible, so:

$$Ax = A_B x_B + A_N x_N = b \quad (x_N = 0)$$

Solving for x_B, we have:

$$x_B = A_B^{-1}(b - A_N x_N) \text{ and } x_N = 0$$

Hence:

$$
\begin{aligned}
c^T(y - x) &= c_B^T(y_B - x_B) + c_N^T(y_N - x_N) \\
&= c_B^T[A_B^{-1}(b - A_N y_N) - A_B^{-1}b] + c_N^T(y_N - 0) \\
&= c_B^T[A_B^{-1}b - A_B^{-1}A_N y_N - A_B^{-1}b] + c_N^T y_N \\
&= -c_B^T A_B^{-1} A_N y_N + c_N^T y_N \\
&= (c_N^T - c_B^T A_B^{-1} A_N) y_N \\
&= (c_N^T - w^T A_N) y_N \\
&= s_N^T y_N
\end{aligned}
$$

From the hypothesis of the theorem we have $s_N \geq 0$. Since y is feasible, we also have $y_N \geq 0$, so:

$$c^T(y - x) \geq 0$$

Hence:

$$c^T x \leq c^T y$$

$\square$

Theorem 8.2 (Feasibility Condition) *Let (B, N) be the current basic partition and (D, H) the next basic partition that is generated by the simplex algorithm. Let x and y be the associated basic solutions. If solution x is feasible, then solution y is also feasible.* '

Proof It is sufficient to prove that $y_D \geq 0$. So:

$$y_D = A_D^{-1}b = (A_B E)^{-1}b = E^{-1}A_B^{-1}b = E^{-1}x_B$$

For each i, we have:

$$\begin{cases} x_{B[i]} - \frac{h_{il}}{h_{rl}}x_{B[r]}, & i \neq r \\ \frac{x_{Br}}{h_{rl}}, & i = r \end{cases}$$

It is obvious that $y_D \geq 0$.   □

The task of solving an LP problem is equivalent to that of searching over basic feasible solutions. If an LP has m constraints and n variables (after adding slack variables), there are at most

$$\binom{n}{m} = \frac{n!}{m!\,(n-m)!}$$

basic solutions.

Extensive computational experience with the revised simplex algorithm applied to problems from various scientific fields has indicated that the algorithm can be expected to converge to an optimal solution in between m and $\frac{3m}{2}$ iterations. Since the simplex algorithm moves from a feasible vertex to an adjacent one, degenerate vertices may cause trouble. The degeneracy is the rule in practical LPs, not the exception.

Klee and Minty [8] showed that there are explicit LPs with n constraints for which the largest coefficient pivot rule can take $2^n - 1$ iterations to reach the optimal solution.

Definition 8.3 A cycle in the simplex algorithm is a sequence of $k + 1$ iterations with corresponding bases

$$B_0, B_1, \cdots, B_k, B_0 \text{ and } k \geq 1$$

The famous Klee-Minty example is:

$$\begin{array}{ll} \max & \sum_{j=1}^{n} 10^{n-j}x_j \\ \text{s.t.} & 2\sum_{j=1}^{i-1} 10^{i-j}x_j + x_i \leq 100^{i-1} \\ & x_j \geq 0, \quad (i,j) = 1, 2, \cdots, n \end{array}$$

For $(i,j) = 1, 2, 3, 4$ we have

$$\begin{array}{ll} \max & 1000x_1 + 100x_2 + 10x_3 + x_4 \\ \text{s.t.} & x_1 \leq 1 \\ & 20x_1 + x_2 \leq 100 \\ & 200x_1 + 20x_2 + x_3 \leq 10,000 \\ & 2,000x_1 + 200x_2 + 20x_3 + x_4 \leq 1,000,000 \\ & x_j \geq 0, \quad (j = 1, 2, 3, 4) \end{array}$$

8.3.1 Phase I

In Phase I, we apply the revised simplex algorithm to an auxiliary LP problem, not the original one, in order to find an initial feasible solution for the original problem. Assuming that the original LP problem is in its standard form (LP.1), the auxiliary LP problem that is solved in Phase I is the following:

$$
\begin{aligned}
\min z = \quad 0x_1 \;+\; 0x_2 \;+\cdots+\; 0x_n \;+\quad x_{n+1} \\
\text{s.t.} \quad A_{11}x_1 + A_{12}x_2 + \cdots + A_{1n}x_n + A_{1,n+1}x_{n+1} &= b_1 \\
A_{21}x_1 + A_{22}x_2 + \cdots + A_{2n}x_n + A_{2,n+1}x_{n+1} &= b_2 \\
\vdots \quad \vdots \quad \vdots \quad \vdots \quad \cdots \quad \vdots \quad \vdots \quad \vdots \qquad \vdots \\
A_{m1}x_1 + A_{m2}x_2 + \cdots + A_{mn}x_n + A_{m,n+1}x_{n+1} &= b_m \\
x_j \geq 0, \quad (j = 1, 2, \cdots, n+1)
\end{aligned}
\tag{8.2}
$$

In matrix notation, we have:

$$
\begin{aligned}
\min \quad & x_{n+1} \\
\text{s.t.} \quad & Ax + dx_{n+1} = b \\
& x_j \geq 0, \quad j = 1, 2, \cdots, n+1
\end{aligned}
\tag{LP.2}
$$

where $A \in \mathbb{R}^{m\times n}$, $(x) \in \mathbb{R}^n$, $b \in \mathbb{R}^m$.

The leaving variable k is the index with the minimum value of x_B. Moreover, the artificial variable x_{n+1} enters the basis. Hence, we update the initial basic list $B = B \cup [n+1] \setminus [k]$. Finally, we update the initial nonbasic list $N = N \cup [k]$.

The last column of matrix A in the auxiliary problem is calculated as:

$$
d = -A_B e
$$

where e is a column vector of ones.

Theorem 8.3 *If basis B is not feasible in problem (LP.1), then basis $B \setminus B[r] \cup [n+1]$ is feasible in problem (LP.2), where r is defined by the following relation:*

$$
(A_B^{-1}b)_r = \min \left\{ (A_B^{-1}b)_i : i = 1, 2, \cdots, m \right\}
$$

Proof Since basis B is not feasible, then $(A_B^{-1}b)_r < 0$. Basic variables x_B are given by

$$
x_B = A_B^{-1}b - A_B^{-1}dx_{n+1} = A_B^{-1}b + ex_{n+1}
$$

where e is a vector of ones. By setting in last relation $x_{n+1} = -(A_B^{-1}b)_r \geq 0$, we have

$$x_j = 0, \, j \in N \text{ and } x_{B[i]} = (A_B^{-1}b)_i - (A_B^{-1}b)_r \geq 0, \, i = 1, 2, \cdots, m, i \neq r$$

because $(A_B^{-1})_r$ is the most negative element of $(A_B^{-1})_i$, $i = 1, 2, \cdots, m, i \neq r$.

□

In the first iteration of Phase I, we calculate the following matrices and vectors:

$$A_B^{-1}, x_B = A_B^{-1}b, w^T = c_B^T A_B^{-1}, \text{ and } s_N^T = c_N^T - w^T A_N$$

Then, we perform the optimality test of Phase I. We proceed with Phase II when one of the following conditions holds true:

- If all the elements of s_N are greater than or equal to 0 ($s_N \geq 0$). If the value x_B for variable $n + 1$ ($x_{B[n+1]}$) is not equal to 0, then the LP problem is infeasible. Otherwise, we proceed with Phase II.
- If the artificial variable left from the basic list, we proceed with Phase II.

If the auxiliary LP problem is not optimal, then we continue with the remaining steps of Phase I. We select the entering variable l according to a specific pivoting rule. In Chapter 6, we presented six pivoting rules. Any of these pivoting rules can be used. Then, we calculate the pivoting column:

$$h_l = A_B^{-1} A_l$$

Next, we perform the minimum ratio test in order to find the leaving variable k:

$$x_k = x_{B[r]} = \frac{x_{B[r]}}{h_{il}} = \min \left\{ \frac{x_{B[i]}}{h_{il}} : h_{il} > 0 \right\}$$

Then, we update the basic and nonbasic lists:

$$B = B \cup [l] \setminus [k], N = N \cup [k] \setminus [l]$$

Next, we update the basis inverse using a basis update method. In Chapter 7, we presented two basis update methods. Any of these basis update methods can be used. Note that we compute the basis inverse from scratch every t iterations. A good value of t for most LPs is 80. There are three reasons for computing the basis inverse periodically from scratch: (i) numerical stability (e.g., elimination of round-off errors), (ii) refactoring the data structures used, and (iii) preserving the degeneracy issues.

Finally, we update vectors x_B, w, and s_N:

$$x_B = A_B^{-1}b, w^T = c_B^T A_B^{-1}, \text{ and } s_N^T = c_N^T - w^T A_N$$

We start again with the optimality test and all the other steps presented in Phase I until we find a feasible basis for (LP.1) or we find that (LP.1) is infeasible.

8.3.2 Phase II

In Phase II, we apply the revised simplex algorithm to (LP.1). If we have performed the Phase I, then the artificial variable has left from the basic list and we also delete it from the nonbasic list.

In the first iteration of Phase II, we calculate the following matrices and vectors:

$$A_B^{-1}, x_B = A_B^{-1}b, w^T = c_B^T A_B^{-1}, \text{ and } s_N^T = c_N^T - w^T A_N$$

Then, we perform the optimality test of Phase II. If all the elements of s_N are greater than or equal to 0 ($s_N \geq 0$), then (LP.1) is optimal. If (LP.1) is not optimal, then we continue with the remaining steps of Phase II. We select the entering variable l according to a pivoting rule. In Chapter 6, we presented six pivoting rules. Any of these pivoting rules can be used. Then, we calculate the pivoting column:

$$h_l = A_B^{-1} A_l$$

If all the elements of h_l are less than or equal to 0 ($h_l \leq 0$), then (LP.1) is unbounded. Otherwise, we perform the minimum ratio test to find the leaving variable k:

$$x_k = x_{B[r]} = \frac{x_{B[r]}}{h_{il}} = \min\left\{ \frac{x_{B[i]}}{h_{il}} : h_{il} > 0 \right\}$$

If two or more nonbasic (basic) variable coefficients are eligible and have the same value, we break ties by selecting the nonbasic (basic) variable with the smallest subscript. Next, we update the basic and nonbasic lists:

$$B = B \cup [l] \setminus [k], N = N \cup [k] \setminus [l]$$

Then, we update the basis inverse using a basis update method. In Chapter 7, we presented two basis update methods. Any of these basis update methods can be used. Note that we compute the basis inverse from scratch every t iterations in order to reduce numerical errors from the basis update method (as already mentioned, a good value of t for most LPs is 80).

Finally, we update vectors x_B, w, and s_N:

$$x_B = A_B^{-1}b, w^T = c_B^T A_B^{-1}, \text{ and } s_N^T = c_N^T - w^T A_N$$

We start again with the optimality test and all the other steps presented in Phase II until we find an optimal basis for (LP.1) or we find that (LP.1) is unbounded.

8.4 Examples

Example 1 The LP problem that will be solved is the following:

$$
\begin{aligned}
\min z = \ & x_1 && - x_3 - 3x_4 \\
\text{s.t.} \quad & 2x_1 && + 2x_3 + 3x_4 = 10 \\
& && - 2x_2 - 2x_3 - 6x_4 = -6 \\
& x_j \geq 0, \quad (j = 1, 2, 3, 4)
\end{aligned}
$$

In matrix notation the above LP problem is written as follows:

$$
A = \begin{bmatrix} 2 & 0 & 2 & 3 \\ 0 & -2 & -2 & -6 \end{bmatrix}, c = \begin{bmatrix} 1 \\ 0 \\ -1 \\ -3 \end{bmatrix}, b = \begin{bmatrix} 10 \\ -6 \end{bmatrix}, Eqin = \begin{bmatrix} 0 \\ 0 \end{bmatrix}
$$

The LP problem is in its standard form. Let's start with the basic list $B = [1, 2]$ and the nonbasic list $N = [3, 4]$. It is easy to find out that this basis is feasible ($x_B \geq 0$):

$$
A_B = \begin{bmatrix} 2 & 0 \\ 0 & -2 \end{bmatrix}
$$

$$
A_B^{-1} = \begin{bmatrix} 1/2 & 0 \\ 0 & -1/2 \end{bmatrix}
$$

$$
x_B = A_B^{-1}b = \begin{bmatrix} 1/2 & 0 \\ 0 & -1/2 \end{bmatrix} \begin{bmatrix} 10 \\ -6 \end{bmatrix} = \begin{bmatrix} 5 \\ 3 \end{bmatrix} \geq 0
$$

That means that we do not have to apply Phase I. Hence, we can proceed with Phase II.

Phase II - 1st Iteration

The initial data of the problem is:

$$
c_B = \begin{bmatrix} 1 \\ 0 \end{bmatrix}, c_N = \begin{bmatrix} -1 \\ -3 \end{bmatrix}
$$

$$
A_B = \begin{bmatrix} 2 & 0 \\ 0 & -2 \end{bmatrix}, A_N = \begin{bmatrix} 2 & 3 \\ -2 & -6 \end{bmatrix}
$$

$$
A_B^{-1} = \begin{bmatrix} 1/2 & 0 \\ 0 & -1/2 \end{bmatrix}
$$

$$x_B = A_B^{-1} b = \begin{bmatrix} 1/2 & 0 \\ 0 & -1/2 \end{bmatrix} \begin{bmatrix} 10 \\ -6 \end{bmatrix} = \begin{bmatrix} 5 \\ 3 \end{bmatrix}$$

$$w^T = c_B^T A_B^{-1} = \begin{bmatrix} 1 & 0 \end{bmatrix} \begin{bmatrix} 1/2 & 0 \\ 0 & -1/2 \end{bmatrix} = \begin{bmatrix} 1/2 & 0 \end{bmatrix}$$

$$s_N^T = c_N^T - w^T A_N = \begin{bmatrix} -1 & -3 \end{bmatrix} - \begin{bmatrix} 1/2 & 0 \end{bmatrix} \begin{bmatrix} 2 & 3 \\ -2 & -6 \end{bmatrix} = \begin{bmatrix} -2 & -9/2 \end{bmatrix}$$

$s_N \not\geq 0$, so the current basis is not optimal. According to Dantzig's rule, the entering variable is x_4, because it is the one with the most negative $\bar{s}_l$ ($-9/2$). Hence, $l = 4$.

We continue calculating the pivoting column:

$$h_4 = A_B^{-1} A_{.4} = \begin{bmatrix} 1/2 & 0 \\ 0 & -1/2 \end{bmatrix} \begin{bmatrix} 3 \\ -6 \end{bmatrix} = \begin{bmatrix} 3/2 \\ 3 \end{bmatrix}$$

There are elements in vector h_4 that are greater than 0, so we perform the minimum ratio test:

$$\min \left\{ \frac{x_{B[i]}}{h_{il}} : h_{il} > 0 \right\} = \min \left\{ \frac{x_{B[1]}}{h_{14}}, \frac{x_{B[2]}}{h_{24}} \right\} = \min \left\{ \frac{5}{1/2}, \frac{3}{3} \right\} = \frac{3}{3} = \frac{x_{B[2]}}{h_{24}}$$

The leaving variable is x_2. Hence, $k = 2$ and $r = 2$.
Next, we update the basic and nonbasic lists:

$$B = \begin{bmatrix} 1 & 4 \end{bmatrix}, N = \begin{bmatrix} 3 & 2 \end{bmatrix}$$

Hence:

$$c_B = \begin{bmatrix} 1 \\ -3 \end{bmatrix}, c_N = \begin{bmatrix} -1 \\ 0 \end{bmatrix}$$

$$A_B = \begin{bmatrix} 2 & 3 \\ 0 & -6 \end{bmatrix}, A_N = \begin{bmatrix} 2 & 0 \\ -2 & -2 \end{bmatrix}$$

Then, we update the basis inverse using the PFI update scheme:

$$E^{-1} = \begin{bmatrix} 1 & -1/2 \\ 0 & 1/3 \end{bmatrix}$$

$$(A_{\bar{B}})^{-1} = E^{-1} A_B^{-1} = \begin{bmatrix} 1 & -1/2 \\ 0 & 1/3 \end{bmatrix} \begin{bmatrix} 1/2 & 0 \\ 0 & -1/2 \end{bmatrix} = \begin{bmatrix} 1/2 & 1/4 \\ 0 & -1/6 \end{bmatrix}$$

Finally, we update vectors x_B, w, and s_N:

$$x_B = A_B^{-1}b = \begin{bmatrix} 1/2 & 1/4 \\ 0 & -1/6 \end{bmatrix} \begin{bmatrix} 10 \\ -6 \end{bmatrix} = \begin{bmatrix} 7/2 \\ 1 \end{bmatrix}$$

$$w^T = c_B^T A_B^{-1} = \begin{bmatrix} 1 & -3 \end{bmatrix} \begin{bmatrix} 1/2 & 1/4 \\ 0 & -1/6 \end{bmatrix} = \begin{bmatrix} 1/2 & 3/4 \end{bmatrix}$$

$$s_N^T = c_N^T - w^T A_N = \begin{bmatrix} -1 & 0 \end{bmatrix} - \begin{bmatrix} 1/2 & 3/4 \end{bmatrix} \begin{bmatrix} 2 & 0 \\ -2 & -2 \end{bmatrix} = \begin{bmatrix} -1/2 & 3/2 \end{bmatrix}$$

Phase II - 2nd Iteration

The data at the beginning of this iteration is:

$$c_B = \begin{bmatrix} 1 \\ -3 \end{bmatrix}, c_N = \begin{bmatrix} -1 \\ 0 \end{bmatrix}$$

$$A_B = \begin{bmatrix} 2 & 3 \\ 0 & -6 \end{bmatrix}, A_N = \begin{bmatrix} 2 & 0 \\ -2 & -2 \end{bmatrix}$$

$$A_B^{-1} = \begin{bmatrix} 1/2 & 1/4 \\ 0 & -1/6 \end{bmatrix}$$

$$x_B = \begin{bmatrix} 7/2 \\ 1 \end{bmatrix}, w^T = \begin{bmatrix} 1/2 & 3/4 \end{bmatrix}, s_N^T = \begin{bmatrix} -1/2 & 3/2 \end{bmatrix}$$

$s_N \not\geq 0$, so the current basis is not optimal. According to Dantzig's rule, the entering variable is x_3, because it is the one with the most negative $\bar{s}_l$ $(-1/2)$. Hence, $l = 3$.

We continue calculating the pivoting column:

$$h_3 = A_B^{-1} A_{.3} = \begin{bmatrix} 1/2 & 1/4 \\ 0 & -1/6 \end{bmatrix} \begin{bmatrix} 2 \\ -2 \end{bmatrix} = \begin{bmatrix} 1/2 \\ 1/3 \end{bmatrix}$$

There are elements in vector h_3 that are greater than 0, so we perform the minimum ratio test:

$$\min \left\{ \frac{x_{B[i]}}{h_{il}} : h_{il} > 0 \right\} = \min \left\{ \frac{x_{B[1]}}{h_{13}}, \frac{x_{B[4]}}{h_{23}} \right\} = \min \left\{ \frac{7/2}{1/2}, \frac{1}{1/3} \right\} = \min \{7, 3\} = \frac{x_{B[4]}}{h_{23}}$$

The leaving variable is x_4. Hence, $k = 4$ and $r = 2$.
Next, we update the basic and nonbasic lists:

$$B = \begin{bmatrix} 1 & 3 \end{bmatrix}, N = \begin{bmatrix} 4 & 2 \end{bmatrix}$$

Hence:

$$c_B = \begin{bmatrix} 1 \\ -1 \end{bmatrix}, c_N = \begin{bmatrix} -3 \\ 0 \end{bmatrix}$$

$$A_B = \begin{bmatrix} 2 & 2 \\ 0 & -2 \end{bmatrix}, A_N = \begin{bmatrix} 3 & 0 \\ -6 & -2 \end{bmatrix}$$

Then, we update the basis inverse using the PFI update scheme:

$$E^{-1} = \begin{bmatrix} 1 & -3/2 \\ 0 & 3 \end{bmatrix}$$

$$(A_{\bar{B}})^{-1} = E^{-1}A_B^{-1} = \begin{bmatrix} 1 & -3/2 \\ 0 & 3 \end{bmatrix}\begin{bmatrix} 1/2 & 1/4 \\ 0 & -1/6 \end{bmatrix} = \begin{bmatrix} 1/2 & 1/2 \\ 0 & -1/2 \end{bmatrix}$$

Finally, we update vectors x_B, w, and s_N:

$$x_B = A_B^{-1}b = \begin{bmatrix} 1/2 & 1/2 \\ 0 & -1/2 \end{bmatrix}\begin{bmatrix} 10 \\ -6 \end{bmatrix} = \begin{bmatrix} 2 \\ 3 \end{bmatrix}$$

$$w^T = c_B^T A_B^{-1} = \begin{bmatrix} 1 & -1 \end{bmatrix}\begin{bmatrix} 1/2 & 1/2 \\ 0 & -1/2 \end{bmatrix} = \begin{bmatrix} 1/2 & 1 \end{bmatrix}$$

$$s_N^T = c_N^T - w^T A_N = \begin{bmatrix} -3 & 0 \end{bmatrix} - \begin{bmatrix} 1/2 & 1 \end{bmatrix}\begin{bmatrix} 3 & 0 \\ -6 & -2 \end{bmatrix} = \begin{bmatrix} 3/2 & 2 \end{bmatrix}$$

Phase II - 3rd Iteration

The data at the beginning of this iteration is:

$$c_B = \begin{bmatrix} 1 \\ -1 \end{bmatrix}, c_N = \begin{bmatrix} -3 \\ 0 \end{bmatrix}$$

$$A_B = \begin{bmatrix} 2 & 2 \\ 0 & -2 \end{bmatrix}, A_N = \begin{bmatrix} 3 & 0 \\ -6 & -2 \end{bmatrix}$$

$$A_B^{-1} = \begin{bmatrix} 1/2 & 1/2 \\ 0 & -1/2 \end{bmatrix}$$

$$x_B = \begin{bmatrix} 2 \\ 3 \end{bmatrix}, w^T = \begin{bmatrix} 1/2 & 1 \end{bmatrix}, s_N^T = \begin{bmatrix} 3/2 & 2 \end{bmatrix}$$

$s_N \geq 0$, so the current basic partition $B = [1, 3]$ and $N = [4, 2]$ is optimal:

$$x_B = \begin{bmatrix} 2 \\ 3 \end{bmatrix}$$

$$z = c_B^T x_B = -1$$

Example 2 Consider the following LP problem:

$$
\begin{aligned}
\min z = {}& 5x_1 + 2x_2 - 4x_3 \\
\text{s.t.} \quad & 6x_1 + x_2 - 2x_3 \geq 5 \\
& x_1 + x_2 + x_3 \leq 4 \\
& 6x_1 + 4x_2 - 2x_3 \geq 10 \\
x_j \geq 0, \quad & (j = 1, 2, 3)
\end{aligned}
$$

In matrix notation the above LP problem is written as follows:

$$
A = \begin{bmatrix} 6 & 1 & -2 \\ 1 & 1 & 1 \\ 6 & 4 & -2 \end{bmatrix}, c = \begin{bmatrix} 5 \\ 2 \\ -4 \end{bmatrix}, b = \begin{bmatrix} 5 \\ 4 \\ 10 \end{bmatrix}, Eqin = \begin{bmatrix} 1 \\ -1 \\ 1 \end{bmatrix}
$$

Initially, we need to convert the LP problem in its standard form:

$$
\begin{aligned}
\min z = {}& 5x_1 + 2x_2 - 4x_3 \\
\text{s.t.} \quad & 6x_1 + x_2 - 2x_3 - x_4 \qquad\qquad = 5 \\
& x_1 + x_2 + x_3 \qquad + x_5 \quad = 4 \\
& 6x_1 + 4x_2 - 2x_3 \qquad\qquad - x_6 = 10 \\
x_j \geq 0, \quad & (j = 1, 2, 3, 4, 5, 6)
\end{aligned}
$$

In matrix notation the above LP problem is written as follows:

$$
A = \begin{bmatrix} 6 & 1 & -2 & -1 & 0 & 0 \\ 1 & 1 & 1 & 0 & 1 & 0 \\ 6 & 4 & -2 & 0 & 0 & -1 \end{bmatrix}, c = \begin{bmatrix} 5 \\ 2 \\ -4 \\ 0 \\ 0 \\ 0 \end{bmatrix}, b = \begin{bmatrix} 5 \\ 4 \\ 10 \end{bmatrix}, Eqin = \begin{bmatrix} 0 \\ 0 \\ 0 \end{bmatrix}
$$

Let's start with the basic list $B = [4, 5, 6]$ and the nonbasic list $N = [1, 2, 3]$. It is easy to find out that this basis is not feasible ($x_B \not\geq 0$):

$$
A_B = \begin{bmatrix} -1 & 0 & 0 \\ 0 & 1 & 0 \\ 0 & 0 & -1 \end{bmatrix} = A_B^{-1}
$$

$$x_B = A_B^{-1}b = \begin{bmatrix} -1 & 0 & 0 \\ 0 & 1 & 0 \\ 0 & 0 & -1 \end{bmatrix} \begin{bmatrix} 5 \\ 4 \\ 10 \end{bmatrix} = \begin{bmatrix} -5 \\ 4 \\ -10 \end{bmatrix} \not\geq 0$$

We apply Phase I. We add the artificial variable x_7 and calculate vector d as follows:

$$d = -A_B e = - \begin{bmatrix} -1 & 0 & 0 \\ 0 & 1 & 0 \\ 0 & 0 & -1 \end{bmatrix} \begin{bmatrix} 1 \\ 1 \\ 1 \end{bmatrix} = \begin{bmatrix} 1 \\ -1 \\ 1 \end{bmatrix}$$

The leaving variable is x_6:

$$\min \left\{ x_{B[4]}, x_{B[5]}, x_{B[6]} \right\} = \min \left\{ -5, 4, -10 \right\} = -10$$

Hence, the auxiliary LP problem that will be solved in Phase I, using one artificial variable, is the following:

$$
\begin{aligned}
\min z = & & & & x_7 \\
\text{s.t.} \quad 6x_1 + & x_2 - 2x_3 - x_4 & & + x_7 = & 5 \\
x_1 + & x_2 + x_3 & + x_5 & - x_7 = & 4 \\
6x_1 + & 4x_2 - 2x_3 & & - x_6 + x_7 = & 10 \\
x_j \geq 0, \quad & (j = 1, 2, 3, 4, 5, 6, 7) & & &
\end{aligned}
$$

In matrix notation the above LP problem is written as follows:

$$
A = \begin{bmatrix} 6 & 1 & -2 & -1 & 0 & 0 & 1 \\ 1 & 1 & 1 & 0 & 1 & 0 & -1 \\ 6 & 4 & -2 & 0 & 0 & -1 & 1 \end{bmatrix}, c = \begin{bmatrix} 0 \\ 0 \\ 0 \\ 0 \\ 0 \\ 0 \\ 1 \end{bmatrix}, b = \begin{bmatrix} 5 \\ 4 \\ 10 \end{bmatrix}, Eqin = \begin{bmatrix} 0 \\ 0 \\ 0 \end{bmatrix}
$$

Phase I - 1st Iteration

The initial basic list of Phase I is $B = [4, 5, 6] \cup [7] \setminus [6] = [4, 5, 7]$ and the nonbasic list is $N = [1, 2, 3] \cup [6] = [1, 2, 3, 6]$. The initial data of the problem is:

$$
c_B = \begin{bmatrix} 0 \\ 0 \\ 1 \end{bmatrix}, c_N = \begin{bmatrix} 0 \\ 0 \\ 0 \\ 0 \end{bmatrix}
$$

$$A_B = \begin{bmatrix} -1 & 0 & 1 \\ 0 & 1 & -1 \\ 0 & 0 & 1 \end{bmatrix}, A_N = \begin{bmatrix} 6 & 1 & -2 & 0 \\ 1 & 1 & 1 & 0 \\ 6 & 4 & -2 & -1 \end{bmatrix}$$

$$A_B^{-1} = \begin{bmatrix} -1 & 0 & 1 \\ 0 & 1 & 1 \\ 0 & 0 & 1 \end{bmatrix}$$

$$x_B = A_B^{-1}b = \begin{bmatrix} -1 & 0 & 1 \\ 0 & 1 & 1 \\ 0 & 0 & 1 \end{bmatrix} \begin{bmatrix} 5 \\ 4 \\ 10 \end{bmatrix} = \begin{bmatrix} 5 \\ 14 \\ 10 \end{bmatrix}$$

$$w^T = c_B^T A_B^{-1} = \begin{bmatrix} 0 & 0 & 1 \end{bmatrix} \begin{bmatrix} -1 & 0 & 1 \\ 0 & 1 & 1 \\ 0 & 0 & 1 \end{bmatrix} = \begin{bmatrix} 0 & 0 & 1 \end{bmatrix}$$

$$s_N^T = c_N^T - w^T A_N = \begin{bmatrix} 0 & 0 & 0 & 0 \end{bmatrix} - \begin{bmatrix} 0 & 0 & 1 \end{bmatrix} \begin{bmatrix} 6 & 1 & -2 & 0 \\ 1 & 1 & 1 & 0 \\ 6 & 4 & -2 & -1 \end{bmatrix} = \begin{bmatrix} -6 & -4 & 2 & 1 \end{bmatrix}$$

$s_N \not\geq 0$, so the current basis is not optimal. According to Dantzig's rule, the entering variable is x_1, because it is the one with the most negative $\bar{s}_l$ (-6). Hence, $l = 1$.

We continue calculating the pivoting column:

$$h_1 = A_B^{-1} A_{.1} = \begin{bmatrix} -1 & 0 & 1 \\ 0 & 1 & 1 \\ 0 & 0 & 1 \end{bmatrix} \begin{bmatrix} 6 \\ 1 \\ 6 \end{bmatrix} = \begin{bmatrix} 0 \\ 7 \\ 6 \end{bmatrix}$$

There are elements in vector h_1 that are greater than 0, so we perform the minimum ratio test (where the letter x is used below to represent that $h_{il} \leq 0$, therefore $\frac{x_{B[i]}}{h_{il}}$ is not defined):

$$\min \left\{ \frac{x_{B[i]}}{h_{il}} : h_{il} > 0 \right\} = \min \left\{ x, \frac{x_{B[5]}}{h_{21}}, \frac{x_{B[7]}}{h_{31}} \right\} = \min \left\{ x, \frac{14}{7}, \frac{10}{6} \right\} = \frac{10}{6} = \frac{x_{B[7]}}{h_{31}}$$

The leaving variable is x_7. Hence, $k = 7$ and $r = 3$.
Next, we update the basic and nonbasic lists:

$$B = \begin{bmatrix} 4 & 5 & 1 \end{bmatrix}, N = \begin{bmatrix} 7 & 2 & 3 & 6 \end{bmatrix}$$

Hence:

$$c_B = \begin{bmatrix} 0 \\ 0 \\ 0 \end{bmatrix}, c_N = \begin{bmatrix} 1 \\ 0 \\ 0 \\ 0 \end{bmatrix}$$

$$A_B = \begin{bmatrix} -1 & 0 & 6 \\ 0 & 1 & 1 \\ 0 & 0 & 6 \end{bmatrix}, A_N = \begin{bmatrix} 1 & 1 & -2 & 0 \\ -1 & 1 & 1 & 0 \\ 1 & 4 & -2 & -1 \end{bmatrix}$$

Then, we update the basis inverse using the PFI update scheme:

$$E^{-1} = \begin{bmatrix} 1 & 0 & 0 \\ 0 & 1 & -7/6 \\ 0 & 0 & 7/6 \end{bmatrix}$$

$$(A_{\overline{B}})^{-1} = E^{-1} A_B^{-1} = \begin{bmatrix} 1 & 0 & 0 \\ 0 & 1 & -7/6 \\ 0 & 0 & 7/6 \end{bmatrix} \begin{bmatrix} -1 & 0 & 1 \\ 0 & 1 & 1 \\ 0 & 0 & 1 \end{bmatrix} = \begin{bmatrix} -1 & 0 & 1 \\ 0 & 1 & -1/6 \\ 0 & 0 & 1/6 \end{bmatrix}$$

Finally, we update vectors x_B, w, and s_N:

$$x_B = A_B^{-1} b = \begin{bmatrix} -1 & 0 & 1 \\ 0 & 1 & -1/6 \\ 0 & 0 & 1/6 \end{bmatrix} \begin{bmatrix} 5 \\ 4 \\ 10 \end{bmatrix} = \begin{bmatrix} 5 \\ 7/3 \\ 5/3 \end{bmatrix}$$

$$w^T = c_B^T A_B^{-1} = \begin{bmatrix} 0 & 0 & 0 \end{bmatrix} \begin{bmatrix} -1 & 0 & 1 \\ 0 & 1 & -1/6 \\ 0 & 0 & 1/6 \end{bmatrix} = \begin{bmatrix} 0 & 0 & 0 \end{bmatrix}$$

$$s_N^T = c_N^T - w^T A_N = \begin{bmatrix} 1 & 0 & 0 & 0 \end{bmatrix} - \begin{bmatrix} 0 & 0 & 0 \end{bmatrix} \begin{bmatrix} 1 & 1 & -2 & 0 \\ -1 & 1 & 1 & 0 \\ 1 & 4 & -2 & -1 \end{bmatrix} = \begin{bmatrix} 1 & 0 & 0 & 0 \end{bmatrix}$$

Phase I - 2nd Iteration

The data at the beginning of this iteration is:

$$c_B = \begin{bmatrix} 0 \\ 0 \\ 0 \end{bmatrix}, c_N = \begin{bmatrix} 1 \\ 0 \\ 0 \\ 0 \end{bmatrix}$$

$$A_B = \begin{bmatrix} -1 & 0 & 6 \\ 0 & 1 & 1 \\ 0 & 0 & 6 \end{bmatrix}, A_N = \begin{bmatrix} 1 & 1 & -2 & 0 \\ -1 & 1 & 1 & 0 \\ 1 & 4 & -2 & -1 \end{bmatrix} A_B^{-1} = \begin{bmatrix} -1 & 0 & 1 \\ 0 & 1 & -1/6 \\ 0 & 0 & 1/6 \end{bmatrix}$$

$$x_B = \begin{bmatrix} 5 \\ 7/3 \\ 5/3 \end{bmatrix}, w^T = \begin{bmatrix} 0 & 0 & 0 \end{bmatrix}, s_N^T = \begin{bmatrix} 1 & 0 & 0 & 0 \end{bmatrix}$$

$s_N \geq 0$, so the current basis is optimal for the auxiliary LP problem of Phase I. We found a feasible basis ($B = [4, 5, 1]$) for the initial LP problem. We delete the artificial variable from the nonbasic list, so the nonbasic list is $N = [2, 3, 6]$. Hence, we proceed to Phase II.

Phase II - 3rd Iteration

The data of the problem is:

$$c_B = \begin{bmatrix} 0 \\ 0 \\ 5 \end{bmatrix}, c_N = \begin{bmatrix} 2 \\ -4 \\ 0 \end{bmatrix}$$

$$A_B = \begin{bmatrix} -1 & 0 & 6 \\ 0 & 1 & 1 \\ 0 & 0 & 6 \end{bmatrix}, A_N = \begin{bmatrix} 1 & -2 & 0 \\ 1 & 1 & 0 \\ 4 & -2 & -1 \end{bmatrix}$$

$$A_B^{-1} = \begin{bmatrix} -1 & 0 & 1 \\ 0 & 1 & -1/6 \\ 0 & 0 & 1/6 \end{bmatrix}$$

$$x_B = A_B^{-1} b = \begin{bmatrix} -1 & 0 & 1 \\ 0 & 1 & -1/6 \\ 0 & 0 & 1/6 \end{bmatrix} \begin{bmatrix} 5 \\ 4 \\ 10 \end{bmatrix} = \begin{bmatrix} 5 \\ 7/3 \\ 5/3 \end{bmatrix}$$

$$w^T = c_B^T A_B^{-1} = \begin{bmatrix} 0 & 0 & 5 \end{bmatrix} \begin{bmatrix} -1 & 0 & 1 \\ 0 & 1 & -1/6 \\ 0 & 0 & 1/6 \end{bmatrix} = \begin{bmatrix} 0 & 0 & 5/6 \end{bmatrix}$$

$$s_N^T = c_N^T - w^T A_N = \begin{bmatrix} 2 & -4 & 0 \end{bmatrix} - \begin{bmatrix} 0 & 0 & 5/6 \end{bmatrix} \begin{bmatrix} 1 & -2 & 0 \\ 1 & 1 & 0 \\ 4 & -2 & -1 \end{bmatrix} = \begin{bmatrix} -4/3 & -7/3 & 5/6 \end{bmatrix}$$

$s_N \ngeq 0$, so the current basis is not optimal. According to Dantzig's rule, the entering variable is x_3, because it is the one with the most negative $\bar{s}_l$ ($-7/3$). Hence, $l = 3$.

We continue calculating the pivoting column:

$$h_3 = A_B^{-1}A_{.3} = \begin{bmatrix} -1 & 0 & 1 \\ 0 & 1 & -1/6 \\ 0 & 0 & 1/6 \end{bmatrix} \begin{bmatrix} -2 \\ 1 \\ -2 \end{bmatrix} = \begin{bmatrix} 0 \\ 4/3 \\ -1/3 \end{bmatrix}$$

There are elements in vector h_3 that are greater than 0, so we perform the minimum ratio test:

$$\min\left\{\frac{x_{B[i]}}{h_{il}} : h_{il} > 0\right\} = \min\left\{x, \frac{x_{B[5]}}{h_{23}}, x\right\} = \min\left\{x, \frac{7/3}{4/3}, x\right\} = \frac{7/3}{4/3} = \frac{x_{B[5]}}{h_{23}}$$

The leaving variable is x_5. Hence, $k = 5$ and $r = 2$.
Next, we update the basic and nonbasic lists:

$$B = \begin{bmatrix} 4 & 3 & 1 \end{bmatrix}, N = \begin{bmatrix} 2 & 5 & 6 \end{bmatrix}$$

Hence:

$$c_B = \begin{bmatrix} 0 \\ -4 \\ 5 \end{bmatrix}, c_N = \begin{bmatrix} 2 \\ 0 \\ 0 \end{bmatrix}$$

$$A_B = \begin{bmatrix} -1 & -2 & 6 \\ 0 & 1 & 1 \\ 0 & -2 & 6 \end{bmatrix}, A_N = \begin{bmatrix} 1 & 0 & 0 \\ 1 & 1 & 0 \\ 4 & 0 & -1 \end{bmatrix}$$

Then, we update the basis inverse using the PFI update scheme:

$$E^{-1} = \begin{bmatrix} 1 & 0 & 0 \\ 0 & 3/4 & 0 \\ 0 & 1/4 & 1 \end{bmatrix}$$

$$(A_{\overline{B}})^{-1} = E^{-1}A_B^{-1} = \begin{bmatrix} 1 & 0 & 0 \\ 0 & 3/4 & 0 \\ 0 & 1/4 & 1 \end{bmatrix} \begin{bmatrix} -1 & 0 & 1 \\ 0 & 1 & -1/6 \\ 0 & 0 & 1/6 \end{bmatrix} = \begin{bmatrix} -1 & 0 & 1 \\ 0 & 3/4 & -1/8 \\ 0 & 1/4 & 1/8 \end{bmatrix}$$

Finally, we update vectors x_B, w, and s_N:

$$x_B = A_B^{-1}b = \begin{bmatrix} -1 & 0 & 1 \\ 0 & 3/4 & -1/8 \\ 0 & 1/4 & 1/8 \end{bmatrix} \begin{bmatrix} 5 \\ 4 \\ 10 \end{bmatrix} = \begin{bmatrix} 5 \\ 7/4 \\ 9/4 \end{bmatrix}$$

$$w^T = c_B^T A_B^{-1} = \begin{bmatrix} 0 & -4 & 5 \end{bmatrix} \begin{bmatrix} -1 & 0 & 1 \\ 0 & 3/4 & -1/8 \\ 0 & 1/4 & 1/8 \end{bmatrix} = \begin{bmatrix} 0 & -7/4 & 9/8 \end{bmatrix}$$

$$s_N^T = c_N^T - w^T A_N = \begin{bmatrix} 2 & 0 & 0 \end{bmatrix} - \begin{bmatrix} 0 & -7/4 & 9/8 \end{bmatrix} \begin{bmatrix} 1 & 0 & 0 \\ 1 & 1 & 0 \\ 4 & 0 & -1 \end{bmatrix} = \begin{bmatrix} -3/4 & 7/4 & 9/8 \end{bmatrix}$$

Phase II - 4th Iteration

The data at the beginning of this iteration is:

$$c_B = \begin{bmatrix} 0 \\ -4 \\ 5 \end{bmatrix}, c_N = \begin{bmatrix} 2 \\ 0 \\ 0 \end{bmatrix}$$

$$A_B = \begin{bmatrix} -1 & -2 & 6 \\ 0 & 1 & 1 \\ 0 & -2 & 6 \end{bmatrix}, A_N = \begin{bmatrix} 1 & 0 & 0 \\ 1 & 1 & 0 \\ 4 & 0 & -1 \end{bmatrix}$$

$$A_B^{-1} = \begin{bmatrix} -1 & 0 & 1 \\ 0 & 3/4 & -1/8 \\ 0 & 1/4 & 1/8 \end{bmatrix}$$

$$x_B = \begin{bmatrix} 5 \\ 7/4 \\ 9/4 \end{bmatrix}, w^T = \begin{bmatrix} 0 & -7/4 & 9/8 \end{bmatrix}, s_N^T = \begin{bmatrix} -3/4 & 7/4 & 9/8 \end{bmatrix}$$

$s_N \not\geq 0$, so the current basis is not optimal. According to Dantzig's rule, the entering variable is x_2, because it is the one with the most negative $\bar{s}_l$ $(-3/4)$. Hence, $l = 2$.

We continue calculating the pivoting column:

$$h_2 = A_B^{-1} A_{.2} = \begin{bmatrix} -1 & 0 & 1 \\ 0 & 3/4 & -1/8 \\ 0 & 1/4 & 1/8 \end{bmatrix} \begin{bmatrix} 1 \\ 1 \\ 4 \end{bmatrix} = \begin{bmatrix} 3 \\ 1/4 \\ 3/4 \end{bmatrix}$$

There are elements in vector h_2 that are greater than 0, so we perform the minimum ratio test:

$$\min \left\{ \frac{x_{B[i]}}{h_{il}} : h_{il} > 0 \right\} = \min \left\{ \frac{x_{B[4]}}{h_{12}}, \frac{x_{B[3]}}{h_{22}}, \frac{x_{B[1]}}{h_{32}} \right\} = \min \left\{ \frac{5}{3}, \frac{7/4}{1/4}, \frac{9/4}{3/4} \right\} =$$

$$\min \left\{ \frac{5}{3}, 7, 3 \right\} = \frac{5}{3} = \frac{x_{B[4]}}{h_{12}}$$

The leaving variable is x_4. Hence, $k = 4$ and $r = 1$.

Next, we update the basic and nonbasic lists:

$$B = \begin{bmatrix} 2 & 3 & 1 \end{bmatrix}, N = \begin{bmatrix} 4 & 5 & 6 \end{bmatrix}$$

Hence:

$$c_B = \begin{bmatrix} 2 \\ -4 \\ 5 \end{bmatrix}, c_N = \begin{bmatrix} 0 \\ 0 \\ 0 \end{bmatrix}$$

$$A_B = \begin{bmatrix} 1 & -2 & 6 \\ 1 & 1 & 1 \\ 4 & -2 & 6 \end{bmatrix}, A_N = \begin{bmatrix} -1 & 0 & 0 \\ 0 & 1 & 0 \\ 0 & 0 & -1 \end{bmatrix}$$

Then, we update the basis inverse using the PFI update scheme:

$$E^{-1} = \begin{bmatrix} 1/3 & 0 & 0 \\ -1/12 & 1 & 0 \\ -1/4 & 0 & 1 \end{bmatrix}$$

$$(A_{\overline{B}})^{-1} = E^{-1}A_B^{-1} = \begin{bmatrix} 1/3 & 0 & 0 \\ -1/12 & 1 & 0 \\ -1/4 & 0 & 1 \end{bmatrix} \begin{bmatrix} -1 & 0 & 1 \\ 0 & 3/4 & -1/8 \\ 0 & 1/4 & 1/8 \end{bmatrix} = \begin{bmatrix} -1/3 & 0 & 1/3 \\ 1/12 & 3/4 & 15/72 \\ 1/4 & 1/4 & -1/8 \end{bmatrix}$$

Finally, we update vectors x_B, w, and s_N:

$$x_B = A_B^{-1}b = \begin{bmatrix} -1/3 & 0 & 1/3 \\ 1/12 & 3/4 & 15/72 \\ 1/4 & 1/4 & -1/8 \end{bmatrix} \begin{bmatrix} 5 \\ 4 \\ 10 \end{bmatrix} = \begin{bmatrix} 5/3 \\ 4/3 \\ 1 \end{bmatrix}$$

$$w^T = c_B^T A_B^{-1} = \begin{bmatrix} 2 & -4 & 5 \end{bmatrix} \begin{bmatrix} -1/3 & 0 & 1/3 \\ 1/12 & 3/4 & 15/72 \\ 1/4 & 1/4 & -1/8 \end{bmatrix} = \begin{bmatrix} 1/4 & -7/4 & 7/8 \end{bmatrix}$$

$$s_N^T = c_N^T - w^T A_N = \begin{bmatrix} 0 & 0 & 0 \end{bmatrix} - \begin{bmatrix} 1/4 & -7/4 & 7/8 \end{bmatrix} \begin{bmatrix} -1 & 0 & 0 \\ 0 & 1 & 0 \\ 0 & 0 & -1 \end{bmatrix} = \begin{bmatrix} 1/4 & 7/4 & 7/8 \end{bmatrix}$$

Phase II - 5th Iteration

The data at the beginning of this iteration is:

$$c_B = \begin{bmatrix} 2 \\ -4 \\ 5 \end{bmatrix}, c_N = \begin{bmatrix} 0 \\ 0 \\ 0 \end{bmatrix}$$

$$A_B = \begin{bmatrix} 1 & -2 & 6 \\ 1 & 1 & 1 \\ 4 & -2 & 6 \end{bmatrix}, A_N = \begin{bmatrix} -1 & 0 & 0 \\ 0 & 1 & 0 \\ 0 & 0 & -1 \end{bmatrix}$$

$$A_B^{-1} = \begin{bmatrix} -1/3 & 0 & 1/3 \\ 1/12 & 3/4 & 15/72 \\ 1/4 & 1/4 & -1/8 \end{bmatrix}$$

$$x_B = \begin{bmatrix} 5/3 \\ 4/3 \\ 1 \end{bmatrix}, w^T = [1/4 \; -7/4 \; 7/8], s_N^T = [1/4 \; 7/4 \; 7/8]$$

$s_N \geq 0$, so the current basis, $B = [2, 3, 1]$ and $N = [4, 5, 6]$, is optimal:

$$x_B = \begin{bmatrix} 5/3 \\ 4/3 \\ 1 \end{bmatrix}$$

$$z = c_B^T x_B = 3$$

Example 3 Consider the following LP problem:

$$\begin{array}{rl}
\min z = & - \; x_2 - 2x_3 \\
\text{s.t.} & x_1 + x_2 + x_3 = 5 \\
& - 2x_2 - 3x_3 = 3 \\
x_j \geq 0, & (j = 1, 2, 3)
\end{array}$$

In matrix notation the above LP problem is written as follows:

$$A = \begin{bmatrix} 1 & 1 & 1 \\ 0 & -2 & -3 \end{bmatrix}, c = \begin{bmatrix} 0 \\ -1 \\ -2 \end{bmatrix}, b = \begin{bmatrix} 5 \\ 3 \end{bmatrix}, Eqin = \begin{bmatrix} 0 \\ 0 \end{bmatrix}$$

The LP problem is in its standard form. Let's start with the basic list $B = [1, 2]$ and the nonbasic list $N = [3]$. It is easy to find out that this basis is not feasible ($x_B \not\geq 0$):

$$A_B = \begin{bmatrix} 1 & 1 \\ 0 & -2 \end{bmatrix}$$

$$A_B^{-1} = \begin{bmatrix} 1 & 1/2 \\ 0 & -1/2 \end{bmatrix}$$

$$x_B = A_B^{-1} b = \begin{bmatrix} 1 & 1/2 \\ 0 & -1/2 \end{bmatrix} \begin{bmatrix} 5 \\ 3 \end{bmatrix} = \begin{bmatrix} 13/2 \\ -3/2 \end{bmatrix}$$

We apply Phase I. We add the artificial variable x_4 and calculate vector d as follows:

$$d = -A_B e = -\begin{bmatrix} 1 & 1 \\ 0 & -2 \end{bmatrix}\begin{bmatrix} 1 \\ 1 \end{bmatrix} = \begin{bmatrix} -2 \\ 2 \end{bmatrix}$$

The leaving variable is x_2:

$$\min\left\{x_{B[1]}, x_{B[2]}\right\} = \min\left\{13/2, -3/2\right\} = -3/2$$

Hence, the auxiliary LP problem that will be solved in Phase I is the following:

$$
\begin{aligned}
\min z = \quad & & & x_4 \\
\text{s.t.} \quad x_1 + & x_2 & x_3 - 2x_4 &= 5 \\
& -2x_2 & -3x_3 + 2x_4 &= 3 \\
x_j \geq 0, \quad & (j = 1, 2, 3, 4)
\end{aligned}
$$

In matrix notation the above LP problem is written as follows:

$$A = \begin{bmatrix} 1 & 1 & 1 & -2 \\ 0 & -2 & -3 & 2 \end{bmatrix}, c = \begin{bmatrix} 0 \\ 0 \\ 0 \\ 1 \end{bmatrix}, b = \begin{bmatrix} 5 \\ 3 \end{bmatrix}, Eqin = \begin{bmatrix} 0 \\ 0 \end{bmatrix}$$

Phase I - 1st Iteration

The initial basic list of Phase I is $B = [1, 2] \cup [4] \setminus [2] = [1, 4]$ and the nonbasic list is $N = [3] \cup [2] = [3, 2]$. The initial data of the problem is:

$$c_B = \begin{bmatrix} 0 \\ 1 \end{bmatrix}, c_N = \begin{bmatrix} 0 \\ 0 \end{bmatrix}$$

$$A_B = \begin{bmatrix} 1 & -2 \\ 0 & 2 \end{bmatrix}, A_N = \begin{bmatrix} 1 & 1 \\ -3 & -2 \end{bmatrix}$$

$$A_B^{-1} = \begin{bmatrix} 1 & 1 \\ 0 & 1/2 \end{bmatrix}$$

$$x_B = A_B^{-1}b = \begin{bmatrix} 1 & 1 \\ 0 & 1/2 \end{bmatrix}\begin{bmatrix} 5 \\ 3 \end{bmatrix} = \begin{bmatrix} 8 \\ 3/2 \end{bmatrix}$$

$$w^T = c_B^T A_B^{-1} = \begin{bmatrix} 0 & 1 \end{bmatrix}\begin{bmatrix} 1 & 1 \\ 0 & 1/2 \end{bmatrix} = \begin{bmatrix} 0 & 1/2 \end{bmatrix}$$

$$s_N^T = c_N^T - w^T A_N = \begin{bmatrix} 0 & 0 \end{bmatrix} - \begin{bmatrix} 0 & 1/2 \end{bmatrix}\begin{bmatrix} 1 & 1 \\ -3 & -2 \end{bmatrix} = \begin{bmatrix} 3/2 & 1 \end{bmatrix}$$

$s_N \geq 0$, but the artificial variable x_4 is in the basic list and $x_{B[4]} = 3/2 > 0$, so the LP problem is infeasible.

Example 4 Consider the following LP problem:

$$
\begin{aligned}
\min z = \quad & 3x_1 - 2x_2 \\
\text{s.t.} \quad & 2x_1 + 4x_2 \geq 8 \\
& x_1 + x_2 \geq 3 \\
& -2x_1 - 7x_2 \leq -15 \\
& x_j \geq 0, \quad (j = 1, 2)
\end{aligned}
$$

In matrix notation the above LP problem is written as follows:

$$
A = \begin{bmatrix} 2 & 4 \\ 1 & 1 \\ -2 & -7 \end{bmatrix}, c = \begin{bmatrix} 3 \\ -2 \end{bmatrix}, b = \begin{bmatrix} 8 \\ 3 \\ -15 \end{bmatrix}, Eqin = \begin{bmatrix} 1 \\ 1 \\ -1 \end{bmatrix}
$$

Initially, we need to convert the LP problem in its standard form:

$$
\begin{aligned}
\min z = \quad & 3x_1 - 2x_2 \\
\text{s.t.} \quad & 2x_1 + 4x_2 - x_3 && = 8 \\
& x_1 + x_2 \quad\;\; - x_4 && = 3 \\
& -2x_1 - 7x_2 && + x_5 = -15 \\
& x_j \geq 0, \quad (j = 1, 2, 3, 4, 5)
\end{aligned}
$$

In matrix notation the above LP problem is written as follows:

$$
A = \begin{bmatrix} 2 & 4 & -1 & 0 & 0 \\ 1 & 1 & 0 & -1 & 0 \\ -2 & -7 & 0 & 0 & 1 \end{bmatrix}, c = \begin{bmatrix} 3 \\ -2 \\ 0 \\ 0 \\ 0 \end{bmatrix}, b = \begin{bmatrix} 8 \\ 3 \\ -15 \end{bmatrix}, Eqin = \begin{bmatrix} 0 \\ 0 \\ 0 \end{bmatrix}
$$

Let's start with the basic list $B = [3, 4, 5]$ and the nonbasic list $N = [1, 2]$. It is easy to find out that this basis is not feasible ($x_B \ngeq 0$):

$$
A_B = \begin{bmatrix} -1 & 0 & 0 \\ 0 & -1 & 0 \\ 0 & 0 & 1 \end{bmatrix}
$$

$$
A_B^{-1} = \begin{bmatrix} -1 & 0 & 0 \\ 0 & -1 & 0 \\ 0 & 0 & 1 \end{bmatrix}
$$

$$x_B = A_B^{-1}b = \begin{bmatrix} -1 & 0 & 0 \\ 0 & -1 & 0 \\ 0 & 0 & 1 \end{bmatrix} \begin{bmatrix} 8 \\ 3 \\ -15 \end{bmatrix} = \begin{bmatrix} -8 \\ -3 \\ -15 \end{bmatrix}$$

We apply Phase I. We add the artificial variable x_6 and calculate vector d as follows:

$$d = -A_B e = - \begin{bmatrix} -1 & 0 & 0 \\ 0 & -1 & 0 \\ 0 & 0 & 1 \end{bmatrix} \begin{bmatrix} 1 \\ 1 \\ 1 \end{bmatrix} = \begin{bmatrix} 1 \\ 1 \\ -1 \end{bmatrix}$$

The leaving variable is x_5:

$$\min \{x_{B[3]}, x_{B[4]}, x_{B[5]}\} = \min \{-8, -3, -15\} = -15$$

Hence, the auxiliary LP problem that will be solved in Phase I is the following:

$$
\begin{aligned}
\min z = \quad & 3x_1 - 2x_2 & & + x_6 \\
\text{s.t.} \quad & 2x_1 + 4x_2 - x_3 & & + x_6 = \quad 8 \\
& x_1 + x_2 & - x_4 & + x_6 = \quad 3 \\
& -2x_1 - 7x_2 & & + x_5 - x_6 = -15 \\
& x_j \geq 0, \quad (j = 1, 2, 3, 4, 5, 6)
\end{aligned}
$$

In matrix notation the above LP problem is written as follows:

$$A = \begin{bmatrix} 2 & 4 & -1 & 0 & 0 & 1 \\ 1 & 1 & 0 & -1 & 0 & 1 \\ -2 & -7 & 0 & 0 & 1 & -1 \end{bmatrix}, c = \begin{bmatrix} 0 \\ 0 \\ 0 \\ 0 \\ 0 \\ 1 \end{bmatrix}, b = \begin{bmatrix} 8 \\ 3 \\ -15 \end{bmatrix}, Eqin = \begin{bmatrix} 0 \\ 0 \\ 0 \end{bmatrix}$$

Phase I - 1st Iteration

The initial basic list of Phase I is $B = [3, 4, 5] \cup [6] \setminus [5] = [3, 4, 6]$ and the nonbasic list is $N = [1, 2] \cup [5] = [1, 2, 5]$. The initial data of the problem is:

$$c_B = \begin{bmatrix} 0 \\ 0 \\ 1 \end{bmatrix}, c_N = \begin{bmatrix} 0 \\ 0 \\ 0 \end{bmatrix}$$

$$A_B = \begin{bmatrix} -1 & 0 & 1 \\ 0 & -1 & 1 \\ 0 & 0 & -1 \end{bmatrix}, A_N = \begin{bmatrix} 2 & 4 & 0 \\ 1 & 1 & 0 \\ -2 & -7 & 1 \end{bmatrix}$$

$$A_B^{-1} = \begin{bmatrix} -1 & 0 & -1 \\ 0 & -1 & -1 \\ 0 & 0 & -1 \end{bmatrix}$$

$$x_B = A_B^{-1}b = \begin{bmatrix} -1 & 0 & -1 \\ 0 & -1 & -1 \\ 0 & 0 & -1 \end{bmatrix} \begin{bmatrix} 8 \\ 3 \\ -15 \end{bmatrix} = \begin{bmatrix} 7 \\ 12 \\ 15 \end{bmatrix}$$

$$w^T = c_B^T A_B^{-1} = \begin{bmatrix} 0 & 0 & 1 \end{bmatrix} \begin{bmatrix} -1 & 0 & -1 \\ 0 & -1 & -1 \\ 0 & 0 & -1 \end{bmatrix} = \begin{bmatrix} 0 & 0 & -1 \end{bmatrix}$$

$$s_N^T = c_N^T - w^T A_N = \begin{bmatrix} 0 & 0 & 0 \end{bmatrix} - \begin{bmatrix} 0 & 0 & -1 \end{bmatrix} \begin{bmatrix} 2 & 4 & 0 \\ 1 & 1 & 0 \\ -2 & -7 & 1 \end{bmatrix} = \begin{bmatrix} -2 & -7 & 1 \end{bmatrix}$$

$s_N \not\ge 0$, so the current basis is not optimal. According to Dantzig's rule, the entering variable is x_2, because it is the one with the most negative $\bar{s}_l$ (-7). Hence, $l = 2$.

We continue calculating the pivoting column:

$$h_2 = A_B^{-1}A_{.2} = \begin{bmatrix} -1 & 0 & -1 \\ 0 & -1 & -1 \\ 0 & 0 & -1 \end{bmatrix} \begin{bmatrix} 4 \\ 1 \\ -7 \end{bmatrix} = \begin{bmatrix} 3 \\ 6 \\ 7 \end{bmatrix}$$

There are elements in vector h_2 that are greater than 0, so we perform the minimum ratio test:

$$\min \left\{ \frac{x_{B[i]}}{h_{il}} : h_{il} > 0 \right\} = \min \left\{ \frac{x_{B[3]}}{h_{12}}, \frac{x_{B[4]}}{h_{22}}, \frac{x_{B[6]}}{h_{32}} \right\} = \min \left\{ \frac{7}{3}, \frac{12}{6}, \frac{15}{7} \right\} = \frac{12}{6} = \frac{x_{B[4]}}{h_{22}}$$

The leaving variable is x_4. Hence, $k = 4$ and $r = 2$.
Next, we update the basic and nonbasic lists:

$$B = \begin{bmatrix} 3 & 2 & 6 \end{bmatrix}, N = \begin{bmatrix} 1 & 4 & 5 \end{bmatrix}$$

Hence:

$$c_B = \begin{bmatrix} 0 \\ 0 \\ 1 \end{bmatrix}, c_N = \begin{bmatrix} 0 \\ 0 \\ 0 \end{bmatrix}$$

$$A_B = \begin{bmatrix} -1 & 4 & 1 \\ 0 & 1 & 1 \\ 0 & -7 & -1 \end{bmatrix}, A_N = \begin{bmatrix} 2 & 0 & 0 \\ 1 & -1 & 0 \\ -2 & 0 & 1 \end{bmatrix}$$

Then, we update the basis inverse using the PFI update scheme:

$$E^{-1} = \begin{bmatrix} 1 & -1/2 & 0 \\ 0 & 1/6 & 0 \\ 0 & -7/6 & 1 \end{bmatrix}$$

$$(A_{\bar{B}})^{-1} = E^{-1}A_B^{-1} = \begin{bmatrix} 1 & 0 & 0 \\ 0 & 1 & -7/6 \\ 0 & 0 & 7/6 \end{bmatrix} \begin{bmatrix} -1 & 0 & -1 \\ 0 & -1 & -1 \\ 0 & 0 & -1 \end{bmatrix} = \begin{bmatrix} -1 & 1/2 & -1/2 \\ 0 & -1/6 & -1/6 \\ 0 & 7/6 & 1/6 \end{bmatrix}$$

Finally, we update vectors x_B, w, and s_N:

$$x_B = A_B^{-1}b = \begin{bmatrix} -1 & 1/2 & -1/2 \\ 0 & -1/6 & -1/6 \\ 0 & 7/6 & 1/6 \end{bmatrix} \begin{bmatrix} 8 \\ 3 \\ -15 \end{bmatrix} = \begin{bmatrix} 1 \\ 2 \\ 1 \end{bmatrix}$$

$$w^T = c_B^T A_B^{-1} = \begin{bmatrix} 0 & 0 & 1 \end{bmatrix} \begin{bmatrix} -1 & 1/2 & -1/2 \\ 0 & -1/6 & -1/6 \\ 0 & 7/6 & 1/6 \end{bmatrix} = \begin{bmatrix} 0 & 7/6 & 1/6 \end{bmatrix}$$

$$s_N^T = c_N^T - w^T A_N = \begin{bmatrix} 0 & 0 & 0 \end{bmatrix} - \begin{bmatrix} 0 & 7/6 & 1/6 \end{bmatrix} \begin{bmatrix} 2 & 0 & 0 \\ 1 & -1 & 0 \\ -2 & 0 & 1 \end{bmatrix} = \begin{bmatrix} -5/6 & 7/6 & -1/6 \end{bmatrix}$$

Phase I - 2nd Iteration

The data at the beginning of this iteration is:

$$c_B = \begin{bmatrix} 0 \\ 0 \\ 1 \end{bmatrix}, c_N = \begin{bmatrix} 0 \\ 0 \\ 0 \end{bmatrix}$$

$$A_B = \begin{bmatrix} -1 & 4 & 1 \\ 0 & 1 & 1 \\ 0 & -7 & -1 \end{bmatrix}, A_N = \begin{bmatrix} 2 & 0 & 0 \\ 1 & -1 & 0 \\ -2 & 0 & 1 \end{bmatrix}$$

$$A_B^{-1} = \begin{bmatrix} -1 & 1/2 & -1/2 \\ 0 & -1/6 & -1/6 \\ 0 & 7/6 & 1/6 \end{bmatrix}$$

$$x_B = \begin{bmatrix} 1 \\ 2 \\ 1 \end{bmatrix}, w^T = \begin{bmatrix} 0 & 7/6 & 1/6 \end{bmatrix}, s_N^T = \begin{bmatrix} -5/6 & 7/6 & -1/6 \end{bmatrix}$$

$s_N \not\geq 0$, so the current basis is not optimal. According to Dantzig's rule, the entering variable is x_1, because it is the one with the most negative $\bar{s}_l$ ($-5/6$). Hence, $l = 1$.

We continue calculating the pivoting column:

$$h_1 = A_B^{-1} A_{.1} = \begin{bmatrix} -1 & 1/2 & -1/2 \\ 0 & -1/6 & -1/6 \\ 0 & 7/6 & 1/6 \end{bmatrix} \begin{bmatrix} 2 \\ 1 \\ -2 \end{bmatrix} = \begin{bmatrix} -1/2 \\ 1/6 \\ 5/6 \end{bmatrix}$$

There are elements in vector h_1 that are greater than 0, so we perform the minimum ratio test:

$$\min \left\{ \frac{x_{B[i]}}{h_{il}} : h_{il} > 0 \right\} = \min \left\{ x, \frac{x_{B[2]}}{h_{21}}, \frac{x_{B[6]}}{h_{31}} \right\} = \min \left\{ x, \frac{2}{1/6}, \frac{1}{5/6} \right\} = \frac{1}{5/6} = \frac{x_{B[6]}}{h_{31}}$$

The leaving variable is x_6. Hence, $k = 6$ and $r = 3$.
Next, we update the basic and nonbasic lists:

$$B = \begin{bmatrix} 3 & 2 & 1 \end{bmatrix}, N = \begin{bmatrix} 6 & 4 & 5 \end{bmatrix}$$

Hence:

$$c_B = \begin{bmatrix} 0 \\ 0 \\ 0 \end{bmatrix}, c_N = \begin{bmatrix} 1 \\ 0 \\ 0 \end{bmatrix}$$

$$A_B = \begin{bmatrix} -1 & 4 & 2 \\ 0 & 1 & 1 \\ 0 & -7 & -2 \end{bmatrix}, A_N = \begin{bmatrix} 1 & 0 & 0 \\ 1 & -1 & 0 \\ -1 & 0 & 1 \end{bmatrix}$$

Then, we update the basis inverse using the PFI update scheme:

$$E^{-1} = \begin{bmatrix} 1 & 0 & 3/5 \\ 0 & 1 & -1/5 \\ 0 & 0 & 6/5 \end{bmatrix}$$

$$(A_{\bar{B}})^{-1} = E^{-1} A_B^{-1} = \begin{bmatrix} 1 & 0 & 3/5 \\ 0 & 1 & -1/5 \\ 0 & 0 & 6/5 \end{bmatrix} \begin{bmatrix} -1 & 1/2 & -1/2 \\ 0 & -1/6 & -1/6 \\ 0 & 7/6 & 1/6 \end{bmatrix} = \begin{bmatrix} -1 & 6/5 & -2/5 \\ 0 & -2/5 & -1/5 \\ 0 & 7/5 & 1/5 \end{bmatrix}$$

Finally, we update vectors x_B, w, and s_N:

$$x_B = A_B^{-1} b = \begin{bmatrix} -1 & 6/5 & -2/5 \\ 0 & -2/5 & -1/5 \\ 0 & 7/5 & 1/5 \end{bmatrix} \begin{bmatrix} 8 \\ 3 \\ -15 \end{bmatrix} = \begin{bmatrix} 8/5 \\ 9/5 \\ 6/5 \end{bmatrix}$$

$$w^T = c_B^T A_B^{-1} = [0\ 0\ 0] \begin{bmatrix} -1 & 6/5 & -2/5 \\ 0 & -2/5 & -1/5 \\ 0 & 7/5 & 1/5 \end{bmatrix} = [0\ 0\ 0]$$

$$s_N^T = c_N^T - w^T A_N = [1\ 0\ 0] - [0\ 0\ 0] \begin{bmatrix} 1 & 0 & 0 \\ 1 & -1 & 0 \\ -1 & 0 & 1 \end{bmatrix} = [1\ 0\ 0]$$

Phase I - 3rd Iteration

The data at the beginning of this iteration is:

$$c_B = \begin{bmatrix} 0 \\ 0 \\ 0 \end{bmatrix}, c_N = \begin{bmatrix} 1 \\ 0 \\ 0 \end{bmatrix}$$

$$A_B = \begin{bmatrix} -1 & 4 & 2 \\ 0 & 1 & 1 \\ 0 & -7 & -2 \end{bmatrix}, A_N = \begin{bmatrix} 1 & 0 & 0 \\ 1 & -1 & 0 \\ -1 & 0 & 1 \end{bmatrix}$$

$$A_B^{-1} = \begin{bmatrix} -1 & 6/5 & -2/5 \\ 0 & -2/5 & -1/5 \\ 0 & 7/5 & 1/5 \end{bmatrix}$$

$$x_B = \begin{bmatrix} 8/5 \\ 9/5 \\ 6/5 \end{bmatrix}, w^T = [0\ 0\ 0], s_N^T = [1\ 0\ 0]$$

$s_N \geq 0$, so the current basis is optimal for the auxiliary LP problem of Phase I. We found a feasible basis ($B = [3, 2, 1]$) for the initial LP problem. We delete the artificial variable from the nonbasic list, so the nonbasic list is $N = [4, 5]$. Hence, we proceed to Phase II.

Phase II - 4th Iteration

The data of the problem is:

$$c_B = \begin{bmatrix} 0 \\ 2 \\ -3 \end{bmatrix}, c_N = \begin{bmatrix} 0 \\ 0 \end{bmatrix}$$

$$A_B = \begin{bmatrix} -1 & 4 & 2 \\ 0 & 1 & 1 \\ 0 & -7 & -2 \end{bmatrix}, A_N = \begin{bmatrix} 0 & 0 \\ -1 & 0 \\ 0 & 1 \end{bmatrix}$$

$$A_B^{-1} = \begin{bmatrix} -1 & 6/5 & -2/5 \\ 0 & -2/5 & -1/5 \\ 0 & 7/5 & 1/5 \end{bmatrix}$$

$$x_B = A_B^{-1}b = \begin{bmatrix} -1 & 6/5 & -2/5 \\ 0 & -2/5 & -1/5 \\ 0 & 7/5 & 1/5 \end{bmatrix} \begin{bmatrix} 8 \\ 3 \\ -15 \end{bmatrix} = \begin{bmatrix} 8/5 \\ 9/5 \\ 6/5 \end{bmatrix}$$

$$w^T = c_B^T A_B^{-1} = \begin{bmatrix} 0 & 2 & -3 \end{bmatrix} \begin{bmatrix} -1 & 6/5 & -2/5 \\ 0 & -2/5 & -1/5 \\ 0 & 7/5 & 1/5 \end{bmatrix} = \begin{bmatrix} 0 & -5 & -1 \end{bmatrix}$$

$$s_N^T = c_N^T - w^T A_N = \begin{bmatrix} 0 & 0 \end{bmatrix} - \begin{bmatrix} 0 & -5 & -1 \end{bmatrix} \begin{bmatrix} 0 & 0 \\ -1 & 0 \\ 0 & 1 \end{bmatrix} = \begin{bmatrix} -5 & 1 \end{bmatrix}$$

$s_N \not\geq 0$, so the current basis is not optimal. According to Dantzig's rule, the entering variable is x_4, because it is the one with the most negative $\bar{s}_l$ (-5). Hence, $l = 4$.

We continue calculating the pivoting column:

$$h_4 = A_B^{-1}A_{.4} = \begin{bmatrix} -1 & 6/5 & -2/5 \\ 0 & -2/5 & -1/5 \\ 0 & 7/5 & 1/5 \end{bmatrix} \begin{bmatrix} 0 \\ -1 \\ 0 \end{bmatrix} = \begin{bmatrix} -6/5 \\ 2/5 \\ -7/5 \end{bmatrix}$$

There are elements in vector h_4 that are greater than 0, so we perform the minimum ratio test:

$$\min\left\{ \frac{x_{B[i]}}{h_{il}} : h_{il} > 0 \right\} = \min\left\{ x, \frac{x_{B[2]}}{h_{24}}, x \right\} = \min\left\{ x, \frac{9/5}{2/5}, x \right\} = \frac{9/5}{2/5} = \frac{x_{B[2]}}{h_{24}}$$

The leaving variable is x_2. Hence, $k = 2$ and $r = 2$.
Next, we update the basic and nonbasic lists:

$$B = \begin{bmatrix} 3 & 4 & 1 \end{bmatrix}, N = \begin{bmatrix} 2 & 5 \end{bmatrix}$$

Hence:

$$c_B = \begin{bmatrix} 0 \\ 0 \\ -3 \end{bmatrix}, c_N = \begin{bmatrix} 2 \\ 0 \end{bmatrix}$$

$$A_B = \begin{bmatrix} -1 & 0 & 2 \\ 0 & -1 & 1 \\ 0 & 0 & -2 \end{bmatrix}, A_N = \begin{bmatrix} 4 & 0 \\ 1 & 0 \\ -7 & 1 \end{bmatrix}$$

Then, we update the basis inverse using the PFI update scheme:

$$E^{-1} = \begin{bmatrix} 1 & 3 & 0 \\ 0 & 5/2 & 0 \\ 0 & 7/2 & 1 \end{bmatrix}$$

$$(A_{\bar{B}})^{-1} = E^{-1}A_B^{-1} = \begin{bmatrix} 1 & 3 & 0 \\ 0 & 5/2 & 0 \\ 0 & 7/2 & 1 \end{bmatrix}\begin{bmatrix} -1 & 6/5 & -2/5 \\ 0 & -2/5 & -1/5 \\ 0 & 7/5 & 1/5 \end{bmatrix} = \begin{bmatrix} -1 & 0 & -1 \\ 0 & -1 & -1/2 \\ 0 & 0 & -1/2 \end{bmatrix}$$

Finally, we update vectors x_B, w, and s_N:

$$x_B = A_B^{-1}b = \begin{bmatrix} -1 & 0 & -1 \\ 0 & -1 & -1/2 \\ 0 & 0 & -1/2 \end{bmatrix}\begin{bmatrix} 8 \\ 3 \\ -15 \end{bmatrix} = \begin{bmatrix} 7 \\ 9/2 \\ 15/2 \end{bmatrix}$$

$$w^T = c_B^T A_B^{-1} = \begin{bmatrix} 0 & 0 & -3 \end{bmatrix}\begin{bmatrix} -1 & 0 & -1 \\ 0 & -1 & -1/2 \\ 0 & 0 & -1/2 \end{bmatrix} = \begin{bmatrix} 0 & 0 & 3/2 \end{bmatrix}$$

$$s_N^T = c_N^T - w^T A_N = \begin{bmatrix} 2 & 0 \end{bmatrix} - \begin{bmatrix} 0 & 0 & 3/2 \end{bmatrix}\begin{bmatrix} 4 & 0 \\ 1 & 0 \\ -7 & 1 \end{bmatrix} = \begin{bmatrix} 25/2 & -3/2 \end{bmatrix}$$

Phase II - 5th Iteration

The data at the beginning of this iteration is:

$$c_B = \begin{bmatrix} 0 \\ 0 \\ -3 \end{bmatrix}, c_N = \begin{bmatrix} 2 \\ 0 \end{bmatrix}$$

$$A_B = \begin{bmatrix} -1 & 0 & 2 \\ 0 & -1 & 1 \\ 0 & 0 & -2 \end{bmatrix}, A_N = \begin{bmatrix} 4 & 0 \\ 1 & 0 \\ -7 & 1 \end{bmatrix}$$

$$A_B^{-1} = \begin{bmatrix} -1 & 0 & -1 \\ 0 & -1 & -1/2 \\ 0 & 0 & = 1/2 \end{bmatrix}$$

$$x_B = \begin{bmatrix} 7 \\ 9/2 \\ 15/2 \end{bmatrix}, w^T = \begin{bmatrix} 0 & 0 & 3/2 \end{bmatrix}, s_N^T = \begin{bmatrix} 25/2 & -3/2 \end{bmatrix}$$

$s_N \not\geq 0$, so the current basis is not optimal. According to Dantzig's rule, the entering variable is x_5, because it is the one with the most negative $\bar{s}_l$ $(-3/2)$. Hence, $l = 5$.

We continue calculating the pivoting column:

$$h_5 = A_B^{-1}A._5 = \begin{bmatrix} -1 & 0 & -1 \\ 0 & -1 & -1/2 \\ 0 & 0 & = 1/2 \end{bmatrix} \begin{bmatrix} 0 \\ 0 \\ 1 \end{bmatrix} = \begin{bmatrix} -1 \\ -1/2 \\ -1/2 \end{bmatrix}$$

There are not any elements in vector h_2 that are greater than 0, so the LP problem is unbounded.

8.5   Implementation in MATLAB

This subsection presents the implementation in MATLAB of the revised primal simplex algorithm. Some necessary notations should be introduced before the presentation of the implementation of the revised primal simplex algorithm. Let A be a $m \times n$ matrix with the coefficients of the constraints, c be a $n \times 1$ vector with the coefficients of the objective function, b be a $m \times 1$ vector with the right-hand side of the constraints, $Eqin$ be a $m \times 1$ vector with the type of the constraints, $MinMaxLP$ be the type of optimization, $c0$ be the constant term of the objective function, $reinv$ be the number of iterations that the basis inverse is recomputed from scratch, $tole1$ be the tolerance for the basic solution, $tole2$ be the tolerance for the reduced costs, $tole3$ be the tolerance for the pivoting column, $scalingTechnique$ be the scaling method (0: no scaling, 1: arithmetic mean, 2: de Buchet for the case p = 1, 3: de Buchet for the case p = 2, 4: de Buchet for the case p = Inf, 5: entropy, 6: equilibration, 7: geometric mean, 8: IBM MPSX, 9: LP-norm for the case p = 1, 10: LP-norm for the case p = 2, 11: LP-norm for the case p = Inf), $pivotingRule$ be the pivoting rule (1: Bland's rule, 2: Dantzig's rule, 3: Greatest Increment Method, 4: Least Recently Considered Method, 5: Partial Pricing Rule, 6: Steepest Edge Rule), $basisUpdateMethod$ be the basis update method (1: PFI, 2: MPFI), $xsol$ be a $m \times 1$ vector with the solution found, $fval$ be the value of the objective function at the solution $xsol$, $exitflag$ be the reason that the algorithm terminated (0: optimal solution found, 1: the LP problem is infeasible, 2: the LP problem is unbounded, -1: the input data is not logically or numerically correct), and $iterations$ be the number of iterations.

The revised primal simplex algorithm is implemented with two functions: (i) one function that implements the revised primal simplex algorithm (filename: rsa.m), and (ii) one function that finds an initial invertible basis (filename: lprref.m).

The function for the implementation of the revised primal simplex algorithm takes as input the matrix with the coefficients of the constraints (matrix A), the vector with the coefficients of the objective function (vector c), the vector with the right-hand side of the constraints (vector b), the vector with the type of the constraints (vector $Eqin$), the type of optimization (variable $MinMaxLP$), the constant term of the objective function (variable $c0$), the number of iterations that the basis inverse is recomputed from scratch (variable *reinv*), the tolerance of the basic solution (variable *tole1*), the tolerance for the reduced costs (variable *tole2*), the tolerance for the pivoting column (variable *tole3*), the scaling method (variable *scalingTechnique*), the pivoting rule (variable *pivotingRule*), and the basis update method (variable *basisUpdateMethod*), and returns as output the solution found (vector *xsol*), the value of the objective function at the solution *xsol* (variable *fval*), the reason that the algorithm terminated (variable *exitflag*), and the number of iterations (variable *iterations*).

In lines 64–67, the output variables are initialized. In lines 69–95, we set default values to the missing inputs (if any). Next, we check if the input data is logically correct (lines 101–132). If the type of optimization is maximization, then we multiply vector c and constant $c0$ by -1 (lines 135–138). Then, the presolve analysis is performed (lines 141–152). Next, we scale the LP problem using the selected scaling technique (lines 159–187). Then, we find an invertible initial basis using the *lprref* function (lines 190–249). If the density of matrix A is less than 20%, then we use sparse algebra for faster computations (lines 256–260). Then, we pre-allocate memory for variables that hold large amounts of data and initialize the necessary variables (lines 263–298). Next, we check if the initial basis is feasible in order to avoid Phase I (lines 300–304).

Phase I is performed in lines 306–539. Initially, we find the index of the minimum element of vector x_B and create a copy of vector c in order to add the artificial variable (lines 309–312). We also calculate vector d and redefine matrix A since the artificial variable needs to be added for Phase I (lines 314–317). Then, we compute the new basic and nonbasic lists and initialize the necessary variables (lines 319–353). At each iteration of Phase I, we perform the optimality test in order to find if the basis found is feasible for the original LP problem or if the original LP problem is infeasible (lines 360–423). If the auxiliary LP problem is not optimal, then we find the entering variable using the selected pivoting rule, find the leaving variable using the minimum ratio test, and update the basis inverse and the necessary variables (lines 424–537).

Phase II is performed in lines 540–741. Initially, we initialize the necessary variables (lines 548–583). At each iteration of Phase II, we perform the optimality test in order to find if the LP problem is optimal (lines 724–739). If the LP problem is not optimal, then we find the entering variable using the selected pivoting rule, find the leaving variable using the minimum ratio test, and update the basis inverse and the necessary variables (lines 593–701). Moreover, we print intermediate results every 100 iterations (lines 704–717) and we also check if the LP problem is unbounded (lines 718–723).

```
1.    function [xsol, fval, exitflag, iterations] = ...
2.        rsa(A, c, b, Eqin, MinMaxLP, c0, reinv, tole1, ...
3.        tole2, tole3, scalingTechnique, pivotingRule, ...
4.        basisUpdateMethod)
5.    % Filename: rsa.m
6.    % Description: the function is an implementation of the
```

```
7.    % revised primal simplex algorithm
8.    % Authors: Ploskas, N., & Samaras, N.
9.    %
10.   % Syntax: [xsol, fval, exitflag, iterations] = ...
11.   %    rsa(A, c, b, Eqin, MinMaxLP, c0, reinv, tole1, ...
12.   %    tole2, tole3, scalingTechnique, pivotingRule, ...
13.   %    basisUpdateMethod)
14.   %
15.   % Input:
16.   % -- A: matrix of coefficients of the constraints
17.   %    (size m x n)
18.   % -- c: vector of coefficients of the objective function
19.   %    (size n x 1)
20.   % -- b: vector of the right-hand side of the constraints
21.   %    (size m x 1)
22.   % -- Eqin: vector of the type of the constraints
23.   %    (size m x 1)
24.   % -- MinMaxLP: the type of optimization (optional:
25.   %    default value -1 - minimization)
26.   % -- c0: constant term of the objective function
27.   %    (optional: default value 0)
28.   % -- reinv: every reinv number of iterations, the basis
29.   %    inverse is re-computed from scratch (optional:
30.   %    default value 80)
31.   % -- tole1: tolerance for the basic solution (optional:
32.   %    default value 1e-07)
33.   % -- tole2: tolerance for the reduced costs (optional:
34.   %    default value 1e-09)
35.   % -- tole3: tolerance for the pivoting column (optional:
36.   %    default value 1e-09)
37.   % -- scalingTechnique: the scaling method to be used
38.   %    (0: no scaling, 1: arithmetic mean, 2: de Buchet for
39.   %    the case p = 1, 3: de Buchet for the case p = 2,
40.   %    4: de Buchet for the case p = Inf, 5: entropy,
41.   %    6: equilibration, 7: geometric mean, 8: IBM MPSX,
42.   %    9: LP-norm for the case p = 1, 10: LP-norm for
43.   %    the case p = 2, 11: LP-norm for the case p = Inf)
44.   %    (optional: default value 6)
45.   % -- pivotingRule: the pivoting rule to be used
46.   %    (1: Bland's rule, 2: Dantzig's rule, 3: Greatest
47.   %    Increment Method, 4: Least Recently Considered
48.   %    Method, 5: Partial Pricing Rule, 6: Steepest Edge
49.   %    Rule) (optional: default value 2)
50.   % -- basisUpdateMethod: the basis update method to be used
51.   %    (1: PFI, 2: MPFI) (optional: default value 1)
52.   %
53.   % Output:
54.   % -- xsol: the solution found (size m x 1)
55.   % -- fval: the value of the objective function at the
56.   %    solution x
57.   % -- exitflag: the reason that the algorithm terminated
58.   %    (0: optimal solution found, 1: the LP problem is
59.   %    infeasible, 2: the LP problem is unbounded, -1: the
60.   %    input data is not logically or numerically correct)
```

```
61.  % -- iterations: the number of iterations
62.
63.  % initialize output variables
64.  xsol = [];
65.  fval = 0;
66.  exitflag = 0;
67.  iterations = 0;
68.  % set default values to missing inputs
69.  if ~exist('MinMaxLP')
70.      MinMaxLP = -1;
71.  end
72.  if ~exist('c0')
73.      c0 = 0;
74.  end
75.  if ~exist('reinv')
76.      reinv = 80;
77.  end
78.  if ~exist('tole1')
79.      tole1 = 1e-7;
80.  end
81.  if ~exist('tole2')
82.      tole2 = 1e-9;
83.  end
84.  if ~exist('tole3')
85.      tole3 = 1e-9;
86.  end
87.  if ~exist('scalingTechnique')
88.      scalingTechnique = 6;
89.  end
90.  if ~exist('pivotingRule')
91.      pivotingRule = 2;
92.  end
93.  if ~exist('basisUpdateMethod')
94.      basisUpdateMethod = 1;
95.  end
96.  [m, n] = size(A); % find the size of matrix A
97.  [m2, n2] = size(c); % find the size of vector c
98.  [m3, n3] = size(Eqin); % find the size of vector Eqin
99.  [m4, n4] = size(b); % find the size of vector b
100. % check if input data is logically correct
101. if n2 ~= 1
102.     disp('Vector c is not a column vector.')
103.     exitflag = -1;
104.     return
105. end
106. if n ~= m2
107.     disp(['The number of columns in matrix A and ' ...
108.          'the number of rows in vector c do not match.'])
109.     exitflag = -1;
110.     return
111. end
112. if m4 ~= m
113.     disp(['The number of the right-hand side values ' ...
114.          'is not equal to the number of constraints.'])
```

```
115.    exitflag = -1;
116.    return
117. end
118. if n3 ~= 1
119.    disp('Vector Eqin is not a column vector')
120.    exitflag = -1;
121.    return
122. end
123. if n4 ~= 1
124.    disp('Vector b is not a column vector')
125.    exitflag = -1;
126.    return
127. end
128. if m4 ~= m3
129.    disp('The size of vectors Eqin and b does not match')
130.    exitflag = -1;
131.    return
132. end
133. % if the type of optimization is maximization, then multiply
134. % vector c and constant c0 by -1
135. if MinMaxLP == 1
136.    c = -c;
137.    c0 = -c0;
138. end
139. % perform the presolve analysis
140. disp('---- P R E S O L V E    A N A L Y S I S ----')
141. [A, c, b, Eqin, c0, infeasible, unbounded] = ...
142.    presolve(A, c, b, Eqin, c0);
143. if infeasible == 1 % the LP problem is infeasible
144.     disp('The LP problem is infeasible')
145.    exitflag = 1;
146.    return
147. end
148. if unbounded == 1 % the LP problem is unbounded
149.     disp('The LP problem is unbounded')
150.    exitflag = 2;
151.    return
152. end
153. [m, n] = size(A); % find the new size of matrix A
154. [m2, ~] = size(c); % find the new size of vector c
155. [m3, ~] = size(Eqin); % find the size of vector Eqin
156. % scale the LP problem using the selected scaling
157. % technique
158. disp('---- S C A L I N G ----')
159. if scalingTechnique == 1 % arithmetic mean
160.     [A, c, b, ~, ~] = arithmeticMean(A, c, b);
161. % de buchet for the case p = 1
162. elseif scalingTechnique == 2
163.     [A, c, b, ~, ~] = debuchet1(A, c, b);
164. % de buchet for the case p = 2
165. elseif scalingTechnique == 3
166.     [A, c, b, ~, ~] = debuchet2(A, c, b);
167. % de buchet for the case p = Inf
168. elseif scalingTechnique == 4
```

```
169.     [A, c, b, ~, ~] = debuchetinf(A, c, b);
170. elseif scalingTechnique == 5 % entropy
171.     [A, c, b, ~, ~] = entropy(A, c, b);
172. elseif scalingTechnique == 6 % equilibration
173.     [A, c, b, ~, ~] = equilibration(A, c, b);
174. elseif scalingTechnique == 7 % geometric mean
175.     [A, c, b, ~, ~] = geometricMean(A, c, b);
176. elseif scalingTechnique == 8 % IBM MPSX
177.     [A, c, b, ~, ~, ~, ~] = ibmmpsx(A, c, b);
178. % LP-norm for the case p = 1
179. elseif scalingTechnique == 9
180.     [A, c, b, ~, ~] = lpnorm1(A, c, b);
181. % LP-norm for the case p = 2
182. elseif scalingTechnique == 10
183.     [A, c, b, ~, ~] = lpnorm2(A, c, b);
184. % LP-norm for the case p = Inf
185. elseif scalingTechnique == 11
186.     [A, c, b, ~, ~] = lpnorminf(A, c, b);
187. end
188. % find an invertible basis
189. disp('---- I N I T I A L    B A S I S ----')
190. flag = isequal(Eqin, zeros(m3, 1));
191. if flag == 1 % all constraints are equalities
192.     % select an initial invertible basis using lprref
193.     % function
194.     [~, ~, jb, out, ~, exitflag] = lprref(A, ...
195.         b, Eqin, 1e-10);
196.     if exitflag == 1 % the LP problem is infeasible
197.         disp('The LP problem is infeasible')
198.         return
199.     end
200.     % create the basic and nonbasic lists
201.     BasicList = jb;
202.     NonBasicList = setdiff(1:n, BasicList);
203.     % delete redundant constraints found by lprref
204.     A(out, :) = [];
205.     b(out, :) = [];
206.     Eqin(out, :) = [];
207. else % some or all constraints are inequalities
208.     % add slack variables
209.     axm = nnz(Eqin);
210.     c(m2 + 1:m2 + axm, :) = 0;
211.     A(:, n + 1:n + axm) = zeros(m, axm);
212.     curcol = 1;
213.     for i = 1:m3
214.         % 'greater than or equal to' inequality constraint
215.         if Eqin(i, 1) == 1
216.             A(i, n + curcol) = -1;
217.             curcol = curcol + 1;
218.         % 'less than or equal to' inequality constraint
219.         elseif Eqin(i, 1) == -1
220.             A(i, n + curcol) = 1;
221.             curcol = curcol + 1;
222.         % unrecognized type of constraint
```

```
223.            elseif Eqin(i,1) ~= 0
224.                disp('Vector Eqin is not numerically correct.')
225.                exitflag = -1;
226.                return
227.            end
228.        end
229.        % select an initial invertible basis using lprref
230.        % function
231.        [~, ~, jb, out, ~, exitflag] = lprref(A, ...
232.            b, Eqin, 1e-10);
233.        if exitflag == 1 % the LP problem is infeasible
234.            disp('The LP problem is infeasible')
235.            return
236.        end
237.        % create the basic and nonbasic lists
238.        [~, y1] = size(A);
239.        temp = n + 1:y1;
240.        for i = 1:length(temp)
241.            jb(length(jb) + 1) = temp(i);
242.        end
243.        BasicList = sort(jb);
244.        NonBasicList = setdiff(1:y1, BasicList);
245.        % delete redundant constraints found by lprref
246.        A(out, :) = [];
247.        b(out, :) = [];
248.        Eqin(out, :) = [];
249. end
250. flag = 0;
251. [m1, n1] = size(A); % new size of matrix A
252. % calculate the density of matrix A
253. density = (nnz(A) / (m1 * n1)) * 100;
254. % if the density of matrix A is less than 20%, then
255. % use sparse algebra for faster computations
256. if density < 20
257.     A = sparse(A);
258.     c = sparse(c);
259.     b = sparse(b);
260. end
261. % preallocate memory for variables that hold large
262. % amounts of data
263. Xb = spalloc(m1, 1, m1); % basic solution
264. h_l = spalloc(m1, 1, m1); % pivoting column
265. optTest = spalloc(m1, 1, m1); % optimality test vector
266. % the matrix of the basic variables
267. Basis = spalloc(m1, length(BasicList), m1 ...
268.     * length(BasicList));
269. % the matrix of the nonbasic variables
270. N = spalloc(m1, length(NonBasicList), m1 ...
271.     * length(NonBasicList));
272. w = spalloc(1, n1, n1); % simplex multiplier
273. Sn = spalloc(1, n1, n1); % reduced costs
274. % initialize data
275. % the matrix of the basic variables
276. Basis = A(:, BasicList);
```

```
277. % the coefficients of the objective function for the
278. % basic variables
279. cb = c(BasicList);
280. % the matrix of the nonbasic variables
281. N = A(:, NonBasicList);
282. % the coefficients of the objective function for the
283. % nonbasic variables
284. cn = c(NonBasicList);
285. % calculate the basis inverse
286. BasisInv = inv(Basis);
287. Xb = BasisInv * b; % basic solution
288. % set to zero, the values of Xb that are less than or
289. % equal to tole1
290. toler = abs(Xb) <= tole1;
291. Xb(toler == 1) = 0;
292. w = cb' * BasisInv; % calculate the simplex multiplier
293. % calculate the reduced costs
294. Sn = sparse(cn' - w * N);
295. % set to zero, the values of Sn that are less than or
296. % equal to tole2
297. toler = abs(Sn) <= tole2;
298. Sn(toler == 1) = 0;
299. % check if the Phase I is needed
300. if all(Xb >= 0) % the solution is feasible, skip Phase I
301.     flag = 1;       % and proceed to Phase II
302. else % the solution is not feasible, proceed to Phase I
303.     flag = 0;
304. end
305. counter = 1;
306. if flag == 0 % Phase I
307.     disp('---- P H A S E    I ----')
308.     % find the index of the minimum element of vector Xb
309.     [~, minindex] = min(Xb);
310.     % create a copy of vector c for the Phase I
311.     c2 = sparse(zeros(length(c), 1));
312.     c2(length(c) + 1) = 1;
313.     % calculate vector d
314.     d = sparse(-(Basis) * ones(m1, 1));
315.     % re-define matrix A since the artificial variable
316.     % needs to be added for Phase I
317.     A(:, n1 + 1) = d;
318.     % compute the new basic and nonbasic variables
319.     NonBasicList(1, length(NonBasicList) + 1) = ...
320.         BasicList(1, minindex);
321.     BasicList(:, minindex) = n1 + 1;
322.     [~, n] = size(A); % the new size of matrix A
323.     flagfase1 = 0;
324.     Basis = A(:, BasicList); % the new basis
325.     % the new coefficients of the objective function for
326.     % the basic variables
327.     fb = c2(BasicList);
328.     % the matrix of the nonbasic variables
329.     N = A(:, NonBasicList);
330.     % the new coefficients of the objective function for
```

```
331.    % the nonbasic variables
332.    fn = c2(NonBasicList);
333.    % re-compute the inverse of the matrix
334.    BasisInv = inv(Basis);
335.    % compute the basic solution
336.    Xb = BasisInv * b;
337.    % set to zero, the values of Xb that are less than or
338.    % equal to tole1
339.    toler = abs(Xb) <= tole1;
340.    Xb(toler == 1) = 0;
341.    % compute the simplex multiplier
342.    w = fb' * BasisInv;
343.    HRN = w * N;
344.    % set to zero, the values of HRN that are less than or
345.    % equal to tole2
346.    toler = abs(HRN) <= tole2;
347.    HRN(toler == 1) = 0;
348.    % compute the reduced costs
349.    Sn = fn' - HRN;
350.    % set to zero, the values of Sn that are less than or
351.    % equal to tole2
352.    toler = abs(Sn) <= tole2;
353.    Sn(toler == 1) = 0;
354.    while flagfase1 == 0 % iterate in Phase I
355.        clear rr;
356.        clear mrt;
357.        clear h_l;
358.        [~, col] = find(NonBasicList == n);
359.        % optimality test for Phase I
360.        if all(Sn >= 0) || ~isempty(col)
361.            % if the artificial variable is leaving from
362.            % the basic list, Phase I must be terminated
363.            % and Phase II must start.
364.            % The basis produced is feasible for Phase II
365.            if ~isempty(col)
366.                NonBasicList(col) = [];
367.                A(:, n) = [];
368.                flag = 1;
369.                iterations = iterations + 1;
370.                break
371.            else
372.                % If the artificial variable is in the
373.                % basic list but Sn >= 0, the artificial
374.                % variable is leaving and as an entering
375.                % one, we pick a random but acceptable
376.                % variable.
377.                % The basis produced is feasible for
378.                % Phase II
379.                [~, col] = find(BasicList == n);
380.                if ~isempty(col)
381.                    if Xb(col) == 0
382.                        A(:, n) = [];
383.                        for i = 1:length(NonBasicList)
384.                            l = NonBasicList(i);
```

```
385.                              h_l = BasisInv * A(:, 1);
386.                              % set to zero, the values of
387.                              % h_l that are less than or
388.                              % equal to tole3
389.                              toler = abs(h_l) <= tole3;
390.                              h_l(toler == 1) = 0;
391.                              if h_l(col) ~= 0
392.                                  % update the basic and
393.                                  % nonbasic lists
394.                                  BasicList(col) = l;
395.                                  NonBasicList(i) = [];
396.                                  % re-compute the basis
397.                                  % inverse
398.                                  BasisInv = inv(A(:, ...
399.                                      BasicList));
400.                                  % re-compute the
401.                                  % reduced costs
402.                                  Sn = c(NonBasicList)' - ...
403.                                      c(BasicList)' * BasisInv ...
404.                                      * A(:, NonBasicList);
405.                                  % set to zero, the values of
406.                                  % Sn that are less than or
407.                                  % equal to tole2
408.                                  toler = abs(Sn) <= tole2;
409.                                  Sn(toler == 1) = 0;
410.                                  flag = 1;
411.                                  iterations = iterations + 1;
412.                                  break
413.                              end
414.                          end
415.                          break
416.                      else % the LP problem is infeasible
417.                          disp('The LP problem is infeasible')
418.                          exitflag = 1;
419.                          iterations = iterations + 1;
420.                          return
421.                      end
422.                  end
423.              end
424.          else % the LP problem of Phase I is not optimal
425.              % find the entering variable using the selected
426.              % pivoting rule
427.              if pivotingRule == 1 % bland
428.                  t = bland(Sn, NonBasicList);
429.              elseif pivotingRule == 2 % dantzig
430.                  t = dantzig(Sn);
431.              % greatest increment
432.              elseif pivotingRule == 3
433.                  t = gim(Sn, NonBasicList, A, ...
434.                      BasisInv, Xb, tole3);
435.              % least recently considered
436.              elseif pivotingRule == 4
437.                  if ~exist('ind') % first time, start at zero
438.                      t = lrcm(Sn, NonBasicList, 0, ...
```

```
439.                         iterations);
440.                 else % continue from the last found index
441.                     t = lrcm(Sn, NonBasicList, ...
442.                         lrcmLast, iterations);
443.                 end
444.                 lrcmLast = t;
445.             elseif pivotingRule == 5 % partial pricing
446.                 % select a segment size according to the
447.                 % size of the problem
448.                 if length(Sn) <= 500
449.                     segmentSize = 20;
450.                 elseif length(Sn) > 1000
451.                     segmentSize = 50;
452.                 else
453.                     segmentSize = 35;
454.                 end
455.                 t = partialPricing(Sn, segmentSize);
456.             elseif pivotingRule == 6 % steepest edge
457.                 t = steepestEdge(Sn, NonBasicList, ...
458.                     A, BasisInv);
459.             end
460.             % index of the entering variable
461.             l = NonBasicList(t);
462.             % calculate the pivoting column
463.             h_l = BasisInv * A(:, l);
464.             % set to zero, the values of h_l that are
465.             % less than or equal to tole3
466.             toler = abs(h_l) <= tole3;
467.             h_l(toler == 1) = 0;
468.             optTest = h_l > 0;
469.             % perform the minimum ratio test to find
470.             % the leaving variable
471.             if sum(optTest) > 0
472.                 mrt = find(h_l > 0);
473.                 Xb_hl_div = Xb(mrt) ./ h_l(mrt);
474.                 [p, r] = min(Xb_hl_div);
475.                 rr = find(Xb_hl_div == p);
476.                 % break the ties in order to avoid stalling
477.                 if length(rr) > 1
478.                     r = mrt(rr);
479.                     % index of the leaving variable
480.                     k = BasicList(r);
481.                     [k, ~] = max(k);
482.                     r = find(BasicList == k);
483.                 else
484.                     r = mrt(r);
485.                     % index of the leaving variable
486.                     k = BasicList(r);
487.                 end
488.                 % pivoting and update vectors and matrices
489.                 a = Sn(t);
490.                 f = Xb(r);
491.                 % update the basic list
492.                 BasicList(r) = l;
```

```
493.                    g = h_l(r); % the pivot element
494.                    % update the nonbasic list
495.                    NonBasicList(t) = k;
496.                    % update the matrix of the nonbasic variables
497.                    N = A(:, NonBasicList);
498.                    iterations = iterations + 1;
499.                    % calculate the new basic solution
500.                    Xb(r) = 0;
501.                    h_12 = h_l;
502.                    h_12(r) = -1;
503.                    Xb = Xb - (f / g) * h_12;
504.                    % set to zero, the values of Xb that
505.                    % are less than or equal to tole1
506.                    toler = abs(Xb) <= tole1;
507.                    Xb(toler == 1) = 0;
508.                    % update the reduced costs
509.                    Sn(t) = 0;
510.                    HRN = BasisInv(r, :) * N;
511.                    % set to zero, the values of HRN that
512.                    % are less than or equal to tole2
513.                    toler = abs(HRN) <= tole2;
514.                    HRN(toler == 1) = 0;
515.                    Sn = Sn - (a / g) * HRN;
516.                    % set to zero, the values of Sn that
517.                    % are less than or equal to tole2
518.                    toler = abs(Sn) <= tole2;
519.                    Sn(toler == 1) = 0;
520.                    % basis inverse
521.                    if iterations == counter * reinv
522.                        % recompute the inverse of the basis
523.                        % from scratch every reinv iterations
524.                        BasisInv = inv(A(:, BasicList));
525.                        counter = counter + 1;
526.                        h_l(r) = g;
527.                    else % basis update according to the
528.                        % selected basis update method
529.                        if basisUpdateMethod == 1 % pfi
530.                            BasisInv = pfi(BasisInv, h_l, r);
531.                        else % mpfi
532.                            BasisInv = mpfi(BasisInv, h_l, r);
533.                        end
534.                        h_l(r) = -1;
535.                    end
536.                end
537.            end
538.      end
539. end
540. if flag == 1 % Phase II
541.     disp('---- P H A S E    II ----')
542.     % matrix A does not contain now the artificial variable.
543.     % Here we solve the original LP problem
544.     flagfase2 = 0;
545.     [m1, n1] = size(A); % new size of matrix A
546.     % calculate the needed variables for the algorithm to
```

```
547.      % begin with the matrix of the basic variables
548.      Basis = spalloc(m1, length(BasicList), m1 * ...
549.           length(BasicList));
550.      % the matrix of the nonbasic variables
551.      N = spalloc(m1, length(NonBasicList), m1 * ...
552.           length(NonBasicList));
553.      w = spalloc(1, n1, n1); % simplex multiplier
554.      Sn = spalloc(1, n1, n1); % reduced costs
555.      Basis = A(:, BasicList); % the new basis
556.      % the coefficients of the objective function for the
557.      % basic variables
558.      cb = c(BasicList);
559.      % the matrix of the basic variables
560.      N = A(:, NonBasicList);
561.      % the coefficients of the objective function for the
562.      % nonbasic variables
563.      cn = c(NonBasicList);
564.      % calculate the basis inverse
565.      BasisInv = inv(Basis);
566.      Xb = BasisInv * b; % basic solution
567.      % set to zero, the values of Xb that are less than or
568.      % equal to tole1
569.      toler = abs(Xb) <= tole1;
570.      Xb(toler == 1) = 0;
571.      % calculate the simplex multiplier
572.      w = cb' * BasisInv;
573.      HRN = w * N;
574.      % set to zero, the values of HRN that are less than or
575.      % equal to tole2
576.      toler = abs(HRN) <= tole2;
577.      HRN(toler == 1) = 0;
578.      % calculate the reduced costs
579.      Sn = cn' - HRN;
580.      % set to zero, the values of Sn that are less than or
581.      % equal to tole2
582.      toler = abs(Sn) <= tole2;
583.      Sn(toler == 1) = 0;
584.      while flagfase2 == 0 % iterate in Phase II
585.           clear optTest
586.           clear rr;
587.           clear mrt;
588.           clear h_1;
589.           % optimality test for Phase II
590.           if ~all(Sn >= 0) == 1
591.                % find the entering variable using the
592.                % selected pivoting rule
593.                if pivotingRule == 1 % bland
594.                     t = bland(Sn, NonBasicList);
595.                elseif pivotingRule == 2 % dantzig
596.                     t = dantzig(Sn);
597.                % greatest increment
598.                elseif pivotingRule == 3
599.                     t = gim(Sn, NonBasicList, A, ...
600.                          BasisInv, Xb, tole3);
```

```
601.              % least recently considered
602.              elseif pivotingRule == 4
603.                  if ~exist('ind') % first time, start at zero
604.                      t = lrcm(Sn, NonBasicList, 0, ...
605.                          iterations);
606.                  else % continue from the last found index
607.                      t = lrcm(Sn, NonBasicList, ...
608.                          lrcmLast, iterations);
609.                  end
610.                  lrcmLast = t;
611.              elseif pivotingRule == 5 % partial pricing
612.                  % select a segment size according to the
613.                  % size of the problem
614.                  if length(Sn) <= 500
615.                      segmentSize = 20;
616.                  elseif length(Sn) > 1000
617.                      segmentSize = 50;
618.                  else
619.                      segmentSize = 35;
620.                  end
621.                  t = partialPricing(Sn, segmentSize);
622.              elseif pivotingRule == 6 % steepest edge
623.                  t = steepestEdge(Sn, NonBasicList, ...
624.                      A, BasisInv);
625.              end
626.              % index of the entering variable
627.              l = NonBasicList(t);
628.              % calculate the pivoting column
629.              h_l = BasisInv * A(:, l);
630.              % set to zero, the values of h_l that are
631.              % less than or equal to tole3
632.              toler = abs(h_l) <= tole3;
633.              h_l(toler == 1) = 0;
634.              optTest = h_l > 0;
635.              % perform the minimum ratio test to find the
636.              % leaving variable
637.              if sum(optTest) > 0
638.                  mrt = find(h_l > 0);
639.                  Xb_hl_div = Xb(mrt) ./ h_l(mrt);
640.                  [p, r] = min(Xb_hl_div);
641.                  rr = find(Xb_hl_div == p);
642.                  % break the ties in order to avoid stalling
643.                  if length(rr) > 1
644.                      r = mrt(rr);
645.                      % index of the leaving variable
646.                      k = BasicList(r);
647.                      [k, ~] = max(k);
648.                      r = find(BasicList == k);
649.                  else
650.                      r = mrt(r);
651.                      % index of the leaving variable
652.                      k = BasicList(r);
653.                  end
654.                  % pivoting and update vectors and
```

```
655.               % matrices
656.               a = Sn(t);
657.               f = Xb(r);
658.               % update the basic list
659.               BasicList(r) = l;
660.               g = h_l(r); % the pivot element
661.               % update the nonbasic list
662.               NonBasicList(t) = k;
663.               % update the matrix of the nonbasic
664.               % variables
665.               N = A(:, NonBasicList);
666.               iterations = iterations + 1;
667.               % calculate the new basic solution
668.               Xb(r) = 0;
669.               h_12 = h_l;
670.               h_12(r) = -1;
671.               Xb = Xb - (f / g) * h_12;
672.               % set to zero, the values of Xb that
673.               % are less than or equal to tole1
674.               toler = abs(Xb) <= tole1;
675.               Xb(toler == 1) = 0;
676.               % update the reduced costs
677.               Sn(t) = 0;
678.               HRN = BasisInv(r, :) * N;
679.               % set to zero, the values of HRN that
680.               % are less than or equal to tole2
681.               toler = abs(HRN) <= tole2;
682.               HRN(toler == 1) = 0;
683.               Sn = Sn - (a / g) * HRN;
684.               % set to zero, the values of Sn that
685.               % are less than or equal to tole2
686.               toler = abs(Sn) <= tole2;
687.               Sn(toler == 1) = 0;
688.               % basis inverse
689.               if iterations == counter * reinv
690.                   BasisInv = inv(A(:, BasicList));
691.                   counter = counter + 1;
692.                   h_l(r) = g;
693.               else % basis update according to the
694.                   % selected basis update method
695.                   if basisUpdateMethod == 1 % pfi
696.                       BasisInv = pfi(BasisInv, h_l, r);
697.                   else % mpfi
698.                       BasisInv = mpfi(BasisInv, h_l, r);
699.                   end
700.                   h_l(r) = -1;
701.               end
702.               % print intermediate results every 100
703.               % iterations
704.               if mod(iterations, 100) == 0
705.                   % calculate the value of the objective
706.                   % function
707.                   if MinMaxLP == 1 % maximization
708.                       fval = full(-(c(BasicList)' * Xb) ...
```

```
709.                                    + c0);
710.                          else % minimization
711.                              fval = full(c(BasicList)' * Xb ...
712.                                  + c0);
713.                          end
714.                          fprintf(['Iteration %i - ' ...
715.                              'objective value: %f\n'], ...
716.                              iterations, fval);
717.                      end
718.                  else % the problem is unbounded
719.                      disp('The LP problem is unbounded')
720.                      exitflag = 2;
721.                      iterations = iterations + 1;
722.                      return
723.                  end
724.              else % the problem is optimal
725.                  % calculate the value of the objective
726.                  % function
727.                  if MinMaxLP == 1 % maximization
728.                      fval = full(-((c(BasicList))' * Xb ...
729.                          + c0));
730.                  else % minimization
731.                      fval = full((c(BasicList))' * Xb ...
732.                          + c0);
733.                  end
734.                  exitflag = 0;
735.                  xsol = Xb;
736.                  iterations = iterations + 1;
737.                  disp('The LP problem is optimal')
738.                  return
739.              end
740.      end
741. end
742. end
```

The function that finds an invertible initial basis (filename: lprref.m) takes as input the matrix with the coefficients of the constraints (matrix A), the vector with the right-hand side of the constraints (vector b), the vector with the type of the constraints (vector $Eqin$), and the tolerance (variable tol), and returns as output the reduced row echelon form of the matrix with the coefficients of the constraints (matrix $AEqin$), the reduced vector with the right-hand side of the constraints (vector $bEqin$), the initial basis (matrix jb), the vector with the redundant indices (vector out), the matrix with the row redundant indices (matrix $rowindex$), and a flag variable showing if the LP problem is infeasible or not (variable $infeasible$).

In lines 30–31, the output variables are initialized. In lines 33–35, we find all the equality constraints. Next, we compute the tolerance if it was not given as input (lines 39–41). Then, we apply Gauss-Jordan elimination with partial pivoting in order to find the initial basis (lines 46–75). Next, we find the indices of the redundant constraints (lines 77–93).

```
1.    function [AEqin, bEqin, jb, out, rowindex, ...
2.        infeasible] = lprref(A, b, Eqin, tol)
3.    % Filename: lprref.m
4.    % Description: the function calculates a row echelon
5.    % form of matrix A for LPs using Gauss-Jordan
6.    % elimination with partial pivoting
7.    % Authors: Ploskas, N., & Samaras, N.
8.    %
9.    % Syntax: [AEqin, bEqin, jb, out, rowindex, ...
10.   %    infeasible] = lprref(A, b, Eqin, tol)
11.   %
12.   % Input:
13.   % -- A: matrix of coefficients of the constraints
14.   %    (size m x n)
15.   % -- b: vector of the right-hand side of the constraints
16.   %    (size m x 1)
17.   % -- Eqin: vector of the type of the constraints
18.   %    (size m x 1)
19.   % -- tol: tolerance
20.   %
21.   % Output:
22.   % -- AEqin: reduced row echelon form of A (size m x n)
23.   % -- bEqin: reduced vector b (size m x 1)
24.   % -- jb: basis of matrix A (size m x m)
25.   % -- rowindex: a matrix with the row redundant indices
26.   %    (size 2 x m)
27.   % -- infeasible: if the problem is infeasible or not
28.
29.   % initialize output variables
30.   infeasible = 0;
31.   out = [];
32.   % find all equality constraints
33.   a0 = find(Eqin == 0);
34.   AEqin = A(a0, :);
35.   bEqin = b(a0, :);
36.   [m, n] = size(AEqin);
37.   rowindex = zeros(2, m);
38.   rowindex(1, 1:m) = a0';
39.   if nargin < 4 % compute tol, if it was not given as input
40.       tol = max(m, n) * eps * norm(A, 'inf');
41.   end
42.   i = 1;
43.   j = 1;
44.   jb = [];
45.   % apply Gauss-Jordan elimination with partial pivoting
46.   while (i <= m) && (j <= n)
47.       [p, k] = max(abs(AEqin(i:m, j)));
48.       if p < tol
49.           AEqin(i:m, j) = zeros(m - i + 1, 1);
50.           j = j + 1;
51.       elseif p ~= 0
52.           k = k + i - 1;
53.           jb = [jb j];
54.           AEqin([i k], :) = AEqin([k i], :);
```

```
55.              bEqin([i k], :) = bEqin([k i], :);
56.              rowindex(:, [i k]) = rowindex(:, [k i]);
57.              bEqin(i, :) = bEqin(i, :) / AEqin(i, j);
58.              AEqin(i, j:n) = AEqin(i, j:n) / AEqin(i, j);
59.              i_nz = find(AEqin(:, j));
60.              i_nz = setdiff(i_nz, i);
61.              for t = i_nz
62.                  if bEqin(i) ~= 0
63.                      bEqin(t) = bEqin(t) - AEqin(t, j) * bEqin(i);
64.                      toler = abs(bEqin) <= tol;
65.                      bEqin(toler == 1) = 0;
66.                  end
67.                  AEqin(t, j:n) = AEqin(t, j:n) - ...
68.                      AEqin(t, j) * AEqin(i, j:n);
69.                  toler = abs(AEqin) <= tol;
70.                  AEqin(toler == 1) = 0;
71.              end
72.              i = i + 1;
73.              j = j + 1;
74.          end
75.      end
76.      % check for redundant and infeasible constraints
77.      i = 1;
78.      for h = [1:i - 1 i + 1:m]
79.          % redundant constraint
80.          if (AEqin(h, :) == 0) & (bEqin(h) == 0)
81.              rowindex(2, h) = 1;
82.          end
83.          % infeasible constraint
84.          if (AEqin(h,:) == 0) & (bEqin(h) ~= 0)
85.              infeasible = 1;
86.              return;
87.          end
88.      end
89.      % find the indices of the redundant constraints
90.      if any(rowindex(2, :) == 1)
91.          y = find(rowindex(2, :) == 1);
92.          out = y;
93.      end
94.  end
```

8.6   Computational Study

In this section, we present a computational study. The aim of the computational study is to compare the efficiency of the proposed implementation with MATLAB's simplex algorithm. The test set used in the computational study is 19 LPs from the Netlib (optimal, Kennington, and infeasible LPs) and Mészáros problem sets that were presented in Section 3.6 and a set of randomly generated sparse LPs. Tables 8.2, 8.3 present the total execution time and the number of iterations of the

Table 8.2 Total execution time and number of iterations of the proposed implementation of the revised primal simplex algorithm and MATLAB's simplex algorithm over the benchmark set

Problem				RSA		MATLAB's SA	
Name	Constraints	Variables	Nonzeros	Time	Iterations	Time	Iterations
beaconfd	173	262	3,375	0.15	26	0.05	19
cari	400	1,200	152,800	10.01	963	0.74	509
farm	7	12	36	0.01	5	0.02	2
itest6	11	8	20	0.01	2	0.01	1
klein2	477	54	4,585	0.46	500	0.32	1
nsic1	451	463	2,853	0.27	369	0.23	333
nsic2	465	463	3,015	0.33	473	0.23	324
osa-07	1,118	23,949	143,694	10.15	997	3.71	213
osa-14	2,337	52,460	314,760	67.22	2,454	13.42	454
osa-30	4,350	100,024	600,138	320.19	4,681	50.68	837
rosen1	520	1,024	23,274	3.33	1,893	2.56	2,574
rosen10	2,056	4,096	62,136	39.71	5,116	10.62	4,937
rosen2	1,032	2,048	46,504	19.29	4,242	6.38	4,670
rosen7	264	512	7,770	0.29	420	0.61	816
rosen8	520	1,024	15,538	1.43	1,045	1.49	1,629
sc205	205	203	551	0.20	165	0.10	126
scfxm1	330	457	2,589	0.74	339	0.35	192
sctap2	1,090	1,880	6,714	1.86	1,001	1.77	660
sctap3	1,480	2,480	8,874	4.37	1,809	2.33	679
Geometric mean				1.47	463.78	0.76	210.96

revised primal simplex algorithm that was presented in this chapter and MATLAB's simplex algorithm that will be presented in Appendix A over the benchmark set and a set of randomly generated sparse LPs, respectively. The following abbreviations are used: (i) RSA—the proposed implementation of the revised primal simplex algorithm, and (ii) MATLAB's SA—MATLAB's simplex algorithm. For each instance, we averaged times over 10 runs. All times are measured in seconds. The execution times of the revised primal simplex algorithm include the time needed to presolve and scale (with arithmetic mean) the LPs.

MATLAB's simplex algorithm is almost two times faster than the proposed implementation over the benchmark set, while requiring almost half of the iterations needed for the revised primal simplex algorithm. On the other hand, the proposed implementation is an order of magnitude faster than MATLAB's simplex algorithm over the set of randomly generated sparse LPs. MATLAB's simplex algorithm fails to solve most of the randomly generated sparse LPs since it cannot find an initial feasible solution.

Table 8.3 Total execution time and number of iterations of the proposed implementation of the revised primal simplex algorithm and MATLAB's simplex algorithm over a set of randomly generated sparse LPs (5% density)

	RSA		MATLAB's SA	
Problem	Time	Iterations	Time	Iterations
100×100	0.13	1	0.18	81
200×200	0.17	236	1.53	902
300×300	0.39	584	7.23	4,710
400×400	0.81	911	21.72	10,715
500×500	1.94	1,565	38.87	9,787
600×600	4.41	2,569	88.58	15,467
700×700	7.97	3,919	189.60	39,549
800×800	16.25	6,016	–	–
900×900	26.14	6,883	–	–
$1,000 \times 1,000$	34.55	7,909	–	–
$1,100 \times 1,100$	70.41	13,483	–	–
$1,200 \times 1,200$	63.55	10,172	–	–
$1,300 \times 1,300$	133.95	18,004	–	–
$1,400 \times 1,400$	160.29	18,805	–	–
$1,500 \times 1,500$	205.85	19,768	–	–
$1,600 \times 1,600$	327.74	28,866	–	–
$1,700 \times 1,700$	463.99	36,016	–	–
$1,800 \times 1,800$	530.54	33,873	–	–
$1,900 \times 1,900$	917.11	52,492	–	–
$2,000 \times 2,000$	926.92	50,926	–	–
Geometric mean	24.97	4,868.59	11.62	4,174.04

8.7   Chapter Review

The simplex algorithm is one of the top ten algorithms with the greatest influence in the twentieth century and the most widely used method for solving linear programming problems (LPs). Nearly all Fortune 500 companies use the simplex algorithm to optimize several tasks. In this chapter, we presented the revised primal simplex algorithm. Numerical examples were presented in order for the reader to understand better the algorithm. Furthermore, an implementation of the algorithm in MATLAB was presented. The implementation is modular allowing the user to select which scaling technique, pivoting rule, and basis update method will use in order to solve LPs. Finally, we performed a computational study over benchmark LPs and randomly generated sparse LPs in order to compare the efficiency of the proposed implementation with MATLAB's simplex algorithm. Computational results showed that MATLAB's simplex algorithm is almost two times faster than the proposed implementation over the set of benchmark problems, while requiring almost half of the iterations needed for the revised primal simplex algorithm. On the other hand, the proposed implementation is an order of magnitude faster than MATLAB's simplex algorithm over the set of randomly generated sparse LPs. MATLAB's simplex algorithm fails to solve most of the randomly generated sparse LPs since it cannot find an initial feasible solution.

References

1. Bixby, R. E. (1992). Implementing the simplex method: The initial basis. *ORSA Journal on Computing, 4*, 267–284.
2. Borgwardt, H. K. (1982). The average number of pivot steps required by the simplex method is polynomial. *Zeitschrift fur Operational Research, 26*(1), 157–177.
3. Carstens, D. M. (1968) Crashing techniques. In W. Orchard-Hays (Ed.), *Advanced linear-programming computing techniques* (pp. 131–139). New York: McGraw-Hill.
4. Dantzig, G. B. (1949). Programming in linear structure. *Econometrica, 17*, 73–74.
5. Dantzig, G. B. (1963). *Linear programming and extensions*. Princeton, NJ: Princeton University Press.
6. Dongarra, J., & Sullivan, F. (2000). Guest editors' introduction: the top 10 algorithms. *Computing in Science & Engineering, 2*(1), 22–23.
7. Gould, N. I. M., & Reid, J. K. (1989). New crash procedures for large systems of linear constraints. *Mathematical Programming, 45*, 475–501.
8. Klee, V., & Minty, G. J. (1972). How good is the simplex algorithm. In O. Shisha (Ed.), *Inequalities - III*. New York and London: Academic Press Inc.
9. Maros, I., & Mitra, G. (1998). Strategies for creating advanced bases for large-scale linear programming problems. *INFORMS Journal on Computing, 10*, 248–260.
10. Smale, S. (1983). On the average number of steps of the simplex method of linear programming. *Mathematical Programming, 27*(3), 241–262.

Chapter 9
Revised Dual Simplex Algorithm

Abstract In Section 2.6, we presented the basic duality concepts. This chapter extends these concepts and presents the dual simplex algorithm. The dual simplex algorithm is an attractive alternative for solving linear programming problems (LPs). The dual simplex algorithm is very efficient on many types of problems and is especially useful in integer linear programming. This chapter presents the revised dual simplex algorithm. Numerical examples are presented in order for the reader to understand better the algorithm. Furthermore, an implementation of the algorithm in MATLAB is presented. The implementation is modular allowing the user to select which scaling technique and basis update method will use in order to solve LPs. Finally, a computational study over benchmark LPs and randomly generated sparse LPs is performed in order to compare the efficiency of the proposed implementation with the revised primal simplex algorithm presented in Chapter 8.

9.1 Chapter Objectives

- Present the revised dual simplex algorithm.
- Present advanced duality concepts.
- Solve linear programming problems using the revised dual simplex algorithm.
- Implement the revised dual simplex algorithm using MATLAB.
- Compare the computational performance of the revised dual simplex algorithm with the revised primal simplex algorithm.

9.2 Introduction

In Section 2.6, we showed that every primal LP problem is associated with a dual LP problem. For a primal LP problem

$$
\begin{aligned}
\min \ z = \ & c^T x \\
\text{s.t.} \quad Ax \ & \leq b \\
x \ & \geq 0
\end{aligned}
\tag{9.1}
$$

© Springer International Publishing AG 2017

N. Ploskas, N. Samaras, *Linear Programming Using MATLAB®*,
Springer Optimization and Its Applications 127, DOI 10.1007/978-3-319-65919-0_9

We can find its dual problem in its canonical form

$$\max z = b^T w$$
$$\text{s.t.} \quad A^T w \geq c \tag{9.2}$$
$$w \geq 0$$

In Section 2.6, we presented the principles of the transformation of a primal LP problem to its dual. We also discussed the duality theorems. The duality theorems establish the relationship between the primal and the dual LP problem.

Theorem 9.1 (Symmetry) *The dual LP problem of a dual LP problem is the primal LP problem.*

Proof Let's convert the dual LP problem (Equation (9.2)) to the form that the primal LP problem is written (Equation (9.1)):

$$\min z = -b^T w$$
$$\text{s.t.} \quad -A^T w \leq -c \tag{9.3}$$
$$w \geq 0$$

The dual of the above problem is

$$\max z = -c^T y$$
$$\text{s.t.} \quad -Ay \geq -b \tag{9.4}$$
$$y \geq 0$$

or equivalently

$$\min z = c^T y$$
$$\text{s.t.} \quad Ay \leq b \tag{9.5}$$
$$y \geq 0$$

Setting $x = y$, we get the primal LP problem (Equation (9.1))

$$\min z = c^T x$$
$$\text{s.t.} \quad Ax \leq b \tag{9.6}$$
$$x \geq 0$$

$\square$

Hence, we proved that the primal and the dual LP problems are symmetric.

Theorem 9.2 (Weak Duality) *If x is a feasible solution to the primal LP problem and w is a feasible solution to the dual LP problem, then $b^T w \leq c^T x$.*

Proof As x and w are solutions of the primal and the dual LP problem, respectively, and $x \geq 0$, we have

$$Ax = b \text{ and } w^T A \leq c^T$$

Then, we multiply the left-hand side equation by w^T, and the right-hand side equation by x

$$w^T A x = w^T b \text{ and } w^T A x \leq c^T x$$

Hence

$$w^T b \leq c^T x$$

☐

The week duality theorem provides some useful information about the relation between the primal and the dual LP problems.

Corollary 9.1 *If the primal or the dual LP problem is unbounded, then the other problem is infeasible.*

Proof Let's assume that the primal LP problem is unbounded and the dual LP problem has a feasible solution (contradiction). According to the week duality theorem, the feasible values of the primal problem are bounded below.

Similarly, let's assume that the dual LP problem is unbounded, and the primal LP problem has a feasible solution (contradiction). According to the week duality theorem, the feasible values of the dual problem are bounded below.   ☐

Corollary 9.2 *Let x and w be feasible solutions to the primal and the dual LP problem, respectively. If $c^T x = b^T w$, then x and w are optimal solutions to the primal and the dual LP problem, respectively.*

Proof Let x and w be feasible solutions to the primal and the dual LP problem, respectively. According to the weak duality theorem, for every primal feasible solution y, it holds that $c^T y \geq b^T w = c^T x$, which proves that x is an optimal solution to the primal LP problem.   ☐

Theorem 9.3 (Strong Duality) *If there exists an optimal solution to an LP problem, then there exists an optimal one to its dual, and the respective optimal values are equal.*

Proof Since the primal LP problem has an optimal solution, it has also a basic optimal solution. Let B and N be the optimal basic and nonbasic partition, respectively. Let x_{n+j} be the slack variable of constraint j in the primal LP problem. Then, we apply the primal simplex algorithm in this modified problem. The optimal basic solution is $x_B = A_B^{-1} b$, $x_N = 0$, and $s_j = c_j - w^T A_j \geq 0$, where $w^T = c_B^T A_B^{-1}$ and $1 \leq j \leq n + m$. In order to prove the theorem, we show that w is a feasible solution of the dual LP problem, and that $c^T x = b^T w$.

Table 9.1 Relationship between the primal and dual LPs

Primal \Dual	Finite optimum	Unbounded	Infeasible
Finite optimum	Yes; values equal	Impossible	Impossible
Unbounded	Impossible	Impossible	Possible
Infeasible	Impossible	Possible	Possible

For $j = n + i$, where $1 \le i \le m$, we have

$$A_j = -e_j \left(e_j \in \mathbb{R}^m\right) \text{ and } c_j = 0$$

Hence

$$0 \le s_j = -w^T A_j + c_j = w^T e_i = w_i$$

Since $s \ge 0$, w is a feasible solution of the dual LP problem. Moreover, x is a basic solution ($x_B = A_B^{-1}b$)

$$c^T x = c_B^T x_B + c_N^T x_N = c_B^T x_B = c_B^T A_B^{-1} b = w^T b$$

$\square$

The strong duality theorem guarantees that the dual has an optimal solution if and only if the primal does. As a result, the objective values of the primal and dual LP problems are the same. Moreover, the primal and dual LP problems are related. As already mentioned in Section 2.6, we have the following relationship between the primal and dual LP problems (Table 9.1):

- If the primal LP problem has a finite solution, then the dual LP problem has also a solution. The values of the objective function of the primal and the dual LPs are equal.
- If the primal LP problem is unbounded, then the dual LP problem is infeasible.
- If the primal LP problem is infeasible, then the dual LP problem may be either unbounded or infeasible.

Similarly:

- If the dual LP problem has a finite solution, then the primal LP problem has also a solution. The values of the objective function of the primal and the dual LPs are equal.
- If the dual LP problem is unbounded, then the primal LP problem is infeasible.
- If the dual LP problem is infeasible, then the primal LP problem may be either unbounded or infeasible.

The duality theorems describe the relationship of the primal and the dual LP problem and they can be used to find out if the feasible solutions x and w are optimal for the primal and the dual LP problem, respectively. However, the duality theorems do not provide a way to compute the solution of the primal (dual) LP problem if we already know the solution of the dual (primal) LP problem.

The complementary slackness theorem is useful, because it helps us to interpret the dual variables. If we already know the solution to the primal LP problem, then we can easily find the solution to the dual LP problem, and vice versa.

Theorem 9.4 (Complementary Slackness) *Let x and w be feasible solutions to the primal and the dual LP problem, respectively. These solutions are optimal for the primal and the dual LP problem, respectively, iff:*

$$\left(c^T - w^T A\right) x = 0 \tag{9.7}$$

and

$$w^T \left(Ax - b\right) = 0 \tag{9.8}$$

Proof From Equations (9.7) and (9.8), we have:

$$c^T x = w^T A x \tag{9.9}$$

$$w^T A x = w^T b \tag{9.10}$$

Hence, $c^T x = w^T b$. According to the strong duality theorem, x is the optimal solution to the primal LP problem. Similarly, we can show that w is the optimal solution to the dual LP problem. □

The dual simplex algorithm is of great importance since it can be more efficient than the primal simplex algorithm in some types of LPs, and is especially useful in integer linear programming. Lemke [6] proposed the dual simplex algorithm in 1954. However, it was not considered as an alternative to the primal simplex algorithm for nearly 40 years. In the early 90s, the contributions of Forrest and Goldfarb [3] and Fourer [4] motivated many researchers and optimization software developers to develop new efficient implementations of the dual simplex algorithm. Nowadays, large-scale LPs can be solved by either a primal simplex algorithm or a dual simplex algorithm (or a combination of a primal and a dual algorithm) or an interior point algorithm (Chapter 11). There exist LPs for which one of these methods outperforms the others.

This chapter presents the revised dual simplex algorithm. Initially, the description of the algorithm is given. The various steps of the algorithm are presented. Numerical examples are also presented in order for the reader to understand better the algorithm. Furthermore, an implementation of the algorithm in MATLAB is presented. The implementation is modular allowing the user to select which scaling technique and basis update method will use in order to solve LPs. Note that controlling the above parameters may affect the execution time and the number of iterations. Tuning the algorithm for a given problem structure can dramatically improve its performance. Finally, a computational study is performed in order to compare the efficiency of the proposed implementation with the revised primal simplex algorithm presented in Chapter 8.

The structure of this chapter is as follows. In Section 9.3, the revised dual simplex algorithm is described. All different steps of the algorithm are presented in detail. Four examples are solved in Section 9.4 using the revised dual simplex algorithm. Section 9.5 presents the MATLAB code of the revised dual simplex algorithm. Section 9.6 presents a computational study that compares the proposed implementation and the revised primal simplex algorithm presented in Chapter 8. Finally, conclusions and further discussion are presented in Section 9.7.

9.3   Description of the Algorithm

A formal description of the revised dual simplex algorithm is given in Table 9.2 and a flow diagram of its major steps in Figure 9.1.

Initially, the LP problem is presolved. We use the presolve routine presented in Section 4.15. This routine presolves the original LP problem using the eleven presolve methods presented in Chapter 4. After the presolve analysis, the modified LP problem is scaled. In Chapter 5, we presented eleven scaling techniques. Any of these scaling techniques can be used to scale the LP problem.

Then, we calculate an initial basis to start with. That means that we seek for an invertible basis. This basis includes a subset of the columns. Some columns are obvious choices for this initial basis. Slack variables only have one nonzero value in the corresponding column, so they are good choices to include them in the initial basis. Deciding on the other columns is not as easy, especially on large LPs. A general procedure to select an initial basis is presented in Section 9.5. Function *lprref* applies Gauss-Jordan elimination with partial pivoting in order to find an initial invertible basis. Many other sophisticated techniques have been proposed for the calculation of the initial basis. Interested readers are referred to [1, 2, 5, 7]

Consider the following LP problem (LP.1) in the standard form:

$$
\begin{aligned}
\min \quad & c^T x \\
\text{s.t.} \quad & Ax = b \qquad\qquad (LP.1) \\
& x \geq 0
\end{aligned}
$$

where $A \in \mathbb{R}^{m \times n}$, $(c, x) \in \mathbb{R}^n$, $b \in \mathbb{R}^m$, and T denotes transposition. We assume that A has full rank, $rank(A) = m$, $m < n$. The dual problem associated with (LP.1) is:

$$
\begin{aligned}
\max \quad & b^T w \\
\text{s.t.} \quad & A^T w + s = c \qquad\qquad (DP.1) \\
& s \geq 0
\end{aligned}
$$

where $w \in \mathbb{R}^m$ and $s \in \mathbb{R}^n$.

Table 9.2 Revised dual simplex algorithm

Step 0. *(Initialization).*

Presolve the LP problem.

Scale the LP problem.

Select an initial basic solution (B, N).

if the initial basic solution is dual feasible then proceed to step 2.

Step 1. *(Dual Algorithm with big-M Method).*

Add an artificial constraint $e^T x_N + x_{n+1} = M, x_{n+1} \geq 0$. Construct the

vector of the coefficients of M in the right-hand side, $\bar{\bar{b}}$.

Set $s_p = \min \{ s_j : j \in N \}$, $\bar{B} = B \cup p$, and $\bar{N} = N \cup \{ n+1 \}$.

Find $\bar{x}_B, \bar{\bar{x}}_B, w$, and s_N. Now, apply the algorithm of Step 2

to the modified big-M problem. The original LP problem can be either

optimal or infeasible or unbounded.

Step 2. *(Dual Simplex Algorithm).*

Step 2.0. *(Initialization).*

Compute $(A_B)^{-1}$ and vectors x_B, w, and s_N.

Step 2.1. *(Test of Optimality).*

if $x_B \geq 0$ then STOP. The primal LP problem is optimal.

else

  Choose the leaving variable k such that $x_{B[r]} = x_k = \min \{ x_{B[i]} : x_{B[i]} \leq 0 \}$.

  Variable x_k leaves the basis.

Step 2.2. *(Pivoting).*

Compute vector $H_{rN} = \left(A_B^{-1} \right)_r A_N$.

if $H_{lN} \geq 0$ then STOP. The primal LP problem is infeasible.

else

  Choose the entering variable $x_{N[t]} = x_l$ using the minimum ratio test:

$x_l = x_{N[t]} = \frac{-s_{N[t]}}{H_{rN}} = \min \left\{ \frac{-s_{N[i]}}{H_{iN}} : H_{iN} < 0 \right\}$

Step 2.3. *(Update).*

Swap indices k and l. Update the new basis inverse $\left(A_{\bar{B}} \right)^{-1}$, using a basis

update scheme. Update vectors x_B, w, and s_N.

Go to Step 2.1.

Let B be a subset of the index set $\{ 1, 2, \cdots, n \}$. If $x \in \mathbb{R}^n$, we denote by x_B the part of x indexed by B. The set of indices corresponding to nonbasic variables is denoted by $N = \{ 1, 2, \cdots, n \} \setminus B$. A_B denotes the submatrix of A containing the columns A_j such that $j \in B$. Similarly, A_N denotes the submatrix of A containing the columns A_j such that $j \in N$. If $A_B \in \mathbb{R}^{m \times m}$ is non-singular (invertible), then B is called a basis and A_B is a basic matrix. Given a basis B the associated solution $x_B = A_B^{-1} b$, $x_N = 0$ is called a basic solution. The row vector $w^T = c_B^T A_B^{-1}$ stands for the simplex multipliers. The solution w is dual feasible iff $s_N^T = c_N^T - w^T A_N \geq 0$, and we proceed to apply the revised dual simplex algorithm to the original problem. Otherwise, it is called dual infeasible. In this case, we can apply a Phase I procedure, similar to the one presented in Chapter 8 for the revised primal simplex algorithm, in order to find

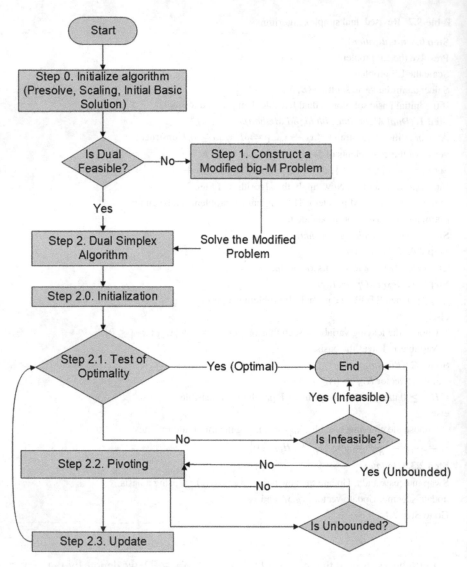

Fig. 9.1 Revised dual simplex algorithm

an initial dual feasible basis. However, in this chapter, we present another procedure, the big-M method, to solve LPs when we do not have an initial dual feasible basis. Initially, we present the dual simplex algorithm and then a modified big-M method for solving LPs using the dual simplex algorithm.

9.3.1 Dual Simplex Algorithm

If we find an initial dual feasible solution ($s_N \geq 0$), we can apply the revised dual simplex algorithm to (LP.1). In the first iteration of the dual simplex algorithm, we calculate the following matrices and vectors:

$$A_B^{-1}, x_B = A_B^{-1}b, w^T = c_B^T A_B^{-1}, \text{ and } s_N^T = c_N^T - w^T A_N$$

Then, we perform the optimality test. If all the elements of x_B are greater than or equal to 0 ($x_B \geq 0$), then the primal LP problem is optimal. If the primal LP problem is not optimal, then we continue with the remaining steps of the algorithm. We select the leaving variable k such that $x_{B[r]} = x_k = \min\{x_{B[i]} : x_{B[i]} \leq 0\}$.

Then, we compute vector $H_{rN} = \left(A_B^{-1}\right)_r A_N$. If $H_{rN} \geq 0$, then the primal LP problem is infeasible. Otherwise, we select the entering variable $x_{N[t]} = x_l$:

$$x_l = x_{N[t]} = \frac{-s_{N[t]}}{H_{rN}} = \min\left\{\frac{-s_{N[i]}}{H_{iN}} : H_{iN} < 0\right\}$$

If two or more nonbasic (basic) variable coefficients are eligible and have the same value, we break ties by selecting the nonbasic (basic) variable with the largest subscript. Then, we calculate the pivoting column:

$$h_l = A_B^{-1} A_l$$

If all the elements of h_l are less than or equal to 0 ($h_l \leq 0$), then the primal LP problem is unbounded. Otherwise, we update the basic and nonbasic lists:

$$B = B \cup [l] \setminus [k], N = N \cup [k] \setminus [l]$$

Then, we update the basis inverse using a basis update method. In Chapter 7, we presented two basis update methods. Any of these basis update methods can be used. Note that we compute the basis inverse from scratch every t iterations in order to reduce numerical errors from the basis update method (a good value of t for most LPs is 80).

Finally, we compute vectors x_B, w, and s_N:

$$x_B = A_B^{-1}b, w^T = c_B^T A_B^{-1}, \text{ and } s_N^T = c_N^T - w^T A_N$$

We start again with the optimality test and all the other steps presented in the algorithm until we find an optimal basis for the primal LP problem or we find that the primal LP problem is infeasible or unbounded.

9.3.2   Modified Big-M Method

Generally, there are two methods that can be used to obtain a primal or dual feasible basic solution in order to initialize the primal or the dual simplex algorithm, respectively: (i) the Phase I method, and (ii) the big-M method. We presented a two-phase method in Chapter 8 in order to find a primal feasible basic solution for the primal simplex algorithm. In this chapter, we present the big-M method.

The big-M method is similar to the two-phase method, except that it essentially attempts to execute Phase I and Phase II in a single execution of the primal or dual simplex algorithm. The big-M method adds an artificial variable to each constraint and modifies the objective function in order to penalize the artificial variables:

$$
\begin{aligned}
\min z = \ & c^T x + M e^T y \\
\text{s.t.} \quad & Ax + I_m y = b \\
& x, y \geq 0
\end{aligned}
\tag{9.11}
$$

where M is a large positive constant, much larger than the largest coefficient in vector c. The value of M is usually chosen to be at least 100 times larger than the largest coefficient in the original objective function. If the original LP is feasible and its optimal value is finite, all artificial variables are eventually driven to zero, and we switch attention to the minimization of the original objective function.

The main disadvantage in implementing the big-M method comes from the following question. How big should M be? It is clear that M should be large enough so that some basic feasible solution with the artificial variable equal to zero has objective value strictly better than the best basic feasible solution having the artificial variable strictly positive. We must be very careful in selecting a value for M.

Definition 9.1 The positive number M is so large as when the problem (LP.1) is feasible, each basic solution verifies the inequality $e^T x_N \leq M$ as strict inequality ($x_{n+1} < M$).

In order to avoid this computational disadvantage, we use a modified big-M method [8].

Definition 9.2 Let α, β, γ, and δ be numbers that are computed from the dual simplex algorithm. Number M is so large that the following relations hold true:

$$
\alpha M + \beta = \gamma M + \delta, \text{ if } \alpha = \gamma \wedge \beta = \delta
$$

$$
\alpha M + \beta > \gamma M + \delta, \text{ if } \alpha > \gamma \vee \alpha = \gamma, \beta > \delta
$$

$$
\alpha M + \beta < \gamma M + \delta, \text{ if } \alpha < \gamma \vee \alpha = \gamma, \beta < \delta
$$

We separate the coefficients of M in the right-hand side, $\bar{\bar{b}}$, from the coefficients of the initial right-hand side after adding an artificial variable and an artificial constraint $e^T x_N + x_{n+1} = M$, $x_{n+1} \geq 0$. The modified LP problem that is solved is the following:

$$
\begin{aligned}
\min z = &\ \bar{c}^T y \\
\text{s.t.} \quad (A, f) y = &\ \left(\bar{b}, \bar{\bar{b}} M\right) \\
y \geq &\ 0
\end{aligned}
$$

(9.12)

where M is a very large positive number, $y^T = (x^T, x_{n+1})$ are the structural variables, $\bar{c} = (\bar{c}^T, 0)$, $\bar{b} = (b, 0)$, and $\bar{\bar{b}} = (0, 0, \cdots, 0, 1) \in \mathbb{R}^{n+1}$ are the coefficients of M in the right-hand side. The vector f is given by the relation

$$ f = -Be $$

where e is the vector of ones. We can add the artificial variable to the nonbasic list and select the entering variable p from $s_p = \min \{s_j : j \in N\}$. Then, we update the basic and nonbasic lists, $\bar{B} = B \cup p$ and $\bar{N} = N \cup \{n + 1\}$. Next, we compute vectors x_B, w, and s_N. Since there is a very large number M in the right-hand side vector, we rewrite vector x_B in terms of the coefficients of the initial right-hand side vector and the coefficients of M, $\bar{x}_B$, and $\bar{\bar{x}}_B$ ($x_B = \bar{x}_B + M\bar{\bar{x}}_B$).

At this point, we have constructed a dual feasible solution for the big-M problem (Equation (9.12)), so the dual simplex algorithm can be applied. A proof of correctness of this modified big-M method is given in [8].

When applying the dual simplex algorithm in this auxiliary problem, we initially use vector $\bar{\bar{x}}_B$ to check for optimality. At some iteration, the artificial variable enters the basis. Then, the artificial variable will never be chosen again as a leaving variable [8]. From that iteration, we use vector $\bar{x}_B$ to check for optimality.

9.4 Examples

Example 1 The LP problem that will be solved is the following:

$$
\begin{aligned}
\min z = &\ x_1 && + \ x_3 + 3x_4 \\
\text{s.t.} \quad &\ 2x_1 - x_2 - 2x_3 - 3x_4 = -4 \\
&\ \quad\quad - 2x_2 - 2x_3 - 6x_4 = -6 \\
&\ x_j \geq 0, \quad (j = 1, 2, 3, 4)
\end{aligned}
$$

In matrix notation the above LP problem is written as follows:

$$
A = \begin{bmatrix} 2 & -1 & -2 & -3 \\ 0 & -2 & -2 & -6 \end{bmatrix}, \ c = \begin{bmatrix} 1 \\ 0 \\ 1 \\ 3 \end{bmatrix}, \ b = \begin{bmatrix} -4 \\ -6 \end{bmatrix}, \ Eqin = \begin{bmatrix} 0 \\ 0 \end{bmatrix}
$$

The LP problem is in its standard form. Let's start with the basic list $B = [1, 2]$ and the nonbasic list $N = [3, 4]$. It is easy to find out that this basis is dual feasible ($s_N \geq 0$):

$$A_B = \begin{bmatrix} 2 & -1 \\ 0 & -2 \end{bmatrix}$$

$$A_B^{-1} = \begin{bmatrix} 1/2 & -1/4 \\ 0 & -1/2 \end{bmatrix}$$

$$x_B = A_B^{-1}b = \begin{bmatrix} 1/2 & -1/4 \\ 0 & -1/2 \end{bmatrix} \begin{bmatrix} -4 \\ -6 \end{bmatrix} = \begin{bmatrix} -1/2 \\ 3 \end{bmatrix}$$

$$w^T = c_B^T A_B^{-1} = \begin{bmatrix} 1 & 0 \end{bmatrix} \begin{bmatrix} 1/2 & -1/4 \\ 0 & -1/2 \end{bmatrix} = \begin{bmatrix} 1/2 & -1/4 \end{bmatrix}$$

$$s_N^T = c_N^T - w^T A_N = \begin{bmatrix} 1 & 3 \end{bmatrix} - \begin{bmatrix} 1/2 & -1/4 \end{bmatrix} \begin{bmatrix} -2 & -3 \\ -2 & -6 \end{bmatrix} = \begin{bmatrix} 3/2 & 3 \end{bmatrix}$$

That means that we do not have to apply the modified big-M method. Hence, we can proceed with the dual simplex algorithm. Note that the initial basic solution x_B is not primal feasible, so we would need to apply Phase I if we were solving this example with the primal simplex algorithm.

1st Iteration

The initial data of the problem is:

$$c_B = \begin{bmatrix} 1 \\ 0 \end{bmatrix}, c_N = \begin{bmatrix} 1 \\ 3 \end{bmatrix}$$

$$A_B = \begin{bmatrix} 2 & -1 \\ 0 & -2 \end{bmatrix}, A_N = \begin{bmatrix} -2 & -3 \\ -2 & -6 \end{bmatrix}$$

$$A_B^{-1} = \begin{bmatrix} 1/2 & -1/4 \\ 0 & -1/2 \end{bmatrix}$$

$$x_B = A_B^{-1}b = \begin{bmatrix} 1/2 & -1/4 \\ 0 & -1/2 \end{bmatrix} \begin{bmatrix} -4 \\ -6 \end{bmatrix} = \begin{bmatrix} -1/2 \\ 3 \end{bmatrix}$$

$$w^T = c_B^T A_B^{-1} = \begin{bmatrix} 1 & 0 \end{bmatrix} \begin{bmatrix} 1/2 & -1/4 \\ 0 & -1/2 \end{bmatrix} = \begin{bmatrix} 1/2 & -1/4 \end{bmatrix}$$

$$s_N^T = c_N^T - w^T A_N = \begin{bmatrix} 1 & 3 \end{bmatrix} - \begin{bmatrix} 1/2 & -1/4 \end{bmatrix} \begin{bmatrix} -2 & -3 \\ -2 & -6 \end{bmatrix} = \begin{bmatrix} 3/2 & 3 \end{bmatrix}$$

$x_B \not\geq 0$, so the current basis is not optimal. We choose the leaving variable k such that $x_{B[r]} = x_k = \min \{x_{B[i]} : x_{B[i]} \leq 0\}$. The leaving variable is x_1 since it is the only one with a negative x_B. Hence, $k = 1$.

Next, we compute vector H_{rN}:

$$H_{rN} = \left(A_B^{-1}\right)_{r.} A_N = [1/2 \; -1/4] \begin{bmatrix} -2 & -3 \\ -2 & -6 \end{bmatrix} = [-1/2 \; 0]$$

$H_{rN} \not\geq 0$. Hence, we choose the entering variable $x_{N[t]} = x_l$ using the minimum ratio test:

$$\min \left\{ \frac{-s_{N[i]}}{H_i} : H_i < 0 \right\} = \min \left\{ \frac{-s_{N[1]}}{H_1} \right\} = \frac{-3/2}{-1/2} = \frac{-s_{N[1]}}{H_1}$$

The entering variable is x_3. Hence, $l = 3$ and $t = 1$.
We continue calculating the pivoting column:

$$h_3 = A_B^{-1} A_{.3} = \begin{bmatrix} 1/2 & -1/4 \\ 0 & -1/2 \end{bmatrix} \begin{bmatrix} -2 \\ -2 \end{bmatrix} = \begin{bmatrix} -1/2 \\ 1 \end{bmatrix}$$

There are elements in vector h_3 that are greater than 0. Next, we update the basic and nonbasic lists:

$$B = [3 \; 2], N = [1 \; 4]$$

Hence:

$$c_B = \begin{bmatrix} 1 \\ 0 \end{bmatrix}, c_N = \begin{bmatrix} 1 \\ 3 \end{bmatrix}$$

$$A_B = \begin{bmatrix} -2 & -1 \\ -2 & -2 \end{bmatrix}, A_N = \begin{bmatrix} 2 & -3 \\ 0 & -6 \end{bmatrix}$$

Then, we update the basis inverse using the PFI update scheme:

$$E^{-1} = \begin{bmatrix} -2 & 0 \\ 2 & 1 \end{bmatrix}$$

$$(A_{\bar{B}})^{-1} = E^{-1} A_B^{-1} = \begin{bmatrix} -2 & 0 \\ 2 & 1 \end{bmatrix} \begin{bmatrix} 1/2 & -1/4 \\ 0 & -1/2 \end{bmatrix} = \begin{bmatrix} -1 & 1/2 \\ 1 & -1 \end{bmatrix}$$

Finally, we update vectors x_B, w, and s_N:

$$x_B = A_B^{-1} b = \begin{bmatrix} -1 & 1/2 \\ 1 & -1 \end{bmatrix} \begin{bmatrix} -4 \\ -6 \end{bmatrix} = \begin{bmatrix} 1 \\ 2 \end{bmatrix}$$

$$w^T = c_B^T A_B^{-1} = \begin{bmatrix} 1 \\ 0 \end{bmatrix} \begin{bmatrix} -1 & 1/2 \\ 1 & -1 \end{bmatrix} = \begin{bmatrix} -1 & 1/2 \end{bmatrix}$$

$$s_N^T = c_N^T - w^T A_N = \begin{bmatrix} 1 & 3 \end{bmatrix} - \begin{bmatrix} -1 & 1/2 \end{bmatrix} \begin{bmatrix} 2 & -3 \\ 0 & -6 \end{bmatrix} = \begin{bmatrix} 3 & 3 \end{bmatrix}$$

2nd Iteration

The data at the beginning of this iteration is:

$$c_B = \begin{bmatrix} 1 \\ 0 \end{bmatrix}, c_N = \begin{bmatrix} 1 \\ 3 \end{bmatrix}$$

$$A_B = \begin{bmatrix} -2 & -1 \\ -2 & -2 \end{bmatrix}, A_N = \begin{bmatrix} 2 & -3 \\ 0 & -6 \end{bmatrix}$$

$$A_B^{-1} = \begin{bmatrix} -1 & 1/2 \\ 1 & -1 \end{bmatrix}$$

$$x_B = \begin{bmatrix} 1 \\ 2 \end{bmatrix}, w^T = \begin{bmatrix} -1 & 1/2 \end{bmatrix}, s_N^T = \begin{bmatrix} 3 & 3 \end{bmatrix}$$

$x_B \geq 0$, so the current basic partition $B = [3, 2]$ and $N = [1, 4]$ is optimal:

$$x_B = \begin{bmatrix} 1 \\ 2 \end{bmatrix}$$

$$z = c_B^T x_B = 1$$

Example 2 Consider the following LP problem:

$$\begin{aligned} \min z = {}& 5x_1 + 2x_2 - 4x_3 \\ \text{s.t.} \quad & 6x_1 + x_2 - 2x_3 \geq 5 \\ & x_1 + x_2 + x_3 \leq 4 \\ & 6x_1 + 4x_2 - 2x_3 \geq 10 \\ x_j \geq 0, \quad & (j = 1, 2, 3) \end{aligned}$$

In matrix notation the above LP problem is written as follows:

$$A = \begin{bmatrix} 6 & 1 & -2 \\ 1 & 1 & 1 \\ 6 & 4 & -2 \end{bmatrix}, c = \begin{bmatrix} 5 \\ 2 \\ -4 \end{bmatrix}, b = \begin{bmatrix} 5 \\ 4 \\ 10 \end{bmatrix}, Eqin = \begin{bmatrix} 1 \\ -1 \\ 1 \end{bmatrix}$$

Initially, we need to convert the LP problem in its standard form:

$$\begin{aligned}
\min z = \; & 5x_1 + 2x_2 - 4x_3 \\
\text{s.t.} \quad & 6x_1 + x_2 - 2x_3 - x_4 && = 5 \\
& x_1 + x_2 + x_3 && + x_5 && = 4 \\
& 6x_1 + 4x_2 - 2x_3 && - x_6 = 10 \\
& x_j \geq 0, \quad (j = 1, 2, 3, 4, 5, 6)
\end{aligned}$$

In matrix notation the above LP problem is written as follows:

$$A = \begin{bmatrix} 6 & 1 & -2 & -1 & 0 & 0 \\ 1 & 1 & 1 & 0 & 1 & 0 \\ 6 & 4 & -2 & 0 & 0 & -1 \end{bmatrix}, c = \begin{bmatrix} 5 \\ 2 \\ -4 \\ 0 \\ 0 \\ 0 \end{bmatrix}, b = \begin{bmatrix} 5 \\ 4 \\ 10 \end{bmatrix}, Eqin = \begin{bmatrix} 0 \\ 0 \\ 0 \end{bmatrix}$$

Let's start with the basic list $B = [4, 5, 6]$ and the nonbasic list $N = [1, 2, 3]$. It is easy to find out that this basis is not dual feasible ($s_N \not\geq 0$):

$$A_B = \begin{bmatrix} -1 & 0 & 0 \\ 0 & 1 & 0 \\ 0 & 0 & -1 \end{bmatrix} = A_B^{-1}$$

$$x_B = A_B^{-1} b = \begin{bmatrix} -1 & 0 & 0 \\ 0 & 1 & 0 \\ 0 & 0 & -1 \end{bmatrix} \begin{bmatrix} 5 \\ 4 \\ 10 \end{bmatrix} = \begin{bmatrix} -5 \\ 4 \\ -10 \end{bmatrix}$$

$$w^T = c_B^T A_B^{-1} = \begin{bmatrix} 0 \\ 0 \\ 0 \end{bmatrix} \begin{bmatrix} -1 & 0 & 0 \\ 0 & 1 & 0 \\ 0 & 0 & -1 \end{bmatrix} = \begin{bmatrix} 0 & 0 & 0 \end{bmatrix}$$

$$s_N^T = c_N^T - w^T A_N = \begin{bmatrix} 5 & 2 & -4 \end{bmatrix} - \begin{bmatrix} 0 & 0 & 0 \end{bmatrix} \begin{bmatrix} 6 & 1 & -2 \\ 1 & 1 & 1 \\ 6 & 4 & -2 \end{bmatrix} = \begin{bmatrix} 5 & 2 & -4 \end{bmatrix}$$

We apply the modified big-M method. We separate the coefficients of M in the right-hand side, $\bar{\bar{b}}$, from the coefficients of the initial right-hand side, $\bar{b}$, after adding the artificial variable x_7 and an artificial constraint $e^T x_N + x_7 = M$, $x_7 \geq 0$. The modified LP problem that is solved is the following:

$$\min z = 5x_1 + 2x_2 - 4x_3$$

$$
\begin{aligned}
\text{s.t.} \quad 6x_1 + x_2 - 2x_3 - x_4 &= 0M + 5 \\
x_1 + x_2 + x_3 + x_5 &= 0M + 4 \\
6x_1 + 4x_2 - 2x_3 - x_6 &= 0M + 10 \\
x_1 + x_2 + x_3 + x_7 &= 1M + 0 \\
\end{aligned}
$$

$$x_j \geq 0, \quad (j = 1, 2, 3, 4, 5, 6, 7)$$

In matrix notation the above LP problem is written as follows:

$$
A = \begin{bmatrix} 6 & 1 & -2 & -1 & 0 & 0 & 0 \\ 1 & 1 & 1 & 0 & 1 & 0 & 0 \\ 6 & 4 & -2 & 0 & 0 & -1 & 0 \\ 1 & 1 & 1 & 0 & 0 & 0 & 1 \end{bmatrix}, c = \begin{bmatrix} 5 \\ 2 \\ -4 \\ 0 \\ 0 \\ 0 \\ 0 \end{bmatrix}, b = \begin{bmatrix} 5 \\ 4 \\ 10 \\ 0 \end{bmatrix}, \bar{b} = \begin{bmatrix} 0 \\ 0 \\ 0 \\ 1 \end{bmatrix}, Eqin = \begin{bmatrix} 0 \\ 0 \\ 0 \\ 0 \end{bmatrix}
$$

The entering variable is x_3:

$$\min \left\{ s_{N[1]}, s_{N[2]}, s_{N[3]} \right\} = \min \{5, 2, -4\} = -4$$

1st Iteration

The initial basic list is $B = [4, 5, 6] \cup [3] = [4, 5, 6, 3]$ and the nonbasic list is $N = [1, 2, 3] \cup [7] \setminus [3] = [1, 2, 7]$. The initial data of the problem is:

$$
c_B = \begin{bmatrix} 0 \\ 0 \\ 0 \\ -4 \end{bmatrix}, c_N = \begin{bmatrix} 5 \\ 2 \\ 0 \end{bmatrix}
$$

$$
A_B = \begin{bmatrix} -1 & 0 & 0 & -2 \\ 0 & 1 & 0 & 1 \\ 0 & 0 & -1 & -2 \\ 0 & 0 & 0 & 1 \end{bmatrix}, A_N = \begin{bmatrix} 6 & 1 & 0 \\ 1 & 1 & 0 \\ 6 & 4 & 0 \\ 1 & 1 & 1 \end{bmatrix}
$$

$$
A_B^{-1} = \begin{bmatrix} -1 & 0 & 0 & -2 \\ 0 & 1 & 0 & -1 \\ 0 & 0 & -1 & -2 \\ 0 & 0 & 0 & 1 \end{bmatrix}
$$

$$\bar{x}_B = A_B^{-1}\bar{b} = \begin{bmatrix} -1 & 0 & 0 & -2 \\ 0 & 1 & 0 & -1 \\ 0 & 0 & -1 & -2 \\ 0 & 0 & 0 & 1 \end{bmatrix} \begin{bmatrix} 5 \\ 4 \\ 10 \\ 0 \end{bmatrix} = \begin{bmatrix} -5 \\ 4 \\ -10 \\ 0 \end{bmatrix}$$

$$\bar{\bar{x}}_B = A_B^{-1}\bar{\bar{b}} = \begin{bmatrix} -1 & 0 & 0 & -2 \\ 0 & 1 & 0 & -1 \\ 0 & 0 & -1 & -2 \\ 0 & 0 & 0 & 1 \end{bmatrix} \begin{bmatrix} 0 \\ 0 \\ 0 \\ 1 \end{bmatrix} = \begin{bmatrix} -2 \\ -1 \\ -2 \\ 1 \end{bmatrix}$$

$$w^T = c_B^T A_B^{-1} = \begin{bmatrix} 0 & 0 & 0 & -4 \end{bmatrix} \begin{bmatrix} -1 & 0 & 0 & -2 \\ 0 & 1 & 0 & -1 \\ 0 & 0 & -1 & -2 \\ 0 & 0 & 0 & 1 \end{bmatrix} = \begin{bmatrix} 0 & 0 & 0 & -4 \end{bmatrix}$$

$$s_N^T = c_N^T - w^T A_N = \begin{bmatrix} 5 & 2 & 0 \end{bmatrix} - \begin{bmatrix} 0 & 0 & 0 & -4 \end{bmatrix} \begin{bmatrix} 6 & 1 & 0 \\ 1 & 1 & 0 \\ 6 & 4 & 0 \\ 1 & 1 & 1 \end{bmatrix} = \begin{bmatrix} 9 & 6 & 4 \end{bmatrix}$$

$\bar{x}_B \not\geq 0$, so the current basis is not dual optimal. The artificial variable x_7 is in the nonbasic list, so we use $\bar{\bar{x}}_B$ to select the leaving variable. Variables x_4 and x_6 are the ones with the most negative $\bar{\bar{x}}_B$. In order to avoid stalling, we break the ties by selecting the variable with the largest index, that is, x_6. Hence, the leaving variable is x_6, $k = 6$ ($r = 3$).

Next, we compute vector H_{rN}:

$$H_{rN} = \left(A_B^{-1}\right)_{r.} A_N = \begin{bmatrix} 0 & 0 & -1 & -2 \end{bmatrix} \begin{bmatrix} 6 & 1 & 0 \\ 1 & 1 & 0 \\ 6 & 4 & 0 \\ 1 & 1 & 1 \end{bmatrix} = \begin{bmatrix} -8 & -6 & -2 \end{bmatrix}$$

$H_{rN} \not\geq 0$. Hence, we choose the entering variable $x_{N[t]} = x_l$ using the minimum ratio test:

$$\min \left\{ \frac{-s_{N[i]}}{H_i} : H_i < 0 \right\} = \min \left\{ \frac{-s_{N[1]}}{H_1}, \frac{-s_{N[2]}}{H_2}, \frac{-s_{N[3]}}{H_3} \right\}$$

$$= \min \left\{ \frac{-9}{-8}, \frac{-6}{-6}, \frac{-4}{-2} \right\} = 1 = \frac{-s_{N[2]}}{H_2}$$

The entering variable is x_2. Hence, $l = 2$ and $t = 2$.

We continue calculating the pivoting column:

$$h_2 = A_B^{-1}A_{.2} = \begin{bmatrix} -1 & 0 & 0 & -2 \\ 0 & 1 & 0 & -1 \\ 0 & 0 & -1 & -2 \\ 0 & 0 & 0 & 1 \end{bmatrix}\begin{bmatrix} 1 \\ 1 \\ 4 \\ 1 \end{bmatrix} = \begin{bmatrix} -3 \\ 0 \\ -6 \\ 1 \end{bmatrix}$$

There are elements in vector h_2 that are greater than 0. Next, we update the basic and nonbasic lists:

$$B = \begin{bmatrix} 4 & 5 & 2 & 3 \end{bmatrix}, N = \begin{bmatrix} 1 & 6 & 7 \end{bmatrix}$$

Hence:

$$c_B = \begin{bmatrix} 0 \\ 0 \\ 2 \\ -4 \end{bmatrix}, c_N = \begin{bmatrix} 5 \\ 0 \\ 0 \end{bmatrix}$$

$$A_B = \begin{bmatrix} -1 & 0 & 1 & -2 \\ 0 & 1 & 1 & 1 \\ 0 & 0 & 4 & -2 \\ 0 & 0 & 1 & 1 \end{bmatrix}, A_N = \begin{bmatrix} 6 & 0 & 0 \\ 1 & 0 & 0 \\ 6 & -1 & 0 \\ 1 & 0 & 1 \end{bmatrix}$$

Then, we update the basis inverse using the PFI update scheme:

$$E^{-1} = \begin{bmatrix} 1 & 0 & -1/2 & 0 \\ 0 & 1 & 0 & 0 \\ 0 & 0 & -1/6 & 0 \\ 0 & 0 & 1/6 & 1 \end{bmatrix}$$

$$(A_{\bar{B}})^{-1} = E^{-1}A_B^{-1} = \begin{bmatrix} 1 & 0 & -1/2 & 0 \\ 0 & 1 & 0 & 0 \\ 0 & 0 & -1/6 & 0 \\ 0 & 0 & 1/6 & 1 \end{bmatrix}\begin{bmatrix} -1 & 0 & 0 & -2 \\ 0 & 1 & 0 & -1 \\ 0 & 0 & -1 & -2 \\ 0 & 0 & 0 & 1 \end{bmatrix} = \begin{bmatrix} -1 & 0 & 1/2 & -1 \\ 0 & 1 & 0 & -1 \\ 0 & 0 & 1/6 & 1/3 \\ 0 & 0 & -1/6 & 2/3 \end{bmatrix}$$

Finally, we update vectors $\bar{x}_B$, $\bar{\bar{x}}_B$, w, and s_N:

$$\bar{x}_B = A_B^{-1}\bar{b} = \begin{bmatrix} -1 & 0 & 1/2 & -1 \\ 0 & 1 & 0 & -1 \\ 0 & 0 & 1/6 & 1/3 \\ 0 & 0 & -1/6 & 2/3 \end{bmatrix}\begin{bmatrix} 5 \\ 4 \\ 10 \\ 0 \end{bmatrix} = \begin{bmatrix} 0 \\ 0 \\ 5/3 \\ -5/3 \end{bmatrix}$$

$$\bar{\bar{x}}_B = A_B^{-1}\bar{\bar{b}} = \begin{bmatrix} -1 & 0 & 1/2 & -1 \\ 0 & 1 & 0 & -1 \\ 0 & 0 & 1/6 & 1/3 \\ 0 & 0 & -1/6 & 2/3 \end{bmatrix} \begin{bmatrix} 0 \\ 0 \\ 0 \\ 1 \end{bmatrix} = \begin{bmatrix} -1 \\ -1 \\ 1/3 \\ 2/3 \end{bmatrix}$$

$$w^T = c_B^T A_B^{-1} = \begin{bmatrix} 0 & 0 & 2 & -4 \end{bmatrix} \begin{bmatrix} -1 & 0 & 1/2 & -1 \\ 0 & 1 & 0 & -1 \\ 0 & 0 & 1/6 & 1/3 \\ 0 & 0 & -1/6 & 2/3 \end{bmatrix} = \begin{bmatrix} 0 & 0 & 1 & -2 \end{bmatrix}$$

$$s_N^T = c_N^T - w^T A_N = \begin{bmatrix} 5 & 0 & 0 \end{bmatrix} - \begin{bmatrix} 0 & 0 & 1 & -2 \end{bmatrix} \begin{bmatrix} 6 & 0 & 0 \\ 1 & 0 & 0 \\ 6 & -1 & 0 \\ 1 & 0 & 1 \end{bmatrix} = \begin{bmatrix} 1 & 1 & 2 \end{bmatrix}$$

2nd Iteration

The data at the beginning of this iteration is:

$$c_B = \begin{bmatrix} 0 \\ 0 \\ 2 \\ -4 \end{bmatrix}, c_N = \begin{bmatrix} 5 \\ 0 \\ 0 \end{bmatrix}$$

$$A_B = \begin{bmatrix} -1 & 0 & 1 & -2 \\ 0 & 1 & 1 & 1 \\ 0 & 0 & 4 & -2 \\ 0 & 0 & 1 & 1 \end{bmatrix}, A_N = \begin{bmatrix} 6 & 0 & 0 \\ 1 & 0 & 0 \\ 6 & -1 & 0 \\ 1 & 0 & 1 \end{bmatrix}$$

$$A_B^{-1} = \begin{bmatrix} -1 & 0 & 1/2 & -1 \\ 0 & 1 & 0 & -1 \\ 0 & 0 & 1/6 & 1/3 \\ 0 & 0 & -1/6 & 2/3 \end{bmatrix}$$

$$\bar{x}_B = A_B^{-1}\bar{b} = \begin{bmatrix} -1 & 0 & 1/2 & -1 \\ 0 & 1 & 0 & -1 \\ 0 & 0 & 1/6 & 1/3 \\ 0 & 0 & -1/6 & 2/3 \end{bmatrix} \begin{bmatrix} 5 \\ 4 \\ 10 \\ 0 \end{bmatrix} = \begin{bmatrix} 0 \\ 0 \\ 5/3 \\ -5/3 \end{bmatrix}$$

$$\bar{\bar{x}}_B = A_B^{-1}\bar{\bar{b}} = \begin{bmatrix} -1 & 0 & 1/2 & -1 \\ 0 & 1 & 0 & -1 \\ 0 & 0 & 1/6 & 1/3 \\ 0 & 0 & -1/6 & 2/3 \end{bmatrix} \begin{bmatrix} 0 \\ 0 \\ 0 \\ 1 \end{bmatrix} = \begin{bmatrix} -1 \\ -1 \\ 1/3 \\ 2/3 \end{bmatrix}$$

$$w^T = c_B^T A_B^{-1} = \begin{bmatrix} 0\ 0\ 2 -4 \end{bmatrix} \begin{bmatrix} -1 & 0 & 1/2 & -1 \\ 0 & 1 & 0 & -1 \\ 0 & 0 & 1/6 & 1/3 \\ 0 & 0 & -1/6 & 2/3 \end{bmatrix} = \begin{bmatrix} 0\ 0\ 1 -2 \end{bmatrix}$$

$$s_N^T = c_N^T - w^T A_N = \begin{bmatrix} 5\ 0\ 0 \end{bmatrix} - \begin{bmatrix} 0\ 0\ 1 -2 \end{bmatrix} \begin{bmatrix} 6 & 0 & 0 \\ 1 & 0 & 0 \\ 6 & -1 & 0 \\ 1 & 0 & 1 \end{bmatrix} = \begin{bmatrix} 1\ 1\ 2 \end{bmatrix}$$

$\bar{\bar{x}}_B \not\geq 0$, so the current basis is not dual optimal. The artificial variable x_7 is in the nonbasic list, so we use $\bar{\bar{x}}_B$ to select the leaving variable. Variables x_4 and x_5 are the ones with the most negative $\bar{\bar{x}}_B$. In order to avoid stalling, we break the ties by selecting the variable with the largest index, that is, x_5. Hence, the leaving variable is x_5, $k = 5$ ($r = 2$).

Next, we compute vector H_{rN}:

$$H_{rN} = \left(A_B^{-1} \right)_{r.} A_N = \begin{bmatrix} 0\ 1\ 0 -1 \end{bmatrix} \begin{bmatrix} 6 & 0 & 0 \\ 1 & 0 & 0 \\ 6 & -1 & 0 \\ 1 & 0 & 1 \end{bmatrix} = \begin{bmatrix} 0\ 0 -1 \end{bmatrix}$$

$H_{rN} \not\geq 0$. Hence, we choose the entering variable $x_{N[t]} = x_l$ using the minimum ratio test:

$$\min \left\{ \frac{-s_{N[i]}}{H_i} : H_i < 0 \right\} = \min \left\{ \frac{-s_{N[3]}}{H_3} \right\} = \frac{-2}{-1} = \frac{-s_{N[3]}}{H_3}$$

The entering variable is x_7. Hence, $l = 7$ and $t = 1$.
We continue calculating the pivoting column:

$$h_7 = A_B^{-1} A_{.7} = \begin{bmatrix} -1 & 0 & 1/2 & -1 \\ 0 & 1 & 0 & -1 \\ 0 & 0 & 1/6 & 1/3 \\ 0 & 0 & -1/6 & 2/3 \end{bmatrix} \begin{bmatrix} 0 \\ 0 \\ 0 \\ 1 \end{bmatrix} = \begin{bmatrix} -1 \\ -1 \\ 1/3 \\ 2/3 \end{bmatrix}$$

There are elements in vector h_7 that are greater than 0. Next, we update the basic and nonbasic lists:

$$B = \begin{bmatrix} 4\ 7\ 2\ 3 \end{bmatrix}, N = \begin{bmatrix} 1\ 6\ 5 \end{bmatrix}$$

Hence:

$$c_B = \begin{bmatrix} 0 \\ 0 \\ 2 \\ -4 \end{bmatrix}, c_N = \begin{bmatrix} 5 \\ 0 \\ 0 \end{bmatrix}$$

$$A_B = \begin{bmatrix} -1 & 0 & 1 & -2 \\ 0 & 0 & 1 & 1 \\ 0 & 0 & 4 & -2 \\ 0 & 1 & 1 & 1 \end{bmatrix}, A_N = \begin{bmatrix} 6 & 0 & 0 \\ 1 & 0 & 1 \\ 6 & -1 & 0 \\ 1 & 0 & 0 \end{bmatrix}$$

Then, we update the basis inverse using the PFI update scheme:

$$E^{-1} = \begin{bmatrix} 1 & -1 & 0 & 0 \\ 0 & -1 & 0 & 0 \\ 0 & 1/3 & 1 & 0 \\ 0 & 2/3 & 0 & 1 \end{bmatrix}$$

$$(A_{\bar{B}})^{-1} = E^{-1}A_B^{-1} = \begin{bmatrix} 1 & -1 & 0 & 0 \\ 0 & -1 & 0 & 0 \\ 0 & 1/3 & 1 & 0 \\ 0 & 2/3 & 0 & 1 \end{bmatrix} \begin{bmatrix} -1 & 0 & 1/2 & -1 \\ 0 & 1 & 0 & -1 \\ 0 & 0 & 1/6 & 1/3 \\ 0 & 0 & -1/6 & 2/3 \end{bmatrix} = \begin{bmatrix} -1 & -1 & 1/2 & 0 \\ 0 & -1 & 0 & 1 \\ 0 & 1/3 & 1/6 & 0 \\ 0 & 2/3 & -1/6 & 0 \end{bmatrix}$$

Finally, we update vectors $\bar{x}_B$, $\bar{\bar{x}}_B$, w, and s_N:

$$\bar{x}_B = A_B^{-1}\bar{b} = \begin{bmatrix} -1 & -1 & 1/2 & 0 \\ 0 & -1 & 0 & 1 \\ 0 & 1/3 & 1/6 & 0 \\ 0 & 2/3 & -1/6 & 0 \end{bmatrix} \begin{bmatrix} 5 \\ 4 \\ 10 \\ 0 \end{bmatrix} = \begin{bmatrix} -4 \\ -4 \\ 3 \\ 1 \end{bmatrix}$$

$$\bar{\bar{x}}_B = A_B^{-1}\bar{\bar{b}} = \begin{bmatrix} -1 & -1 & 1/2 & 0 \\ 0 & -1 & 0 & 1 \\ 0 & 1/3 & 1/6 & 0 \\ 0 & 2/3 & -1/6 & 0 \end{bmatrix} \begin{bmatrix} 0 \\ 0 \\ 0 \\ 1 \end{bmatrix} = \begin{bmatrix} 0 \\ 1 \\ 0 \\ 0 \end{bmatrix}$$

$$w^T = c_B^T A_B^{-1} = [0\ 0\ 2\ -4] \begin{bmatrix} -1 & -1 & 1/2 & 0 \\ 0 & -1 & 0 & 1 \\ 0 & 1/3 & 1/6 & 0 \\ 0 & 2/3 & -1/6 & 0 \end{bmatrix} = [0\ -2\ 1\ 0]$$

$$s_N^T = c_N^T - w^T A_N = [5\ 0\ 0] - [0\ -2\ 1\ 0] \begin{bmatrix} 6 & 0 & 0 \\ 1 & 0 & 1 \\ 6 & -1 & 0 \\ 1 & 0 & 0 \end{bmatrix} = [1\ 1\ 2]$$

3rd Iteration

The data at the beginning of this iteration is:

$$c_B = \begin{bmatrix} 0 \\ 0 \\ 2 \\ -4 \end{bmatrix}, c_N = \begin{bmatrix} 5 \\ 0 \\ 0 \end{bmatrix}$$

$$A_B = \begin{bmatrix} -1 & 0 & 1 & -2 \\ 0 & 0 & 1 & 1 \\ 0 & 0 & 4 & -2 \\ 0 & 1 & 1 & 1 \end{bmatrix}, A_N = \begin{bmatrix} 6 & 0 & 0 \\ 1 & 0 & 1 \\ 6 & -1 & 0 \\ 1 & 0 & 0 \end{bmatrix}$$

$$A_B^{-1} = \begin{bmatrix} -1 & -1 & 1/2 & 0 \\ 0 & -1 & 0 & 1 \\ 0 & 1/3 & 1/6 & 0 \\ 0 & 2/3 & -1/6 & 0 \end{bmatrix}$$

$$\bar{x}_B = A_B^{-1}\bar{b} = \begin{bmatrix} -1 & -1 & 1/2 & 0 \\ 0 & -1 & 0 & 1 \\ 0 & 1/3 & 1/6 & 0 \\ 0 & 2/3 & -1/6 & 0 \end{bmatrix} \begin{bmatrix} 5 \\ 4 \\ 10 \\ 0 \end{bmatrix} = \begin{bmatrix} -4 \\ -4 \\ 3 \\ 1 \end{bmatrix}$$

$$\bar{\bar{x}}_B = A_B^{-1}\bar{\bar{b}} = \begin{bmatrix} -1 & -1 & 1/2 & 0 \\ 0 & -1 & 0 & 1 \\ 0 & 1/3 & 1/6 & 0 \\ 0 & 2/3 & -1/6 & 0 \end{bmatrix} \begin{bmatrix} 0 \\ 0 \\ 0 \\ 1 \end{bmatrix} = \begin{bmatrix} 0 \\ 1 \\ 0 \\ 0 \end{bmatrix}$$

$$w^T = c_B^T A_B^{-1} = \begin{bmatrix} 0 & 0 & 2 & -4 \end{bmatrix} \begin{bmatrix} -1 & -1 & 1/2 & 0 \\ 0 & -1 & 0 & 1 \\ 0 & 1/3 & 1/6 & 0 \\ 0 & 2/3 & -1/6 & 0 \end{bmatrix} = \begin{bmatrix} 0 & -2 & 1 & 0 \end{bmatrix}$$

$$s_N^T = c_N^T - w^T A_N = \begin{bmatrix} 5 & 0 & 0 \end{bmatrix} - \begin{bmatrix} 0 & -2 & 1 & 0 \end{bmatrix} \begin{bmatrix} 6 & 0 & 0 \\ 1 & 0 & 1 \\ 6 & -1 & 0 \\ 1 & 0 & 0 \end{bmatrix} = \begin{bmatrix} 1 & 1 & 2 \end{bmatrix}$$

$\bar{\bar{x}}_B \geq 0$, since the artificial variable x_7 is in the basic list. However, the current basis is not dual optimal, since $\bar{x}_B \not\geq 0$. So, we use $\bar{x}_B$ to select the leaving variable. Variable x_4 is the only one with a negative $\bar{x}_B$ (note that we do not consider selecting the artificial variable x_7). Hence, the leaving variable is x_4, $k = 4$ ($r = 1$).

Next, we compute vector H_{rN}:

$$H_{rN} = \left(A_B^{-1}\right)_{r.} A_N = \begin{bmatrix} -1 & -1 & 1/2 & 0 \end{bmatrix} \begin{bmatrix} 6 & 0 & 0 \\ 1 & 0 & 1 \\ 6 & -1 & 0 \\ 1 & 0 & 0 \end{bmatrix} = \begin{bmatrix} -4 & -1/2 & -1 \end{bmatrix}$$

$H_{rN} \not> 0$. Hence, we choose the entering variable $x_{N[t]} = x_l$ using the minimum ratio test:

$$\min \left\{ \frac{-s_{N[i]}}{H_i} : H_i < 0 \right\} = \min \left\{ \frac{-s_{N[1]}}{H_1}, \frac{-s_{N[2]}}{H_2}, \frac{-s_{N[3]}}{H_3} \right\}$$

$$= \min \left\{ \frac{-1}{-4}, \frac{-1}{-1/2}, \frac{-2}{-1} \right\} = \frac{-1}{-4} = \frac{-s_{N[1]}}{H_1}$$

The entering variable is x_1. Hence, $l = 1$ and $t = 1$.
We continue calculating the pivoting column:

$$h_1 = A_B^{-1} A_{.1} = \begin{bmatrix} -1 & -1 & 1/2 & 0 \\ 0 & -1 & 0 & 1 \\ 0 & 1/3 & 1/6 & 0 \\ 0 & 2/3 & -1/6 & 0 \end{bmatrix} \begin{bmatrix} 6 \\ 1 \\ 6 \\ 1 \end{bmatrix} = \begin{bmatrix} -4 \\ 0 \\ 4/3 \\ -1/3 \end{bmatrix}$$

There are elements in vector h_1 that are greater than 0. Next, we update the basic and nonbasic lists:

$$B = \begin{bmatrix} 1 & 7 & 2 & 3 \end{bmatrix}, N = \begin{bmatrix} 4 & 6 & 5 \end{bmatrix}$$

Hence:

$$c_B = \begin{bmatrix} 5 \\ 0 \\ 2 \\ -4 \end{bmatrix}, c_N = \begin{bmatrix} 0 \\ 0 \\ 0 \end{bmatrix}$$

$$A_B = \begin{bmatrix} 6 & 0 & 1 & -2 \\ 1 & 0 & 1 & 1 \\ 6 & 0 & 4 & -2 \\ 1 & 1 & 1 & 1 \end{bmatrix}, A_N = \begin{bmatrix} -1 & 0 & 0 \\ 0 & 0 & 1 \\ 0 & -1 & 0 \\ 0 & 0 & 0 \end{bmatrix}$$

Then, we update the basis inverse using the PFI update scheme:

$$E^{-1} = \begin{bmatrix} -1/4 & 0 & 0 & 0 \\ 0 & 1 & 0 & 0 \\ 1/3 & 0 & 1 & 0 \\ -1/12 & 0 & 0 & 1 \end{bmatrix}$$

$$(A_{\bar{B}})^{-1} = E^{-1}A_B^{-1} = \begin{bmatrix} -1/4 & 0 & 0 & 0 \\ 0 & 1 & 0 & 0 \\ 1/3 & 0 & 1 & 0 \\ -1/12 & 0 & 0 & 1 \end{bmatrix} \begin{bmatrix} -1 & -1 & 1/2 & 0 \\ 0 & -1 & 0 & 1 \\ 0 & 1/3 & 1/6 & 0 \\ 0 & 2/3 & -1/6 & 0 \end{bmatrix} = \begin{bmatrix} 1/4 & 1/4 & -1/8 & 0 \\ 0 & -1 & 0 & 1 \\ -1/3 & 0 & 1/3 & 0 \\ 1/12 & 3/4 & -5/24 & 0 \end{bmatrix}$$

Finally, we update vectors $\bar{x}_B$, w, and s_N:

$$\bar{x}_B = A_B^{-1}\bar{b} = \begin{bmatrix} 1/4 & 1/4 & -1/8 & 0 \\ 0 & -1 & 0 & 1 \\ -1/3 & 0 & 1/3 & 0 \\ 1/12 & 3/4 & -5/24 & 0 \end{bmatrix} \begin{bmatrix} 5 \\ 4 \\ 10 \\ 0 \end{bmatrix} = \begin{bmatrix} 1 \\ -4 \\ 5/3 \\ 4/3 \end{bmatrix}$$

$$w^T = c_B^T A_B^{-1} = \begin{bmatrix} 5 & 0 & 2 & -4 \end{bmatrix} \begin{bmatrix} 1/4 & 1/4 & -1/8 & 0 \\ 0 & -1 & 0 & 1 \\ -1/3 & 0 & 1/3 & 0 \\ 1/12 & 3/4 & -5/24 & 0 \end{bmatrix} = \begin{bmatrix} 1/4 & -7/4 & 7/8 & 0 \end{bmatrix}$$

$$s_N^T = c_N^T - w^T A_N = \begin{bmatrix} 0 & 0 & 0 \end{bmatrix} - \begin{bmatrix} 1/4 & -7/4 & 7/8 & 0 \end{bmatrix} \begin{bmatrix} -1 & 0 & 0 \\ 0 & 0 & 1 \\ 0 & -1 & 0 \\ 0 & 0 & 0 \end{bmatrix} = \begin{bmatrix} 1/4 & 7/8 & 7/4 \end{bmatrix}$$

4th Iteration

The data at the beginning of this iteration is:

$$c_B = \begin{bmatrix} 5 \\ 0 \\ 2 \\ -4 \end{bmatrix}, c_N = \begin{bmatrix} 0 \\ 0 \\ 0 \end{bmatrix}$$

$$A_B = \begin{bmatrix} 6 & 0 & 1 & -2 \\ 1 & 0 & 1 & 1 \\ 6 & 0 & 4 & -2 \\ 1 & 1 & 1 & 1 \end{bmatrix}, A_N = \begin{bmatrix} -1 & 0 & 0 \\ 0 & 0 & 1 \\ 0 & -1 & 0 \\ 0 & 0 & 0 \end{bmatrix}$$

$$A_B^{-1} = \begin{bmatrix} 1/4 & 1/4 & -1/8 & 0 \\ 0 & -1 & 0 & 1 \\ -1/3 & 0 & 1/3 & 0 \\ 1/12 & 3/4 & -5/24 & 0 \end{bmatrix}$$

$$\bar{x}_B = A_B^{-1}\bar{b} = \begin{bmatrix} 1/4 & 1/4 & -1/8 & 0 \\ 0 & -1 & 0 & 1 \\ -1/3 & 0 & 1/3 & 0 \\ 1/12 & 3/4 & -5/24 & 0 \end{bmatrix} \begin{bmatrix} 5 \\ 4 \\ 10 \\ 0 \end{bmatrix} = \begin{bmatrix} 1 \\ -4 \\ 5/3 \\ 4/3 \end{bmatrix}$$

$$w^T = c_B^T A_B^{-1} = \begin{bmatrix} 5 & 0 & 2 & -4 \end{bmatrix} \begin{bmatrix} 1/4 & 1/4 & -1/8 & 0 \\ 0 & -1 & 0 & 1 \\ -1/3 & 0 & 1/3 & 0 \\ 1/12 & 3/4 & -5/24 & 0 \end{bmatrix} = \begin{bmatrix} 1/4 & -7/4 & 7/8 & 0 \end{bmatrix}$$

$$s_N^T = c_N^T - w^T A_N = \begin{bmatrix} 0 & 0 & 0 \end{bmatrix} - \begin{bmatrix} 1/4 & -7/4 & 7/8 & 0 \end{bmatrix} \begin{bmatrix} -1 & 0 & 0 \\ 0 & 0 & 1 \\ 0 & -1 & 0 \\ 0 & 0 & 0 \end{bmatrix} = \begin{bmatrix} 1/4 & 7/8 & 7/4 \end{bmatrix}$$

$\bar{x}_B \geq 0$, since the artificial variable x_7 is in the basic list. In addition, $\bar{x}_{B_i} \geq 0$, where i is a set of indices of the elements of $\bar{x}_B$ that are equal to 0. Hence, the current basic partition $B = [1, 7, 2, 3]$ and $N = [4, 6, 5]$ is optimal. We can delete the artificial variable x_7 from vector x_B:

$$x_B = \begin{bmatrix} 1 \\ 5/3 \\ 4/3 \end{bmatrix}$$

$$z = c_B^T x_B = 3$$

Example 3 Consider the following LP problem:

$$\begin{aligned} \min z = \quad & - x_2 - 2x_3 \\ \text{s.t.} \quad & x_1 + x_2 + x_3 = 5 \\ & - 2x_2 - 3x_3 = 3 \\ x_j \geq 0, \quad & (j = 1, 2, 3) \end{aligned}$$

In matrix notation the above LP problem is written as follows:

$$A = \begin{bmatrix} 1 & 1 & 1 \\ 0 & -2 & -3 \end{bmatrix}, c = \begin{bmatrix} 0 \\ -1 \\ -2 \end{bmatrix}, b = \begin{bmatrix} 5 \\ 3 \end{bmatrix}, Eqin = \begin{bmatrix} 0 \\ 0 \end{bmatrix}$$

The LP problem is in its standard form. Let's start with the basic list $B = [1, 2]$ and the nonbasic list $N = [3]$. It is easy to find out that this basis is not dual feasible ($s_N \not\geq 0$):

$$A_B = \begin{bmatrix} 1 & 1 \\ 0 & -2 \end{bmatrix}$$

$$A_B^{-1} = \begin{bmatrix} 1 & 1/2 \\ 0 & -1/2 \end{bmatrix}$$

$$x_B = A_B^{-1} b = \begin{bmatrix} 1 & 1/2 \\ 0 & -1/2 \end{bmatrix} \begin{bmatrix} 5 \\ 3 \end{bmatrix} = \begin{bmatrix} 13/2 \\ -3/2 \end{bmatrix}$$

$$w^T = c_B^T A_B^{-1} = \begin{bmatrix} 0 & -1 \end{bmatrix} \begin{bmatrix} 1 & 1/2 \\ 0 & -1/2 \end{bmatrix} = \begin{bmatrix} 0 & 1/2 \end{bmatrix}$$

$$s_N^T = c_N^T - w^T A_N = \begin{bmatrix} -2 \end{bmatrix} - \begin{bmatrix} 0 & 1/2 \end{bmatrix} \begin{bmatrix} 1 \\ -3 \end{bmatrix} = \begin{bmatrix} -1/2 \end{bmatrix}$$

We apply the modified big-M method. We separate the coefficients of M in the right-hand side, $\bar{\bar{b}}$, from the coefficients of the initial right-hand side after adding the artificial variable x_4 and an artificial constraint $e^T x_N + x_4 = M$, $x_4 \geq 0$. The modified LP problem that is solved is the following:

$$
\begin{aligned}
\min z = \quad & - x_2 - 2x_3 \\
\text{s.t.} \quad & x_1 + x_2 + x_3 && = 0M + 5 \\
& - 2x_2 - 3x_3 && = 0M + 3 \\
& x_3 + x_4 && = 1M + 0 \\
& x_j \geq 0, \quad (j = 1, 2, 3, 4)
\end{aligned}
$$

In matrix notation the above LP problem is written as follows:

$$A = \begin{bmatrix} 1 & 1 & 1 & 0 \\ 0 & -2 & -3 & 0 \\ 0 & 0 & 1 & 1 \end{bmatrix}, c = \begin{bmatrix} 0 \\ -1 \\ -2 \\ 0 \end{bmatrix}, \bar{b} = \begin{bmatrix} 5 \\ 3 \\ 0 \end{bmatrix}, \bar{\bar{b}} = \begin{bmatrix} 0 \\ 0 \\ 1 \end{bmatrix}, Eqin = \begin{bmatrix} 0 \\ 0 \\ 0 \end{bmatrix}$$

The entering variable is x_3:

$$\min \{s_{N[1]}, s_{N[2]}, s_{N[3]}\} = \min \{-1/2, 0, 0\} = -1/2$$

1st Iteration

The initial basic list is $B = [1, 2] \cup [3] = [1, 2, 3]$ and the nonbasic list is $N = [3] \cup [4] \setminus [3] = [4]$. The initial data of the problem is:

$$c_B = \begin{bmatrix} 0 \\ -1 \\ -2 \end{bmatrix}, c_N = \begin{bmatrix} 0 \end{bmatrix}$$

$$A_B = \begin{bmatrix} 1 & 1 & 1 \\ 0 & -2 & -3 \\ 0 & 0 & 1 \end{bmatrix}, A_N = \begin{bmatrix} 0 \\ 0 \\ 1 \end{bmatrix}$$

$$A_B^{-1} = \begin{bmatrix} 1 & 1/2 & 1/2 \\ 0 & -1/2 & -3/2 \\ 0 & 0 & 1 \end{bmatrix}$$

$$\bar{x}_B = A_B^{-1} b = \begin{bmatrix} 1 & 1/2 & 1/2 \\ 0 & -1/2 & -3/2 \\ 0 & 0 & 1 \end{bmatrix} \begin{bmatrix} 5 \\ 3 \\ 0 \end{bmatrix} = \begin{bmatrix} 13/2 \\ -3/2 \\ 0 \end{bmatrix}$$

$$\bar{\bar{x}}_B = A_B^{-1} \bar{\bar{b}} = \begin{bmatrix} 1 & 1/2 & 1/2 \\ 0 & -1/2 & -3/2 \\ 0 & 0 & 1 \end{bmatrix} \begin{bmatrix} 0 \\ 0 \\ 1 \end{bmatrix} = \begin{bmatrix} 1/2 \\ -3/2 \\ 1 \end{bmatrix}$$

$$w^T = c_B^T A_B^{-1} = \begin{bmatrix} 0 & -1 & -2 \end{bmatrix} \begin{bmatrix} 1 & 1/2 & 1/2 \\ 0 & -1/2 & -3/2 \\ 0 & 0 & 1 \end{bmatrix} = \begin{bmatrix} 0 & 1/2 & -1/2 \end{bmatrix}$$

$$s_N^T = c_N^T - w^T A_N = \begin{bmatrix} 0 \end{bmatrix} - \begin{bmatrix} 0 & 1/2 & -1/2 \end{bmatrix} \begin{bmatrix} 0 \\ 0 \\ 1 \end{bmatrix} = \begin{bmatrix} 1/2 \end{bmatrix}$$

$\bar{\bar{x}}_B \not\geq 0$, so the current basis is not dual optimal. The artificial variable x_4 is in the nonbasic list, so we use $\bar{\bar{x}}_B$ to select the leaving variable. Variable x_2 is the only one with a negative $\bar{\bar{x}}_B$. Hence, the leaving variable is x_2, $k = 2$ ($r = 3$).

Next, we compute vector H_{rN}:

$$H_{rN} = \left(A_B^{-1} \right)_{r.} A_N = \begin{bmatrix} 0 & -1/2 & -3/2 \end{bmatrix} \begin{bmatrix} 0 \\ 0 \\ 1 \end{bmatrix} = \begin{bmatrix} -3/2 \end{bmatrix}$$

$H_{rN} \not\geq 0$. Hence, we choose the entering variable $x_{N[t]} = x_l$ using the minimum ratio test:

$$\min \left\{ \frac{-s_{N[i]}}{H_i} : H_i < 0 \right\} = \min \left\{ \frac{-s_{N[1]}}{H_1} \right\} = \min \left\{ \frac{-1/2}{-3/2} \right\} = \frac{-s_{N[1]}}{H_1}$$

The entering variable is x_4. Hence, $l = 4$ and $t = 1$.
We continue calculating the pivoting column:

$$h_4 = A_B^{-1} A_{.4} = \begin{bmatrix} 1 & 1/2 & 1/2 \\ 0 & -1/2 & -3/2 \\ 0 & 0 & 1 \end{bmatrix} \begin{bmatrix} 0 \\ 0 \\ 1 \end{bmatrix} = \begin{bmatrix} 1/2 \\ -3/2 \\ 1 \end{bmatrix}$$

There are elements in vector h_4 that are greater than 0. Next, we update the basic and nonbasic lists:

$$B = \begin{bmatrix} 1 & 4 & 3 \end{bmatrix}, N = \begin{bmatrix} 2 \end{bmatrix}$$

Hence:

$$c_B = \begin{bmatrix} 0 \\ 0 \\ -2 \end{bmatrix}, c_N = \begin{bmatrix} -1 \end{bmatrix}$$

$$A_B = \begin{bmatrix} 1 & 0 & 1 \\ 0 & 0 & -3 \\ 0 & 1 & 1 \end{bmatrix}, A_N = \begin{bmatrix} 1 \\ -2 \\ 0 \end{bmatrix}$$

Then, we update the basis inverse using the PFI update scheme:

$$E^{-1} = \begin{bmatrix} 1 & 1/3 & 0 \\ 0 & -2/3 & 0 \\ 0 & 2/3 & 1 \end{bmatrix}$$

$$(A_{\bar{B}})^{-1} = E^{-1}A_B^{-1} = \begin{bmatrix} 1 & 1/3 & 0 \\ 0 & -2/3 & 0 \\ 0 & 2/3 & 1 \end{bmatrix}\begin{bmatrix} 1 & 1/2 & 1/2 \\ 0 & -1/2 & -3/2 \\ 0 & 0 & 1 \end{bmatrix} = \begin{bmatrix} 1 & 1/3 & 0 \\ 0 & 1/3 & 1 \\ 0 & -1/3 & 0 \end{bmatrix}$$

Finally, we update vectors $\bar{x}_B$, $\bar{\bar{x}}_B$, w, and s_N:

$$\bar{x}_B = A_B^{-1}\bar{b} = \begin{bmatrix} 1 & 1/3 & 0 \\ 0 & 1/3 & 1 \\ 0 & -1/3 & 0 \end{bmatrix}\begin{bmatrix} 5 \\ 3 \\ 0 \end{bmatrix} = \begin{bmatrix} 6 \\ 1 \\ -1 \end{bmatrix}$$

$$\bar{\bar{x}}_B = A_B^{-1}\bar{\bar{b}} = \begin{bmatrix} 1 & 1/3 & 0 \\ 0 & 1/3 & 1 \\ 0 & -1/3 & 0 \end{bmatrix}\begin{bmatrix} 0 \\ 0 \\ 1 \end{bmatrix} = \begin{bmatrix} 0 \\ 1 \\ 0 \end{bmatrix}$$

$$w^T = c_B^T A_B^{-1} = \begin{bmatrix} 0 & 0 & -2 \end{bmatrix}\begin{bmatrix} 1 & 1/3 & 0 \\ 0 & 1/3 & 1 \\ 0 & -1/3 & 0 \end{bmatrix} = \begin{bmatrix} 0 & 2/3 & 0 \end{bmatrix}$$

$$s_N^T = c_N^T - w^T A_N = \begin{bmatrix} -1 \end{bmatrix} - \begin{bmatrix} 0 & 2/3 & 0 \end{bmatrix}\begin{bmatrix} 1 \\ -2 \\ 0 \end{bmatrix} = \begin{bmatrix} 1/3 \end{bmatrix}$$

2nd Iteration

The data at the beginning of this iteration is:

$$c_B = \begin{bmatrix} 0 \\ 0 \\ -2 \end{bmatrix}, c_N = \begin{bmatrix} -1 \end{bmatrix}$$

$$A_B = \begin{bmatrix} 1 & 0 & 1 \\ 0 & 0 & -3 \\ 0 & 1 & 1 \end{bmatrix}, A_N = \begin{bmatrix} 1 \\ -2 \\ 0 \end{bmatrix}$$

$$A_B^{-1} = \begin{bmatrix} 1 & 1/3 & 0 \\ 0 & 1/3 & 1 \\ 0 & -1/3 & 0 \end{bmatrix}$$

$$\bar{x}_B = A_B^{-1}\bar{b} = \begin{bmatrix} 1 & 1/3 & 0 \\ 0 & 1/3 & 1 \\ 0 & -1/3 & 0 \end{bmatrix} \begin{bmatrix} 5 \\ 3 \\ 0 \end{bmatrix} = \begin{bmatrix} 6 \\ 1 \\ -1 \end{bmatrix}$$

$$\bar{\bar{x}}_B = A_B^{-1}\bar{\bar{b}} = \begin{bmatrix} 1 & 1/3 & 0 \\ 0 & 1/3 & 1 \\ 0 & -1/3 & 0 \end{bmatrix} \begin{bmatrix} 0 \\ 0 \\ 1 \end{bmatrix} = \begin{bmatrix} 0 \\ 1 \\ 0 \end{bmatrix}$$

$$w^T = c_B^T A_B^{-1} = \begin{bmatrix} 0 & 0 & -2 \end{bmatrix} \begin{bmatrix} 1 & 1/3 & 0 \\ 0 & 1/3 & 1 \\ 0 & -1/3 & 0 \end{bmatrix} = \begin{bmatrix} 0 & 2/3 & 0 \end{bmatrix}$$

$$s_N^T = c_N^T - w^T A_N = \begin{bmatrix} -1 \end{bmatrix} - \begin{bmatrix} 0 & 2/3 & 0 \end{bmatrix} \begin{bmatrix} 1 \\ -2 \\ 0 \end{bmatrix} = \begin{bmatrix} 1/3 \end{bmatrix}$$

$\bar{\bar{x}}_B \geq 0$, since the artificial variable x_4 is in the basic list. However, the current basis is not dual optimal, since $\bar{x}_B \ngeq 0$. So, we use $\bar{x}_B$ to select the leaving variable. Variable x_3 is the only one with a negative $\bar{x}_B$. Hence, the leaving variable is x_3, $k = 3$ $(r = 3)$.

Next, we compute vector H_{rN}:

$$H_{rN} = \left(A_B^{-1}\right)_{r.} A_N = \begin{bmatrix} 0 & -1/3 & 0 \end{bmatrix} \begin{bmatrix} 1 \\ -2 \\ 0 \end{bmatrix} = \begin{bmatrix} 2/3 \end{bmatrix}$$

$H_{rN} \geq 0$, so the LP problem is infeasible.

Example 4 Consider the following LP problem:

$$
\begin{aligned}
\min z = \quad & 3x_1 + 2x_2 \\
\text{s.t.} \quad & 2x_1 + 4x_2 \geq 8 \\
& x_1 + x_2 \geq 3 \\
& -2x_1 - 7x_2 \leq -15 \\
& x_j \geq 0, \quad (j = 1, 2)
\end{aligned}
$$

In matrix notation the above LP problem is written as follows:

$$
A = \begin{bmatrix} 2 & 4 \\ 1 & 1 \\ -2 & -7 \end{bmatrix}, c = \begin{bmatrix} 3 \\ -2 \end{bmatrix}, b = \begin{bmatrix} 8 \\ 3 \\ -15 \end{bmatrix}, Eqin = \begin{bmatrix} 1 \\ 1 \\ -1 \end{bmatrix}
$$

Initially, we need to convert the LP problem in its standard form:

$$
\begin{aligned}
\min z = \quad & 3x_1 + 2x_2 \\
\text{s.t.} \quad & 2x_1 + 4x_2 - x_3 \qquad\qquad\;\; = 8 \\
& x_1 + x_2 \qquad - x_4 \qquad\;\; = 3 \\
& -2x_1 - 7x_2 \qquad\qquad + x_5 = -15 \\
& x_j \geq 0, \quad (j = 1, 2, 3, 4, 5)
\end{aligned}
$$

In matrix notation the above LP problem is written as follows:

$$
A = \begin{bmatrix} 2 & 4 & -1 & 0 & 0 \\ 1 & 1 & 0 & -1 & 0 \\ -2 & -7 & 0 & 0 & 1 \end{bmatrix}, c = \begin{bmatrix} 3 \\ 2 \\ 0 \\ 0 \\ 0 \end{bmatrix}, b = \begin{bmatrix} 8 \\ 3 \\ -15 \end{bmatrix}, Eqin = \begin{bmatrix} 0 \\ 0 \\ 0 \end{bmatrix}
$$

The LP problem is in its standard form. Let's start with the basic list $B = [3, 4, 5]$ and the nonbasic list $N = [1, 2]$. It is easy to find out that this basis is dual feasible $(s_N \geq 0)$:

$$
A_B = \begin{bmatrix} -1 & 0 & 0 \\ 0 & -1 & 0 \\ 0 & 0 & 1 \end{bmatrix}
$$

$$
A_B^{-1} = \begin{bmatrix} -1 & 0 & 0 \\ 0 & -1 & 0 \\ 0 & 0 & 1 \end{bmatrix}
$$

$$x_B = A_B^{-1}b = \begin{bmatrix} -1 & 0 & 0 \\ 0 & -1 & 0 \\ 0 & 0 & 1 \end{bmatrix} \begin{bmatrix} 8 \\ 3 \\ -15 \end{bmatrix} = \begin{bmatrix} -8 \\ -3 \\ -15 \end{bmatrix}$$

$$w^T = c_B^T A_B^{-1} = \begin{bmatrix} 0 & 0 & 0 \end{bmatrix} \begin{bmatrix} -1 & 0 & 0 \\ 0 & -1 & 0 \\ 0 & 0 & 1 \end{bmatrix} = \begin{bmatrix} 0 & 0 & 0 \end{bmatrix}$$

$$s_N^T = c_N^T - w^T A_N = \begin{bmatrix} 3 & 2 \end{bmatrix} - \begin{bmatrix} 0 & 0 & 0 \end{bmatrix} \begin{bmatrix} 2 & 4 \\ 1 & 1 \\ -2 & -7 \end{bmatrix} = \begin{bmatrix} 3 & 2 \end{bmatrix}$$

That means that we do not have to apply the modified big-M method. Hence, we can proceed with the dual simplex algorithm. Note that the initial basic solution x_B is not primal feasible, so we would need to apply Phase I if we were solving this example with the primal simplex algorithm.

1st Iteration

The initial data of the problem is:

$$c_B = \begin{bmatrix} 0 \\ 0 \\ 0 \end{bmatrix}, c_N = \begin{bmatrix} 3 \\ 2 \end{bmatrix}$$

$$A_B = \begin{bmatrix} -1 & 0 & 0 \\ 0 & -1 & 0 \\ 0 & 0 & 1 \end{bmatrix}, A_N = \begin{bmatrix} 2 & 4 \\ 1 & 1 \\ -2 & -7 \end{bmatrix}$$

$$A_B^{-1} = \begin{bmatrix} -1 & 0 & 0 \\ 0 & -1 & 0 \\ 0 & 0 & 1 \end{bmatrix}$$

$$x_B = A_B^{-1}b = \begin{bmatrix} -1 & 0 & 0 \\ 0 & -1 & 0 \\ 0 & 0 & 1 \end{bmatrix} \begin{bmatrix} 8 \\ 3 \\ -15 \end{bmatrix} = \begin{bmatrix} -8 \\ -3 \\ -15 \end{bmatrix}$$

$$w^T = c_B^T A_B^{-1} = \begin{bmatrix} 0 & 0 & 0 \end{bmatrix} \begin{bmatrix} -1 & 0 & 0 \\ 0 & -1 & 0 \\ 0 & 0 & 1 \end{bmatrix} = \begin{bmatrix} 0 & 0 & 0 \end{bmatrix}$$

$$s_N^T = c_N^T - w^T A_N = \begin{bmatrix} 3 & 2 \end{bmatrix} - \begin{bmatrix} 0 & 0 & 0 \end{bmatrix} \begin{bmatrix} 2 & 4 \\ 1 & 1 \\ -2 & -7 \end{bmatrix} = \begin{bmatrix} 3 & 2 \end{bmatrix}$$

$x_B \ngeq 0$, so the current basis is not optimal. We choose the leaving variable k such that $x_{B[r]} = x_k < 0$. The leaving variable is x_5 since it has the largest negative x_B. Hence, $k = 5$ ($r = 3$).

Next, we compute vector H_{rN}:

$$H_{rN} = \left(A_B^{-1}\right)_r A_N = \begin{bmatrix} 0 & 0 & 1 \end{bmatrix} \begin{bmatrix} 2 & 4 \\ 1 & 1 \\ -2 & -7 \end{bmatrix} = \begin{bmatrix} -2 & -7 \end{bmatrix}$$

$H_{rN} \ngeq 0$. Hence, we choose the entering variable $x_{N[t]} = x_l$ using the minimum ratio test:

$$\min \left\{ \frac{-s_{N[i]}}{H_i} : H_i < 0 \right\} = \min \left\{ \frac{-s_{N[1]}}{H_1}, \frac{-s_{N[2]}}{H_2} \right\} = \min \left\{ \frac{-3}{-2}, \frac{-2}{-7} \right\} = \frac{2}{7} = \frac{-s_{N[2]}}{H_2}$$

The entering variable is x_2. Hence, $l = 2$ and $t = 2$.
We continue calculating the pivoting column:

$$h_2 = A_B^{-1} A_{.2} = \begin{bmatrix} -1 & 0 & 0 \\ 0 & -1 & 0 \\ 0 & 0 & 1 \end{bmatrix} \begin{bmatrix} 4 \\ 1 \\ -7 \end{bmatrix} = \begin{bmatrix} -4 \\ -1 \\ -7 \end{bmatrix}$$

There are not any elements in vector h_3 that are greater than 0, so the LP problem is unbounded.

9.5   Implementation in MATLAB

This subsection presents the implementation in MATLAB of the revised dual simplex algorithm. Some necessary notations should be introduced before the presentation of the implementation of the revised dual simplex algorithm. Let A be a $m \times n$ matrix with the coefficients of the constraints, c be a $n \times 1$ vector with the coefficients of the objective function, b be a $m \times 1$ vector with the right-hand side of the constraints, *Eqin* be a $m \times 1$ vector with the type of the constraints, *MinMaxLP* be the type of optimization, $c0$ be the constant term of the objective function, *reinv* be the number of iterations that the basis inverse is recomputed from scratch, *tole*1 be the tolerance for the basic solution, *tole*2 be the tolerance for the reduced costs, *tole*3 be the tolerance for the pivoting column, *scalingTechnique* be the scaling method (0: no scaling, 1: arithmetic mean, 2: de Buchet for the case p = 1, 3: de Buchet for the case p = 2, 4: de Buchet for the case p = Inf, 5: entropy, 6: equilibration, 7: geometric mean, 8: IBM MPSX, 9: LP-norm for the case p = 1, 10: LP-norm for the case p = 2, 11: LP-norm for the case p = Inf), *basisUpdateMethod* be the basis update method (1: PFI, 2: MPFI), *xsol* be a $m \times 1$ vector with the solution found, *fval* be the value of the objective function at the solution *xsol*, *exitflag* be the reason that the algorithm terminated (0: optimal solution found, 1: the LP problem is infeasible, 2: the LP problem is unbounded, -1: the input data is not logically or numerically correct), and *iterations* be the number of iterations.

The revised dual simplex algorithm is implemented with two functions: (i) one function that implements the revised dual simplex algorithm (filename: `rdsa.m`), and (ii) one function that finds an initial invertible basis (filename: `lprref.m`).

The function for the implementation of the revised dual simplex algorithm takes as input the matrix with the coefficients of the constraints (matrix A), the vector with the coefficients of the objective function (vector c), the vector with the right-hand side of the constraints (vector b), the vector with the type of the constraints (vector $Eqin$), the type of optimization (variable $MinMaxLP$), the constant term of the objective function (variable $c0$), the number of iterations that the basis inverse is recomputed from scratch (variable $reinv$), the tolerance of the basic solution (variable $tole1$), the tolerance for the reduced costs (variable $tole2$), the tolerance for the pivoting column (variable $tole3$), the scaling method (variable $scalingTechnique$), and the basis update method (variable $basisUpdateMethod$), and returns as output the solution found (vector $xsol$), the value of the objective function at the solution $xsol$ (variable $fval$), the reason that the algorithm terminated (variable $exitflag$), and the number of iterations (variable $iterations$).

In lines 57–60, the output variables are initialized. In lines 62–85, we set default values to the missing inputs (if any). Next, we check if the input data is logically correct (lines 91–122). If the type of optimization is maximization, then we multiply vector c and constant $c0$ by -1 (lines 125–128). Then, the presolve analysis is performed (lines 131–142). Next, we scale the LP problem using the selected scaling technique (lines 149–177). Then, we find an invertible initial basis using the *lprref* function (lines 180–239). If the density of matrix A is less than 20%, then we use sparse algebra for faster computations (lines 246–250). Then, we pre-allocate memory for variables that hold large amounts of data and initialize the necessary variables (lines 253–287). Next, we check if the initial basis is dual feasible (lines 289–294).

If the initial basis is not dual feasible, then we apply the modified big-M method (lines 296–636). Initially, we find the index of the minimum element of vector s_N and create the auxiliary LP problem (lines 300–316). Then, we compute the new basic and nonbasic lists and initialize the necessary variables (lines 318–351). At each iteration of the dual simplex algorithm, we perform the optimality test in order to find if the basis found is optimal for the original LP problem or if the original LP problem is infeasible or unbounded (lines 352–635). If the auxiliary LP problem is not optimal, then we find the leaving variable (lines 489–525), find the entering variable using the minimum ratio test (lines 542–552), and update the basis inverse and the necessary variables (lines 554–621).

If the initial basis is dual feasible, then we apply the dual simplex algorithm to the original problem (lines 637–769). At each iteration, we perform the optimality test in order to find if the LP problem is optimal (lines 641–654). If the LP problem is not optimal, then we find the leaving variable (lines 656–670), find the entering variable using the minimum ratio test (lines 687–697), and update the basis inverse and the necessary variables (lines 699–768).

```
1.    function [xsol, fval, exitflag, iterations] = ...
2.        rdsa(A, c, b, Eqin, MinMaxLP, c0, reinv, tole1, ...
3.        tole2, tole3, scalingTechnique, basisUpdateMethod)
4.    % Filename: rdsa.m
5.    % Description: the function is an implementation of the
6.    % revised dual simplex algorithm
7.    % Authors: Ploskas, N., & Samaras, N.
8.    %
9.    % Syntax: [xsol, fval, exitflag, iterations] = ...
10.   %    rdsa(A, c, b, Eqin, MinMaxLP, c0, reinv, tole1, ...
11.   %    tole2, tole3, scalingTechnique, basisUpdateMethod)
12.   %
13.   % Input:
14.   % -- A: matrix of coefficients of the constraints
15.   %    (size m x n)
16.   % -- c: vector of coefficients of the objective function
17.   %    (size n x 1)
18.   % -- b: vector of the right-hand side of the constraints
19.   %    (size m x 1)
20.   % -- Eqin: vector of the type of the constraints
21.   %    (size m x 1)
22.   % -- MinMaxLP: the type of optimization (optional:
23.   %    default value -1 - minimization)
24.   % -- c0: constant term of the objective function
25.   %    (optional: default value 0)
26.   % -- reinv: every reinv number of iterations, the basis
27.   %    inverse is re-computed from scratch (optional:
28.   %    default value 80)
29.   % -- tole1: tolerance for the basic solution (optional:
30.   %    default value 1e-07)
31.   % -- tole2: tolerance for the reduced costs (optional:
32.   %    default value 1e-09)
33.   % -- tole3: tolerance for the pivoting column (optional:
34.   %    default value 1e-09)
35.   % -- scalingTechnique: the scaling method to be used
36.   %    (0: no scaling, 1: arithmetic mean, 2: de Buchet for
37.   %    the case p = 1, 3: de Buchet for the case p = 2,
38.   %    4: de Buchet for the case p = Inf, 5: entropy,
39.   %    6: equilibration, 7: geometric mean, 8: IBM MPSX,
40.   %    9: LP-norm for the case p = 1, 10: LP-norm for
41.   %    the case p = 2, 11: LP-norm for the case p = Inf)
42.   %    (optional: default value 6)
43.   % -- basisUpdateMethod: the basis update method to be used
44.   %    (1: PFI, 2: MPFI) (optional: default value 1)
45.   %
46.   % Output:
47.   % -- xsol: the solution found (size m x 1)
48.   % -- fval: the value of the objective function at the
49.   %    solution x
50.   % -- exitflag: the reason that the algorithm terminated
51.   %    (0: optimal solution found, 1: the LP problem is
52.   %    infeasible, 2: the LP problem is unbounded, -1: the
53.   %    input data is not logically or numerically correct)
54.   % -- iterations: the number of iterations
```

```
55.
56.  % initialize output variables
57.  xsol = [];
58.  fval = 0;
59.  exitflag = 0;
60.  iterations = 0;
61.  % set default values to missing inputs
62.  if ~exist('MinMaxLP')
63.     MinMaxLP = -1;
64.  end
65.  if ~exist('c0')
66.     c0 = 0;
67.  end
68.  if ~exist('reinv')
69.     reinv = 80;
70.  end
71.  if ~exist('tole1')
72.     tole1 = 1e-7;
73.  end
74.  if ~exist('tole2')
75.     tole2 = 1e-9;
76.  end
77.  if ~exist('tole3')
78.     tole3 = 1e-9;
79.  end
80.  if ~exist('scalingTechnique')
81.     scalingTechnique = 6;
82.  end
83.  if ~exist('basisUpdateMethod')
84.     basisUpdateMethod = 1;
85.  end
86.  [m, n] = size(A); % find the size of matrix A
87.  [m2, n2] = size(c); % find the size of vector c
88.  [m3, n3] = size(Eqin); % find the size of vector Eqin
89.  [m4, n4] = size(b); % find the size of vector b
90.  % check if input data is logically correct
91.  if n2 ~= 1
92.     disp('Vector c is not a column vector.')
93.     exitflag = -1;
94.     return
95.  end
96.  if n ~= m2
97.     disp(['The number of columns in matrix A and ' ...
98.           'the number of rows in vector c do not match.'])
99.     exitflag = -1;
100.    return
101. end
102. if m4 ~= m
103.    disp(['The number of the right-hand side values ' ...
104.          'is not equal to the number of constraints.'])
105.    exitflag = -1;
106.    return
107. end
108. if n3 ~= 1
```

```
109.    disp('Vector Eqin is not a column vector')
110.    exitflag = -1;
111.    return
112. end
113. if n4 ~= 1
114.    disp('Vector b is not a column vector')
115.    exitflag = -1;
116.    return
117. end
118. if m4 ~= m3
119.    disp('The size of vectors Eqin and b does not match')
120.    exitflag = -1;
121.    return
122. end
123. % if the type of optimization is maximization, then multiply
124. % vector c and constant c0 by -1
125. if MinMaxLP == 1
126.    c = -c;
127.    c0 = -c0;
128. end
129. % perform the presolve analysis
130. disp('---- P R E S O L V E    A N A L Y S I S ----')
131. [A, c, b, Eqin, c0, infeasible, unbounded] = ...
132.     presolve(A, c, b, Eqin, c0);
133. if infeasible == 1 % the LP problem is infeasible
134.    disp('The LP problem is infeasible')
135.    exitflag = 1;
136.    return
137. end
138. if unbounded == 1 % the LP problem is unbounded
139.    disp('The LP problem is unbounded')
140.    exitflag = 2;
141.    return
142. end
143. [m, n] = size(A); % find the new size of matrix A
144. [m2, ~] = size(c); % find the new size of vector c
145. [m3, ~] = size(Eqin); % find the size of vector Eqin
146. % scale the LP problem using the selected scaling
147. % technique
148. disp('---- S C A L I N G ----')
149. if scalingTechnique == 1 % arithmetic mean
150.    [A, c, b, ~, ~] = arithmeticMean(A, c, b);
151. % de buchet for the case p = 1
152. elseif scalingTechnique == 2
153.    [A, c, b, ~, ~] = debuchet1(A, c, b);
154. % de buchet for the case p = 2
155. elseif scalingTechnique == 3
156.    [A, c, b, ~, ~] = debuchet2(A, c, b);
157. % de buchet for the case p = Inf
158. elseif scalingTechnique == 4
159.    [A, c, b, ~, ~] = debuchetinf(A, c, b);
160. elseif scalingTechnique == 5 % entropy
161.    [A, c, b, ~, ~] = entropy(A, c, b);
162. elseif scalingTechnique == 6 % equilibration
```

```
163.     [A, c, b, ~, ~] = equilibration(A, c, b);
164. elseif scalingTechnique == 7 % geometric mean
165.     [A, c, b, ~, ~] = geometricMean(A, c, b);
166. elseif scalingTechnique == 8 % IBM MPSX
167.     [A, c, b, ~, ~, ~, ~] = ibmmpsx(A, c, b);
168. % LP-norm for the case p = 1
169. elseif scalingTechnique == 9
170.     [A, c, b, ~, ~] = lpnorm1(A, c, b);
171. % LP-norm for the case p = 2
172. elseif scalingTechnique == 10
173.     [A, c, b, ~, ~] = lpnorm2(A, c, b);
174. % LP-norm for the case p = Inf
175. elseif scalingTechnique == 11
176.     [A, c, b, ~, ~] = lpnorminf(A, c, b);
177. end
178. % find an invertible basis
179. disp('---- I N I T I A L    B A S I S ----')
180. flag = isequal(Eqin, zeros(m3, 1));
181. if flag == 1 % all constraints are equalities
182.     % select an initial invertible basis using lprref
183.     % function
184.     [~, ~, jb, out, ~, exitflag] = lprref(A, ...
185.         b, Eqin, 1e-10);
186.     if exitflag == 1 % the LP problem is infeasible
187.         disp('The LP problem is infeasible')
188.         return
189.     end
190.     % create the basic and nonbasic lists
191.     BasicList = jb;
192.     NonBasicList = setdiff(1:n, BasicList);
193.     % delete redundant constraints found by lprref
194.     A(out, :) = [];
195.     b(out, :) = [];
196.     Eqin(out, :) = [];
197. else % some or all constraints are inequalities
198.     % add slack variables
199.     axm = nnz(Eqin);
200.     c(m2 + 1:m2 + axm, :) = 0;
201.     A(:, n + 1:n + axm) = sparse(m, axm);
202.     curcol = 1;
203.     for i = 1:m3
204.         % 'greater than or equal to' inequality constraint
205.         if Eqin(i, 1) == 1
206.             A(i, n + curcol) = -1;
207.             curcol = curcol + 1;
208.         % 'less than or equal to' inequality constraint
209.         elseif Eqin(i, 1) == -1
210.             A(i, n + curcol) = 1;
211.             curcol = curcol + 1;
212.         % unrecognized type of constraint
213.         elseif Eqin(i,1) ~= 0
214.             disp('Vector Eqin is not numerically correct.')
215.             exitflag = -1;
216.             return
```

```
217.            end
218.       end
219.       % select an initial invertible basis using lprref
220.       % function
221.       [~, ~, jb, out, ~, exitflag] = lprref(A, ...
222.            b, Eqin, 1e-10);
223.       if exitflag == 1 % the LP problem is infeasible
224.            disp('The LP problem is infeasible')
225.            return
226.       end
227.       % create the basic and nonbasic lists
228.       [~, y1] = size(A);
229.       temp = n + 1:y1;
230.       for i = 1:length(temp)
231.            jb(length(jb) + 1) = temp(i);
232.       end
233.       BasicList = sort(jb);
234.       NonBasicList = setdiff(1:y1, BasicList);
235.       % delete redundant constraints found by lprref
236.       A(out, :) = [];
237.       b(out, :) = [];
238.       Eqin(out, :) = [];
239. end
240. flag = 0;
241. [m1, n1] = size(A); % new size of matrix A
242. % calculate the density of matrix A
243. density = (nnz(A) / (m1 * n1)) * 100;
244. % if the density of matrix A is less than 20%, then
245. % use sparse algebra for faster computations
246. if density < 20
247.     A = sparse(A);
248.     c = sparse(c);
249.     b = sparse(b);
250. end
251. % preallocate memory for variables that hold large
252. % amounts of data
253. Xb = spalloc(m1, 1, m1); % basic solution
254. h_l = spalloc(m1, 1, m1); % pivoting column
255. % the matrix of the basic variables
256. Basis = spalloc(m1, length(BasicList), m1 ...
257.      * length(BasicList));
258. % the matrix of the nonbasic variables
259. N = spalloc(m1, length(NonBasicList), m1 ...
260.      * length(NonBasicList));
261. w = spalloc(1, n1, n1); % simplex multiplier
262. Sn = spalloc(1, n1, n1); % reduced costs
263. % initialize data
264. % the matrix of the basic variables
265. Basis = A(:, BasicList);
266. % the coefficients of the objective function for the
267. % basic variables
268. cb = c(BasicList);
269. % the matrix of the nonbasic variables
270. N = A(:, NonBasicList);
```

```
271. % the coefficients of the objective function for the
272. % nonbasic variables
273. cn = c(NonBasicList);
274. % calculate the basis inverse
275. BasisInv = inv(Basis);
276. Xb = BasisInv * b; % basic solution
277. % set to zero, the values of Xb that are less than or
278. % equal to tole1
279. toler = abs(Xb) <= tole1;
280. Xb(toler == 1) = 0;
281. w = cb' * BasisInv; % calculate the simplex multiplier
282. % calculate the reduced costs
283. Sn = sparse(cn' - w * N);
284. % set to zero, the values of Sn that are less than or
285. % equal to tole2
286. toler = abs(Sn) <= tole2;
287. Sn(toler == 1) = 0;
288. % check if the current basis is dual feasible
289. if all(Sn >= 0) % the solution is dual feasible, skip big-M
290.     % method and proceed to the dual simplex algorithm
291.     flag = 1;
292. else % the solution is not dual feasible, apply big-M method
293.     flag = 0;
294. end
295. counter = 1;
296. if flag == 0 % modified big-M method
297.     disp(['---- D U A L     W I T H     B I G - M    ' ...
298.         'M E T H O D ----'])
299.     % find the entering variable
300.     [p, t] = min(Sn);
301.     rr = find(Sn == p);
302.     % break the ties in order to avoid stalling
303.     if length(rr) > 1
304.         l = NonBasicList(rr);
305.         [l, ~] = max(l);
306.     else
307.         l = NonBasicList(t);
308.     end
309.     % add a constraint and an artificial variable
310.     A(m1 + 1, :) = sparse(1, n1);
311.     A(m1 + 1, NonBasicList) = 1;
312.     A = [A [sparse(m1, 1); 1]];
313.     c = [c; 0];
314.     b = [b; 0];
315.     % the right-hand side of the big-M method
316.     bM = [sparse(m1, 1); 1];
317.     % compute the new basic and nonbasic variables
318.     BasicList(m1 + 1) = l;
319.     NonBasicList(NonBasicList == l) = n1 + 1;
320.     % the artificial variable is in the nonbasic list
321.     artificialVariableInN = 1;
322.     Basis = A(:, BasicList); % the new basis
323.     % the matrix of the nonbasic variables
324.     N = A(:, NonBasicList);
```

```
325.    % re-compute the inverse of the matrix
326.    BasisInv = inv(Basis);
327.    % compute the basic solution for the original problem
328.    % and the big-M problem
329.    Xb = BasisInv * b;
330.    % set to zero, the values of Xb that are less than or
331.    % equal to tole1
332.    toler = abs(Xb) <= tole1;
333.    Xb(toler == 1) = 0;
334.    XbM = BasisInv * bM;
335.    % set to zero, the values of XbM that are less than or
336.    % equal to tole1
337.    toler = abs(XbM) <= tole1;
338.    XbM(toler == 1) = 0;
339.    % compute the simplex multiplier
340.    w = c(BasicList)' * BasisInv;
341.    HRN = w * N;
342.    % set to zero, the values of HRN that are less than or
343.    % equal to tole2
344.    toler = abs(HRN) <= tole2;
345.    HRN(toler == 1) = 0;
346.    % compute the reduced costs
347.    Sn = c(NonBasicList)' - HRN;
348.    % set to zero, the values of Sn that are less than or
349.    % equal to tole2
350.    toler = abs(Sn) <= tole2;
351.    Sn(toler == 1) = 0;
352.    while flag == 0 % iterate in big-M method
353.        % optimality test
354.        if all(XbM > 0)
355.            col = find(BasicList == (n1 + 1), 1);
356.            if ~isempty(col)
357.                % if the artificial variable is in the
358.                % basic list, the problem is optimal
359.                % calculate the value of the objective
360.                % function
361.                if MinMaxLP == 1 % maximization
362.                    fval = full(-((c(BasicList))' * Xb ...
363.                        + c0));
364.                else % minimization
365.                    fval = full((c(BasicList))' * Xb ...
366.                        + c0);
367.                end
368.                exitflag = 0;
369.                xsol = Xb;
370.                iterations = iterations + 1;
371.                disp('The LP problem is optimal')
372.                return
373.            else
374.                % if the artificial variable is not in
375.                % the basic list, check if the reduced
376.                % cost of the artificial variable is
377.                % equal to zero
378.                % compute the simplex multiplier
```

```
379.                    w = c(BasicList)' * BasisInv;
380.                    HRN = w * A;
381.                    % set to zero, the values of HRN that are
382.                    % less than or equal to tole2
383.                    toler = abs(HRN) <= tole2;
384.                    HRN(toler == 1) = 0;
385.                    % compute the reduced costs
386.                    S = c' - HRN;
387.                    % set to zero, the values of S that are
388.                    % less than or equal to tole2
389.                    toler = abs(S) <= tole2;
390.                    S(toler == 1) = 0;
391.                    if S(n1 + 1) == 0 % if the reduced cost of
392.                        % the artificial variable is equal to
393.                        % zero, then the problem is optimal
394.                        % calculate the value of the objective
395.                        % function
396.                        if MinMaxLP == 1 % maximization
397.                            fval = full(-((c(BasicList))' * Xb ...
398.                                + c0));
399.                        else % minimization
400.                            fval = full((c(BasicList))' * Xb ...
401.                                + c0);
402.                        end
403.                        exitflag = 0;
404.                        xsol = Xb;
405.                        iterations = iterations + 1;
406.                        disp('The LP problem is optimal')
407.                        return
408.                    else % the problem is unbounded
409.                        disp('The LP problem is unbounded')
410.                        exitflag = 2;
411.                        iterations = iterations + 1;
412.                        return
413.                    end
414.                end
415.        elseif all(XbM >= 0) % optimality test
416.            row = find(XbM == 0);
417.            if ~isempty(row)
418.                if all(Xb(row) >= 0)
419.                    col = find(BasicList == (n1 + 1));
420.                    if (~isempty(col) && XbM(col) == 0 && ...
421.                        Xb(col) == 0) || isempty(col)
422.                        % if the artificial variable is in
423.                        % the basic list, check if the reduced
424.                        % cost of the artificial variable is
425.                        % equal to zero
426.                        % compute the simplex multiplier
427.                        w = c(BasicList)' * BasisInv;
428.                        HRN = w * A;
429.                        % set to zero, the values of HRN that
430.                        % are less than or equal to tole2
431.                        toler = abs(HRN) <= tole2;
432.                        HRN(toler == 1) = 0;
```

```
433.                             % compute the reduced costs
434.                             S = c' - HRN;
435.                             % set to zero, the values of S that
436.                             % are less than or equal to tole2
437.                             toler = abs(S) <= tole2;
438.                             S(toler == 1) = 0;
439.                             if S(n1 + 1) == 0 % if the reduced
440.                                 % cost of the artificial variable
441.                                 % is equal to zero, then the
442.                                 % problem is optimal
443.                                 % calculate the value of the
444.                                 % objective function
445.                                 if MinMaxLP == 1 % maximization
446.                                     fval = full(-((c(BasicList))' ...
447.                                         * Xb + c0));
448.                                 else % minimization
449.                                     fval = full((c(BasicList))' ...
450.                                         * Xb + c0);
451.                                 end
452.                                 exitflag = 0;
453.                                 xsol = Xb;
454.                                 iterations = iterations + 1;
455.                                 disp('The LP problem is optimal')
456.                                 return
457.                             else % the problem is unbounded
458.                                 disp('The LP problem is unbounded')
459.                                 exitflag = 2;
460.                                 iterations = iterations + 1;
461.                                 return
462.                             end
463.                         elseif (~isempty(col) && XbM(col) > 0) ...
464.                                 || Xb(col) > 0
465.                             % the problem is optimal
466.                             % calculate the value of the objective
467.                             % function
468.                             if MinMaxLP == 1 % maximization
469.                                 fval = full(-((c(BasicList))' * Xb ...
470.                                     + c0));
471.                             else % minimization
472.                                 fval = full((c(BasicList))' * Xb ...
473.                                     + c0);
474.                             end
475.                             exitflag = 0;
476.                             xsol = Xb;
477.                             iterations = iterations + 1;
478.                             disp('The LP problem is optimal')
479.                             return
480.                         end
481.                     else % the artificial variable left the
482.                         % nonbasic list
483.                         artificialVariableInN = 0;
484.                     end
485.             end
486.     end
```

```
487.            % if the artificial variable is in the nonbasic list,
488.            % use XbM to select the leaving variable
489.            if artificialVariableInN == 1
490.                % find the leaving variable
491.                mrt = find(XbM < 0);
492.                [a, r] = min(XbM(mrt));
493.                rr = find((XbM(mrt)) == a);
494.                % break the ties in order to avoid stalling
495.                if length(rr) > 1
496.                    r = mrt(rr);
497.                    % index of the leaving variable
498.                    k = BasicList(r);
499.                    [k, ~] = max(k);
500.                    r = find(BasicList == k);
501.                else
502.                    r = mrt(r);
503.                    % index of the leaving variable
504.                    k = BasicList(r);
505.                end
506.            else % if the artificial variable is not in the
507.                 % nonbasic list, use Xb to select the
508.                 % leaving variable
509.                mrt = Xb(row) < 0;
510.                mrt = row(mrt);
511.                [a, r] = min(Xb(mrt));
512.                rr = find((Xb(mrt)) == a);
513.                % break the ties in order to avoid stalling
514.                if length(rr) > 1
515.                    r = mrt(rr);
516.                    k = BasicList(r);
517.                    [k, ~] = max(k);
518.                    % index of the leaving variable
519.                    r = find(BasicList == k);
520.                else
521.                    r = mrt(r);
522.                    % index of the leaving variable
523.                    k = BasicList(r);
524.                end
525.            end
526.            % compute HRN vector
527.            HRN = BasisInv(r, :) * N;
528.            % set to zero, the values of HRN that are less than
529.            % orequal to tole2
530.            toler = abs(HRN) <= tole2;
531.            HRN(toler == 1) = 0;
532.            mrt = find(HRN < 0);
533.            if isempty(mrt) % if there is not any candidate
534.                % to enter the basic list, then the problem
535.                % is infeasible
536.                exitflag = 1;
537.                disp('The problem is infeasible')
538.                return;
539.            end
540.            % perform the minimum ratio test to select
```

```
541.        % the entering variable
542.        [a, t] = min(-Sn(mrt) ./ HRN(mrt));
543.        rr = find(-Sn(mrt) ./ HRN(mrt) == a);
544.        % break the ties in order to avoid stalling
545.        if length(rr)>1
546.            l = NonBasicList(mrt(rr));
547.            % index of the entering variable
548.            [l, ~] = max(l);
549.        else
550.            % index of the entering variable
551.            l = NonBasicList(mrt(t));
552.        end
553.        % calculate the pivoting column
554.        h_l = BasisInv * A(:, l);
555.        % set to zero, the values of h_l that are less than
556.        % or equal to tole3
557.        toler = abs(h_l) <= tole3;
558.        h_l(toler == 1) = 0;
559.        % check if the problem is unbounded
560.        if all(h_l <= 0)
561.            disp('The LP problem is unbounded')
562.            exitflag = 2;
563.            iterations = iterations + 1;
564.            return
565.        end
566.        % pivoting and update vectors and matrices
567.        f = Xb(r);
568.        fM = XbM(r);
569.        t = find(NonBasicList == l);
570.        g = h_l(r);
571.        v = Sn(t);
572.        % update the basic and the nonbasic lists
573.        BasicList(r) = l;
574.        NonBasicList(t) = k;
575.        % update the matrix of the nonbasic variables
576.        N = A(:, NonBasicList);
577.        iterations = iterations + 1;
578.        % calculate the new basic solution
579.        Xb(r) = 0;
580.        XbM(r) = 0;
581.        h_l2 = h_l;
582.        h_l2(r) = -1;
583.        if artificialVariableInN == 1 % the artificial
584.            % variable is in the nonbasic list, so
585.            % update XbM
586.            XbM = XbM - (fM / g) * h_l2;
587.            % set to zero, the values of XbM that are less
588.            % than or equal to tole1
589.            toler = abs(XbM) <= tole1;
590.            XbM(toler == 1) = 0;
591.        end
592.        Xb = Xb - (f / g) * h_l2;
593.        % set to zero, the values of Xb that are less than
594.        % or equal to tole1
```

```
595.          toler = abs(Xb) <= tole1;
596.          Xb(toler == 1) = 0;
597.          % update the reduced costs
598.          Sn(t) = 0;
599.          HRN(t) = 1;
600.          Sn = Sn - (v / g) * HRN;
601.          % set to zero, the values of Sn that are less than
602.          % or equal to tole2
603.          toler = abs(Sn) <= tole2;
604.          Sn(toler == 1) = 0;
605.          % basis inverse
606.          if iterations == counter * reinv
607.              % recompute the inverse of the basis
608.              % from scratch every reinv iterations
609.              BasisInv = inv(A(:, BasicList));
610.              counter = counter + 1;
611.              h_l(r) = g;
612.          else
613.              % basis update according to the
614.              % selected basis update method
615.              if basisUpdateMethod == 1 % pfi
616.                  BasisInv = pfi(BasisInv, h_l, r);
617.              else % mpfi
618.                  BasisInv = mpfi(BasisInv, h_l, r);
619.              end
620.              h_l(r) = -1;
621.          end
622.          % print intermediate results every 100
623.          % iterations
624.          if mod(iterations, 100) == 0
625.              % calculate the value of the objective
626.              % function
627.              if MinMaxLP == 1 % maximization
628.                  fval = full(-(c(BasicList)' * Xb) + c0);
629.              else % minimization
630.                  fval = full(c(BasicList)' * Xb + c0);
631.              end
632.              fprintf(['Iteration %i - objective value: ' ...
633.                  '%f\n'], iterations, fval);
634.          end
635.      end
636. end
637. if flag == 1 % dual simplex method
638.     disp('---- D U A L ----')
639.     while flag == 1
640.         % optimality test
641.         if all(Xb >= 0) % the problem is optimal
642.             % calculate the value of the objective
643.             % function
644.             if MinMaxLP == 1 % maximization
645.                 fval = full(-((c(BasicList))' * Xb + c0));
646.             else % minimization
647.                 fval = full((c(BasicList))' * Xb + c0);
648.             end
```

```
649.            exitflag = 0;
650.            xsol = Xb;
651.            iterations = iterations + 1;
652.            disp('The LP problem is optimal')
653.            return
654.        end
655.        % find the leaving variable
656.        mrt = find(Xb < 0);
657.        [a, r] = min(Xb(mrt));
658.        rr = find(Xb(mrt) == a);
659.        % break the ties in order to avoid stalling
660.        if length(rr) > 1
661.            r = mrt(rr);
662.            % index of the leaving variable
663.            k = BasicList(r);
664.            [k, ~] = max(k);
665.            r = find(BasicList == k);
666.        else
667.            r = mrt(r);
668.            % index of the leaving variable
669.            k = BasicList(r);
670.        end
671.        % compute HRN vector
672.        HRN = BasisInv(r, :) * N;
673.        % set to zero, the values of HRN that are less than
674.        % or equal to tole3
675.        toler = abs(HRN) <= tole2;
676.        HRN(toler == 1) = 0;
677.        mrt = find(HRN < 0);
678.        if isempty(mrt) % if there is not any candidate
679.            % to enter the basic list, then the problem
680.            % is infeasible
681.            exitflag = 1;
682.            disp('The problem is infeasible')
683.            return;
684.        end
685.        % perform the minimum ratio test to select
686.        % the entering variable
687.        [a, t] = min(-Sn(mrt) ./ HRN(mrt));
688.        rr = find(-Sn(mrt) ./ HRN(mrt) == a);
689.        % break the ties in order to avoid stalling
690.        if length(rr) >1
691.            l = NonBasicList(mrt(rr));
692.            % index of the entering variable
693.            [l, ~] = max(l);
694.        else
695.            % index of the entering variable
696.            l = NonBasicList(mrt(t));
697.        end
698.        % calculate the pivoting column
699.        h_l = BasisInv * A(:, l);
700.        % set to zero, the values of h_l that are less than
701.        % or equal to tole3
702.        toler = abs(h_l) <= tole3;
```

```
703.            h_1(toler == 1) = 0;
704.            % check if the problem is unbounded
705.            if all(h_1 <= 0)
706.                disp('The LP problem is unbounded')
707.                exitflag = 2;
708.                iterations = iterations + 1;
709.                return
710.            end
711.            % pivoting and update vectors and matrices
712.            f = Xb(r);
713.            t = find(NonBasicList == 1);
714.            g = h_1(r);
715.            % update the basic and the nonbasic lists
716.            BasicList(r) = 1;
717.            NonBasicList(t) = k;
718.            % update the matrix of the nonbasic variables
719.            N = A(:, NonBasicList);
720.            iterations = iterations + 1;
721.            % calculate the new basic solution
722.            Xb(r) = 0;
723.            h_12 = h_1;
724.            h_12(r) = -1;
725.            Xb = Xb - (f / g) * h_12;
726.            % set to zero, the values of Xb that are less than
727.            % or equal to tole1
728.            toler = abs(Xb) <= tole1;
729.            Xb(toler == 1) = 0;
730.            % update the reduced costs
731.            Sn(t) = 0;
732.            HRN(t) = 1;
733.            Sn = Sn + a * HRN;
734.            % set to zero, the values of Sn that are less than
735.            % or equal to tole2
736.            toler = abs(Sn) <= tole2;
737.            Sn(toler == 1) = 0;
738.            % basic inverse
739.            if iterations == counter * reinv
740.                % recompute the inverse of the basis
741.                % from scratch every reinv iterations
742.                BasisInv = inv(A(:, BasicList));
743.                counter = counter + 1;
744.                h_1(r) = g;
745.            else
746.                % basis update according to the
747.                % selected basis update method
748.                if basisUpdateMethod == 1 % pfi
749.                    BasisInv = pfi(BasisInv, h_1, r);
750.                else % mpfi
751.                    BasisInv = mpfi(BasisInv, h_1, r);
752.                end
753.                h_1(r) = -1;
754.            end
755.            % print intermediate results every 100
756.            % iterations
```

```
757.            if mod(iterations, 100) == 0
758.                % calculate the value of the objective
759.                % function
760.                if MinMaxLP == 1 % maximization
761.                    fval = full(-(c(BasicList)' * Xb) + c0);
762.                else % minimization
763.                    fval = full(c(BasicList)' * Xb + c0);
764.                end
765.                fprintf(['Iteration %i - objective value: ' ...
766.                    '%f\n'], iterations, fval);
767.            end
768.       end
769.  end
770.  end
```

The function that finds an invertible initial basis (filename: lprref.m) takes as input the matrix with the coefficients of the constraints (matrix *A*), the vector with the right-hand side of the constraints (vector *b*), the vector with the type of the constraints (vector *Eqin*), and the tolerance (variable *tol*), and returns as output the reduced row echelon form of the matrix with the coefficients of the constraints (matrix *AEqin*), the reduced vector with the right-hand side of the constraints (vector *bEqin*), the initial basis (matrix *jb*), the vector with the redundant indices (vector *out*), the matrix with the row redundant indices (matrix *rowindex*), and a flag variable showing if the LP problem is infeasible or not (variable *infeasible*).

In lines 30–31, the output variables are initialized. In lines 33–35, we find all the equality constraints. Next, we compute the tolerance if it was not given as input (lines 39–41). Then, we apply Gauss-Jordan elimination with partial pivoting in order to find the initial basis (lines 46–75). Next, we find the indices of the redundant constraints (lines 77–93).

```
1.    function [AEqin, bEqin, jb, out, rowindex, ...
2.         infeasible] = lprref(A, b, Eqin, tol)
3.    % Filename: lprref.m
4.    % Description: the function calculates a row echelon
5.    % form of matrix A for LPs using Gauss-Jordan
6.    % elimination with partial pivoting
7.    % Authors: Ploskas, N., & Samaras, N.
8.    %
9.    % Syntax: [AEqin, bEqin, jb, out, rowindex, ...
10.   %    infeasible] = lprref(A, b, Eqin, tol)
11.   %
12.   % Input:
13.   % -- A: matrix of coefficients of the constraints
14.   %    (size m x n)
15.   % -- b: vector of the right-hand side of the constraints
16.   %    (size m x 1)
17.   % -- Eqin: vector of the type of the constraints
18.   %    (size m x 1)
```

```
19.   % -- tol: tolerance
20.   %
21.   % Output:
22.   % -- AEqin: reduced row echelon form of A (size m x n)
23.   % -- bEqin: reduced vector b (size m x 1)
24.   % -- jb: basis of matrix A (size m x m)
25.   % -- rowindex: a matrix with the row redundant indices
26.   %    (size 2 x m)
27.   % -- infeasible: if the problem is infeasible or not
28.
29.   % initialize output variables
30.   infeasible = 0;
31.   out = [];
32.   % find all equality constraints
33.   a0 = find(Eqin == 0);
34.   AEqin = A(a0, :);
35.   bEqin = b(a0, :);
36.   [m, n] = size(AEqin);
37.   rowindex = zeros(2, m);
38.   rowindex(1, 1:m) = a0';
39.   if nargin < 4 % compute tol, if it was not given as input
40.       tol = max(m, n) * eps * norm(A, 'inf');
41.   end
42.   i = 1;
43.   j = 1;
44.   jb = [];
45.   % apply Gauss-Jordan elimination with partial pivoting
46.   while (i <= m) && (j <= n)
47.       [p, k] = max(abs(AEqin(i:m, j)));
48.       if p < tol
49.           AEqin(i:m, j) = zeros(m - i + 1, 1);
50.           j = j + 1;
51.       elseif p ~= 0
52.           k = k + i - 1;
53.           jb = [jb j];
54.           AEqin([i k], :) = AEqin([k i], :);
55.           bEqin([i k], :) = bEqin([k i], :);
56.           rowindex(:, [i k]) = rowindex(:, [k i]);
57.           bEqin(i, :) = bEqin(i, :) / AEqin(i, j);
58.           AEqin(i, j:n) = AEqin(i, j:n) / AEqin(i, j);
59.           i_nz = find(AEqin(:, j));
60.           i_nz = setdiff(i_nz, i);
61.           for t = i_nz
62.               if bEqin(i) ~= 0
63.                   bEqin(t) = bEqin(t) - AEqin(t, j) * bEqin(i);
64.                   toler = abs(bEqin) <= tol;
65.                   bEqin(toler == 1) = 0;
66.               end
67.               AEqin(t, j:n) = AEqin(t, j:n) - ...
68.                   AEqin(t, j) * AEqin(i, j:n);
69.               toler = abs(AEqin) <= tol;
70.               AEqin(toler == 1) = 0;
71.           end
72.           i = i + 1;
```

```
73.                j = j + 1;
74.        end
75.    end
76.    % check for redundant and infeasible constraints
77.    i = 1;
78.    for h = [1:i - 1 i + 1:m]
79.        % redundant constraint
80.        if (AEqin(h, :) == 0) & (bEqin(h) == 0)
81.            rowindex(2, h) = 1;
82.        end
83.        % infeasible constraint
84.        if (AEqin(h,:) == 0) & (bEqin(h) ~= 0)
85.            infeasible = 1;
86.            return;
87.        end
88.    end
89.    % find the indices of the redundant constraints
90.    if any(rowindex(2, :) == 1)
91.        y = find(rowindex(2, :) == 1);
92.        out = y;
93.    end
94.    end
```

9.6   Computational Study

In this section, we present a computational study. The aim of the computational study is to compare the efficiency of the proposed implementation with the revised primal simplex algorithm presented in Chapter 8. The test set used in the computational study is 19 LPs from the Netlib (optimal, Kennington, and infeasible LPs) and Mészáros problem sets that were presented in Section 3.6 and a set of randomly generated sparse LPs. Tables 9.3 and 9.4 present the total execution time and the number of iterations of the revised primal simplex algorithm that was presented in Chapter 8 and the revised dual simplex algorithm that was presented in this chapter over the benchmark set and a set of randomly generated sparse LPs, respectively. The following abbreviations are used: (i) RSA—the proposed implementation of the revised primal simplex algorithm, and (ii) RDSA—the proposed implementation of the revised dual simplex algorithm. For each instance, we averaged times over 10 runs. All times are measured in seconds. The execution times of both algorithms include the time needed to presolve and scale (with arithmetic mean) the LPs.

The revised dual simplex algorithm is almost 1.6 times faster than the revised primal simplex algorithm over the benchmark set, while requiring almost 1.6 less iterations than the iterations needed for the revised primal simplex algorithm. On the other hand, the revised primal simplex algorithm is almost 1.2 times faster than

Table 9.3 Total execution time and number of iterations of the proposed implementation of the revised primal simplex algorithm and the revised dual simplex algorithm over the benchmark set

Problem				RSA		RDSA	
Name	Constraints	Variables	Nonzeros	Time	Iterations	Time	Iterations
beaconfd	173	262	3,375	0.15	26	0.10	25
cari	400	1,200	152,800	10.01	963	10.00	745
farm	7	12	36	0.01	5	0.01	2
itest6	11	8	20	0.01	2	0.01	1
klein2	477	54	4,585	0.46	500	1.23	1,520
nsic1	451	463	2,853	0.27	369	0.24	335
nsic2	465	463	3,015	0.33	473	0.39	571
osa-07	1,118	23,949	143,694	10.15	997	6.27	427
osa-14	2,337	52,460	314,760	67.22	2,454	31.20	963
osa-30	4,350	100,024	600,138	320.19	4,681	118.77	1,941
rosen1	520	1,024	23,274	3.33	1,893	0.91	621
rosen10	2,056	4,096	62,136	39.71	5,116	12.95	2,281
rosen2	1,032	2,048	46,504	19.29	4,242	3.53	1,197
rosen7	264	512	7,770	0.29	420	0.14	183
rosen8	520	1,024	15,538	1.43	1,045	0.63	492
sc205	205	203	551	0.20	165	0.29	354
scfxm1	330	457	2,589	0.74	339	0.72	303
sctap2	1,090	1,880	6,714	1.86	1,001	1.19	570
sctap3	1,480	2,480	8,874	4.37	1,809	2.04	740
Geometric mean				1.47	463.78	0.93	289.66

the revised dual simplex algorithm over the set of randomly generated sparse LPs, while requiring almost 1.2 times more iterations than the iterations needed for the revised dual simplex algorithm.

9.7 Chapter Review

The dual simplex algorithm is an attractive alternative for solving linear programming problems. The dual simplex algorithm is very efficient on many types of problems and is especially useful in integer linear programming. This chapter presented the revised dual simplex algorithm. Numerical examples were presented in order for the reader to understand better the algorithm. Furthermore, an implementation of the algorithm in MATLAB was presented. The implementation is modular allowing the user to select which scaling technique and basis update method will use in order to solve LPs. Finally, we performed a computational study over benchmark LPs and randomly generated sparse LPs in order to compare the efficiency of the proposed implementation with the revised primal simplex

Table 9.4 Total execution time and number of iterations of the proposed implementation of the revised primal simplex algorithm and the revised dual simplex algorithm over a set of randomly generated sparse LPs (5% density)

Problem	RSA		RDSA	
	Time	Iterations	Time	Iterations
100x100	0.13	1	0.05	1
200x200	0.17	236	0.09	72
300x300	0.39	584	0.41	491
400x400	0.81	911	1.21	1,012
500x500	1.94	1,565	0.82	561
600x600	4.41	2,569	1.82	998
700x700	7.97	3,919	25.36	6,554
800x800	16.25	6,016	53.40	10,147
900x900	26.14	6,883	13.47	3,125
1,000x1,000	34.55	7,909	15.01	3,254
1,100x1,100	70.41	13,483	179.02	17,966
1,200x1,200	63.55	10,172	33.63	4,829
1,300x1,300	133.95	18,004	340.12	25,135
1,400x1,400	160.29	18,805	516.93	31,676
1,500x1,500	205.85	19,768	126.77	10,973
1,600x1,600	327.74	28,866	967.64	44,734
1,700x1,700	463.99	36,016	1,216.43	50,589
1,800x1,800	530.54	33,873	283.06	16,931
1,900x1,900	917.11	52,492	1,754.04	59,657
2,000x2,000	926.92	50,926	2,980.28	83,448
Geometric mean	24.97	4,868.59	29.32	4,117.13

algorithm. Computational results showed that the revised dual simplex algorithm is almost 1.6 times faster than the revised primal simplex algorithm over the benchmark set, while requiring almost 1.6 less iterations than the iterations needed for the revised primal simplex algorithm. On the other hand, the revised primal simplex algorithm is almost 1.2 times faster than the revised dual simplex algorithm over the set of randomly generated sparse LPs, while requiring almost 1.2 times more iterations than the iterations needed for the revised dual simplex algorithm.

References

1. Bixby, R. E. (1992). Implementing the simplex method: The initial basis. *ORSA Journal on Computing, 4*, 267–284.
2. Carstens, D. M. (1968) Crashing techniques. In W. Orchard-Hays (Ed.), *Advanced linear-programming computing techniques* (pp. 131–139). New York: McGraw-Hill.
3. Forrest, J. J., & Goldfarb, D. (1992). Steepest-edge simplex algorithms for linear programming. *Mathematical Programming, 57*(1–3), 341–374.

4. Fourer, R. (1994). *Notes on the dual simplex method.* Draft report.
5. Gould, N. I. M., & Reid, J. K. (1989). New crash procedures for large systems of linear constraints. *Mathematical Programming, 45,* 475–501.
6. Lemke, C. E. (1954). The dual method of solving the linear programming problem. *Naval Research Logistics Quarterly, 1*(1), 36–47.
7. Maros, I., & Mitra, G. (1998). Strategies for creating advanced bases for large-scale linear programming problems. *INFORMS Journal on Computing, 10,* 248–260.
8. Paparrizos, K., Samaras, N., & Stephanides, G. (2003). A new efficient primal dual simplex algorithm. *Computers & Operations Research, 30*(9), 1383–1399. ISO 690

Chapter 10
Exterior Point Simplex Algorithm

Abstract The exterior point simplex algorithm is a simplex-type algorithm that moves in the exterior of the feasible solution and constructs basic infeasible solutions instead of constructing feasible solutions like simplex algorithm does. This chapter presents the exterior point simplex algorithm. Numerical examples are presented in order for the reader to understand better the algorithm. Furthermore, an implementation of the algorithm in MATLAB is presented. The implementation is modular allowing the user to select which scaling technique and basis update method will use in order to solve LPs. Finally, a computational study over benchmark LPs and randomly generated sparse LPs is performed in order to compare the efficiency of the proposed implementation with the revised primal simplex algorithm presented in Chapter 8.

10.1 Chapter Objectives

- Present the exterior point simplex algorithm.
- Understand all different methods that can be used in each step of the exterior point simplex algorithm.
- Solve linear programming problems using the exterior point simplex algorithm.
- Implement the exterior point simplex algorithm using MATLAB.
- Compare the computational performance of the exterior point simplex algorithm with the revised primal simplex algorithm.

10.2 Introduction

The implementation of pivoting algorithms that will be more efficient than simplex algorithm is still an active field of research. An approach to enhance the performance of simplex algorithm is to move in the exterior of the feasible solution and construct basic infeasible solutions instead of constructing feasible solutions like simplex

---

The original version of this chapter was revised. An erratum to this chapter can be found at https://doi.org/10.1007/978-3-319-65919-0_13

N. Ploskas, N. Samaras, *Linear Programming Using MATLAB*®,
Springer Optimization and Its Applications 127, DOI 10.1007/978-3-319-65919-0_10

algorithm does. Such an algorithm is called exterior point simplex algorithm and was proposed by Paparrizos initially for the assignment problem [5] and then for the solution of LPs [6, 8]. Paparrizos et al. [10] have pointed out that the geometry of the exterior point simplex algorithm reveals that this algorithm is faster than primal simplex algorithm. Exterior point simplex algorithm relies on the idea that making steps in directions that are linear combinations of attractive descent directions can lead to faster practical convergence than that available from simplex algorithm. Exterior point simplex algorithm constructs two paths to the optimal solution. One path consists of basic but not feasible solutions, while the second path is feasible. Furthermore, in [13] new variations of exterior point simplex algorithm using three different strategies to enter Phase II are presented. Another approach is the transformation of the exterior path into a dual feasible simplex path. This algorithm is called Primal-Dual Exterior Point Simplex Algorithm (PDEPSA) and its revised form can be found in [9]. A more efficient approach is the Primal-Dual Interior Point Simplex Algorithm (PDIPSA), which is proposed by Samaras [12]. PDIPSA can deal with the problems of stalling and cycling more effectively and as a result improves the performance of the primal-dual exterior point algorithms. The advantage of PDIPSA stems from the fact that it uses an interior point in order to compute the leaving variable in contrast to primal-dual exterior point algorithms which use a boundary point. A complete review of exterior point simplex algorithms for linear and network programming can be found in [7]. Ploskas and Samaras [11] proposed a GPU-based implementation of PDEPSA achieving a maximum speedup of 181 on dense LPs and 20 on sparse LPs over MATLAB's interior point method. Moreover, the GPU-based implementation was $2.3\times$ faster than MATLAB's interior point method on a set of benchmark LPs.

This chapter presents the exterior point simplex algorithm. Initially, the description of the algorithm is given. The various steps of the algorithm are presented. Numerical examples are also presented in order for the reader to understand better the algorithm. Furthermore, an implementation of the algorithm in MATLAB is presented. The implementation is modular allowing the user to select which scaling technique and basis update method will use in order to solve LPs. Finally, a computational study is performed in order to compare the efficiency of the proposed implementation with the revised primal simplex algorithm.

The structure of this chapter is as follows. In Section 10.3, the exterior point simplex algorithm is described. All different steps of the algorithm are presented in detail. Four examples are solved in Section 10.4 using the exterior point simplex algorithm. Section 10.5 presents the MATLAB code of the exterior point simplex algorithm. Section 10.6 presents a computational study that compares the proposed implementation and the revised primal simplex algorithm. Finally, conclusions and further discussion are presented in Section 10.7.

10.3   Description of the Algorithm

A formal description of the exterior point simplex algorithm is given in Table 10.1 and a flow diagram of its major steps in Figure 10.1.

Table 10.1 Exterior point simplex algorithm

Step 0. *(Initialization).*

Presolve the LP problem.

Scale the LP problem.

Select an initial basic solution (B, N).

Find the set of indices $P = \{j \in N : s_j < 0\}$.

If $P \neq \emptyset$ and the direction d_B crosses the feasible region then proceed to step 2.

Step 1. *(Phase I).*

Construct an auxiliary problem by adding an artificial variable x_{n+1} with a

coefficient vector equal to $-A_B e$, where $e = (1, 1, \cdots, 1)^T \in \mathbb{R}^m$.

Apply the steps of the revised simplex algorithm in Phase I. If the final basic

solution (B, N) is feasible, then proceed to step 2 in order to solve (LP.1).

(LP.1) can be either optimal or infeasible.

Step 2. *(Phase II).*

Step 2.0. *(Initialization).*

Compute $(A_B)^{-1}$ and vectors x_B, w, and s_N.

Find the sets of indices $P = \{j \in N : s_j < 0\}$ and $Q = \{j \in N : s_j \geq 0\}$.

Define an arbitrary vector $\lambda = (\lambda_1, \lambda_2, \cdots, \lambda_{|P|}) > 0$ and compute s_0 as follows:

$s_0 = \sum_{j \in P} \lambda_j s_j$

and the direction

$d_B = -\sum_{j \in P} \lambda_j h_j$, where $h_j = A_B^{-1} A_j$.

Step 2.1. *(Test of Optimality).*

if $P = \emptyset$ then STOP. (LP.1) is optimal.

else

   if $d_B \geq 0$ then

      if $s_0 = 0$ then STOP. (LP.1) is optimal.

   else

      choose the leaving variable $x_{B[r]} = x_k$ using the following relation:

$a = \frac{x_{B[r]}}{-d_{B[r]}} = \min\left\{\frac{x_{B[i]}}{-d_{B[i]}} : d_{B[i]} < 0\right\}$

      If $a = \infty$, then (LP.1) is unbounded.

Step 2.2. *(Pivoting).*

Compute the row vectors: $H_{rP} = \left(A_B^{-1}\right)_{r.} A_P$ and $H_{rQ} = \left(A_B^{-1}\right)_{r.} A_Q$.

Compute the ratios θ_1 and θ_2 using the following relations:

$\theta_1 = \frac{-s_P}{H_{rP}} = \min\left\{\frac{-s_j}{H_{rj}} : H_{rj} > 0 \wedge j \in P\right\}$ and

$\theta_2 = \frac{-s_Q}{H_{rQ}} = \min\left\{\frac{-s_j}{H_{rj}} : H_{rj} < 0 \wedge j \in Q\right\}$

Determine the indices t_1 and t_2 such that $P[t_1] = p$ and $Q[t_2] = q$.

if $\theta_1 \leq \theta_2$ then set $l = p$

else set $l = q$.

Step 2.3. *(Update).*

Swap indices k and l. Update the new basis inverse $\left(A_{\overline{B}}\right)^{-1}$, using a basis

update scheme. Update sets P and Q. Set $B(r) = l$. If $\theta_1 \leq \theta_2$, set $P = P \setminus [l]$

and $Q = Q \cup [k]$. Otherwise, set $Q(t_2) = k$. Update vectors x_B, w, and s_N.

Update the new direction $d_{\overline{B}}$ using the relation

$d_{\overline{B}} = E^{-1} d_B$. If $l \in P$, set $d_{\overline{B}[r]} = d_{B[r]} + \lambda_l$.

Go to Step 2.1.

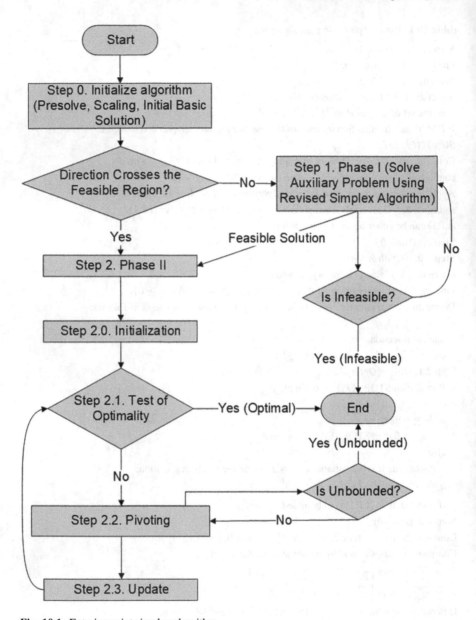

Fig. 10.1 Exterior point simplex algorithm

Initially, the LP problem is presolved. We use the presolve routine presented in Section 4.15. This routine presolves the original LP problem using the eleven presolve methods presented in Chapter 4. After the presolve analysis, the modified LP problem is scaled. In Chapter 5, we presented eleven scaling techniques. Any of these scaling techniques can be used to scale the LP problem.

Then, we calculate an initial basis to start with. That means that we seek for an invertible basis. This basis includes a subset of the columns. Some columns are obvious choices for this initial basis. Slack variables only have one nonzero value in the corresponding column, so they are good choices to include them in the initial basis. Deciding on the other columns is not as easy, especially on large LPs. A general procedure to select an initial basis is presented in Section 10.5. Function *lprref* applies Gauss-Jordan elimination with partial pivoting in order to find an initial invertible basis. Many other sophisticated techniques have been proposed for the calculation of the initial basis. Many other sophisticated techniques have been proposed for the calculation of the initial basis. Interested readers are referred to [1–4].

Consider the following LP problem (LP.1) in the standard form:

$$\min \quad c^T x$$

$$\text{s.t.} \quad Ax = b \qquad (LP.1)$$

$$x \geq 0$$

where $A \in \mathbb{R}^{m \times n}$, $(c, x) \in \mathbb{R}^n$, $b \in \mathbb{R}^m$, and T denotes transposition. We assume that the coefficient matrix A is of full rank ($rank(A) = m$, $m < n$). The dual problem associated with (LP.1) is:

$$\max \quad b^T w$$

$$\text{s.t.} \quad A^T w + s = c \qquad (DP.1)$$

$$s \geq 0$$

where $w \in \mathbb{R}^m$ and $s \in \mathbb{R}^n$.

Let B be a subset of the index set $\{1, 2, \cdots, n\}$ with cardinality number equal to m ($|B| = m$). The partition of the coefficient matrix A as (A_B, A_N), where $|B| = m$ and $|N| = n - m$, is called a basic partition. A_B denotes the submatrix of A containing the columns A_j such that $j \in B$. Similarly, A_N denotes the submatrix of A containing the columns A_j such that $j \in N$. Partitioning matrix A as $A = \begin{bmatrix} A_B & A_N \end{bmatrix}$ and with a corresponding partitioning and ordering of $x^T = \begin{bmatrix} x_B & x_N \end{bmatrix}$ and $c^T = \begin{bmatrix} c_B & c_N \end{bmatrix}$, (LP.1) is written as:

$$\min \quad c_B^T x_B + c_N^T x_N$$

$$\text{s.t.} \quad A_B x_B + A_N x_N = b$$

$$x_B, x_N \geq 0$$

Matrix A_B is the basis and B is the basic list iff A_B is an $m \times n$ non-singular (invertible) submatrix of A. Given a basis B, the associated solution $x_B = A_B^{-1} b$, $x_N = 0$, is called a basic solution. If $x_B \geq 0$, then the above solution is feasible.

Otherwise, it is infeasible. The solution w is feasible to (DP.1) iff $s_N^T = c_N^T - w^T A_N \geq 0$. With s we denote the reduced costs. A given solution is dual feasible iff $s \geq 0$. Notation $x_{B[i]}$ refers to the ith basic variable.

Initially, we determine if the direction d_B crosses the feasible region using the following relation:

$$\beta = \max \left\{ \frac{x_{B(i)}}{-d_{B(i)}} : x_{B(i)} < 0 \right\} < \alpha = \min \left\{ \frac{x_{B(i)}}{-d_{B(i)}} : d_{B(i)} < 0 \right\}$$

where $1 \leq i \leq m$, $d_B = -\sum_{j \in P} \lambda_j h_j$, $h_j = A_B^{-1} A_j$, and λ is an arbitrary vector such that $\lambda = (\lambda_1, \lambda_2, \cdots, \lambda_{|P|}) > 0$. If $P \neq \emptyset$ and $\beta < \alpha$, then the direction d_B crosses the feasible region and we proceed to apply the exterior point simplex algorithm to (LP.1) (Phase II). If $P = \emptyset$ or the direction d_B does not cross the feasible region, then we construct an auxiliary problem to find an initial feasible basis (Phase I) using the revised simplex algorithm. In Phase I, it is possible to find out that the LP problem is infeasible.

10.3.1   Phase I

Phase I has been already presented in Chapter 8. We also describe it here for the sake of completeness. In Phase I, we apply the revised simplex algorithm to an auxiliary problem, not the original one (LP.1), in order to find an initial feasible solution for (LP.1). Assuming that the original problem is in its standard form (LP.1), the auxiliary LP problem that is solved in Phase I is the following:

$$
\begin{aligned}
\min z = \quad & 0x_1 + 0x_2 + \quad \cdots \quad + 0x_n + x_{n+1} \\
\text{s.t.} \quad & A_{11}x_1 + A_{12}x_2 + \quad \cdots \quad + A_{1n}x_n + A_{1,n+1}x_{n+1} = b_1 \\
& A_{21}x_1 + A_{22}x_2 + \quad \cdots \quad + A_{2n}x_n + A_{2,n+1}x_{n+1} = b_2 \\
& \vdots \quad \vdots \quad \vdots \quad \vdots \qquad \cdots \qquad \vdots \quad \vdots \quad \vdots \qquad \vdots \\
& A_{m1}x_1 + A_{m2}x_2 + \quad \cdots \quad + A_{mn}x_n + A_{m,n+1}x_{n+1} = b_m \\
& x_j \geq 0, \qquad j = 1, 2, \cdots, n+1
\end{aligned}
$$

$$(10.1)$$

In matrix notation, we have:

$$
\begin{aligned}
\min \quad & x_{n+1} \\
\text{s.t.} \quad & Ax + dx_{n+1} = b \\
& x_j \geq 0, \quad j = 1, 2, \cdots, n+1
\end{aligned}
$$

$$(LP.2)$$

where $A \in \mathbb{R}^{m \times n}$, $(x) \in \mathbb{R}^n$, $b \in \mathbb{R}^m$.

The leaving variable k is the index with the minimum value of x_B. Moreover, the artificial variable x_{n+1} enters the basis. Hence, we update the initial basic list $B = B \cup [n+1] \setminus [k]$. Finally, $N = N \cup [k]$. It is obvious now that the corresponding basic solution is feasible. Since $x_{B[r]} = -b_r > 0$, then $x_{B[i]} = b_i - b_r \geq 0$, $i = 1, 2, \cdots, m$, $i \neq r$, $x_j = 0$, and $j \in N$.

The last column of matrix A in the auxiliary problem is calculated as:

$$d - -A_B e$$

where e is a column vector of ones.

In the first iteration of Phase I, we calculate the following matrices and vectors:

$$A_B^{-1}, x_B = A_B^{-1}b, w^T = c_B^T A_B^{-1}, \text{ and } s_N^T = c_N^T - w^T A_N$$

Then, we perform the optimality test of Phase I. We proceed with Phase II when one of the following conditions is met:

- If all the elements of s_N are greater than or equal to 0. If the value x_B for the variable $n + 1$ ($x_{B[n+1]}$) is not equal to 0, then the LP problem is infeasible. Otherwise, we proceed with Phase II.
- If the artificial variable left from the basic list.

If the auxiliary LP problem is not optimal, then we continue with the remaining steps of Phase I. We select the entering variable l according to a pivoting rule. In Chapter 6, we presented six pivoting rules. Any of these pivoting rules can be used. Then, we calculate the pivoting column:

$$h_l = A_B^{-1}A_l$$

Then, we perform the minimum ratio test to find the leaving variable k:

$$x_k = x_{B[r]} = \frac{x_{B[r]}}{h_{il}} = \min\left\{ \frac{x_{B[i]}}{h_{il}} : h_{il} > 0 \right\}$$

Next, we update the basic and nonbasic lists:

$$B = B \cup [l] \setminus [k], N = N \cup [k] \setminus [l]$$

Then, we update the basis inverse using a basis update method. In Chapter 7, we presented two basis update methods. Any of these basis update methods can be used.

Finally, we compute vectors x_B, w, and s_N:

$$x_B = A_B^{-1}b, w^T = c_B^T A_B^{-1}, \text{ and } s_N^T = c_N^T - w^T A_N$$

We start again with the optimality test and all the other steps presented in Phase I until we find a feasible basis for the original LP problem (LP.1) or we find that the original LP problem (LP.1) is infeasible.

Note that instead of using the revised primal simplex algorithm to solve the auxiliary problem in Phase I, we can also use the exterior point simplex algorithm. Interested readers are referred to [13].

10.3.2   Phase II

In Phase II, we apply the exterior point simplex algorithm to the original problem (LP.1). If we have applied Phase I, then the artificial variable has left from the basic list and we also delete it from the nonbasic list.

In the first iteration of Phase II, we calculate the following matrices and vectors:

$$A_B^{-1}, x_B = A_B^{-1}b, w = c_B^T A_B^{-1}, \text{ and } s_N = c_N^T - wA_N$$

Then, we find the sets of indices P and Q:

$$P = \{j \in N : s_j < 0\} \text{ and } Q = \{j \in N : s_j \geq 0\}$$

Next, we define an arbitrary vector $\lambda = (\lambda_1, \lambda_2, \cdots, \lambda_{|P|}) > 0$ and compute s_0 as follows:

$$s_0 = \sum_{j \in P} \lambda_j s_j$$

and the direction $d_B = -\sum_{j \in P} \lambda_j h_j$, where $h_j = A_B^{-1} A_P$.

Then, we perform the optimality test of Phase II. We find an optimal solution when one of the following conditions is met:

- If set P is empty.
- If $d_B \geq 0$ and $s_0 = 0$ (if $d_B \geq 0$ and $s_0 < 0$, then (LP.1) is unbounded).

If (LP.1) is not optimal, then we continue with the remaining steps of Phase II. We select the leaving variable k using the following relation:

$$a = \frac{x_{B[r]}}{-d_{B[r]}} = \min\left\{ \frac{x_{B[i]}}{-d_{B[i]}} : d_{B[i]} < 0, i = 1, 2, \cdots, m \right\}$$

If $a = \infty$, problem (LP.1) is unbounded.

Then, we compute the row vectors:

$$H_{rP} = \left(A_B^{-1}\right)_{r.} A_P \text{ and } H_{rQ} = \left(A_B^{-1}\right)_{r.} A_Q$$

Next, we compute the ratios θ_1 and θ_2 using the following relations:

$$\theta_1 = \frac{-s_P}{H_{rP}} = \min\left\{\frac{-s_j}{H_{rj}} : H_{rj} > 0 \wedge j \in P\right\} \text{ and }$$

$$\theta_2 = \frac{-s_Q}{H_{rQ}} = \min\left\{\frac{-s_j}{H_{rj}} : H_{rj} < 0 \wedge j \in Q\right\}$$

Then, we determine the indices t_1 and t_2 such that $P[t_1] = p$ and $Q[t_2] = q$. If $\theta_1 \leq \theta_2$ then $l = p$ and we update sets P and Q:

$$P = P \setminus [l] \text{ and } Q = Q \cup [k]$$

Otherwise ($\theta_1 > \theta_2$), $l = q$ and we update set Q:

$$Q = Q \cup [k] \setminus [l]$$

We also update the basic and nonbasic lists:

$$B = B \cup [l] \setminus [k] \text{ and } N = P \cup Q$$

Then, we update the basis inverse using a basis update method. In Chapter 7, we presented two basis update methods. Any of these basis update methods can be used.

Finally, we update vectors x_B, w, s_N, and d_B:

$$x_B = A_B^{-1}b, w^T = c_B^T A_B^{-1}, s_N^T = c_N^T - w^T A_N, \text{ and } d_{\overline{B}} = E^{-1}d_B$$

where E^{-1} is the eta matrix used in PFI basis update method (see Section 7.6). If $\theta_1 \leq \theta_2$ then $d_{\overline{B}[r]} = d_{\overline{B}[r]} + \lambda_l$.

We start again with the optimality test and all the other steps presented in Phase II until we find an optimal basis for (LP.1) or we find that (LP.1) is unbounded.

Lemma 10.1 *If the algorithm stops because of the optimality test at step 2.1, then the last basic solution is optimal.*

Proof The proof is by induction on the number of iterations. Assume that $s_j \geq 0$ for $j \in Q$. It is easily seen that this induction hypothesis is satisfied by the initial basis. Let $\overline{B}$ be the next basis, $\overline{Q}$ be the updated set Q, and $\overline{s}_j$ the corresponding reduced costs. Then, we have:

$$\overline{s}_j = s_j - \frac{s_l}{h_{rl}}h_{rj} \geq 0, j \neq l, j \in \overline{Q} \tag{10.2}$$

Because of the choice of the entering variable x_l, we have $\frac{s_l}{h_{rl}} \geq 0$. If $h_{rj} \leq 0$, then we have $\overline{s}_l \geq 0$. If $h_{rj} > 0$, relation (10.2) is equivalent to:

$$\frac{s_j}{h_{rj}} \geq \frac{s_l}{h_{rl}}$$

which holds because of the choice of the entering variable. If $j = l$, then $\bar{s}_l = \frac{s_l}{h_{rl}} \geq 0$. Hence, $\bar{s}_j \geq 0, \forall j \in \overline{Q}$.

Let $\overline{B}$ be the basis just before the last basis B. From the stopping condition $P = \emptyset$, we conclude that $\overline{P} = \{l\}$. A simple induction on the number of iterations shows that $\overline{\overline{d}}_{\overline{Q}} = 0$ and $\overline{\overline{d}}_l > 0$. Let $\overline{\overline{d}}$ be the direction corresponding to the increase after the entering variable x_l, $l \in \overline{P}$. Then, $\overline{d} = \lambda \overline{\overline{d}}$, $\lambda > 0$. Let y be the boundary point from which $\overline{d}$ exits the feasible region. It is easily concluded that $x = y \geq 0$. Hence, the last basic solution x is feasible. By the duality theorem, we conclude that x is optimal. $\qquad\square$

Lemma 10.2 *If the algorithm stops because of the unboundedness test at step 2.1, (LP.1) is unbounded.*

Proof If x and y are feasible solutions, from $d = y - x$ we conclude that $Ad = A(y - x) = Ay - Ax = b - b = 0$. From d_Q we have $c^T d = \sum_{j \in P} s_j d_j$. From relation (10.2) and a simple induction, we conclude that $s_j < 0, \forall j \in P$. As $d_j \geq 0$, we have $c^T d < 0$. This concludes the proof. $\qquad\square$

Theorem 10.1 *The algorithm solves (LP.1) correctly after a finite number of iterations.*

Proof The correctness of the algorithm is a consequence of Lemmas 10.1 and 10.2. Let B be the current basis and $\overline{B}$ the previous one. Let $\overline{d}$ correspond to $\overline{B}$ and $\overline{y}$ be the boundary point from which $\overline{d}$ exits the feasible region. Then, $d = \overline{y} - x$, where x is the basic solution corresponding to B. Moreover, $\overline{x} = x + t\overline{d} \geq 0$ for $1 \leq t \leq \frac{x_{B[r]}}{-d_{B[r]}}$. Indeed, for $t = 1$, we have $\overline{x} = \overline{y} \geq 0$. We conclude that $x_{B[r]} \geq 0$. From the nondegeneracy assumption, we have $s_l \neq 0$. Hence, the objective function decreases from iteration to iteration. $\qquad\square$

10.4   Examples

Example 1 The LP problem that will be solved is the following:

$$\begin{aligned}
\min z = \ & x_1 && - \ x_3 - 3x_4 \\
\text{s.t.} \quad & 2x_1 && + 2x_3 + 3x_4 = 10 \\
& && - 2x_2 - 2x_3 - 6x_4 = -6 \\
& x_j \geq 0, \quad (j = 1, 2, 3, 4)
\end{aligned}$$

In matrix notation the above LP problem is written as follows:

$$A = \begin{bmatrix} 2 & 0 & 2 & 3 \\ 0 & -2 & -2 & -6 \end{bmatrix}, c = \begin{bmatrix} 1 \\ 0 \\ -1 \\ -3 \end{bmatrix}, b = \begin{bmatrix} 10 \\ -6 \end{bmatrix}, Eqin = \begin{bmatrix} 0 \\ 0 \end{bmatrix}$$

The LP problem is in its standard form. Let's start with the basic list $B = [1,2]$ and the nonbasic list $N = [3,4]$. Initially, we check if we must apply the Phase I of the algorithm:

$$c_B = \begin{bmatrix} 1 \\ 0 \end{bmatrix}, c_N = \begin{bmatrix} -1 \\ -3 \end{bmatrix}$$

$$A_B = \begin{bmatrix} 2 & 0 \\ 0 & -2 \end{bmatrix}, A_N = \begin{bmatrix} 2 & 3 \\ -2 & -6 \end{bmatrix}$$

$$A_B^{-1} = \begin{bmatrix} 1/2 & 0 \\ 0 & -1/2 \end{bmatrix}$$

$$x_B = A_B^{-1}b = \begin{bmatrix} 1/2 & 0 \\ 0 & -1/2 \end{bmatrix} \begin{bmatrix} 10 \\ -6 \end{bmatrix} = \begin{bmatrix} 5 \\ 3 \end{bmatrix}$$

$$w^T = c_B^T A_B^{-1} = \begin{bmatrix} 1 & 0 \end{bmatrix} \begin{bmatrix} 1/2 & 0 \\ 0 & -1/2 \end{bmatrix} = \begin{bmatrix} 1/2 & 0 \end{bmatrix}$$

$$s_N^T = c_N^T - w^T A_N = \begin{bmatrix} -1 & -3 \end{bmatrix} - \begin{bmatrix} 1/2 & 0 \end{bmatrix} \begin{bmatrix} 2 & 3 \\ -2 & -6 \end{bmatrix} = \begin{bmatrix} -2 & -9/2 \end{bmatrix}$$

$$P = \{j \in N : s_j < 0\} = [3,4], Q = N \setminus P = \emptyset, \lambda = (\lambda_3, \lambda_4) = (1,1)$$

$$h_j = A_B^{-1} A_P = \begin{bmatrix} 1/2 & 0 \\ 0 & -1/2 \end{bmatrix} \begin{bmatrix} 2 & 3 \\ -2 & -6 \end{bmatrix} = \begin{bmatrix} 1 & 3/2 \\ 1 & 3 \end{bmatrix}$$

$$d_B = -\sum_{j \in P} \lambda_j h_j = \begin{bmatrix} -(1+3/2) \\ -(1+3) \end{bmatrix} = \begin{bmatrix} -5/2 \\ -4 \end{bmatrix}$$

Then, we calculate the ratios a and β:

$$\beta = \max \left\{ \frac{x_{B[i]}}{-d_{B[i]}} : x_{B[i]} < 0 \right\} = -\infty$$

$$\alpha = \min \left\{ \frac{x_{B[i]}}{-d_{B[i]}} : d_{B[i]} < 0 \right\} = \min \left\{ \frac{x_{B[1]}}{-d_{B[1]}}, \frac{x_{B[2]}}{-d_{B[2]}} \right\} = \min \left\{ \frac{5}{5/2}, \frac{3}{4} \right\} = \frac{3}{4}$$

Since $P \neq \emptyset$ and $\beta < \alpha$, the direction d_B is feasible. Hence, we can proceed with Phase II.

Phase II - 1st Iteration

The initial data of the problem is:

$$c_B = \begin{bmatrix} 1 \\ 0 \end{bmatrix}, c_N = \begin{bmatrix} -1 \\ -3 \end{bmatrix}$$

$$A_B = \begin{bmatrix} 2 & 0 \\ 0 & -2 \end{bmatrix}, A_N = \begin{bmatrix} 2 & 3 \\ -2 & -6 \end{bmatrix}$$

$$A_B^{-1} = \begin{bmatrix} 1/2 & 0 \\ 0 & -1/2 \end{bmatrix}$$

$$x_B = A_B^{-1} b = \begin{bmatrix} 1/2 & 0 \\ 0 & -1/2 \end{bmatrix} \begin{bmatrix} 10 \\ -6 \end{bmatrix} = \begin{bmatrix} 5 \\ 3 \end{bmatrix}$$

$$w^T = c_B^T A_B^{-1} = \begin{bmatrix} 1 & 0 \end{bmatrix} \begin{bmatrix} 1/2 & 0 \\ 0 & -1/2 \end{bmatrix} = \begin{bmatrix} 1/2 & 0 \end{bmatrix}$$

$$s_N^T = c_N^T - w^T A_N = \begin{bmatrix} -1 & -3 \end{bmatrix} - \begin{bmatrix} 1/2 & 0 \end{bmatrix} \begin{bmatrix} 2 & 3 \\ -2 & -6 \end{bmatrix} = \begin{bmatrix} -2 & -9/2 \end{bmatrix}$$

Then, we find the sets of indices P and Q:

$$P = \{ j \in N : s_j < 0 \} = [3, 4], Q = N \setminus P = \emptyset$$

We define an arbitrary vector $(\lambda_3, \lambda_4) = (1, 1) > 0$ and compute s_0:

$$s_0 = \sum_{j \in P} \lambda_j s_j = \lambda_3 s_3 + \lambda_4 s_4 = -13/2$$

Then, we calculate the direction d_B:

$$h_j = B^{-1} A_P = \begin{bmatrix} 1/2 & 0 \\ 0 & -1/2 \end{bmatrix} \begin{bmatrix} 2 & 3 \\ -2 & -6 \end{bmatrix} = \begin{bmatrix} 1 & 3/2 \\ 1 & 3 \end{bmatrix}$$

$$d_B = -\sum_{j \in P} \lambda_j h_j = \begin{bmatrix} -(1 + 3/2) \\ -(1 + 3) \end{bmatrix} = \begin{bmatrix} -5/2 \\ -4 \end{bmatrix}$$

$P \neq \emptyset$, so the current basis is not optimal. $d_B \not\geq 0$, so we continue to find the leaving variable using the following relation:

$$a = \min \left\{ \frac{x_{B[i]}}{-d_{B[i]}} : d_{B[i]} < 0 \right\} = \min \left\{ \frac{x_{B[1]}}{-d_{B[1]}}, \frac{x_{B[2]}}{-d_{B[2]}} \right\}$$

$$= \min \left\{ \frac{5}{5/2}, \frac{3}{4} \right\} = \frac{3}{4} = \frac{x_{B[2]}}{-d_{B[2]}}$$

The leaving variable is x_2. Hence, $r = 2$ and $k = 2$.

Then, we compute the row vectors:

$$H_{rP} = \left(A_B^{-1}\right)_{r.} A_P = \begin{bmatrix} 0 & -1/2 \end{bmatrix} \begin{bmatrix} 2 & 3 \\ -2 & -6 \end{bmatrix} = \begin{bmatrix} 1 & 3 \end{bmatrix}$$

$$H_{rQ} = \left(A_B^{-1}\right)_{r.} A_Q = \begin{bmatrix} 0 & -1/2 \end{bmatrix} \begin{bmatrix} \, \end{bmatrix} = \begin{bmatrix} \, \end{bmatrix}$$

Next, we compute the ratios θ_1 and θ_2:

$$\theta_1 = \frac{-s_P}{H_{rP}} = \min\left\{ \frac{-s_j}{H_{rj}} : H_{rj} > 0 \wedge j \in P \right\} = \min\left\{ \frac{-s_3}{H_{23}}, \frac{-s_4}{H_{24}} \right\} = \min\left\{ \frac{2}{1}, \frac{9/2}{3} \right\}$$

$$= \frac{9/2}{3} = \frac{3}{2} = \frac{-s_4}{H_{24}}$$

$$\theta_2 = \frac{-s_Q}{H_{rQ}} = \min\left\{ \frac{-s_j}{H_{rj}} : H_{rj} < 0 \wedge j \in Q \right\} = \infty$$

$\theta_1 \leq \theta_2$, so the entering variable is x_4. Hence, $l = 4$. We update sets P and Q and the basic and nonbasic lists:

$$P = P \setminus l = [3, 4] \setminus 4 = [3], Q = Q \cup k = \emptyset \cup 2 = [2]$$

$$B = [1, 4], N = [3, 2]$$

Hence:

$$c_B = \begin{bmatrix} 1 \\ -3 \end{bmatrix}, c_N = \begin{bmatrix} -1 \\ 0 \end{bmatrix}$$

$$A_B = \begin{bmatrix} 2 & 3 \\ 0 & -6 \end{bmatrix}, A_N = \begin{bmatrix} 2 & 0 \\ -2 & -2 \end{bmatrix}$$

We continue calculating the pivoting column:

$$h_4 = A_B^{-1} A_{.4} = \begin{bmatrix} 1/2 & 0 \\ 0 & -1/2 \end{bmatrix} \begin{bmatrix} 3 \\ -6 \end{bmatrix} = \begin{bmatrix} 3/2 \\ 3 \end{bmatrix}, h_4(2) = -1, h_4 = \begin{bmatrix} 3/2 \\ -1 \end{bmatrix}$$

Then, we update the basis inverse using the PFI update scheme:

$$E^{-1} = \begin{bmatrix} 1 & -1/2 \\ 0 & 1/3 \end{bmatrix}$$

$$(A_{\bar{B}})^{-1} = E^{-1} A_B^{-1} = \begin{bmatrix} 1 & -1/2 \\ 0 & 1/3 \end{bmatrix} \begin{bmatrix} 1/2 & 0 \\ 0 & -1/2 \end{bmatrix} = \begin{bmatrix} 1/2 & 1/4 \\ 0 & -1/6 \end{bmatrix}$$

Finally, we update vectors x_B, w, s_N, and d_B:

$$x_B = A_B^{-1}b = \begin{bmatrix} 1/2 & 1/4 \\ 0 & -1/6 \end{bmatrix} \begin{bmatrix} 10 \\ -6 \end{bmatrix} = \begin{bmatrix} 7/2 \\ 1 \end{bmatrix}$$

$$w^T = c_B^T A_B^{-1} = \begin{bmatrix} 1 & -3 \end{bmatrix} \begin{bmatrix} 1/2 & 1/4 \\ 0 & -1/6 \end{bmatrix} = \begin{bmatrix} 1/2 & 3/4 \end{bmatrix}$$

$$s_N^T = c_N^T - w^T A_N = \begin{bmatrix} -1 & 0 \end{bmatrix} - \begin{bmatrix} 1/2 & 3/4 \end{bmatrix} \begin{bmatrix} 2 & 0 \\ -2 & -2 \end{bmatrix} = \begin{bmatrix} -1/2 & 3/2 \end{bmatrix}$$

$$d_{\bar{B}} = E^{-1}d_B = \begin{bmatrix} 1 & -1/2 \\ 0 & 1/3 \end{bmatrix} \begin{bmatrix} -5/2 \\ -4 \end{bmatrix} = \begin{bmatrix} -1/2 \\ -4/3 \end{bmatrix},$$

$$d_{\bar{B}[2]} = d_{\bar{B}[2]} + 1 = -4/3 + 1 = -1/3$$

Phase II - 2nd Iteration

The data at the beginning of this iteration is:

$$c_B = \begin{bmatrix} 1 \\ -3 \end{bmatrix}, c_N = \begin{bmatrix} -1 \\ 0 \end{bmatrix}$$

$$A_B = \begin{bmatrix} 2 & 3 \\ 0 & -6 \end{bmatrix}, A_N = \begin{bmatrix} 2 & 0 \\ -2 & -2 \end{bmatrix}$$

$$A_B^{-1} = \begin{bmatrix} 1/2 & 1/4 \\ 0 & -1/6 \end{bmatrix}$$

$$x_B = \begin{bmatrix} 7/2 \\ 1 \end{bmatrix}, w^T = \begin{bmatrix} 1/2 & 3/4 \end{bmatrix}, s_N^T = \begin{bmatrix} -1/2 & 3/2 \end{bmatrix}, d_B = \begin{bmatrix} -1/2 \\ -1/3 \end{bmatrix}$$

$P \neq \emptyset$, so the current basis is not optimal. $d_B \not\geq 0$, so we continue to find the leaving variable using the following relation:

$$a = \min\left\{ \frac{x_{B[i]}}{-d_{B[i]}} : d_{B[i]} < 0 \right\} = \min\left\{ \frac{x_{B[1]}}{-d_{B[1]}}, \frac{x_{B[4]}}{-d_{B[4]]}} \right\} = \min\left\{ \frac{7/2}{1/2}, \frac{1}{1/3} \right\}$$

$$= \frac{1}{1/3} = 3 = \frac{x_{B[4]}}{-d_{B[4]}}$$

The leaving variable is x_4. Hence, $r = 2$ and $k = 4$.
Then, we compute the row vectors:

$$H_{rP} = \left(A_B^{-1}\right)_r A_P = \begin{bmatrix} 0 & -1/6 \end{bmatrix} \begin{bmatrix} 2 \\ -2 \end{bmatrix} = \begin{bmatrix} 1/3 \end{bmatrix}$$

$$H_{rQ} = \left(A_B^{-1}\right)_{r.} A_Q \begin{bmatrix} 0 & -1/6 \end{bmatrix} \begin{bmatrix} 0 \\ -2 \end{bmatrix} = [1/3]$$

Next, we compute the ratios θ_1 and θ_2:

$$\theta_1 = \frac{-s_P}{H_{rP}} = \min\left\{ \frac{-s_j}{H_{rj}} : H_{rj} > 0 \wedge j \in P \right\} = \min\left\{ \frac{-s_3}{H_{23}} \right\}$$

$$= \min\left\{ \frac{1/2}{1/3} \right\} = \frac{1/2}{1/3} = \frac{3}{2} = \frac{-s_3}{H_{13}}$$

$$\theta_2 = \frac{-s_Q}{H_{rQ}} = \min\left\{ \frac{-s_j}{H_{rj}} : H_{rj} < 0 \wedge j \in Q \right\} = \infty$$

$\theta_1 \le \theta_2$, so the entering variable is x_3. Hence, $l = 3$. We update sets P and Q and the basic and nonbasic lists:

$$P = P \setminus l = [3] \setminus 3 = \emptyset, Q = Q \cup k = [2] \cup 4 = [2, 4]$$

$$B = [1, 3], N = [2, 4]$$

Hence:

$$c_B = \begin{bmatrix} 1 \\ -1 \end{bmatrix}, c_N = \begin{bmatrix} 0 \\ -3 \end{bmatrix}$$

$$A_B = \begin{bmatrix} 2 & 2 \\ 0 & -2 \end{bmatrix}, A_N = \begin{bmatrix} 0 & 3 \\ -2 & -6 \end{bmatrix}$$

We continue calculating the pivoting column:

$$h_3 = A_B^{-1} A_{.3} = \begin{bmatrix} 1/2 & 1/4 \\ 0 & -1/6 \end{bmatrix} \begin{bmatrix} 2 \\ -2 \end{bmatrix} = \begin{bmatrix} 1/2 \\ 1/3 \end{bmatrix}, h_3(2) = -1, h_3 = \begin{bmatrix} 1/2 \\ -1 \end{bmatrix}$$

Then, we update the basis inverse using the PFI update scheme:

$$E^{-1} = \begin{bmatrix} 1 & -3/2 \\ 0 & 3 \end{bmatrix}$$

$$(A_{\overline{B}})^{-1} = E^{-1} A_B^{-1} = \begin{bmatrix} 1 & -3/2 \\ 0 & 3 \end{bmatrix} \begin{bmatrix} 1/2 & 1/4 \\ 0 & -1/6 \end{bmatrix} = \begin{bmatrix} 1/2 & 1/2 \\ 0 & -1/2 \end{bmatrix}$$

Finally, we update vectors x_B, w, and s_N:

$$x_B = A_B^{-1} b = \begin{bmatrix} 1/2 & 1/2 \\ 0 & -1/2 \end{bmatrix} \begin{bmatrix} 10 \\ -6 \end{bmatrix} = \begin{bmatrix} 2 \\ 3 \end{bmatrix}$$

$$w^T = c_B^T A_B^{-1} = \begin{bmatrix} 1 & -1 \end{bmatrix} \begin{bmatrix} 1/2 & 1/2 \\ 0 & -1/2 \end{bmatrix} = \begin{bmatrix} 1/2 & 1 \end{bmatrix}$$

$$s_N^T = c_N^T - w^T A_N = \begin{bmatrix} -3 & 0 \end{bmatrix} - \begin{bmatrix} 1/2 & 1 \end{bmatrix} \begin{bmatrix} 3 & 0 \\ -6 & -2 \end{bmatrix} = \begin{bmatrix} 3/2 & 2 \end{bmatrix}$$

$$d_{\bar{B}} = E^{-1} d_B = \begin{bmatrix} 1 & -3/2 \\ 0 & 3 \end{bmatrix} \begin{bmatrix} -1/2 \\ -1/3 \end{bmatrix} = \begin{bmatrix} 0 \\ -1 \end{bmatrix}, d_{\bar{B}[2]} = d_{\bar{B}[2]} + 1 = -1 + 1 = 0$$

Phase II - 3rd Iteration

The data at the beginning of this iteration is:

$$c_B = \begin{bmatrix} 1 \\ -1 \end{bmatrix}, c_N = \begin{bmatrix} -3 \\ 0 \end{bmatrix}$$

$$A_B = \begin{bmatrix} 2 & 2 \\ 0 & -2 \end{bmatrix}, A_N = \begin{bmatrix} 3 & 0 \\ -6 & -2 \end{bmatrix}$$

$$A_B^{-1} = \begin{bmatrix} 1/2 & 1/2 \\ 0 & -1/2 \end{bmatrix}$$

$$x_B = \begin{bmatrix} 2 \\ 3 \end{bmatrix}, w^T = \begin{bmatrix} 1/2 & 1 \end{bmatrix}, s_N^T = \begin{bmatrix} 3/2 & 2 \end{bmatrix}, d_B = \begin{bmatrix} 0 \\ 0 \end{bmatrix}$$

$P = \emptyset$, so the current basis ($B = [1, 3], N = [2, 4]$) is optimal:

$$x_B = \begin{bmatrix} 2 \\ 3 \end{bmatrix}$$

$$z = c_B^T x_B = -1$$

Example 2 The LP problem that will be solved is the following:

$$\min z = 5x_1 + 2x_2 - 4x_3$$
$$\text{s.t.} \quad 6x_1 + x_2 - 2x_3 \geq 5$$
$$x_1 + x_2 + x_3 \leq 4$$
$$6x_1 + 4x_2 - 2x_3 \geq 10$$
$$x_j \geq 0, \quad (j = 1, 2, 3)$$

In matrix notation the above LP problem is written as follows:

$$A = \begin{bmatrix} 6 & 1 & -2 \\ 1 & 1 & 1 \\ 6 & 4 & -2 \end{bmatrix}, c = \begin{bmatrix} 5 \\ 2 \\ -4 \end{bmatrix}, b = \begin{bmatrix} 5 \\ 4 \\ 10 \end{bmatrix}, Eqin = \begin{bmatrix} 1 \\ -1 \\ 1 \end{bmatrix}$$

Initially, we need to convert the LP problem in its standard form:

$$\begin{aligned}
\min z = {} & 5x_1 + 2x_2 - 4x_3 \\
\text{s.t.} \quad & 6x_1 + x_2 - 2x_3 - x_4 && = 5 \\
& x_1 + x_2 + x_3 + x_5 && = 4 \\
& 6x_1 + 4x_2 - 2x_3 - x_6 && = 10 \\
& x_j \geq 0, \quad (j = 1, 2, 3, 4, 5, 6)
\end{aligned}$$

In matrix notation the above LP problem is written as follows:

$$A = \begin{bmatrix} 6 & 1 & -2 & -1 & 0 & 0 \\ 1 & 1 & 1 & 0 & 1 & 0 \\ 6 & 4 & -2 & 0 & 0 & -1 \end{bmatrix}, c = \begin{bmatrix} 5 \\ 2 \\ -4 \\ 0 \\ 0 \\ 0 \end{bmatrix}, b = \begin{bmatrix} 5 \\ 4 \\ 10 \end{bmatrix}, Eqin = \begin{bmatrix} 0 \\ 0 \\ 0 \end{bmatrix}$$

Let's start with the basic list $B = [4, 5, 6]$ and the nonbasic list $N = [1, 2, 3]$. Initially, we check if we must apply the Phase I of the algorithm:

$$c_B = \begin{bmatrix} 0 \\ 0 \\ 0 \end{bmatrix}, c_N = \begin{bmatrix} 5 \\ 2 \\ -4 \end{bmatrix}$$

$$A_B = \begin{bmatrix} -1 & 0 & 0 \\ 0 & 1 & 0 \\ 0 & 0 & -1 \end{bmatrix}, A_N = \begin{bmatrix} 6 & 1 & -2 \\ 1 & 1 & 1 \\ 6 & 4 & -2 \end{bmatrix}$$

$$A_B^{-1} = \begin{bmatrix} -1 & 0 & 0 \\ 0 & 1 & 0 \\ 0 & 0 & -1 \end{bmatrix}$$

$$x_B = A_B^{-1}b = \begin{bmatrix} -1 & 0 & 0 \\ 0 & 1 & 0 \\ 0 & 0 & -1 \end{bmatrix} \begin{bmatrix} 5 \\ 4 \\ 10 \end{bmatrix} = \begin{bmatrix} -5 \\ 4 \\ -10 \end{bmatrix}$$

$$w^T = c_B^T A_B^{-1} = \begin{bmatrix} 0 & 0 & 0 \end{bmatrix} \begin{bmatrix} -1 & 0 & 0 \\ 0 & 1 & 0 \\ 0 & 0 & -1 \end{bmatrix} = \begin{bmatrix} 0 & 0 & 0 \end{bmatrix}$$

$$s_N^T = c_N^T - w^T A_N = \begin{bmatrix} 5 & 2 & -4 \end{bmatrix} - \begin{bmatrix} 0 & 0 & 0 \end{bmatrix} \begin{bmatrix} 6 & 1 & -2 \\ 1 & 1 & 1 \\ 6 & 4 & -2 \end{bmatrix} = \begin{bmatrix} 5 & 2 & -4 \end{bmatrix}$$

$$P = \{j \in N : s_j < 0\} = [3], Q = N \setminus P = [1, 2, 3] \setminus [3] = [1, 2]$$

$$h_j = A_B^{-1} A_P = \begin{bmatrix} -1 & 0 & 0 \\ 0 & 1 & 0 \\ 0 & 0 & -1 \end{bmatrix} \begin{bmatrix} -2 \\ 1 \\ -2 \end{bmatrix} = \begin{bmatrix} 2 \\ 1 \\ 2 \end{bmatrix}$$

$$\lambda = (\lambda_1, \lambda_2, \lambda_3) = (1, 1, 1), d_B = -\sum_{j \in P} \lambda_j h_j = \begin{bmatrix} -2 \\ -1 \\ -2 \end{bmatrix}$$

Then, we calculate the ratios α and β:

$$\beta = \max \left\{ \frac{x_{B[i]}}{-d_{B[i]}} : x_{B[i]} < 0 \right\} = \max \left\{ \frac{x_{B[4]}}{-d_{B[4]}}, \frac{x_{B[6]}}{-d_{B[6]}} \right\} = \max \left\{ \frac{-5}{2}, \frac{-10}{2} \right\} = -5/2$$

$$\alpha = \min \left\{ \frac{x_{B[i]}}{-d_{B[i]}} : d_{B[i]} < 0 \right\} = \min \left\{ \frac{x_{B[4]}}{-d_{B[4]}}, \frac{x_{B[5]}}{-d_{B[5]}}, \frac{x_{B[6]}}{-d_{B[6]}} \right\} = \min \left\{ \frac{-5}{2}, \frac{4}{2}, \frac{-10}{2} \right\} =$$

$$\frac{-10}{2} = -5$$

$P \neq \emptyset$. However, the first direction d_B is not feasible ($\beta > \alpha$), so we apply Phase I.

The leaving variable is x_6:

$$\min \{x_{B[4]}, x_{B[5]}, x_{B[6]}\} = \min \{-5, 4, -10\} = -10$$

We add the artificial variable x_7 and calculate vector d as follows:

$$d = -A_B e = - \begin{bmatrix} -1 & 0 & 0 \\ 0 & 1 & 0 \\ 0 & 0 & -1 \end{bmatrix} \begin{bmatrix} 1 \\ 1 \\ 1 \end{bmatrix} = \begin{bmatrix} 1 \\ -1 \\ 1 \end{bmatrix}$$

Hence, the auxiliary LP problem that will be solved in Phase I is the following:

$$
\begin{aligned}
\min z = & & & & & x_7 \\
\text{s.t.} \quad & 6x_1 + x_2 - 2x_3 - x_4 & & & & + x_7 = 5 \\
& x_1 + x_2 + x_3 & & + x_5 & & - x_7 = 4 \\
& 6x_1 + 4x_2 - 2x_3 & & & - x_6 & + x_7 = 10 \\
& x_j \geq 0, \quad (j = 1, 2, 3, 4, 5, 6, 7)
\end{aligned}
$$

In matrix notation the above LP problem is written as follows:

$$A = \begin{bmatrix} 6 & 1 & -2 & -1 & 0 & 0 & 1 \\ 1 & 1 & 1 & 0 & 1 & 0 & -1 \\ 6 & 4 & -2 & 0 & 0 & -1 & 1 \end{bmatrix}, c = \begin{bmatrix} 0 \\ 0 \\ 0 \\ 0 \\ 0 \\ 0 \\ 1 \end{bmatrix}, b = \begin{bmatrix} 5 \\ 4 \\ 10 \end{bmatrix}, Eqin = \begin{bmatrix} 0 \\ 0 \\ 0 \end{bmatrix}$$

Phase I - 1st Iteration

Apply the revised simplex algorithm in Phase I. The initial basic list of Phase I is $B = [4, 5, 6] \cup [7] \setminus [6] = [4, 5, 7]$ and the nonbasic list is $N = [1, 2, 3] \cup [6] = [1, 2, 3, 6]$. The initial data of the problem is:

$$c_B = \begin{bmatrix} 0 \\ 0 \\ 1 \end{bmatrix}, c_N = \begin{bmatrix} 0 \\ 0 \\ 0 \\ 0 \end{bmatrix}$$

$$A_B = \begin{bmatrix} -1 & 0 & 1 \\ 0 & 1 & -1 \\ 0 & 0 & 1 \end{bmatrix}, A_N = \begin{bmatrix} 6 & 1 & -2 & 0 \\ 1 & 1 & 1 & 0 \\ 6 & 4 & -2 & -1 \end{bmatrix}$$

$$A_B^{-1} = \begin{bmatrix} -1 & 0 & 1 \\ 0 & 1 & 1 \\ 0 & 0 & 1 \end{bmatrix}$$

$$x_B = A_B^{-1}b = \begin{bmatrix} -1 & 0 & 1 \\ 0 & 1 & 1 \\ 0 & 0 & 1 \end{bmatrix}\begin{bmatrix} 5 \\ 4 \\ 10 \end{bmatrix} = \begin{bmatrix} 5 \\ 14 \\ 10 \end{bmatrix}$$

$$w^T = c_B^T A_B^{-1} = \begin{bmatrix} 0 & 0 & 1 \end{bmatrix}\begin{bmatrix} -1 & 0 & 1 \\ 0 & 1 & 1 \\ 0 & 0 & 1 \end{bmatrix} = \begin{bmatrix} 0 & 0 & 1 \end{bmatrix}$$

$$s_N^T = c_N^T - w^T A_N = \begin{bmatrix} 0 & 0 & 0 & 0 \end{bmatrix} - \begin{bmatrix} 0 & 0 & 1 \end{bmatrix}\begin{bmatrix} 6 & 1 & -2 & 0 \\ 1 & 1 & 1 & 0 \\ 6 & 4 & -2 & -1 \end{bmatrix} = \begin{bmatrix} -6 & -4 & 2 & 1 \end{bmatrix}$$

$s_N \not\geq 0$, so the current basis is not optimal. According to Dantzig's rule, the entering variable is x_1, because it is the one with the most negative $\bar{s}_l$ (-6). Hence, $l = 1$.

We continue calculating the pivoting column:

$$h_1 = A_B^{-1}A_{.1} = \begin{bmatrix} -1 & 0 & 1 \\ 0 & 1 & 1 \\ 0 & 0 & 1 \end{bmatrix} \begin{bmatrix} 6 \\ 1 \\ 6 \end{bmatrix} = \begin{bmatrix} 0 \\ 7 \\ 6 \end{bmatrix}$$

There are elements in vector h_1 that are greater than 0, so we perform the minimum ratio test:

$$\min\left\{ \frac{x_{B[i]}}{h_{il}} : h_{il} > 0 \right\} = \min\left\{ x, \frac{x_{B[5]}}{h_{21}}, \frac{x_{B[7]}}{h_{31}} \right\} = \min\left\{ x, \frac{14}{7}, \frac{10}{6} \right\} = \frac{10}{6} = \frac{x_{B[7]}}{h_{31}}$$

The leaving variable is x_7. Hence, $k = 7$ and $r = 3$.
Next, we update the basic and nonbasic lists:

$$B = \begin{bmatrix} 4 & 5 & 1 \end{bmatrix}, N = \begin{bmatrix} 7 & 2 & 3 & 6 \end{bmatrix}$$

Hence:

$$c_B = \begin{bmatrix} 0 \\ 0 \\ 0 \end{bmatrix}, c_N = \begin{bmatrix} 1 \\ 0 \\ 0 \\ 0 \end{bmatrix}$$

$$A_B = \begin{bmatrix} -1 & 0 & 6 \\ 0 & 1 & 1 \\ 0 & 0 & 6 \end{bmatrix}, A_N = \begin{bmatrix} 1 & 1 & -2 & 0 \\ -1 & 1 & 1 & 0 \\ 1 & 4 & -2 & -1 \end{bmatrix}$$

Then, we update the basis inverse using the PFI update scheme:

$$E^{-1} = \begin{bmatrix} 1 & 0 & 0 \\ 0 & 1 & -7/6 \\ 0 & 0 & 7/6 \end{bmatrix}$$

$$(A_{\bar{B}})^{-1} = E^{-1}A_B^{-1} = \begin{bmatrix} 1 & 0 & 0 \\ 0 & 1 & -7/6 \\ 0 & 0 & 7/6 \end{bmatrix} \begin{bmatrix} -1 & 0 & 1 \\ 0 & 1 & 1 \\ 0 & 0 & 1 \end{bmatrix} = \begin{bmatrix} -1 & 0 & 1 \\ 0 & 1 & -1/6 \\ 0 & 0 & 1/6 \end{bmatrix}$$

Finally, we update vectors x_B, w, and s_N:

$$x_B = A_B^{-1}b = \begin{bmatrix} -1 & 0 & 1 \\ 0 & 1 & -1/6 \\ 0 & 0 & 1/6 \end{bmatrix} \begin{bmatrix} 5 \\ 4 \\ 10 \end{bmatrix} = \begin{bmatrix} 5 \\ 7/3 \\ 5/3 \end{bmatrix}$$

$$w^T = c_B^T A_B^{-1} = \begin{bmatrix} 0 & 0 & 0 \end{bmatrix} \begin{bmatrix} -1 & 0 & 1 \\ 0 & 1 & -1/6 \\ 0 & 0 & 1/6 \end{bmatrix} = \begin{bmatrix} 0 & 0 & 0 \end{bmatrix}$$

$$s_N^T = c_N^T - w^T A_N = \begin{bmatrix} 1 & 0 & 0 & 0 \end{bmatrix} - \begin{bmatrix} 0 & 0 & 0 \end{bmatrix} \begin{bmatrix} 1 & 1 & -2 & 0 \\ -1 & 1 & 1 & 0 \\ 1 & 4 & -2 & -1 \end{bmatrix} = \begin{bmatrix} 1 & 0 & 0 & 0 \end{bmatrix}$$

Phase I - 2nd Iteration

The data at the beginning of this iteration is:

$$c_B = \begin{bmatrix} 0 \\ 0 \\ 0 \end{bmatrix}, c_N = \begin{bmatrix} 1 \\ 0 \\ 0 \\ 0 \end{bmatrix}$$

$$A_B = \begin{bmatrix} -1 & 0 & 6 \\ 0 & 1 & 1 \\ 0 & 0 & 6 \end{bmatrix}, A_N = \begin{bmatrix} 1 & 1 & -2 & 0 \\ -1 & 1 & 1 & 0 \\ 1 & 4 & -2 & -1 \end{bmatrix} A_B^{-1} = \begin{bmatrix} -1 & 0 & 1 \\ 0 & 1 & -1/6 \\ 0 & 0 & 1/6 \end{bmatrix}$$

$$x_B = \begin{bmatrix} 5 \\ 7/3 \\ 5/3 \end{bmatrix}, w^T = \begin{bmatrix} 0 & 0 & 0 \end{bmatrix}, s_N^T = \begin{bmatrix} 1 & 0 & 0 & 0 \end{bmatrix}$$

$s_N \geq 0$, so the current basis is optimal for the auxiliary LP problem of Phase I. We found a feasible basis ($B = [4, 5, 1]$) for the initial LP problem. The nonbasic list is $N = [2, 3, 6]$. Hence, we can restore the original LP problem and proceed to Phase II.

Phase II - 3rd Iteration

The data of the problem is:

$$c_B = \begin{bmatrix} 0 \\ 0 \\ 5 \end{bmatrix}, c_N = \begin{bmatrix} 2 \\ -4 \\ 0 \end{bmatrix}$$

$$A_B = \begin{bmatrix} -1 & 0 & 6 \\ 0 & 1 & 1 \\ 0 & 0 & 6 \end{bmatrix}, A_N = \begin{bmatrix} 1 & -2 & 0 \\ 1 & 1 & 0 \\ 4 & -2 & -1 \end{bmatrix}$$

$$A_B^{-1} = \begin{bmatrix} -1 & 0 & 1 \\ 0 & 1 & -1/6 \\ 0 & 0 & 1/6 \end{bmatrix}$$

$$x_B = A_B^{-1}b = \begin{bmatrix} -1 & 0 & 1 \\ 0 & 1 & -1/6 \\ 0 & 0 & 1/6 \end{bmatrix} \begin{bmatrix} 5 \\ 4 \\ 10 \end{bmatrix} = \begin{bmatrix} 5 \\ 7/3 \\ 5/3 \end{bmatrix}$$

$$w^T = c_B^T A_B^{-1} = [0\ 0\ 5] \begin{bmatrix} -1 & 0 & 1 \\ 0 & 1 & -1/6 \\ 0 & 0 & 1/6 \end{bmatrix} = [0\ 0\ 5/6]$$

$$s_N^T = c_N^T - w^T A_N = [2\ -4\ 0] - [0\ 0\ 5/6] \begin{bmatrix} 1 & -2 & 0 \\ 1 & 1 & 0 \\ 4 & -2 & -1 \end{bmatrix} = [-4/3\ -7/3\ 5/6]$$

Then, we find the sets of indices P and Q:

$$P = \{j \in N : s_j < 0\} = [2, 3], Q = N \setminus P = [6]$$

We define an arbitrary vector $\lambda = (\lambda_2, \lambda_3) = (1, 1) > 0$ and compute s_0:

$$s_0 = \sum_{j \in P} \lambda_j s_j = \lambda_2 s_2 + \lambda_3 s_3 = -4/3 - 7/3 = -11/3$$

Then, we calculate the direction d_B:

$$h_j = B^{-1}A_P = \begin{bmatrix} -1 & 0 & 1 \\ 0 & 1 & -1/6 \\ 0 & 0 & 1/6 \end{bmatrix} \begin{bmatrix} 1 & -2 \\ 1 & 1 \\ 4 & -2 \end{bmatrix} = \begin{bmatrix} 3 & 0 \\ 1/3 & 4/3 \\ 2/3 & -1/3 \end{bmatrix}$$

$$d_B = -\sum_{j \in P} \lambda_j h_j = \begin{bmatrix} -(3 + 0) \\ -(1/3 + 4/3) \\ -(2/3 - 1/3) \end{bmatrix} = \begin{bmatrix} -3 \\ -5/3 \\ -1/3 \end{bmatrix}$$

$P \neq \emptyset$, so the current basis is not optimal. $d_B \not\geq 0$, so we continue to find the leaving variable using the following relation:

$$a = \min \left\{ \frac{x_{B[i]}}{-d_{B[i]}} : d_{B[i]} < 0 \right\} = \min \left\{ \frac{x_{B[4]}}{-d_{B[4]}}, \frac{x_{B[5]}}{-d_{B[5]}}, \frac{x_{B[1]}}{-d_{B[1]}} \right\}$$

$$= \min \left\{ \frac{5}{3}, \frac{7/3}{5/3}, \frac{5/3}{1/3} \right\} = \frac{7}{5} = \frac{x_{B[5]}}{-d_{B[5]}}$$

The leaving variable is x_5. Hence, $r = 2$ and $k = 5$.

Then, we compute the row vectors:

$$H_{rP} = \left(A_B^{-1}\right)_{r.} A_P = \begin{bmatrix} 0 & 1 & -1/6 \end{bmatrix} \begin{bmatrix} 1 & -2 \\ 1 & 1 \\ 4 & -2 \end{bmatrix} = \begin{bmatrix} 1/3 & 4/3 \end{bmatrix}$$

$$H_{rQ} = \left(A_B^{-1}\right)_{r.} A_Q \begin{bmatrix} 0 & 1 & -1/6 \end{bmatrix} \begin{bmatrix} 0 \\ 0 \\ -1 \end{bmatrix} = \begin{bmatrix} 1/6 \end{bmatrix}$$

Next, we compute the ratios θ_1 and θ_2:

$$\theta_1 = \frac{-s_P}{H_{rP}} = \min\left\{\frac{-s_j}{H_{rj}} : H_{rj} > 0 \wedge j \in P\right\} = \min\left\{\frac{-s_2}{H_{12}}, \frac{-s_3}{H_{13}}\right\}$$

$$= \min\left\{\frac{4/3}{1/3}, \frac{7/3}{4/3}\right\} = \frac{7}{4} = \frac{-s_3}{H_{13}}$$

$$\theta_2 = \frac{-s_Q}{H_{rQ}} = \min\left\{\frac{-s_j}{H_{rj}} : H_{rj} < 0 \wedge j \in Q\right\} = \infty$$

$\theta_1 \leq \theta_2$, so the entering variable is x_3. Hence, $l = 3$. We update sets P and Q and the basic and nonbasic lists:

$$P = P \setminus l = [2,3] \setminus 3 = [2], Q = Q \cup k = [6] \cup 5 = [6,5]$$

$$B = [4,3,1], N = [2,6,5]$$

Hence:

$$c_B = \begin{bmatrix} 0 \\ -4 \\ 5 \end{bmatrix}, c_N = \begin{bmatrix} 2 \\ 0 \\ 0 \end{bmatrix}$$

$$A_B = \begin{bmatrix} -1 & -2 & 6 \\ 0 & 1 & 1 \\ 0 & -2 & 6 \end{bmatrix}, A_N = \begin{bmatrix} 1 & 0 & 0 \\ 1 & 0 & 1 \\ 4 & -1 & 0 \end{bmatrix}$$

We continue calculating the pivoting column:

$$h_3 = A_B^{-1} A_{.3} = \begin{bmatrix} -1 & 0 & 1 \\ 0 & 1 & -1/6 \\ 0 & 0 & 1/6 \end{bmatrix} \begin{bmatrix} -2 \\ 1 \\ -2 \end{bmatrix} = \begin{bmatrix} 0 \\ 4/3 \\ -1/3 \end{bmatrix}, h_3(2) = -1, h_3 = \begin{bmatrix} 0 \\ -1 \\ -1/3 \end{bmatrix}$$

Then, we update the basis inverse using the PFI update scheme:

$$E^{-1} = \begin{bmatrix} 1 & 0 & 0 \\ 0 & 3/4 & 0 \\ 0 & 1/4 & 1 \end{bmatrix}$$

$$(A_{\overline{B}})^{-1} = E^{-1}A_B^{-1} = \begin{bmatrix} 1 & 0 & 0 \\ 0 & 3/4 & 0 \\ 0 & 1/4 & 1 \end{bmatrix} \begin{bmatrix} -1 & 0 & 1 \\ 0 & 1 & -1/6 \\ 0 & 0 & 1/6 \end{bmatrix} = \begin{bmatrix} -1 & 0 & 1 \\ 0 & 3/4 & -1/8 \\ 0 & 1/4 & 1/8 \end{bmatrix}$$

Finally, we update vectors x_B, w, s_N, and d_B:

$$x_B = A_B^{-1}b = \begin{bmatrix} -1 & 0 & 1 \\ 0 & 3/4 & -1/8 \\ 0 & 1/4 & 1/8 \end{bmatrix} \begin{bmatrix} 5 \\ 4 \\ 10 \end{bmatrix} = \begin{bmatrix} 5 \\ 7/4 \\ 9/4 \end{bmatrix}$$

$$w^T = c_B^T A_B^{-1} = \begin{bmatrix} 0 & -4 & 5 \end{bmatrix} \begin{bmatrix} -1 & 0 & 1 \\ 0 & 3/4 & -1/8 \\ 0 & 1/4 & 1/8 \end{bmatrix} = \begin{bmatrix} 0 & -7/4 & 9/8 \end{bmatrix}$$

$$s_N^T = c_N^T - w^T A_N = \begin{bmatrix} 2 & 0 & 0 \end{bmatrix} - \begin{bmatrix} 0 & -7/4 & 9/8 \end{bmatrix} \begin{bmatrix} 1 & 0 & 0 \\ 1 & 0 & 1 \\ 4 & -1 & 0 \end{bmatrix} = \begin{bmatrix} -3/4 & 7/4 & 9/8 \end{bmatrix}$$

$$d_{\overline{B}} = E^{-1}d_B = \begin{bmatrix} 1 & 0 & 0 \\ 0 & 3/4 & 0 \\ 0 & 1/4 & 1 \end{bmatrix} \begin{bmatrix} -3 \\ -5/3 \\ -1/3 \end{bmatrix} = \begin{bmatrix} -3 \\ -5/4 \\ -3/4 \end{bmatrix},$$

$$d_{\overline{B}[2]} = d_{\overline{B}[2]} + 1 = -5/4 + 1 = -1/4$$

Phase II - 4th Iteration

The data at the beginning of this iteration is:

$$c_B = \begin{bmatrix} 0 \\ -4 \\ 5 \end{bmatrix}, c_N = \begin{bmatrix} 2 \\ 0 \\ 0 \end{bmatrix}$$

$$A_B = \begin{bmatrix} -1 & -2 & 6 \\ 0 & 1 & 1 \\ 0 & -2 & 6 \end{bmatrix}, A_N = \begin{bmatrix} 1 & 0 & 0 \\ 1 & 0 & 1 \\ 4 & -1 & 0 \end{bmatrix}$$

$$A_B^{-1} = \begin{bmatrix} -1 & 0 & 1 \\ 0 & 3/4 & -1/8 \\ 0 & 1/4 & 1/8 \end{bmatrix}$$

$$x_B = \begin{bmatrix} 5 \\ 7/4 \\ 9/4 \end{bmatrix}, w^T = \begin{bmatrix} 0 & -7/4 & 9/8 \end{bmatrix}, s_N^T = \begin{bmatrix} -3/4 & 7/4 & 9/8 \end{bmatrix}, d_B = \begin{bmatrix} -3 \\ -1/4 \\ -3/4 \end{bmatrix}$$

$P \neq \emptyset$, so the current basis is not optimal. $d_B \ngeq 0$, so we continue to find the leaving variable using the following relation:

$$a = \min\left\{ \frac{x_{B[i]}}{-d_{B[i]}} : d_{B[i]} < 0 \right\} = \min\left\{ \frac{x_{B[4]}}{-d_{B[4]}}, \frac{x_{B[3]}}{-d_{B[3]}}, \frac{x_{B[1]}}{-d_{B[1]}} \right\}$$

$$= \min\left\{ \frac{5}{3}, \frac{7/4}{1/4}, \frac{9/4}{3/4} \right\} = \frac{5}{3} = \frac{x_{B[4]}}{-d_{B[4]}}$$

The leaving variable is x_4. Hence, $r = 1$ and $k = 4$.
Then, we compute the row vectors:

$$H_{rP} = \left(A_B^{-1}\right)_{r.} A_P = \begin{bmatrix} -1 & 0 & 1 \end{bmatrix} \begin{bmatrix} 1 \\ 1 \\ 4 \end{bmatrix} = [3]$$

$$H_{rQ} = \left(A_B^{-1}\right)_{r.} A_Q \begin{bmatrix} -1 & 0 & 1 \end{bmatrix} \begin{bmatrix} 0 & 0 \\ 0 & 1 \\ -1 & 0 \end{bmatrix} = \begin{bmatrix} -1 & 0 \end{bmatrix}$$

Next, we compute the ratios θ_1 and θ_2:

$$\theta_1 = \frac{-s_P}{H_{rP}} = \min\left\{ \frac{-s_j}{H_{rj}} : H_{rj} > 0 \wedge j \in P \right\} = \min\left\{ \frac{-s_2}{H_{12}} \right\}$$

$$= \min\left\{ \frac{3/4}{3} \right\} = \frac{3/4}{3} = \frac{-s_2}{H_{12}}$$

$$\theta_2 = \frac{-s_Q}{H_{rQ}} = \min\left\{ \frac{-s_j}{H_{rj}} : H_{rj} < 0 \wedge j \in Q \right\} = \min\left\{ \frac{-s_6}{H_{16}}, x \right\}$$

$$= \min\left\{ \frac{-9/8}{-1}, x \right\} = \frac{9}{8} = \frac{-s_6}{H_{16}}$$

$\theta_1 \leq \theta_2$, so the entering variable is x_2. Hence, $l = 2$. We update sets P and Q and the basic and nonbasic lists:

$$P = P \setminus l = [2] \setminus 2 = \emptyset, Q = Q \cup k = [6,5] \cup 4 = [6,5,4]$$

$$B = [2,3,1], N = [6,5,4]$$

Hence:

$$c_B = \begin{bmatrix} 2 \\ -4 \\ 5 \end{bmatrix}, c_N = \begin{bmatrix} 0 \\ 0 \\ 0 \end{bmatrix}$$

$$A_B = \begin{bmatrix} 1 & -2 & 6 \\ 1 & 1 & 1 \\ 4 & -2 & 6 \end{bmatrix}, A_N = \begin{bmatrix} 0 & 0 & -1 \\ 0 & 1 & 0 \\ -1 & 0 & 0 \end{bmatrix}$$

Then, we update the basis inverse using the PFI update scheme:

$$E^{-1} = \begin{bmatrix} 1/3 & 0 & 0 \\ -1/12 & 1 & 0 \\ -1/4 & 0 & 1 \end{bmatrix}$$

$$(A_{\bar{B}})^{-1} = E^{-1}A_B^{-1} = \begin{bmatrix} 1/3 & 0 & 0 \\ -1/12 & 1 & 0 \\ -1/4 & 0 & 1 \end{bmatrix}\begin{bmatrix} -1 & 0 & 1 \\ 0 & 3/4 & -1/8 \\ 0 & 1/4 & 1/8 \end{bmatrix} = \begin{bmatrix} -1/3 & 0 & 1/3 \\ 1/12 & 3/4 & 15/72 \\ 1/4 & 1/4 & -1/8 \end{bmatrix}$$

Finally, we update vectors x_B, w, s_N, and d_B:

$$x_B = A_B^{-1}b = \begin{bmatrix} -1/3 & 0 & 1/3 \\ 1/12 & 3/4 & 15/72 \\ 1/4 & 1/4 & -1/8 \end{bmatrix}\begin{bmatrix} 5 \\ 4 \\ 10 \end{bmatrix} = \begin{bmatrix} 5/3 \\ 4/3 \\ 1 \end{bmatrix}$$

$$w^T = c_B^T A_B^{-1} = \begin{bmatrix} 2 & -4 & 5 \end{bmatrix}\begin{bmatrix} -1/3 & 0 & 1/3 \\ 1/12 & 3/4 & 15/72 \\ 1/4 & 1/4 & -1/8 \end{bmatrix} = \begin{bmatrix} 1/4 & -7/4 & 7/8 \end{bmatrix}$$

$$s_N^T = c_N^T - w^T A_N = \begin{bmatrix} 0 & 0 & 0 \end{bmatrix} - \begin{bmatrix} 1/4 & -7/4 & 7/8 \end{bmatrix}\begin{bmatrix} 0 & 0 & -1 \\ 0 & 1 & 0 \\ -1 & 0 & 0 \end{bmatrix} = \begin{bmatrix} 7/8 & 7/4 & 1/4 \end{bmatrix}$$

$$d_{\bar{B}} = E^{-1}d_B = \begin{bmatrix} 1/3 & 0 & 0 \\ -1/12 & 1 & 0 \\ -1/4 & 0 & 1 \end{bmatrix}\begin{bmatrix} -3 \\ -1/4 \\ -3/4 \end{bmatrix} = \begin{bmatrix} -1 \\ 0 \\ 0 \end{bmatrix}, d_{\bar{B}[1]} = d_{\bar{B}[1]} + 1 = -1 + 1 = 0$$

Phase II - 5th Iteration

The data at the beginning of this iteration is:

$$c_B = \begin{bmatrix} 2 \\ -4 \\ 5 \end{bmatrix}, c_N = \begin{bmatrix} 0 \\ 0 \\ 0 \end{bmatrix}$$

$$A_B = \begin{bmatrix} 1 & -2 & 6 \\ 1 & 1 & 1 \\ 4 & -2 & 6 \end{bmatrix}, A_N = \begin{bmatrix} 0 & 0 & -1 \\ 0 & 1 & 0 \\ -1 & 0 & 0 \end{bmatrix}$$

$$A_B^{-1} = \begin{bmatrix} -1/3 & 0 & 1/3 \\ 1/12 & 3/4 & 15/72 \\ 1/4 & 1/4 & -1/8 \end{bmatrix}$$

$$x_B = \begin{bmatrix} 5/3 \\ 4/3 \\ 1 \end{bmatrix}, w^T = [1/4 \ -7/4 \ 7/8], s_N^T = [7/8 \ 7/4 \ 1/4]$$

$P = \emptyset$, so the current basis $(B = [2, 3, 1], N = [6, 5, 4])$ is optimal:

$$x_B = \begin{bmatrix} 5/3 \\ 4/3 \\ 1 \end{bmatrix}$$

$$z = c_B^T x_B = 3$$

Example 3 The LP problem that will be solved is the following:

$$
\begin{aligned}
\min z = \quad & -\ x_2 - 2x_3 \\
\text{s.t.} \quad & x_1 + x_2 + x_3 = 5 \\
& -\ 2x_2 - 3x_3 = 3 \\
& x_j \geq 0, \quad (j = 1, 2, 3)
\end{aligned}
$$

In matrix notation the above LP problem is written as follows:

$$A = \begin{bmatrix} 1 & 1 & 1 \\ 0 & -2 & -3 \end{bmatrix}, c = \begin{bmatrix} 0 \\ -1 \\ -2 \end{bmatrix}, b = \begin{bmatrix} 5 \\ 3 \end{bmatrix}, Eqin = \begin{bmatrix} 0 \\ 0 \end{bmatrix}$$

The LP problem is in its standard form. Let's start with the basic list $B = [1, 2]$ and the nonbasic list $N = [3]$. Initially, we check if we must apply the Phase I of the algorithm:

$$c_B = \begin{bmatrix} 0 \\ -1 \end{bmatrix}, c_N = [-2]$$

$$A_B = \begin{bmatrix} 1 & 1 \\ 0 & -2 \end{bmatrix}, A_N = \begin{bmatrix} 1 \\ -3 \end{bmatrix}$$

$$A_B^{-1} = \begin{bmatrix} 1 & 1/2 \\ 0 & -1/2 \end{bmatrix}$$

$$x_B = A_B^{-1} b = \begin{bmatrix} 1 & 1/2 \\ 0 & -1/2 \end{bmatrix} \begin{bmatrix} 5 \\ 3 \end{bmatrix} = \begin{bmatrix} 13/2 \\ -3/2 \end{bmatrix}$$

$$w^T = c_B^T A_B^{-1} = \begin{bmatrix} 0 & -1 \end{bmatrix} \begin{bmatrix} 1 & 1/2 \\ 0 & -1/2 \end{bmatrix} = \begin{bmatrix} 0 & 1/2 \end{bmatrix}$$

$$s_N^T = c_N^T - w^T A_N = \begin{bmatrix} -2 \end{bmatrix} - \begin{bmatrix} 0 & 1/2 \end{bmatrix} \begin{bmatrix} 1 \\ -3 \end{bmatrix} = \begin{bmatrix} -1/2 \end{bmatrix}$$

$$P = \{j \in N : s_j < 0\} = [3], Q = N \setminus P = [1, 2, 3] \setminus [3] = [1, 2]$$

$$h_j = A_B^{-1} A_P = \begin{bmatrix} 1 & 1/2 \\ 0 & -1/2 \end{bmatrix} \begin{bmatrix} 1 \\ -3 \end{bmatrix} = \begin{bmatrix} -1/2 \\ 3/2 \end{bmatrix}$$

$$\lambda = (\lambda_3) = (1) > 0, d_B = -\sum_{j \in P} \lambda_j h_j = \begin{bmatrix} 1/2 \\ -3/2 \end{bmatrix}$$

Then, we calculate the ratios α and β:

$$\beta = \max \left\{ \frac{x_{B[i]}}{-d_{B[i]}} : x_{B[i]} < 0 \right\} = \max \left\{ x, \frac{x_{B[2]}}{-d_{B[2]}} \right\} = \max \left\{ x, \frac{-3/2}{3/2} \right\} = \frac{-1}{1}$$

$$\alpha = \min \left\{ \frac{x_{B[i]}}{-d_{B[i]}} : d_{B[i]} < 0 \right\} = \min \left\{ x, \frac{x_{B[2]}}{-d_{B[2]}} \right\} = \min \left\{ x, \frac{-3/2}{3/2} \right\} = \frac{-1}{1}$$

$P \neq \emptyset$. However, $\beta = \alpha$, so we apply Phase I.
The leaving variable is x_2:

$$\min \{x_{B[1]}, x_{B[2]}\} = \min \{13/2, -3/2\} = -3/2$$

We add the artificial variable x_4 and calculate vector d as follows:

$$d = -A_B e = - \begin{bmatrix} 1 & 1/2 \\ 0 & -1/2 \end{bmatrix} \begin{bmatrix} 1 \\ 1 \end{bmatrix} = \begin{bmatrix} -2 \\ 2 \end{bmatrix}$$

Hence, the auxiliary LP problem that will be solved in Phase I is the following:

$$
\begin{array}{rl}
\min z = & x_4 \\
\text{s.t.} \quad x_1 + \ x_2 \quad x_3 - 2x_4 = 5 & \\
-2x_2 - 3x_3 + 2x_4 = 3 & \\
x_j \geq 0, \quad (j = 1, 2, 3, 4) &
\end{array}
$$

In matrix notation the above LP problem is written as follows:

$$A = \begin{bmatrix} 1 & 1 & 1 & -2 \\ 0 & -2 & -3 & 2 \end{bmatrix}, c = \begin{bmatrix} 0 \\ 0 \\ 0 \\ 1 \end{bmatrix}, b = \begin{bmatrix} 5 \\ 3 \end{bmatrix}, Eqin = \begin{bmatrix} 0 \\ 0 \end{bmatrix}$$

Phase I - 1st Iteration

The initial basic list of Phase I is $B = [1, 2] \cup [4] \setminus [2] = [1, 4]$ and the nonbasic list is $N = [3] \cup [2] = [3, 2]$. The initial data of the problem is:

$$c_B = \begin{bmatrix} 0 \\ 1 \end{bmatrix}, c_N = \begin{bmatrix} 0 \\ 0 \end{bmatrix}$$

$$A_B = \begin{bmatrix} 1 & -2 \\ 0 & 2 \end{bmatrix}, A_N = \begin{bmatrix} 1 & 1 \\ -3 & -2 \end{bmatrix}$$

$$A_B^{-1} = \begin{bmatrix} 1 & 1 \\ 0 & 1/2 \end{bmatrix}$$

$$x_B = A_B^{-1} b = \begin{bmatrix} 1 & 1 \\ 0 & 1/2 \end{bmatrix} \begin{bmatrix} 5 \\ 3 \end{bmatrix} = \begin{bmatrix} 8 \\ 3/2 \end{bmatrix}$$

$$w^T = c_B^T A_B^{-1} = \begin{bmatrix} 0 & 1 \end{bmatrix} \begin{bmatrix} 1 & 1 \\ 0 & 1/2 \end{bmatrix} = \begin{bmatrix} 0 & 1/2 \end{bmatrix}$$

$$s_N^T = c_N^T - w^T A_N = \begin{bmatrix} 0 & 0 \end{bmatrix} - \begin{bmatrix} 0 & 1/2 \end{bmatrix} \begin{bmatrix} 1 & 1 \\ -3 & -2 \end{bmatrix} = \begin{bmatrix} 3/2 & 1 \end{bmatrix}$$

$s_N \geq 0$, but the artificial variable x_4 is in the basic list and $x_{B[4]} = 3/2 > 0$, so the LP problem is infeasible.

Example 4 The LP problem that will be solved is the following:

$$\begin{array}{rl} \min z = & 3x_1 - 2x_2 \\ \text{s.t.} & 2x_1 + 4x_2 \geq 8 \\ & x_1 + x_2 \geq 3 \\ & -2x_1 - 7x_2 \leq -15 \\ x_j \geq 0, & (j = 1, 2) \end{array}$$

In matrix notation the above LP problem is written as follows:

$$A = \begin{bmatrix} 2 & 4 \\ 1 & 1 \\ -2 & -7 \end{bmatrix}, c = \begin{bmatrix} 3 \\ -2 \end{bmatrix}, b = \begin{bmatrix} 8 \\ 3 \\ -15 \end{bmatrix}, Eqin = \begin{bmatrix} 1 \\ 1 \\ -1 \end{bmatrix}$$

Initially, we need to convert the LP problem in its standard form:

$$
\begin{aligned}
\min z = \quad & 3x_1 - 2x_2 \\
\text{s.t.} \quad & 2x_1 + 4x_2 - x_3 &&&= 8 \\
& x_1 + x_2 && - x_4 && = 3 \\
& -2x_1 - 7x_2 && + x_5 && = -15 \\
& x_j \geq 0, \quad (j = 1,2,3,4,5)
\end{aligned}
$$

In matrix notation the above LP problem is written as follows:

$$
A = \begin{bmatrix} 2 & 4 & -1 & 0 & 0 \\ 1 & 1 & 0 & -1 & 0 \\ -2 & -7 & 0 & 0 & 1 \end{bmatrix}, c = \begin{bmatrix} 3 \\ -2 \\ 0 \\ 0 \\ 0 \end{bmatrix}, b = \begin{bmatrix} 8 \\ 3 \\ -15 \end{bmatrix}, Eqin = \begin{bmatrix} 0 \\ 0 \\ 0 \end{bmatrix}
$$

Let's start with the basic list $B = [3, 4, 5]$ and the nonbasic list $N = [1, 2]$. Initially, we check if we must apply the Phase I of the algorithm:

$$
c_B = \begin{bmatrix} 0 \\ 0 \\ 0 \end{bmatrix}, c_N = \begin{bmatrix} -3 \\ 2 \end{bmatrix}
$$

$$
A_B = \begin{bmatrix} -1 & 0 & 0 \\ 0 & -1 & 0 \\ 0 & 0 & 1 \end{bmatrix}, A_N = \begin{bmatrix} 2 & 4 \\ 1 & 1 \\ -2 & -7 \end{bmatrix}
$$

$$
A_B^{-1} = \begin{bmatrix} -1 & 0 & 0 \\ 0 & -1 & 0 \\ 0 & 0 & 1 \end{bmatrix}
$$

$$
x_B = A_B^{-1}b = \begin{bmatrix} -1 & 0 & 0 \\ 0 & -1 & 0 \\ 0 & 0 & 1 \end{bmatrix}\begin{bmatrix} 8 \\ 3 \\ -15 \end{bmatrix} = \begin{bmatrix} -8 \\ -3 \\ -15 \end{bmatrix}
$$

$$
w^T = c_B^T A_B^{-1} = [0\ 0\ 0]\begin{bmatrix} -1 & 0 & 0 \\ 0 & -1 & 0 \\ 0 & 0 & 1 \end{bmatrix} = [0\ 0\ 0]
$$

$$
s_N^T = c_N^T - w^T A_N = [-3\ 2] - [0\ 0\ 0]\begin{bmatrix} 2 & 4 \\ 1 & 1 \\ -2 & -7 \end{bmatrix} = [-3\ 2]
$$

$$
P = \{ j \in N : s_j < 0 \} = [1], Q = N \setminus P = [1, 2] \setminus [1] = [2]
$$

$$h_j = A_B^{-1} A_P = \begin{bmatrix} -1 & 0 & 0 \\ 0 & -1 & 0 \\ 0 & 0 & 1 \end{bmatrix} \begin{bmatrix} 2 \\ 1 \\ -2 \end{bmatrix} = \begin{bmatrix} -2 \\ -1 \\ -2 \end{bmatrix}$$

$$\lambda = (\lambda_1) = (1) > 0, d_B = -\sum_{j \in P} \lambda_j h_j = \begin{bmatrix} 2 \\ 1 \\ 2 \end{bmatrix}$$

Then, we calculate the ratios α and β:

$$\beta = \max \left\{ \frac{x_{B[i]}}{-d_{B[i]}} : x_{B[i]} < 0 \right\} = \max \left\{ \frac{x_{B[3]}}{-d_{B[3]}}, \frac{x_{B[4]}}{-d_{B[4]}}, \frac{x_{B[5]}}{-d_{B[5]}} \right\}$$

$$= \max \left\{ \frac{8}{2}, \frac{3}{1}, \frac{15}{2} \right\} = \frac{15}{2}$$

$$\alpha = \min \left\{ \frac{x_{B[i]}}{-d_{B[i]}} : d_{B[i]} < 0 \right\} = \infty$$

$P \neq \emptyset$ and $\beta < \alpha$, so we can proceed with Phase II.

Phase II - 1st Iteration

The initial data of the problem is:

$$c_B = \begin{bmatrix} 0 \\ 0 \\ 0 \end{bmatrix}, c_N = \begin{bmatrix} -3 \\ 2 \end{bmatrix}$$

$$A_B = \begin{bmatrix} -1 & 0 & 0 \\ 0 & -1 & 0 \\ 0 & 0 & 1 \end{bmatrix}, A_N = \begin{bmatrix} 2 & 4 \\ 1 & 1 \\ -2 & -7 \end{bmatrix}$$

$$A_B^{-1} = \begin{bmatrix} -1 & 0 & 0 \\ 0 & -1 & 0 \\ 0 & 0 & 1 \end{bmatrix}$$

$$x_B = A_B^{-1} b = \begin{bmatrix} -1 & 0 & 0 \\ 0 & -1 & 0 \\ 0 & 0 & 1 \end{bmatrix} \begin{bmatrix} 8 \\ 3 \\ -15 \end{bmatrix} = \begin{bmatrix} -8 \\ -3 \\ -15 \end{bmatrix}$$

$$w^T = c_B^T A_B^{-1} = [0 \ 0 \ 0] \begin{bmatrix} -1 & 0 & 0 \\ 0 & -1 & 0 \\ 0 & 0 & 1 \end{bmatrix} = [0 \ 0 \ 0]$$

$$s_N^T = c_N^T - w^T A_N = \begin{bmatrix} -3 & 2 \end{bmatrix} - \begin{bmatrix} 0 & 0 & 0 \end{bmatrix} \begin{bmatrix} 2 & 4 \\ 1 & 1 \\ -2 & -7 \end{bmatrix} = \begin{bmatrix} -3 & 2 \end{bmatrix}$$

Then, we find the sets of indices P and Q:

$$P = \{j \in N : s_j < 0\} = [1], Q = N \setminus P = [2]$$

Next, we compute s_0:

$$s_0 = \sum_{j \in P} \lambda_j s_j = \lambda_1 s_1 = -3$$

Then, we calculate the direction d_B:

$$h_j = B^{-1} A_P = \begin{bmatrix} -1 & 0 & 0 \\ 0 & -1 & 0 \\ 0 & 0 & 1 \end{bmatrix} \begin{bmatrix} 2 \\ 1 \\ -2 \end{bmatrix} = \begin{bmatrix} -2 \\ -1 \\ -2 \end{bmatrix}$$

$$d_B = -\sum_{j \in P} \lambda_j h_j = \begin{bmatrix} -(-2) \\ -(-1) \\ -(-2) \end{bmatrix} = \begin{bmatrix} 2 \\ 1 \\ 2 \end{bmatrix}$$

$P \neq \emptyset$, so the current basis is not optimal. $d_B \geq 0$ and $s_0 = -3$, so the LP problem is unbounded.

10.5   Implementation in MATLAB

This subsection presents the implementation in MATLAB of the exterior point simplex algorithm. Some necessary notations should be introduced before the presentation of the implementation of the exterior point simplex algorithm. Let A be a $m \times n$ matrix with the coefficients of the constraints, c be a $n \times 1$ vector with the coefficients of the objective function, b be a $m \times 1$ vector with the right-hand side of the constraints, *Eqin* be a $m \times 1$ vector with the type of the constraints, *MinMaxLP* be the type of optimization, $c0$ be the constant term of the objective function, *reinv* be the number of iterations that the basis inverse is recomputed from scratch, *tole1* be the tolerance for the basic solution, *tole2* be the tolerance for the reduced costs, *tole3* be the tolerance for the pivoting column, *scalingTechnique* be the scaling method (0: no scaling, 1: arithmetic mean, 2: de Buchet for the case p = 1, 3: de Buchet for the case p = 2, 4: de Buchet for the case p = Inf, 5: entropy, 6: equilibration, 7: geometric mean, 8: IBM MPSX, 9: LP-norm for the case p = 1, 10: LP-norm for the case p = 2, 11: LP-norm for the case p = Inf), *basisUpdateMethod* be the basis update method (1: PFI, 2: MPFI), *xsol* be a $m \times 1$ vector with the solution found, $fval$ be the value of the objective function at the solution *xsol*, *exitflag* be the reason that the algorithm terminated (0: optimal solution found, 1: the LP problem is infeasible, 2: the LP problem is unbounded, -1: the input data is not logically or numerically correct), and *iterations* be the number of iterations.

The exterior point simplex algorithm is implemented with two functions: (i) one function that implements the exterior point simplex algorithm (filename: `epsa.m`), and (ii) one function that finds an initial invertible basis (filename: `lprref.m`).

The function for the implementation of the exterior point simplex algorithm takes as input the matrix with the coefficients of the constraints (matrix A), the vector with the coefficients of the objective function (vector c), the vector with the right-hand side of the constraints (vector b), the vector with the type of the constraints (vector *Eqin*), the type of optimization (variable *MinMaxLP*), the constant term of the objective function (variable $c0$), the number of iterations that the basis inverse is recomputed from scratch (variable *reinv*), the tolerance of the basic solution (variable *tole1*), the tolerance for the reduced costs (variable *tole2*), the tolerance for the pivoting column (variable *tole3*), the scaling method (variable *scalingTechnique*), and the basis update method (variable *basisUpdateMethod*), and returns as output the solution found (vector *xsol*), the value of the objective function at the solution *xsol* (variable $fval$), the reason that the algorithm terminated (variable *exitflag*), and the number of iterations (variable *iterations*).

In lines 57–60, the output variables are initialized. In lines 62–88, we set default values to the missing inputs (if any). Next, we check if the input data is logically correct (lines 94–125). If the type of optimization is maximization, then we multiply vector c and constant $c0$ by -1 (lines 128–131). Then, the presolve analysis is performed (lines 134–145). Next, we scale the LP problem using the selected scaling technique (lines 152–180). Then, we find an invertible initial basis using the *lprref* function (lines 183–242). If the density of matrix A is less than 20%, then we use sparse storage for faster computations (lines 249–253). Then, we pre-allocate memory for variables that hold large amounts of data and initialize the necessary variables (lines 256–293). Next, we check if we can avoid Phase I (lines 294–341).

Phase I is performed in lines 343–544. Initially, we find the index of the minimum element of vector x_B and create a copy of vector c in order to add the artificial variable (lines 346–349). We also calculate vector d and redefine matrix A since the artificial variable needs to be added for Phase I (lines 351–354). Then, we compute the new basic and nonbasic lists and initialize the necessary variables (lines 356–390). At each iteration of Phase I, we perform the optimality test in order to find if the basis found is feasible for the original LP problem or if the original LP problem is infeasible (lines 397–460). If the auxiliary LP problem is not optimal, then we find the entering variable using Dantzig's rule, find the leaving variable using the minimum ratio test, and update the basis inverse and the necessary variables (lines 461–542).

Phase II is performed in lines 545–809. Initially, we initialize the necessary variables (lines 553–607). At each iteration of Phase II, we perform the optimality test in order to find if the LP problem is optimal (lines 611–648). If the LP problem

is not optimal, then we find the leaving and the entering variable and update
the basis inverse and the necessary variables (lines 650–793). Moreover, we print
intermediate results every 100 iterations (lines 796–807).

```
1.    function [xsol, fval, exitflag, iterations] = ...
2.        epsa(A, c, b, Eqin, MinMaxLP, c0, reinv, tole1, ...
3.        tole2, tole3, scalingTechnique, basisUpdateMethod)
4.    % Filename: epsa.m
5.    % Description: the function is an implementation of the
6.    % primal exterior point simplex algorithm
7.    % Authors: Ploskas, N., Samaras, N., & Triantafyllidis, Ch.
8.    %
9.    % Syntax: [xsol, fval, exitflag, iterations] = ...
10.   %    epsa(A, c, b, Eqin, MinMaxLP, c0, reinv, tole1, ...
11.   %    tole2, tole3, scalingTechnique, basisUpdateMethod)
12.   %
13.   % Input:
14.   % -- A: matrix of coefficients of the constraints
15.   %    (size m x n)
16.   % -- c: vector of coefficients of the objective function
17.   %    (size n x 1)
18.   % -- b: vector of the right-hand side of the constraints
19.   %    (size m x 1)
20.   % -- Eqin: vector of the type of the constraints
21.   %    (size m x 1)
22.   % -- MinMaxLP: the type of optimization (optional:
23.   %    default value -1 - minimization)
24.   % -- c0: constant term of the objective function
25.   %    (optional: default value 0)
26.   % -- reinv: every reinv number of iterations, the basis
27.   %    inverse is re-computed from scratch (optional:
28.   %    default value 80)
29.   % -- tole1: tolerance for the basic solution (optional:
30.   %    default value 1e-07)
31.   % -- tole2: tolerance for the reduced costs (optional:
32.   %    default value 1e-09)
33.   % -- tole3: tolerance for the pivot column (optional:
34.   %    default value 1e-09)
35.   % -- scalingTechnique: the scaling method to be used
36.   %    (0: no scaling, 1: arithmetic mean, 2: de Buchet for
37.   %    the case p = 1, 3: de Buchet for the case p = 2,
38.   %    4: de Buchet for the case p = Inf, 5: entropy,
39.   %    6: equilibration, 7: geometric mean, 8: IBM MPSX,
40.   %    9: LP-norm for the case p = 1, 10: LP-norm for
41.   %    the case p = 2, 11: LP-norm for the case p = Inf)
42.   %    (optional: default value 6)
43.   % -- basisUpdateMethod: the basis update method to be used
44.   %    (1: PFI, 2: MPFI) (optional: default value 1)
45.   %
46.   % Output:
47.   % -- xsol: the solution found (size m x 1)
48.   % -- fval: the value of the objective function at the
49.   %    solution x
50.   % -- exitflag: the reason that the algorithm terminated
```

```
51.  %      (0: optimal solution found, 1: the LP problem is
52.  %      infeasible, 2: the LP problem is unbounded, -1: the
53.  %      input data is not logically or numerically correct)
54.  % -- iterations: the number of iterations
55.
56.  % initialize output variables
57.  xsol = [];
58.  fval = 0;
59.  exitflag = 0;
60.  iterations = 0;
61.  % set default values to missing inputs
62.  if ~exist('MinMaxLP')
63.      MinMaxLP = -1;
64.  end
65.  if ~exist('c0')
66.      c0 = 0;
67.  end
68.  if ~exist('reinv')
69.      reinv = 80;
70.  end
71.  if ~exist('tole1')
72.      tole1 = 1e-7;
73.  end
74.  if ~exist('tole2')
75.      tole2 = 1e-9;
76.  end
77.  if ~exist('tole3')
78.      tole3 = 1e-9;
79.  end
80.  if ~exist('scalingTechnique')
81.      scalingTechnique = 6;
82.  end
83.  if ~exist('pivotingRule')
84.      pivotingRule = 2;
85.  end
86.  if ~exist('basisUpdateMethod')
87.      basisUpdateMethod = 1;
88.  end
89.  [m, n] = size(A); % find the size of matrix A
90.  [m2, n2] = size(c); % find the size of vector c
91.  [m3, n3] = size(Eqin); % find the size of vector Eqin
92.  [m4, n4] = size(b); % find the size of vector b
93.  % check if input data is logically correct
94.  if n2 ~= 1
95.      disp('Vector c is not a column vector.')
96.      exitflag = -1;
97.      return
98.  end
99.  if n ~= m2
100.     disp(['The number of columns in matrix A and ' ...
101.          'the number of rows in vector c do not match.'])
102.     exitflag = -1;
103.     return
104. end
```

```
105. if m4 ~= m
106.     disp(['The number of the right-hand side values ' ...
107.          'is not equal to the number of constraints.'])
108.     exitflag = -1;
109.     return
110. end
111. if n3 ~= 1
112.     disp('Vector Eqin is not a column vector')
113.     exitflag = -1;
114.     return
115. end
116. if n4 ~= 1
117.     disp('Vector b is not a column vector')
118.     exitflag = -1;
119.     return
120. end
121. if m4 ~= m3
122.     disp('The size of vectors Eqin and b does not match')
123.     exitflag = -1;
124.     return
125. end
126. % if the type of optimization is maximization, then multiply
127. % vector c and constant c0 by -1
128. if MinMaxLP == 1
129.     c = -c;
130.     c0 = -c0;
131. end
132. % perform the presolve analysis
133. disp('---- P R E S O L V E     A N A L Y S I S ----')
134. [A, c, b, Eqin, c0, infeasible, unbounded] = ...
135.     presolve(A, c, b, Eqin, c0);
136. if infeasible == 1 % the LP problem is infeasible
137.     disp('The LP problem is infeasible')
138.     exitflag = 1;
139.     return
140. end
141. if unbounded == 1 % the LP problem is unbounded
142.     disp('The LP problem is unbounded')
143.     exitflag = 2;
144.     return
145. end
146. [m, n] = size(A); % find the new size of matrix A
147. [m2, ~] = size(c); % find the new size of vector c
148. [m3, ~] = size(Eqin); % find the size of vector Eqin
149. % scale the LP problem using the selected scaling
150. % technique
151. disp('---- S C A L I N G ----')
152. if scalingTechnique == 1 % arithmetic mean
153.     [A, c, b, ~, ~] = arithmeticMean(A, c, b);
154. % de buchet for the case p = 1
155. elseif scalingTechnique == 2
156.     [A, c, b, ~, ~] = debuchet1(A, c, b);
157. % de buchet for the case p = 2
158. elseif scalingTechnique == 3
```

```
159.      [A, c, b, ~, ~] = debuchet2(A, c, b);
160. % de buchet for the case p = Inf
161. elseif scalingTechnique == 4
162.      [A, c, b, ~, ~] = debuchetinf(A, c, b);
163. elseif scalingTechnique == 5 % entropy
164.      [A, c, b, ~, ~] = entropy(A, c, b);
165. elseif scalingTechnique == 6 % equilibration
166.      [A, c, b, ~, ~] = equilibration(A, c, b);
167. elseif scalingTechnique == 7 % geometricMean
168.      [A, c, b, ~, ~] = geometricMean(A, c, b);
169. elseif scalingTechnique == 8 % IBM MPSX
170.      [A, c, b, ~, ~, ~, ~] = ibmmpsx(A, c, b);
171. % LP-norm for the case p = 1
172. elseif scalingTechnique == 9
173.      [A, c, b, ~, ~] = lpnorm1(A, c, b);
174. % LP-norm for the case p = 2
175. elseif scalingTechnique == 10
176.      [A, c, b, ~, ~] = lpnorm2(A, c, b);
177. % LP-norm for the case p = Inf
178. elseif scalingTechnique == 11
179.      [A, c, b, ~, ~] = lpnorminf(A, c, b);
180. end
181. % find an invertible basis
182. disp('---- I N I T I A L    B A S I S ----')
183. flag = isequal(Eqin, zeros(m3, 1));
184. if flag == 1 % all constraints are equalities
185.      % select an initial invertible basis using lprref
186.      % function
187.      [~, ~, jb, out, ~, exitflag] = lprref(A, ...
188.          b, Eqin, 1e-10);
189.      if exitflag == 1 % the LP problem is infeasible
190.          disp('The LP problem is infeasible')
191.          return
192.      end
193.      % create the basic and nonbasic lists
194.      BasicList = jb;
195.      NonBasicList = setdiff(1:n, BasicList);
196.      % delete redundant constraints found by lprref
197.      A(out, :) = [];
198.      b(out, :) = [];
199.      Eqin(out, :) = [];
200. else % some or all constraints are inequalities
201.      % add slack variables
202.      axm = nnz(Eqin);
203.      c(m2 + 1:m2 + axm, :) = 0;
204.      A(:, n + 1:n + axm) = zeros(m, axm);
205.      curcol = 1;
206.      for i = 1:m3
207.          % 'greater than or equal to' inequality constraint
208.          if Eqin(i, 1) == 1
209.              A(i, n + curcol) = -1;
210.              curcol = curcol + 1;
211.          % 'less than or equal to' inequality constraint
212.          elseif Eqin(i, 1) == -1
```

```
213.              A(i, n + curcol) = 1;
214.              curcol = curcol + 1;
215.         % unrecognized type of constraint
216.         elseif Eqin(i,1) ~= 0
217.             disp('Vector Eqin is not numerically correct.')
218.             exitflag = -1;
219.             return
220.         end
221.    end
222.    % select an initial invertible basis using lprref
223.    % function
224.    [~, ~, jb, out, ~, exitflag] = lprref(A, ...
225.         b, Eqin, 1e-10);
226.    if exitflag == 1 % the LP problem is infeasible
227.         disp('The LP problem is infeasible')
228.         return
229.    end
230.    % create the basic and nonbasic lists
231.    [~, y1] = size(A);
232.    temp = n + 1:y1;
233.    for i = 1:length(temp)
234.         jb(length(jb) + 1) = temp(i);
235.    end
236.    BasicList = sort(jb);
237.    NonBasicList = setdiff(1:y1, BasicList);
238.    % delete redundant constraints found by lprref
239.    A(out, :) = [];
240.    b(out, :) = [];
241.    Eqin(out, :) = [];
242. end
243. flag = 0;
244. [m1, n1] = size(A); % new size of matrix A
245. % calculate the density of matrix A
246. density = (nnz(A) / (m1 * n1)) * 100;
247. % if the density of matrix A is less than 20%, then
248. % use sparse algebra for faster computations
249. if density < 20
250.    A = sparse(A);
251.    c = sparse(c);
252.    b = sparse(b);
253. end
254. % preallocate memory for variables that hold large
255. % amounts of data
256. Xb = spalloc(m1, 1, m1); % basic solution
257. h_l = spalloc(m1, 1, m1); % pivot column
258. optTest = spalloc(m1, 1, m1); % optimality test vector
259. % the matrix of the basic variables
260. Basis = spalloc(m1, length(BasicList), m1 ...
261.     * length(BasicList));
262. % the matrix of the nonbasic variables
263. N = spalloc(m1, length(NonBasicList), m1 ...
264.     * length(NonBasicList));
265. w = spalloc(1, n1, n1); % simplex multiplier
266. Sn = spalloc(1, n1, n1); % reduced costs
```

```
267. % initialize data
268. % the matrix of the basic variables
269. Basis = A(:, BasicList);
270. % the coefficients of the objective function for the
271. % basic variables
272. cb = c(BasicList);
273. % the matrix of the nonbasic variables
274. N = A(:, NonBasicList);
275. % the coefficients of the objective function for the
276. % nonbasic variables
277. cn = c(NonBasicList);
278. % calculate the basis inverse
279. BasisInv = inv(Basis);
280. Xb = BasisInv * b; % basic solution
281. % set to zero, the values of Xb that are less than or
282. % equal to tole1
283. toler = abs(Xb) <= tole1;
284. Xb(toler == 1) = 0;
285. w = cb' * BasisInv; % simplex multiplier
286. Sn = sparse(cn' - w * N); % reduced costs
287. % set to zero, the values of Xb that are less than or
288. % equal to tole2
289. toler = abs(Sn) <= tole2;
290. Sn(toler == 1) = 0;
291. % check if the direction crosses the feasible region
292. % to avoid Phase I
293. P = NonBasicList(find(Sn < 0));
294. if ~isempty(P)
295.     h_j = BasisInv * A(:, P);
296.     dB = -(sum(h_j, 2));
297.     % set to zero, the values of dB that are less than
298.     % or equal to tole3
299.     toler = (abs(dB)) <= tole3;
300.     dB(toler == 1) = 0;
301.     % calculate ratio a
302.     mrtarat = find(dB < 0);
303.     if isempty(mrtarat)
304.         arat = inf;
305.     else
306.         mrtarat2 = find(-dB([mrtarat]) ~= 0);
307.         mrtarat = mrtarat(mrtarat2);
308.         aratios = Xb([mrtarat]) ./ (-dB([mrtarat]));
309.         % set to zero, the values of aratios that are less
310.         % than or equal to tole1
311.         toler = (abs(aratios)) <= tole1;
312.         aratios(toler==1) = 0;
313.         arat = min(aratios);
314.     end
315.     % calculate ratio b
316.     mrtbrat = find(Xb < 0);
317.     if isempty(mrtbrat)
318.         brat = -inf;
319.     else
320.         mrtbrat2 =find(-dB([mrtbrat]) ~=0 );
```

```
321.          mrtbrat = mrtbrat(mrtbrat2);
322.          bratios = Xb([mrtbrat]) ./ (-dB([mrtbrat]));
323.          % set to zero, the values of bratios that are less
324.          % than or equal to tole1
325.          toler = (abs(bratios)) <= tole1;
326.          bratios(toler==1) = 0;
327.          idinf = find(bratios == -inf);
328.          if ~isempty(idinf)
329.              brat = inf;
330.          else
331.              brat = max(bratios);
332.          end
333.    end
334. end
335. if isempty(P)
336.     flag = 0;
337. elseif brat < arat % proceed to Phase II
338.     flag = 1;
339. else
340.     flag = 0;
341. end
342. counter = 1;
343. if flag == 0 % Phase I
344.     disp('---- P H A S E    I ----')
345.     % find the index of the minimum element of vector Xb
346.     [~, minindex] = min(Xb);
347.     % create a copy of vector c for the Phase I
348.     c2 = sparse(zeros(length(c), 1));
349.     c2(length(c) + 1) = 1;
350.     % calculate vector d
351.     d = sparse(-(Basis) * ones(m1, 1));
352.     % re-define matrix A since artificial variable needs
353.     % to be added for fase 1
354.     A(:, n1 + 1) = d;
355.     % compute the new basic and nonbasic variables
356.     NonBasicList(1, length(NonBasicList) + 1) = ...
357.             BasicList(1, minindex);
358.     BasicList(:, minindex) = n1 + 1;
359.     [~, n] = size(A); % the new size of matrix A
360.     flagfase1 = 0;
361.     Basis = A(:, BasicList); % the new basis
362.     % the new coefficients of the objective function for
363.     % the basic variables
364.     fb = c2(BasicList);
365.     % the matrix of the nonbasic variables
366.     N = A(:, NonBasicList);
367.     % the new coefficients of the objective function for
368.     % the nonbasic variables
369.     fn = c2(NonBasicList);
370.     % re-compute the inverse of the matrix
371.     BasisInv = inv(Basis);
372.     % compute the basic solution
373.     Xb = BasisInv * b;
374.     % set to zero, the values of Xb that are less than or
```

```
375.     % equal to tole1
376.     toler = abs(Xb) <= tole1;
377.     Xb(toler == 1) = 0;
378.     % compute the simplex multiplier
379.     w = fb' * BasisInv;
380.     HRN = w * N;
381.     % set to zero, the values of HRN that are less than or
382.     % equal to tole2
383.     toler = abs(HRN) <= tole2;
384.     HRN(toler == 1) = 0;
385.     % compute the reduced costs
386.     Sn = fn' - HRN;
387.     % set to zero, the values of Sn that are less than or
388.     % equal to tole2
389.     toler = abs(Sn) <= tole2;
390.     Sn(toler == 1) = 0;
391.     while flagfase1 == 0 % iterate in Phase I
392.         clear rr;
393.         clear mrt;
394.         clear h_l;
395.         [~, col] = find(NonBasicList == n);
396.         % optimality test for Phase I
397.         if all(Sn >= 0) || ~isempty(col)
398.             % if the artificial variable is leaving from
399.             % the basic list, Phase I must be terminated
400.             % and Phase II must start.
401.             % The basis produced is feasible for Phase II
402.             if ~isempty(col)
403.                 NonBasicList(col) = [];
404.                 A(:, n) = [];
405.                 flag = 1;
406.                 iterations = iterations + 1;
407.                 break
408.         else
409.                 % If the artificial variable is in the
410.                 % basic list but Sn >= 0, the artificial
411.                 % variable is leaving and as an entering
412.                 % one, we pick a random but acceptable
413.                 % variable.
414.                 % The basis produced is feasible for
415.                 % Phase II
416.                 [~, col] = find(BasicList == n);
417.                 if ~isempty(col)
418.                     if Xb(col) == 0
419.                         A(:, n) = [];
420.                         for i = 1:length(NonBasicList)
421.                             l = NonBasicList(i);
422.                             h_l = BasisInv * A(:, l);
423.                             % set to zero, the values of
424.                             % h_l that are less than or
425.                             % equal to tole3
426.                             toler = abs(h_l) <= tole3;
427.                             h_l(toler == 1) = 0;
428.                             if h_l(col) ~= 0
```

```
429.                                  % update the basic and
430.                                  % nonbasic list
431.                                  BasicList(col) = 1;
432.                                  NonBasicList(i) = [];
433.                                  % re-compute the basis
434.                                  % inverse
435.                                  BasisInv = inv(A(:, ...
436.                                      BasicList));
437.                                  % re-compute the
438.                                  % reduced costs
439.                                  Sn = c(NonBasicList)' - ...
440.                                      c(BasicList)' * BasisInv ...
441.                                      * A(:, NonBasicList);
442.                                  % set to zero, the values of
443.                                  % Sn that are less than or
444.                                  % equal to tole2
445.                                  toler = abs(Sn) <= tole2;
446.                                  Sn(toler == 1) = 0;
447.                                  flag = 1;
448.                                  iterations = iterations + 1;
449.                                  break
450.                              end
451.                          end
452.                          break
453.                      else % the LP problem is infeasible
454.                          disp('The LP problem is infeasible')
455.                          exitflag = 1;
456.                          iterations = iterations + 1;
457.                          return
458.                      end
459.                  end
460.              end
461.          else % the LP problem of Phase I is not optimal
462.              % find the entering variable using Dantzig's
463.              % rule
464.              [~, t] = min(Sn);
465.              % index of the entering variable
466.              l = NonBasicList(t);
467.              % calculate the pivoting column
468.              h_l = BasisInv * A(:, l);
469.              % set to zero, the values of h_l that are
470.              % less than or equal to tole3
471.              toler = abs(h_l) <= tole3;
472.              h_l(toler == 1) = 0;
473.              optTest = h_l > 0;
474.              % perform the minimum ratio test to find
475.              % the leaving variable
476.              if sum(optTest) > 0
477.                  mrt = find(h_l > 0);
478.                  Xb_hl_div = Xb(mrt) ./ h_l(mrt);
479.                  [p, r] = min(Xb_hl_div);
480.                  rr = find(Xb_hl_div == p);
481.                  % avoid stalling because of ties
482.                  if length(rr) > 1
```

```
483.                          r = mrt(rr);
484.                          % index of the leaving variable
485.                          k = BasicList(r);
486.                          [k, ~] = max(k);
487.                          r = find(BasicList == k);
488.                      else
489.                          r = mrt(r);
490.                          % index of the leaving variable
491.                          k = BasicList(r);
492.                      end
493.                      % pivoting and update vectors and matrices
494.                      a = Sn(t);
495.                      f = Xb(r);
496.                      % update the basic list
497.                      BasicList(r) = 1;
498.                      g = h_1(r); % the pivot element
499.                      % update the nonbasic list
500.                      NonBasicList(t) = k;
501.                      % update the matrix of the nonbasic variables
502.                      N = A(:, NonBasicList);
503.                      iterations = iterations + 1;
504.                      % calculate the new basic solution
505.                      Xb(r) = 0;
506.                      h_12 = h_1;
507.                      h_12(r) = -1;
508.                      Xb = Xb - (f / g) * h_12;
509.                      % set to zero, the values of Xb that
510.                      % are less than or equal to tole1
511.                      toler = abs(Xb) <= tole1;
512.                      Xb(toler == 1) = 0;
513.                      % update the reduced costs
514.                      Sn(t) = 0;
515.                      HRN = BasisInv(r, :) * N;
516.                      % set to zero, the values of HRN that
517.                      % are less than or equal to tole2
518.                      toler = abs(HRN) <= tole2;
519.                      HRN(toler == 1) = 0;
520.                      Sn = Sn - (a / g) * HRN;
521.                      % set to zero, the values of Sn that
522.                      % are less than or equal to tole2
523.                      toler = abs(Sn) <= tole2;
524.                      Sn(toler == 1) = 0;
525.                      % basis inverse
526.                      if iterations == counter * reinv
527.                          % recompute the inverse of the basis
528.                          % from scratch every reinv iterations
529.                          BasisInv = inv(A(:, BasicList));
530.                          counter = counter + 1;
531.                          h_1(r) = g;
532.                      else % basis update according to the
533.                          % selected basis update method
534.                          if basisUpdateMethod == 1 % pfi
535.                              BasisInv = pfi(BasisInv, h_1, r);
536.                          else % mpfi
```

```
537.                              BasisInv = mpfi(BasisInv, h_l, r);
538.                    end
539.                    h_l(r) = -1;
540.               end
541.          end
542.       end
543.    end
544. end
545. if flag == 1 %  Phase II
546.    disp('---- P H A S E    II ----')
547.    % matrix A does not contain the artificial variable.
548.    % Here we solve the original LP problem
549.    flagfase2 = 0;
550.    [m1, n1] = size(A); % new size of matrix A
551.    % calculate the needed variables for the algorithm to
552.    % begin with the matrix of the basic variables
553.    Basis = spalloc(m1, length(BasicList), m1 * ...
554.         length(BasicList));
555.    % the matrix of the nonbasic variables
556.    N = spalloc(m1, length(NonBasicList), m1 * ...
557.         length(NonBasicList));
558.    w = spalloc(1, n1, n1); % simplex multiplier
559.    Sn = spalloc(1, n1, n1); % reduced costs
560.    Basis = A(:, BasicList); % the new basis
561.    % the coefficients of the objective function for the
562.    % basic variables
563.    cb = c(BasicList);
564.    % the matrix of the basic variables
565.    N = A(:, NonBasicList);
566.    % the coefficients of the objective function for the
567.    % nonbasic variables
568.    cn = c(NonBasicList);
569.    % calculate the basis inverse
570.    BasisInv = inv(Basis);
571.    Xb = BasisInv * b; % basic solution
572.    % set to zero, the values of Xb that are less than or
573.    % equal to tole1
574.    toler = abs(Xb) <= tole1;
575.    Xb(toler == 1) = 0;
576.    % calculate the simplex multiplier
577.    w = cb' * BasisInv;
578.    HRN = w * N;
579.    % set to zero, the values of HRN that are less than or
580.    % equal to tole2
581.    toler = abs(HRN) <= tole2;
582.    HRN(toler == 1) = 0;
583.    % calculate the reduced costs
584.    Sn = cn' - HRN;
585.    % set to zero, the values of Sn that are less than or
586.    % equal to tole2
587.    toler = abs(Sn) <= tole2;
588.    Sn(toler == 1) = 0;
589.    % compute set P
590.    P = NonBasicList(Sn < 0);
```

```
591.      % compute set Q
592.      Q = NonBasicList(Sn >= 0);
593.      % find direction dB
594.      h_j = BasisInv * A(:, P);
595.      dB = -(sum(h_j, 2));
596.      % find vector SP
597.      SP = c(P)' - w * A(:, P);
598.      % set to zero, the values of SP that are less than or
599.      % equal to tole2
600.      toler = abs(SP) <= tole2;
601.      SP(toler == 1) = 0;
602.      % find vector SQ
603.      SQ = c(Q)' - w * A(:, Q);
604.      % set to zero, the values of SQ that are less than or
605.      % equal to tole2
606.      toler = abs(SQ) <= tole2;
607.      SQ(toler == 1) = 0;
608.      while flagfase2 == 0 % iterate in Phase II
609.          flag = 0;
610.          % optimality test for Phase II
611.          if isempty(P) % the problem is optimal
612.              % calculate the value of the objective function
613.              if MinMaxLP == 1 % maximization
614.                  fval = full(-((c(BasicList))' * Xb ...
615.                      - c0));
616.              else % minimization
617.                  fval = full((c(BasicList))' * Xb ...
618.                      - c0);
619.              end
620.              exitflag = 0;
621.              xsol = Xb;
622.              disp('The LP problem is optimal')
623.              iterations = iterations + 1;
624.              return
625.          end
626.          if all(dB >= 0)
627.              so = sum(SP);
628.              if so == 0 % the problem is optimal
629.                  % calculate the value of the objective function
630.                  if MinMaxLP == 1 % maximization
631.                      fval = full(-((c(BasicList))' * Xb ...
632.                          - c0));
633.                  else % minimization
634.                      fval = full((c(BasicList))' * Xb ...
635.                          - c0);
636.                  end
637.                  exitflag = 0;
638.                  xsol = Xb;
639.                  iterations = iterations + 1;
640.                  disp('The LP problem is optimal')
641.                  return
642.              elseif so < 0 % the problem is unbounded
643.                  disp('The LP problem is unbounded')
644.                  exitflag = 2;
```

```
645.                     iterations = iterations + 1;
646.                     return
647.                 end
648.         end
649.         % find the leaving variable
650.         mrt = find(dB < 0);
651.         Xb_dB_div = Xb(mrt) ./ (-dB(mrt));
652.         [p, r] = min(Xb_dB_div);
653.         if p == +Inf % the problem is unbounded
654.                 disp('The LP problem is unbounded')
655.                 exitflag = 2;
656.                 iterations = iterations + 1;
657.                 return
658.         end
659.         rr = find(Xb_dB_div == p);
660.         if length(rr) > 1 % avoid stalling because of ties
661.                 r = mrt(rr);
662.                 % index of the leaving variable
663.                 k = BasicList(r);
664.                 [k, ~] = min(k);
665.                 r = find(BasicList == k);
666.         else
667.                 r = mrt(r);
668.                 % index of the leaving variable
669.                 k = BasicList(r);
670.         end
671.         % find the entering variable
672.         HrP = BasisInv(r, :) * A(:, P);
673.         % set to zero, the values of HrP that are
674.         % less than or equal to tole3
675.         toler = abs(HrP) <= tole3;
676.         HrP(toler == 1) = 0;
677.         HrQ = BasisInv(r, :) * A(:, Q);
678.         % set to zero, the values of HrQ that are
679.         % less than or equal to tole3
680.         toler = abs(HrQ) <= tole3;
681.         HrQ(toler == 1) = 0;
682.         mrtP = find(HrP > 0);
683.         mrtQ = find(HrQ < 0);
684.         % find theta1
685.         if ~isempty(mrtP)
686.                 Ratios1 = (-SP(mrtP)) ./ (HrP(mrtP));
687.                 % set to zero, the values of Ratios1 that
688.                 % are less than or equal to tole3
689.                 toler = abs(Ratios1) <= tole3;
690.                 Ratios1(toler == 1) = 0;
691.                 % avoid ties
692.                 [Rat1, ~] = min(Ratios1);
693.                 t1 = mrtP(Ratios1 == Rat1);
694.                 m_i = P(t1);
695.                 for i = 1:length(m_i)
696.                         indexes(i) = (find(P == m_i(i)));
697.                 end
698.                 t1 = min(indexes);
```

```
699.                   clear indexes;
700.                   clear m_i;
701.             else
702.                   Rat1 = Inf;
703.             end
704.             % find theta2
705.             if ~isempty(mrtQ)
706.                   Ratios2 = (-SQ(mrtQ)) ./ (HrQ(mrtQ));
707.                   % set to zero, the values of Ratios2 that
708.                   % are less than or equal to tole3
709.                   toler = abs(Ratios2) <= tole3;
710.                   Ratios2(toler == 1) = 0;
711.                   % avoid ties
712.                   [Rat2, ~] = min(Ratios2);
713.                   t2 = mrtQ(Ratios2 == Rat2);
714.                   m_i = Q(t2);
715.                   for i = 1:length(m_i)
716.                         indexes(i) = (find(Q == m_i(i)));
717.                   end
718.                   t2 = min(indexes);
719.                   clear indexes;
720.                   clear m_i;
721.             else
722.                   Rat2 = Inf;
723.             end
724.             % compare theta1 and theta2 to find the
725.             % entering variable
726.             if Rat1 <= Rat2
727.                   l = P(t1);
728.                   P(t1) = [];
729.                   Q = [Q k];
730.                   flag = 1;
731.             else
732.                   l = Q(t2);
733.                   Q(t2) = k;
734.             end
735.             % pivoting and update vectors and matrices
736.             h_l = BasisInv * A(:, l); % pivoting column
737.             g = h_l(r); % the pivot element
738.             % update the basic list
739.             BasicList(r) = l;
740.             % compute dB
741.             EInv = speye(length(BasisInv));
742.             EInv(:, r) = -h_l(1:length(h_l)) / g;
743.             EInv(r, r) = 1 / g;
744.             dB = EInv * dB;
745.             clear EInv;
746.             if flag == 1
747.                   dB(r) = dB(r) + 1;
748.             end
749.             % set to zero, the values of dB that are less
750.             % than or equal to tole1
751.             toler = abs(dB) <= tole1;
752.             dB(toler == 1) = 0;
```

```
753.            % basis inverse
754.            if iterations == counter * reinv
755.                % recompute the inverse of the basis
756.                % from scratch every reinv iterations
757.                BasisInv = inv(A(:, BasicList));
758.                counter = counter + 1;
759.                h_l(r) = g;
760.            else % basis update according to the
761.                % selected basis update method
762.                if basisUpdateMethod == 1 % pfi
763.                    BasisInv = pfi(BasisInv, h_l, r);
764.                else % mpfi
765.                    BasisInv = mpfi(BasisInv, h_l, r);
766.                end
767.                h_l(r) = -1;
768.            end
769.            % calculate the new basic solution
770.            Xb = BasisInv * b;
771.            % set to zero, the values of Xb that are less
772.            % than or equal to tole1
773.            toler = abs(Xb) <= tole1;
774.            Xb(toler == 1) = 0;
775.            % update the nonbasic list
776.            NonBasicList = [P Q];
777.            iterations = iterations + 1;
778.            % calculate the coefficients of the objective
779.            % function for the basic variables
780.            cb = c(BasicList);
781.            % calculate the simplex multiplier
782.            w = cb' * BasisInv;
783.            % calculate vectors SP and SQ
784.            SP = c(P)' - w * A(:, P);
785.            % set to zero, the values of SP that are less
786.            % than or equal to tole2
787.            toler = abs(SP) <= tole2;
788.            SP(toler == 1) = 0;
789.            SQ = c(Q)' - w * A(:, Q);
790.            % set to zero, the values of SQ that are less
791.            % than or equal to tole2
792.            toler = abs(SQ) <= tole2;
793.            SQ(toler == 1) = 0;
794.            % print intermediate results every 100
795.            % iterations
796.            if mod(iterations, 100) == 0
797.                % calculate the value of the objective
798.                % function
799.                if MinMaxLP == 1 % maximization
800.                    fval = full(-(c(BasicList)' * Xb - c0));
801.                else % minimization
802.                    fval = full(c(BasicList)' * Xb - c0);
803.                end
804.                fprintf(['Iteration %i - ' ...
805.                    'objective value: %f\n'], ...
806.                    iterations, fval);
```

```
807.              end
808.      end
809. end
810. end
```

The function that finds an invertible initial basis takes as input the matrix with the coefficients of the constraints (matrix *A*), the vector with the right-hand side of the constraints (vector *b*), the vector with the type of the constraints (vector *Eqin*), and the tolerance (variable *tol*), and returns as output the reduced row echelon form of the matrix with the coefficients of the constraints (matrix *AEqin*), the reduced vector with the right-hand side of the constraints (vector *bEqin*), the initial basis (matrix *jb*), the vector with the redundant indices (vector *out*), the matrix with the row redundant indices (matrix *rowindex*), and a flag variable showing if the LP problem is infeasible or not (variable *infeasible*).

In lines 30–31, the output variables are initialized. In lines 33–35, we find all the equality constraints. Next, we compute the tolerance if it was not given as input (lines 39–41). Then, we apply Gauss-Jordan elimination with partial pivoting in order to find the initial basis (lines 46–75). Next, we find the indices of the redundant constraints (lines 77–93).

```
1.    function [AEqin, bEqin, jb, out, rowindex, ...
2.         infeasible] = lprref(A, b, Eqin, tol)
3.    % Filename: lprref.m
4.    % Description: the function calculates a row echelon
5.    % form of matrix A for LPs using Gauss-Jordan
6.    % elimination with partial pivoting
7.    % Authors: Ploskas, N., & Samaras, N.
8.    %
9.    % Syntax: [AEqin, bEqin, jb, out, rowindex, ...
10.   %    infeasible] = lprref(A, b, Eqin, tol)
11.   %
12.   % Input:
13.   % -- A: matrix of coefficients of the constraints
14.   %    (size m x n)
15.   % -- b: vector of the right-hand side of the constraints
16.   %    (size m x 1)
17.   % -- Eqin: vector of the type of the constraints
18.   %    (size m x 1)
19.   % -- tol: tolerance
20.   %
21.   % Output:
22.   % -- AEqin: reduced row echelon form of A (size m x n)
23.   % -- bEqin: reduced vector b (size m x 1)
24.   % -- jb: basis of matrix A (size m x m)
25.   % -- rowindex: a matrix with the row redundant indices
26.   %    (size 2 x m)
27.   % -- infeasible: if the problem is infeasible or not
28.
29.   % initialize output variables
30.   infeasible = 0;
31.   out = [];
32.   % find all equality constraints
```

```
33.    a0 = find(Eqin == 0);
34.    AEqin = A(a0, :);
35.    bEqin = b(a0, :);
36.    [m, n] = size(AEqin);
37.    rowindex = zeros(2, m);
38.    rowindex(1, 1:m) = a0';
39.    if nargin < 4 % compute tol, if it was not given as input
40.        tol = max(m, n) * eps * norm(A, 'inf');
41.    end
42.    i = 1;
43.    j = 1;
44.    jb = [];
45.    % apply Gauss-Jordan elimination with partial pivoting
46.    while (i <= m) && (j <= n)
47.        [p, k] = max(abs(AEqin(i:m, j)));
48.        if p < tol
49.            AEqin(i:m, j) = zeros(m - i + 1, 1);
50.            j = j + 1;
51.        elseif p ~= 0
52.            k = k + i - 1;
53.            jb = [jb j];
54.            AEqin([i k], :) = AEqin([k i], :);
55.            bEqin([i k], :) = bEqin([k i], :);
56.            rowindex(:, [i k]) = rowindex(:, [k i]);
57.            bEqin(i, :) = bEqin(i, :) / AEqin(i, j);
58.            AEqin(i, j:n) = AEqin(i, j:n) / AEqin(i, j);
59.            i_nz = find(AEqin(:, j));
60.            i_nz = setdiff(i_nz, i);
61.            for t = i_nz
62.                if bEqin(i) ~= 0
63.                    bEqin(t) = bEqin(t) - AEqin(t, j) * bEqin(i);
64.                    toler = abs(bEqin) <= tol;
65.                    bEqin(toler == 1) = 0;
66.                end
67.                AEqin(t, j:n) = AEqin(t, j:n) - ...
68.                    AEqin(t, j) * AEqin(i, j:n);
69.                toler = abs(AEqin) <= tol;
70.                AEqin(toler == 1) = 0;
71.            end
72.            i = i + 1;
73.            j = j + 1;
74.        end
75.    end
76.    % check for redundant and infeasible constraints
77.    i = 1;
78.    for h = [1:i - 1 i + 1:m]
79.        % redundant constraint
80.        if (AEqin(h, :) == 0) & (bEqin(h) == 0)
81.            rowindex(2, h) = 1;
82.        end
83.        % infeasible constraint
84.        if (AEqin(h,:) == 0) & (bEqin(h) ~= 0)
85.            infeasible = 1;
86.            return;
```

```
87.        end
88.    end
89.    % find the indices of the redundant constraints
90.    if any(rowindex(2, :) == 1)
91.        y = find(rowindex(2, :) == 1);
92.        out = y;
93.    end
94.    end
```

10.6 Computational Study

In this section, we present a computational study. The aim of the computational study is to compare the efficiency of the proposed implementation with the revised primal simplex algorithm presented in Chapter 8. The test set used in the computational study is 19 LPs from the Netlib (optimal, Kennington, and infeasible LPs) and Mészáros problem sets that were presented in Section 3.6 and a set of randomly generated sparse LPs. Tables 10.2 and 10.3 present the total execution time and the number of iterations of the revised primal simplex algorithm that was presented in Chapter 8 and the exterior point simplex algorithm that was presented in this chapter over a set of benchmark LPs and over a set of randomly generated sparse LPs, respectively. The following abbreviations are used: (i) RSA— the proposed implementation of the revised primal simplex algorithm, and (ii) EPSA—the proposed implementation of the exterior point simplex algorithm. For each instance, we averaged times over 10 runs. All times are measured in seconds. The execution times of both algorithms include the time needed to presolve and scale (with arithmetic mean) the LPs.

The revised primal simplex algorithm is almost 1.4 times faster than the exterior point simplex algorithm over the benchmark set, while requiring almost 1.4 less iterations than the iterations needed for the exterior point simplex algorithm. On the other hand, the exterior point simplex algorithm is almost 3 times faster than the revised primal simplex algorithm over the set of randomly generated sparse LPs, while requiring almost 3 times less iterations than the iterations needed for the revised primal simplex algorithm.

10.7 Chapter Review

The exterior point simplex algorithm is a simplex-type algorithm that moves in the exterior of the feasible solution and constructs basic infeasible solutions instead of constructing feasible solutions like simplex algorithm does. In this chapter, we presented the exterior point simplex algorithm. Numerical examples were presented in order for the reader to understand better the algorithm. Furthermore, an

Table 10.2 Total execution time and number of iterations of the proposed implementation of the revised primal simplex algorithm and the exterior point simplex algorithm over the benchmark set

Problem				RSA		EPSA	
Name	Constraints	Variables	Nonzeros	Time	Iterations	Time	Iterations
beaconfd	173	262	3,375	0.15	26	0.16	41
cari	400	1,200	152,800	10.01	963	11.09	914
farm	7	12	36	0.01	5	0.01	5
itest6	11	8	20	0.01	2	0.01	2
klein2	477	54	4,585	0.46	500	0.47	500
nsic1	451	463	2,853	0.27	369	0.38	384
nsic2	465	463	3,015	0.33	473	0.54	535
osa-07	1,118	23,949	143,694	10.15	997	68.30	8,720
osa-14	2,337	52,460	314,760	67.22	2,454	375.52	19,720
osa-30	4,350	100,024	600,138	320.19	4,681	1,607.20	39,315
rosen1	520	1,024	23,274	3.33	1,893	3.23	1,661
rosen10	2,056	4,096	62,136	39.71	5,116	25.54	3,893
rosen2	1,032	2,048	46,504	19.29	4,242	15.31	3,581
rosen7	264	512	7,770	0.29	420	0.44	505
rosen8	520	1,024	15,538	1.43	1,045	1.83	1,171
sc205	205	203	551	0.20	165	0.26	246
scfxm1	330	457	2,589	0.74	339	0.86	440
sctap2	1,090	1,880	6,714	1.86	1,001	2.07	782
sctap3	1,480	2,480	8,874	4.37	1,809	3.95	1,157
Geometric mean				1.47	463.78	2.11	657.72

implementation of the algorithm in MATLAB was presented. The implementation is modular allowing the user to select which scaling technique and basis update method will use in order to solve LPs. Finally, a computational study was performed over benchmark LPs and randomly generated sparse LPs in order to compare the efficiency of the proposed implementation with the revised primal simplex algorithm. Computational results showed that the revised primal simplex algorithm is almost 1.4 times faster than the exterior point simplex algorithm over the benchmark set, while requiring almost 1.4 less iterations than the iterations needed for the exterior point simplex algorithm. On the other hand, the exterior point simplex algorithm is almost 3 times faster than the revised primal simplex algorithm over the set of randomly generated sparse LPs, while requiring almost 3 times less iterations than the iterations needed for the revised primal simplex algorithm.

Table 10.3 Total execution time and number of iterations of the proposed implementation of the revised primal simplex algorithm and the exterior point simplex algorithm over a set of randomly generated sparse LPs (5% density)

Problem	RSA		EPSA	
	Time	Iterations	Time	Iterations
100x100	0.13	1	0.04	1
200x200	0.17	236	0.22	305
300x300	0.39	584	0.79	865
400x400	0.81	911	1.67	1,323
500x500	1.94	1,565	2.16	1,364
600x600	4.41	2,569	5.03	2,340
700x700	7.97	3,919	2.88	1,306
800x800	16.25	6,016	3.98	1,522
900x900	26.14	6,883	15.11	3,770
1,000x1,000	34.55	7,909	17.93	3,959
1,100x1,100	70.41	13,483	9.91	2,302
1,200x1,200	63.55	10,172	34.00	5,034
1,300x1,300	133.95	18,004	16.18	2,818
1,400x1,400	160.29	18,805	18.79	2,880
1,500x1,500	205.85	19,768	78.52	7,430
1,600x1,600	327.74	28,866	30.73	3,656
1,700x1,700	463.99	36,016	96.93	7,660
1,800x1,800	530.54	33,873	135.37	8,793
1,900x1,900	917.11	52,492	47.94	4,001
2,000x2,000	926.92	50,926	192.25	10,651
Geometric mean	24.97	4,868.59	8.93	1,858.76

References

1. Bixby, R. E. (1992). Implementing the simplex method: The initial basis. *ORSA Journal on Computing, 4*, 267–284.
2. Carstens, D. M. (1968) Crashing techniques. In W. Orchard-Hays (Ed.), *Advanced linear-programming computing techniques* (pp. 131–139). New York: McGraw-Hill.
3. Gould, N. I. M., & Reid, J. K. (1989). New crash procedures for large systems of linear constraints. *Mathematical Programming, 45*, 475–501.
4. Maros, I., & Mitra, G. (1998). Strategies for creating advanced bases for large-scale linear programming problems. *INFORMS Journal on Computing, 10*, 248–260.
5. Paparrizos, K. (1991). An infeasible exterior point simplex algorithm for assignment problems. *Mathematical Programming, 51*(1–3), 45–54.
6. Paparrizos, K. (1993). An exterior point simplex algorithm for (general) linear programming problems. *Annals of Operations Research, 47*, 497–508.
7. Paparrizos, K., Samaras, N., & Sifaleras, A. (2015). Exterior point simplex-type algorithms for linear and network optimization problems. *Annals of Operations Research, 229*(1), 607–633.
8. Paparrizos, K., Samaras, N., & Stephanides, G. (2003). An efficient simplex type algorithm for sparse and dense linear programs. *European Journal of Operational Research, 148*(2), 323–334.

9. Paparrizos, K., Samaras, N., & Stephanides, G. (2003). A new efficient primal dual simplex algorithm. *Computers & Operations Research, 30*(9), 1383–1399.
10. Paparrizos, K., Samaras, N., & Tsiplidis, K. (2009). Pivoting algorithms for (LP) generating two paths. *Encyclopedia of optimization, 2nd edition*, 2965–2969.
11. Ploskas, N., & Samaras, N. (2015). Efficient GPU-based implementations of simplex type algorithms. *Applied Mathematics and Computation, 250*, 552–570.
12. Samaras, N. (2001). *Computational improvements and efficient implementation of two path pivoting algorithms*. Ph.D. dissertation, Department of Applied Informatics, University of Macedonia.
13. Triantafyllidis, C., & Samaras, N. (2014). Three nearly scaling-invariant versions of an exterior point algorithm for linear programming. *Optimization, 64*(10), 2163–2181.

Chapter 11
Interior Point Methods

Abstract Nowadays, much attention is focused on primal-dual Interior Point Methods (IPMs) due to their great computational performance. IPMs have permanently changed the landscape of mathematical programming theory and computation. Most primal-dual IPMs are based on Mehrotra's Predictor-Corrector (MPC) method. In this chapter, a presentation of the basic concepts of primal-dual IPMs is performed. Next, we present the MPC method. The various steps of the algorithm are presented. Numerical examples are also presented in order for the reader to understand better the algorithm. Furthermore, an implementation of the algorithm in MATLAB is presented. Finally, a computational study over benchmark LPs and randomly generated sparse LPs is performed in order to compare the efficiency of the proposed implementation with MATLAB's IPM solver.

11.1 Chapter Objectives

- Present the basic concepts of primal-dual interior point methods.
- Present Mehrotra's Predictor-Corrector method.
- Solve linear programming problems using Mehrotra's Predictor-Corrector method.
- Implement Mehrotra's Predictor-Corrector method using MATLAB.
- Compare the computational performance of Mehrotra's Predictor-Corrector method with MATLAB's implementation.

11.2 Introduction

Since Dantzig's initial contribution [2], researchers have made many efforts in order to enhance the performance of the simplex algorithm. At the same time when Dantzig presented the simplex algorithm, other researchers proposed IPMs that traverse across the interior of the feasible region [3, 8, 15]. However, these efforts did not compete with the simplex algorithm in practice due to the expensive execution time per iteration and the possibility of numerical instability. The first polynomial algorithm for LP was proposed by Khachiyan [10]. The development

© Springer International Publishing AG 2017   491
N. Ploskas, N. Samaras, *Linear Programming Using MATLAB®*,
Springer Optimization and Its Applications 127, DOI 10.1007/978-3-319-65919-0_11

of the ellipsoid method had a great impact on the theory of LP, but the proposed algorithm did not achieve to compete with the simplex algorithm in practice. The first IPM that outperformed simplex algorithm was proposed by Karmarkar [9] with a lower bound on the computational complexity of $O(nL)$, requiring a total of $O(n^{3.5}L)$ bit operations, where L is the length of a binary coding of the input data:

$$\sum_{i=0}^{m} \sum_{j=0}^{n} \lceil log_2 \left(|A_{ij}| + 1 \right) + 1 \rceil, \text{ and}$$

$$A_{i0} = b_i, A_{0j} = c_j, i = 1, 2, \cdots, m, j = 1, 2, \cdots, n$$

Since Karmarkar's algorithm, many improvements have been made both in theory and in practice of IPMs. Since then, many IPMs have been proposed [4, 5, 7, 11, 12]. Computational results showed that IPMs outperform the simplex algorithm for large-scale LPs [1]. The simplex algorithm tends to perform poorly on large-scale degenerate LPs. IPMs are superior to the simplex algorithm for certain important classes of LPs. IPMs seek to approach the optimal solution through a sequence of points that are always strongly feasible (interior points). In 1992, Mehrotra [13] presented a practical algorithm for linear programming that remains the basis of most current software. Nowadays, much attention is focused on primal-dual IPMs due to their great computational performance. Most primal-dual IPMs are based on Mehrotra's Predictor-Corrector (MPC) method.

Most IPMs fall into one of the three main categories: (i) affine scaling methods, (ii) potential reduction methods, and (iii) central trajectory methods. The affine scaling algorithm is an attractive choice due to its simplicity and its good performance in practice. However, its performance is sensitive to the starting (initial) point. Potential reduction methods do not have the simplicity of affine scaling methods, but they are more attractive than affine scaling methods. IPMs based on the central trajectory are the most useful in theory and the most used in practice. A historical review on the IPMs, including algorithms for linear programming, semi-definite programming, and comments on the complexity theory can be found in [6, 14].

In this chapter, a presentation of the basic concepts of primal-dual IPMs is performed. Next, we present the MPC method. The various steps of the algorithm are presented. Numerical examples are also presented in order for the reader to understand better the algorithm. Furthermore, an implementation of the algorithm in MATLAB is presented. Finally, a computational study is performed in order to compare the efficiency of the proposed implementation with MATLAB's IPM solver.

The structure of this chapter is as follows. In Section 11.3, the background of IPMs is presented. Section 11.4 presents Mehrotra's Predictor-Corrector method. All different steps of the algorithm are presented in detail. Two examples are solved in Section 11.5 using the MPC method. Section 11.6 presents the MATLAB code of MPC method. Section 11.7 presents a computational study that compares the proposed implementation and MATLAB's IPM solver. Finally, conclusions and further discussion are presented in Section 11.8.

11.3 Background

Consider the primal LP problem in its standard form:

$$
\begin{aligned}
\min z = {}& c^T x \\
\text{s.t.} \quad & Ax = b \\
& x \geq 0
\end{aligned}
\tag{11.1}
$$

where $A \in \mathbb{R}^{m \times n}$, $(c, x) \in \mathbb{R}^n$, $b \in \mathbb{R}^m$, and T denotes transposition. As already described in Chapter 2, the equivalent dual LP problem is the following:

$$
\begin{aligned}
\max z = {}& b^T w \\
\text{s.t.} \quad & A^T w \leq c
\end{aligned}
\tag{11.2}
$$

Adding nonnegative slack variables into the dual constraints, we can reformulate the dual LP problem as follows:

$$
\begin{aligned}
\max z = {}& b^T w \\
\text{s.t.} \quad & A^T w + s = c \\
& s \geq 0
\end{aligned}
\tag{11.3}
$$

Together, LPs (11.1) and (11.3) are referred to as the primal-dual pair. Solving the primal LP problem is equivalent to solving the dual LP problem.

The Karush-Kuhn-Tucker (KKT) conditions for the primal-dual pair is a mapping F from $\mathbb{R}^{2n+m}$ to $\mathbb{R}^{2n+m}$:

$$
F(x, w, s) =
\begin{bmatrix}
A^T w + s - c \\
Ax - b \\
XSe
\end{bmatrix}
= 0, (x, s \geq 0)
\tag{11.4}
$$

where $X, S \in \mathbb{R}^{n \times n}$ are diagonal matrices with diagonal x and s, respectively, and $e \in \mathbb{R}^n$ is a vector of ones. KKT conditions state that:

- Any vector x that satisfies the conditions $Ax - b = 0$ and $x \geq 0$ is a feasible point for the primal LP problem (primal feasibility).
- Any pair (w, s) that satisfies the conditions $A^T w + s - c = 0$ and $s \geq 0$ is a feasible point for the dual LP problem (dual feasibility).
- Any point that is both primal and dual feasible and satisfies the condition $XSe = 0$ is optimal for both LPs (primal and dual feasibility).

- The condition $X^*S^*e = 0$ implies that at least one of the pair x_j^* or s_j^* must be zero for all $j = 1, 2, \cdots, n$ (complementary slackness).

According to the duality theory, if x and (w, s) are feasible for the primal and dual LPs, respectively, then $b^T w \le c^T x$. The duality gap ($|c^T x - b^T w|$) is the difference between the objective function of the primal LP problem and the dual LP problem.

Let $u = (x, w, s)$ be the current approximation to the solution and $x > 0$ and $s > 0$. If u is not optimal, we can apply Newton's method to solve the following equation for a fixed barrier parameter $\mu > 0$:

$$F_\mu(u) = \begin{bmatrix} A^T w + s - c \\ Ax - b \\ Xs - \mu e \end{bmatrix} = 0, (x, s \ge 0) \tag{11.5}$$

After performing a Newton iteration, we get a new point $\bar{u}$:

$$\bar{u} = (\bar{x}, \bar{w}, \bar{s}) = (x, w, s) - (\Delta x, \Delta w, \Delta s) \tag{11.6}$$

or using the shortened notation:

$$\bar{u} = u - \Delta u \tag{11.7}$$

Let us define the dual residual (r_d), the primal residual (p_d), and the complementarity residual (r_c) at u as:

$$\begin{aligned} r_d &= A^T w + s - c \\ r_p &= Ax - b \\ r_c &= Xs - \mu e \end{aligned} \tag{11.8}$$

Then, the Newton direction Δu at u is given by:

$$J_\mu(u)\Delta u = F_\mu(u) \tag{11.9}$$

where J is the Jacobian of F. For $\mu > 0$, the Jacobian is nonsingular. Hence, $x > 0$ and $s > 0$. That gives the following linear system:

$$\begin{bmatrix} 0 & A^T & I \\ A & 0 & 0 \\ S & 0 & X \end{bmatrix} \begin{bmatrix} \Delta x \\ \Delta w \\ \Delta s \end{bmatrix} = \begin{bmatrix} r_d \\ r_p \\ r_c \end{bmatrix} \tag{11.10}$$

The above system is of size $2n + m$. We can symmetrize $J_\mu(u)$ by multiplying the last block row from the left by S^{-1}:

$$\begin{bmatrix} 0 & A^T & I \\ A & 0 & 0 \\ I & 0 & S^{-1}X \end{bmatrix} \begin{bmatrix} \Delta x \\ \Delta w \\ \Delta s \end{bmatrix} = \begin{bmatrix} r_d \\ r_p \\ S^{-1}r_c \end{bmatrix} \tag{11.11}$$

The solution of the above system gives as the solution:

$$\begin{aligned} \Delta w &= (AXS^{-1}A^T)^{-1}(r_p - AS^{-1}(r_c - Xr_d)) \\ \Delta s &= r_d - A^T \Delta w \\ \Delta x &= S^{-1}(r_c - X\Delta s) \end{aligned} \tag{11.12}$$

If we define the matrix $M = AXS^{-1}A^T$ and the right-hand side $rhs = r_p + AS^{-1}(-r_c + Xr_d)$, then we get:

$$M\Delta w = rhs \tag{11.13}$$

The solution of the above systems of linear equations yields Δw. Then, we can calculate:

$$\begin{aligned} \Delta s &= r_d - A^T \Delta w \\ \Delta x &= S^{-1}(r_c - X\Delta s) \end{aligned} \tag{11.14}$$

After computing the search direction Δu, we can update the current solution as:

$$\bar{u} = u + \alpha \Delta u \tag{11.15}$$

where $\alpha \in (0, 1]$ is a parameter that prevents $\bar{u}$ from violating the condition $(\bar{x}, \bar{s}) > 0$. Then, we continue applying the Newton method until a termination criterion is satisfied.

MPC method generates a sequence of approximate solutions, $u(\mu) = (x(\mu), w(\mu), s(\mu))$, to the perturbed KKT conditions:

$$\begin{aligned} A^T w(\mu) + s(\mu) - c &= 0 \\ Ax(\mu) - b &= 0 \\ X(\mu)S(\mu)e &= \sigma\mu e \\ (x(\mu), s(\mu) &> 0) \end{aligned} \tag{11.16}$$

where $0 \leq \sigma \leq 1$ and $\mu = x^T s/n > 0$. The parameter σ is called the centering parameter. The above conditions differ from the KKT conditions in the complementary slackness condition, because the solution $u(\mu)$ is uniquely defined for each $\mu > 0$ and the pairwise products $x_j(\mu)s_j(\mu) = \sigma\mu$ for each $j = 1, 2, \cdots, n$.

We can define the central path as follows:

$$C = u(\mu)|\mu > 0 \tag{11.17}$$

The central path is a trajectory of feasible solutions in the interior of the feasible set which leads to an optimal point at the boundary. As μ decreases to 0, the central path C converges to the primal-dual solution.

MPC is a predictor-corrector method and solves two linear systems at each iteration: (i) Equation (11.10) to obtain the predictor direction $\Delta u = (\Delta x^p, \Delta w^p, \Delta s^p)$, and (ii) Equation (11.18) to obtain the direction $\Delta u = (\Delta x, \Delta w, \Delta s)$.

$$
\begin{bmatrix} 0 & A^T & I \\ A & 0 & 0 \\ S & 0 & X \end{bmatrix} \begin{bmatrix} \Delta x \\ \Delta w \\ \Delta s \end{bmatrix} = \begin{bmatrix} A^T w + s - c \\ Ax - b \\ Xs - \sigma \mu e \end{bmatrix} + \begin{bmatrix} 0 \\ 0 \\ \Delta X^p \Delta s^p \end{bmatrix} \tag{11.18}
$$

11.4 Mehrotra's Predictor-Corrector Method

A formal description of the MPC method is given in Table 11.1 and a flow diagram of its major steps in Figure 11.1.

Most primal-dual IPMs need a strictly feasible interior point as a starting point. For some LPs, a strictly feasible initial point is difficult to find. However, MPC is an infeasible primal-dual IPM and it just requires that $(x^0, s^0) > 0$ for the starting point. At each iteration of the algorithm, a point (x, w, s) is calculated. This point is permitted to be infeasible with $(x, s) > 0$.

Table 11.1 Mehrotra's predictor-corrector method

Step 0. *(Initialization).*
Presolve the LP problem.
Scale the LP problem.
Find an initial interior point (x^0, w^0, s^0).
Step 1. *(Test of Optimality).*
Calculate the primal (r_p), dual (r_d), and complementarity (r_c) residuals.
Calculate the duality measure (μ).
if $\max(\mu, ||r_p||, ||r_d||) \leq tol$ then STOP. The LP problem (11.1) is optimal.
Step 2. *(Predictor Step).*
Solve the system in (11.21) for $(\Delta x^p, \Delta w^p, \Delta s^p)$.
Calculate the largest possible step lengths α_p^p, α_d^p.
Step 3. *(Centering Parameter Step).*
Compute the centering parameter σ.
Step 4. *(Corrector Step).*
Solve the system in (11.25) for $(\Delta x, \Delta w, \Delta s)$.
Calculate the primal and dual step lengths α_p, α_d.
Step 5. *(Update Step).*
Update the solution (x, w, s).
Go to Step 1.

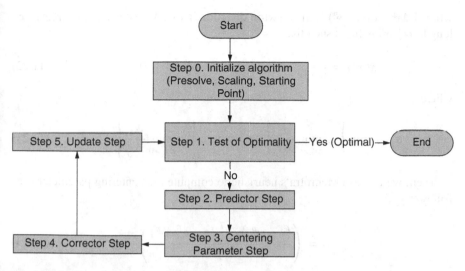

Fig. 11.1 Mehrotra's predictor-corrector method

Mehrotra proposed the following heuristic to obtain a starting point:

$$
\begin{aligned}
\bar{x} &= A^T (AA^T)^{-1} b \\
\bar{w} &= (AA^T)^{-1} Ac \\
\bar{s} &= c - A^T \bar{w} \\
\delta x &= \max(-1.5 \min(\bar{x}), 0) \\
\delta s &= \max(-1.5 \min(\bar{s}), 0) \\
\bar{\delta}_x &= \delta_x + 0.5 \frac{(\bar{x} + \delta_x e)^T (\bar{s} + \delta s e)}{\sum_{i=1}^{n} \bar{s}_i + \delta_s} \\
\bar{\delta}_s &= \delta_s + 0.5 \frac{(\bar{x} + \delta_x e)^T (\bar{s} + \delta s e)}{\sum_{i=1}^{n} \bar{x}_i + \delta_x} \\
x^0 &= \bar{x} + \bar{\delta}_x e \\
w^0 &= \bar{w} \\
s^0 &= \bar{s} + \bar{\delta}_s e
\end{aligned}
\tag{11.19}
$$

Then, we continue generating approximate solutions until a termination criterion is satisfied. The termination criterion can be the following:

$$
\max(\mu, \|Ax - b\|, \|A^T w + s - c\|) \le tol \tag{11.20}
$$

where tol is the tolerance, a very small positive number (e.g., $tol = 10^{-8}$ or $tol = 10^{-10}$).

MPC method uses two directions at each iteration, the predictor and the corrector. In the predictor step, we solve:

$$
\begin{bmatrix} 0 & A^T & I \\ A & 0 & 0 \\ S & 0 & X \end{bmatrix} \begin{bmatrix} \Delta x^p \\ \Delta w^p \\ \Delta s^p \end{bmatrix} = \begin{bmatrix} A^T w + s - c \\ Ax - b \\ Xs \end{bmatrix} = \begin{bmatrix} r_d \\ r_p \\ r_c \end{bmatrix} \tag{11.21}
$$

where $(\Delta x^p, \Delta w^p, \Delta s^p)$ is the Newton direction. Next, we calculate the largest step lengths $\alpha_p^p, \alpha_d^p \in (0, 1]$ such that:

$$x(\alpha_p^p) = x - \alpha_p^p \Delta x^p, s(\alpha_d^p) = s - \alpha_d^p \Delta s^p \geq 0 \qquad (11.22)$$

where:

$$\alpha_p^p = \min \left\{ 1, \min_{\Delta x_i^p > 0} \frac{x_i}{\Delta x_i^p} \right\}, \alpha_d^p = \min \left\{ 1, \min_{\Delta s_i^p > 0} \frac{s_i}{\Delta s_i^p} \right\} \qquad (11.23)$$

Then, we can use Mehrotra's heuristic to compute the centering parameter σ as follows:

$$\sigma = \left(\frac{\left(x - \alpha_p^p \Delta x^p \right)^T \left(s - \alpha_d^p \Delta s^p \right)}{n\mu} \right)^3 \qquad (11.24)$$

where $\mu = x^T s / n$. In the corrector step, we solve:

$$\begin{bmatrix} 0 & A^T & I \\ A & 0 & 0 \\ S & 0 & X \end{bmatrix} \begin{bmatrix} \Delta x \\ \Delta w \\ \Delta s \end{bmatrix} = \begin{bmatrix} A^T w + s - c \\ Ax - b \\ Xs - \sigma \mu e \end{bmatrix} + \begin{bmatrix} 0 \\ 0 \\ \Delta X^p \Delta s^p \end{bmatrix} \qquad (11.25)$$

where we approximate the central path. The coefficient matrix is the same as in the predictor step, so we can use only one factorization. However, the right-hand side is different. Using a value $\eta \in (0, 1)$ such that $\eta \approx 1$, e.g., $\eta = 0.995$ or $\eta = 0.999$, we calculate the primal and dual step lengths:

$$\alpha_p = \min \left\{ 1, \eta \min_{\Delta x_i > 0} \frac{x_i}{\Delta x_i} \right\}, \alpha_d = \min \left\{ 1, \eta \min_{\Delta s_i > 0} \frac{s_i}{\Delta s_i} \right\} \qquad (11.26)$$

Finally, we update the solution as:

$$\begin{aligned} \bar{x} &= x - \alpha_p \Delta x \\ \bar{w} &= w - \alpha_d \Delta w \\ \bar{s} &= s - \alpha_d \Delta s \end{aligned} \qquad (11.27)$$

We continue applying the same procedure until the termination criterion in (11.20) is satisfied.

11.5 Examples

Example 1 The LP problem that will be solved is the following:

$$\begin{aligned}
\min z = & \; x_1 & - \; x_3 - 3x_4 & \\
\text{s.t.} \quad & 2x_1 & + \; 2x_3 + 3x_4 & = 10 \\
& -2x_2 - 2x_3 - 6x_4 & = -6 \\
x_j \geq 0, \quad & (j = 1, 2, 3, 4) &
\end{aligned}$$

In matrix notation the above LP problem is written as follows:

$$A = \begin{bmatrix} 2 & 0 & 2 & 3 \\ 0 & -2 & -2 & -6 \end{bmatrix}, c = \begin{bmatrix} 1 \\ 0 \\ -1 \\ -3 \end{bmatrix}, b = \begin{bmatrix} 10 \\ -6 \end{bmatrix}, Eqin = \begin{bmatrix} 0 \\ 0 \end{bmatrix}$$

The LP problem is in its standard form. Using Equation (11.19), we compute a starting point:

$$\bar{x} = A^T (AA^T)^{-1} b = \begin{bmatrix} 2 & 0 \\ 0 & -2 \\ 2 & -2 \\ 3 & -6 \end{bmatrix} \left(\begin{bmatrix} 2 & 0 & 2 & 3 \\ 0 & -2 & -2 & -6 \end{bmatrix} \begin{bmatrix} 2 & 0 \\ 0 & -2 \\ 2 & -2 \\ 3 & -6 \end{bmatrix} \right)^{-1} \begin{bmatrix} 10 \\ -6 \end{bmatrix}$$

$$= \begin{bmatrix} 2.3333 \\ -0.8939 \\ 1.4394 \\ 0.8182 \end{bmatrix},$$

$$\bar{w} = (AA^T)^{-1} Ac = \left(\begin{bmatrix} 2 & 0 & 2 & 3 \\ 0 & -2 & -2 & -6 \end{bmatrix} \begin{bmatrix} 2 & 0 \\ 0 & -2 \\ 2 & -2 \\ 3 & -6 \end{bmatrix} \right)^{-1} \begin{bmatrix} 2 & 0 & 2 & 3 \\ 0 & -2 & -2 & -6 \end{bmatrix} \begin{bmatrix} 1 \\ 0 \\ -1 \\ -3 \end{bmatrix} = \begin{bmatrix} 0.1667 \\ 0.5379 \end{bmatrix},$$

$$\bar{s} = c - A^T \bar{w} = \begin{bmatrix} 1 \\ 0 \\ -1 \\ -3 \end{bmatrix} - \begin{bmatrix} 2 & 0 \\ 0 & -2 \\ 2 & -2 \\ 3 & -6 \end{bmatrix} \begin{bmatrix} 0.1667 \\ 0.5379 \end{bmatrix} = \begin{bmatrix} 0.6667 \\ 1.0758 \\ -0.2576 \\ -0.2727 \end{bmatrix},$$

$$\delta x = \max(-1.5 \min(\bar{x}), 0) = \max(-1.5 \times -0.8939), 0) = 1.3409,$$

$$\delta s = \max(-1.5 \min(\bar{s}), 0) = \max(-1.5 \times -0.2727, 0) = 0.4091,$$

$$\bar{\delta}_x = \delta_x + 0.5 \frac{(\bar{x} + \delta_x e)^T (\bar{s} + \delta se)}{\sum_{i=1}^n \bar{s}_i + \delta_s} = 2.2768,$$

$$\bar{\delta}_s = \delta_s + 0.5 \frac{(\bar{x} + \delta_x e)^T (\bar{s} + \delta se)}{\sum_{i=1}^n \bar{x}_i + \delta_x} = 0.7033,$$

$$x = \bar{x} + \bar{\delta}_x e = \begin{bmatrix} 4.6102 \\ 1.3829 \\ 3.7162 \\ 3.0950 \end{bmatrix},$$

$$w = \bar{w} = \begin{bmatrix} 0.1667 \\ 0.5379 \end{bmatrix},$$

$$s = \bar{s} + \bar{\delta}_s e = \begin{bmatrix} 1.3700 \\ 1.7791 \\ 0.4458 \\ 0.4306 \end{bmatrix}$$

1st Iteration

Initially, we calculate the primal, dual, and complementarity residuals and the duality measure:

$$r_d = A^T w + s - c = \begin{bmatrix} 2 & 0 \\ 0 & -2 \\ 2 & -2 \\ 3 & -6 \end{bmatrix} \begin{bmatrix} 0.1667 \\ 0.5379 \end{bmatrix} + \begin{bmatrix} 1.3700 \\ 1.7791 \\ 0.4458 \\ 0.4306 \end{bmatrix} - \begin{bmatrix} 1 \\ 0 \\ -1 \\ -3 \end{bmatrix} = \begin{bmatrix} 0.7033 \\ 0.7033 \\ 0.7033 \\ 0.7033 \end{bmatrix},$$

$$r_p = Ax - b = \begin{bmatrix} 2 & 0 & 2 & 3 \\ 0 & -2 & -2 & -6 \end{bmatrix} \begin{bmatrix} 4.6102 \\ 1.3829 \\ 3.7162 \\ 3.0950 \end{bmatrix} - \begin{bmatrix} 10 \\ -6 \end{bmatrix} = \begin{bmatrix} 15.9379 \\ -22.7684 \end{bmatrix},$$

$$r_c = Xs = \begin{bmatrix} 4.6102 & 0 & 0 & 0 \\ 0 & 1.3829 & 0 & 0 \\ 0 & 0 & 3.7162 & 0 \\ 0 & 0 & 0 & 3.0950 \end{bmatrix} \begin{bmatrix} 1.3700 \\ 1.7791 \\ 0.4458 \\ 0.4306 \end{bmatrix} = \begin{bmatrix} 6.3159 \\ 2.4603 \\ 1.6565 \\ 1.3327 \end{bmatrix},$$

$$\mu = x^T s / n = \begin{bmatrix} 4.6102 & 1.3829 & 3.7162 & 3.0950 \end{bmatrix} \begin{bmatrix} 1.3700 \\ 1.7791 \\ 0.4458 \\ 0.4306 \end{bmatrix} / 4 = 2.9414$$

Then, we perform the test of optimality:

$$\max(\mu, ||r_p||, ||r_d||) = \max(2.9414, 27.7923, 1.4067) = 27.7923 > 10^{-8}$$

The LP is not optimal. Next, we compute the coefficient matrix M of the systems (11.21) and (11.25):

$$M = AXS^{-1}A^T = \begin{bmatrix} 111.4972 & -162.7257 \\ -162.7257 & 295.2127 \end{bmatrix}$$

Then, we compute the right-hand side rhs of the system (11.21):

$$rhs = r_p + AS^{-1}(-r_c + Xr_d)) = \begin{bmatrix} 21.6267 \\ -37.1524 \end{bmatrix}$$

In the predictor step, we solve the system in (11.21) for $(\Delta x^p, \Delta w^p, \Delta s^p)$:

$$\Delta w^p = M^{-1}rhs = \begin{bmatrix} 0.0526 \\ -0.0968 \end{bmatrix},$$

$$\Delta s^p = r_d - A^T \Delta w^p = \begin{bmatrix} 0.5980 \\ 0.5097 \\ 0.4044 \\ -0.0356 \end{bmatrix},$$

$$\Delta x^p = S^{-1}(r_c - X\Delta s^p) = \begin{bmatrix} 2.5977 \\ 0.9867 \\ 0.3450 \\ 3.3508 \end{bmatrix}$$

Next, we calculate the largest possible step lengths α_p^p, α_d^p:

$$\alpha_p^p = \min \left\{ 1, \min_{\Delta x_i^p > 0} \frac{x_i}{\Delta x_i^p} \right\} = 0.9237,$$

$$\alpha_d^p = \min \left\{ 1, \min_{\Delta s_i^p > 0} \frac{s_i}{\Delta s_i^p} \right\} = 1$$

Then, we compute the centering parameter σ:

$$\sigma = \left(\frac{(x - \alpha_p^p \Delta x^p)^T (s - \alpha_d^p \Delta s^p)}{n\mu} \right)^3 = 0.0090$$

In the corrector step, we solve the system in (11.25) for $(\Delta x, \Delta w, \Delta s)$:

$$r_c = r_c - \sigma\mu + \Delta X^p \Delta s^p = \begin{bmatrix} 7.8430 \\ 2.9368 \\ 1.7696 \\ 1.1870 \end{bmatrix},$$

$$rhs = r_p + AS^{-1}(-r_c + Xr_d)) = \begin{bmatrix} 19.9049 \\ -38.1393 \end{bmatrix},$$

$$\Delta w = M^{-1}rhs = \begin{bmatrix} -0.0513 \\ -0.1575 \end{bmatrix},$$

$$\Delta s = r_d - A^T \Delta w = \begin{bmatrix} 0.8059 \\ 0.3884 \\ 0.4910 \\ -0.0876 \end{bmatrix},$$

$$\Delta x = S^{-1}(r_c - X\Delta s) = \begin{bmatrix} 3.0129 \\ 1.3488 \\ -0.1234 \\ 3.3862 \end{bmatrix}.$$

Next, we calculate the primal and dual step lengths α_p, α_d:

$$\alpha_p = \min\left\{1, \eta \min_{\Delta x_i > 0} \frac{x_i}{\Delta x_i}\right\} = 0.9094,$$

$$\alpha_d = \min\left\{1, \eta \min_{\Delta s_i > 0} \frac{s_i}{\Delta s_i}\right\} = 0.9034$$

where $\eta = \max(0.995, 1 - \mu) = 0.995$.
Finally, we update the solution (x, w, s):

$$x = x - \alpha_p \Delta x = \begin{bmatrix} 1.8701 \\ 0.1563 \\ 3.8284 \\ 0.0155 \end{bmatrix},$$

$$w = w - \alpha_d \Delta w = \begin{bmatrix} 0.2130 \\ 0.6801 \end{bmatrix},$$

$$s = s - \alpha_d \Delta s = \begin{bmatrix} 0.6420 \\ 1.4282 \\ 0.0022 \\ 0.5097 \end{bmatrix}.$$

2nd Iteration

Initially, we calculate the primal, dual, and complementarity residuals and the duality measure:

$$r_d = A^T w + s - c = \begin{bmatrix} 2 & 0 \\ 0 & -2 \\ 2 & -2 \\ 3 & -6 \end{bmatrix} \begin{bmatrix} 0.2130 \\ 0.6801 \end{bmatrix} + \begin{bmatrix} 0.6420 \\ 1.4282 \\ 0.0022 \\ 0.5097 \end{bmatrix} - \begin{bmatrix} 1 \\ 0 \\ -1 \\ -3 \end{bmatrix} = \begin{bmatrix} 0.0680 \\ 0.0680 \\ 0.0680 \\ 0.0680 \end{bmatrix},$$

$$r_p = Ax - b = \begin{bmatrix} 2 & 0 & 2 & 3 \\ 0 & -2 & -2 & -6 \end{bmatrix} \begin{bmatrix} 1.8701 \\ 0.1563 \\ 3.8284 \\ 0.0155 \end{bmatrix} - \begin{bmatrix} 10 \\ -6 \end{bmatrix} = \begin{bmatrix} 1.4435 \\ -2.0622 \end{bmatrix},$$

$$r_c = Xs = \begin{bmatrix} 1.8701 & 0 & 0 & 0 \\ 0 & 0.1563 & 0 & 0 \\ 0 & 0 & 3.8284 & 0 \\ 0 & 0 & 0 & 0.0155 \end{bmatrix} \begin{bmatrix} 0.6420 \\ 1.4282 \\ 0.0022 \\ 0.5097 \end{bmatrix} = \begin{bmatrix} 1.2006 \\ 0.2232 \\ 0.0085 \\ 0.0079 \end{bmatrix},$$

$$\mu = x^T s / n = \begin{bmatrix} 1.8701 & 0.1563 & 3.8284 & 0.0155 \end{bmatrix} \begin{bmatrix} 0.6420 \\ 1.4282 \\ 0.0022 \\ 0.5097 \end{bmatrix} / 4 = 0.36$$

Then, we perform the test of optimality:

$$\max(\mu, ||r_p||, ||r_d||) = \max(0.36, 2.5172, 0.1359) = 2.5172 > 10^{-8}$$

The LP is not optimal. Next, we compute the coefficient matrix M of the systems (11.21) and (11.25):

$$M = AXS^{-1}A^T = \begin{bmatrix} 6,882.8284 & -6,871.4494 \\ -6,871.4494 & 6,872.4335 \end{bmatrix}$$

Then, we compute the right-hand side *rhs* of the system (11.21):

$$rhs = r_p + AS^{-1}(-r_c + Xr_d)) = \begin{bmatrix} 223.9263 \\ -227.5513 \end{bmatrix}$$

In the predictor step, we solve the system in (11.21) for $(\Delta x^p, \Delta w^p, \Delta s^p)$:

$$\Delta w^p = M^{-1} rhs = \begin{bmatrix} -0.2906 \\ -0.3237 \end{bmatrix},$$

$$\Delta s^p = r_d - A^T \Delta w^p = \begin{bmatrix} 0.6491 \\ -0.5793 \\ 0.0018 \\ -1.0022 \end{bmatrix},$$

$$\Delta x^p = S^{-1}(r_c - X \Delta s^p) = \begin{bmatrix} -0.0208 \\ 0.2196 \\ 0.6738 \\ 0.0459 \end{bmatrix}.$$

Next, we calculate the largest possible step lengths α_p^p, α_d^p:

$$\alpha_p^p = \min \left\{ 1, \min_{\Delta x_i^p > 0} \frac{x_i}{\Delta x_i^p} \right\} = 0.3371,$$

$$\alpha_d^p = \min \left\{ 1, \min_{\Delta s_i^p > 0} \frac{s_i}{\Delta s_i^p} \right\} = 0.9890$$

Then, we compute the centering parameter σ:

$$\sigma = \left(\frac{\left(x - \alpha_p^p \Delta x^p \right)^T \left(s - \alpha_d^p \Delta s^p \right)}{n\mu} \right)^3 = 0.0015$$

In the corrector step, we solve the system in (11.25) for $(\Delta x, \Delta w, \Delta s)$:

$$r_c = r_c - \sigma\mu + \Delta X^p \Delta s^p = \begin{bmatrix} 1.1865 \\ 0.0954 \\ 0.0092 \\ -0.0387 \end{bmatrix},$$

$$rhs = r_p + AS^{-1}(-r_c + Xr_d)) = \begin{bmatrix} 223.6284 \\ -227.6625 \end{bmatrix},$$

$$\Delta w = M^{-1} rhs = \begin{bmatrix} -0.3237 \\ -0.3567 \end{bmatrix},$$

$$\Delta s = r_d - A^T \Delta w = \begin{bmatrix} 0.7153 \\ -0.6455 \\ 0.0018 \\ -1.1015 \end{bmatrix},$$

$$\Delta x = S^{-1}(r_c - X\Delta s) = \begin{bmatrix} -0.2355 \\ 0.1374 \\ 1.0209 \\ -0.0424 \end{bmatrix}$$

Next, we calculate the primal and dual step lengths α_p, α_d:

$$\alpha_p = \min\left\{1, \eta \min_{\Delta x_i > 0} \frac{x_i}{\Delta x_i}\right\} = 1,$$

$$\alpha_d = \min\left\{1, \eta \min_{\Delta s_i > 0} \frac{s_i}{\Delta s_i}\right\} = 0.8930$$

where $\eta = \max(0.995, 1 - \mu) = 0.995$.

Finally, we update the solution (x, w, s):

$$x = x - \alpha_p \Delta x = \begin{bmatrix} 2.1057 \\ 0.0189 \\ 2.8075 \\ 0.0579 \end{bmatrix},$$

$$w = w - \alpha_d \Delta w = \begin{bmatrix} 0.5020 \\ 0.9987 \end{bmatrix},$$

$$s = s - \alpha_d \Delta s = \begin{bmatrix} 0.0032 \\ 2.0047 \\ 0.0006 \\ 1.4934 \end{bmatrix}$$

3rd Iteration

Initially, we calculate the primal, dual, and complementarity residuals and the duality measure:

$$r_d = A^T w + s - c = \begin{bmatrix} 2 & 0 \\ 0 & -2 \\ 2 & -2 \\ 3 & -6 \end{bmatrix} \begin{bmatrix} 0.5020 \\ 0.9987 \end{bmatrix} + \begin{bmatrix} 0.0032 \\ 2.0047 \\ 0.0006 \\ 1.4934 \end{bmatrix} - \begin{bmatrix} 1 \\ 0 \\ -1 \\ -3 \end{bmatrix} = \begin{bmatrix} 0.0073 \\ 0.0073 \\ 0.0073 \\ 0.0073 \end{bmatrix},$$

$$r_p = Ax - b = \begin{bmatrix} 2 & 0 & 2 & 3 \\ 0 & -2 & -2 & -6 \end{bmatrix} \begin{bmatrix} 2.1057 \\ 0.0189 \\ 2.8075 \\ 0.0579 \end{bmatrix} - \begin{bmatrix} 10 \\ -6 \end{bmatrix} = \begin{bmatrix} 2.0250e-13 \\ -1.1546e-14 \end{bmatrix},$$

$$r_c = Xs = \begin{bmatrix} 2.1057 & 0 & 0 & 0 \\ 0 & 0.0189 & 0 & 0 \\ 0 & 0 & 2.8075 & 0 \\ 0 & 0 & 0 & 0.0579 \end{bmatrix} \begin{bmatrix} 0.0032 \\ 2.0047 \\ 0.0006 \\ 1.4934 \end{bmatrix} = \begin{bmatrix} 0.0068 \\ 0.0378 \\ 0.0017 \\ 0.0864 \end{bmatrix},$$

$$\mu = x^T s/n = \begin{bmatrix} 2.1057 & 0.0189 & 2.8075 & 0.0579 \end{bmatrix} \begin{bmatrix} 0.0032 \\ 2.0047 \\ 0.0006 \\ 1.4934 \end{bmatrix} /4 = 0.0332$$

Then, we perform the test of optimality:

$$\max(\mu, ||r_p||, ||r_d||) = \max(0.0332, 2.0283e - 13, 0.0145) = 0.0332 > 10^{-8}$$

The LP is not optimal. Next, we compute the coefficient matrix M of the systems (11.21) and (11.25):

$$M = AXS^{-1}A^T = \begin{bmatrix} 21,058.4373 & -18,434.8141 \\ -18,434.8141 & 18,435.5495 \end{bmatrix}$$

Then, we compute the right-hand side rhs of the system (11.21):

$$rhs = r_p + AS^{-1}(-r_c + Xr_d)) = \begin{bmatrix} 66.5825 \\ -61.0409 \end{bmatrix}$$

In the predictor step, we solve the system in (11.21) for $(\Delta x^p, \Delta w^p, \Delta s^p)$:

$$\Delta w^p = M^{-1} rhs = \begin{bmatrix} 0.0021 \\ -0.0012 \end{bmatrix},$$

$$\Delta s^p = r_d - A^T \Delta w^p = \begin{bmatrix} 0.0030 \\ 0.0049 \\ 0.0007 \\ -0.0063 \end{bmatrix},$$

$$\Delta x^p = S^{-1}(r_c - X\Delta s^p) = \begin{bmatrix} 0.1060 \\ 0.0188 \\ -0.1932 \\ 0.0581 \end{bmatrix}$$

Next, we calculate the largest possible step lengths α_p^p, α_d^p:

$$\alpha_p^p = \min\left\{1, \min_{\Delta x_i^p > 0} \frac{x_i}{\Delta x_i^p}\right\} = 0.9958,$$

$$\alpha_d^p = \min\left\{1, \min_{\Delta s_i^p > 0} \frac{s_i}{\Delta s_i^p}\right\} = 0.9356$$

Then, we compute the centering parameter σ:

$$\sigma = \left(\frac{\left(x - \alpha_p^p \Delta x^p\right)^T \left(s - \alpha_d^p \Delta s^p\right)}{n\mu}\right)^3 = 3.8374e - 07$$

In the corrector step, we solve the system in (11.25) for $(\Delta x, \Delta w, \Delta s)$:

$$r_c = r_c - \sigma\mu + \Delta X^p \Delta s^p = \begin{bmatrix} 0.0071 \\ 0.0379 \\ 0.0016 \\ 0.0861 \end{bmatrix},$$

$$rhs = r_p + AS^{-1}(-r_c + Xr_d)) = \begin{bmatrix} 66.7949 \\ -61.4553 \end{bmatrix},$$

$$\Delta w = M^{-1} rhs = \begin{bmatrix} 0.0020 \\ -0.0013 \end{bmatrix},$$

$$\Delta s = r_d - A^T \Delta w = \begin{bmatrix} 0.0032 \\ 0.0047 \\ 0.0006 \\ -0.0066 \end{bmatrix},$$

$$\Delta x = S^{-1}(r_c - X\Delta s) = \begin{bmatrix} 0.1057 \\ 0.0189 \\ -0.1926 \\ 0.0579 \end{bmatrix}$$

Next, we calculate the primal and dual step lengths α_p, α_d:

$$\alpha_p = \min\left\{1, \eta \min_{\Delta x_i > 0} \frac{x_i}{\Delta x_i}\right\} = 0.9948,$$

$$\alpha_d = \min\left\{1, \eta \min_{\Delta s_i > 0} \frac{s_i}{\Delta s_i}\right\} = 0.9974$$

where $\eta = \max(0.995, 1 - \mu) = 0.995$.

Finally, we update the solution (x, w, s):

$$x = x - \alpha_p \Delta x = \begin{bmatrix} 2.0005 \\ 0.0001 \\ 2.9990 \\ 0.0003 \end{bmatrix},$$

$$w = w - \alpha_d \Delta w = \begin{bmatrix} 0.5000 \\ 1.0000 \end{bmatrix},$$

$$s = s - \alpha_d \Delta s = \begin{bmatrix} 0.0000 \\ 2.0000 \\ 0.0000 \\ 1.5000 \end{bmatrix}$$

4th Iteration

Initially, we calculate the primal, dual, and complementarity residuals and the duality measure:

$$r_d = A^T w + s - c = \begin{bmatrix} 2 & 0 \\ 0 & -2 \\ 2 & -2 \\ 3 & -6 \end{bmatrix} \begin{bmatrix} 0.5000 \\ 1.0000 \end{bmatrix} + \begin{bmatrix} 0.0000 \\ 2.0000 \\ 0.0000 \\ 1.5000 \end{bmatrix} - \begin{bmatrix} 1 \\ 0 \\ -1 \\ -3 \end{bmatrix} = \begin{bmatrix} 1.8990e - 05 \\ 1.8990e - 05 \\ 1.8990e - 05 \\ 1.8990e - 05 \end{bmatrix},$$

$$r_p = Ax - b = \begin{bmatrix} 2 & 0 & 2 & 3 \\ 0 & -2 & -2 & -6 \end{bmatrix} \begin{bmatrix} 2.0005 \\ 0.0001 \\ 2.9990 \\ 0.0003 \end{bmatrix} - \begin{bmatrix} 10 \\ -6 \end{bmatrix} = \begin{bmatrix} -1.0658e - 14 \\ 5.3291e - 15 \end{bmatrix},$$

$$r_c = Xs = \begin{bmatrix} 2.0005 & 0 & 0 & 0 \\ 0 & 0.0001 & 0 & 0 \\ 0 & 0 & 2.9990 & 0 \\ 0 & 0 & 0 & 0.0003 \end{bmatrix} \begin{bmatrix} 0.0000 \\ 2.0000 \\ 0.0000 \\ 1.5000 \end{bmatrix} = \begin{bmatrix} 3.2108e - 05 \\ 0.0002 \\ 1.3829e - 05 \\ 0.0004 \end{bmatrix},$$

$$\mu = x^T s/n = \begin{bmatrix} 2.0005 & 0.0001 & 2.9990 & 0.0003 \end{bmatrix} \begin{bmatrix} 0.0000 \\ 2.0000 \\ 0.0000 \\ 1.5000 \end{bmatrix} /4 = 1.6841e - 04$$

Then, we perform the test of optimality:

$$\max(\mu, ||r_p||, ||r_d||) = \max(1.6841e - 04, 1.1916e - 14, 3.7980e - 05)$$
$$= 1.6841e - 04 > 10^{-8}$$

The LP is not optimal. Next, we compute the coefficient matrix M of the systems (11.21) and (11.25):

$$M = AXS^{-1}A^T = \begin{bmatrix} 3,100,214.2529 & -2,601,627.0223 \\ -2,601,627.0223 & 2,601,627.0260 \end{bmatrix}$$

Then, we compute the right-hand side rhs of the system (11.21):

$$rhs = r_p + AS^{-1}(-r_c + Xr_d)) = \begin{bmatrix} 19.4367 \\ -18.7026 \end{bmatrix}$$

In the predictor step, we solve the system in (11.21) for $(\Delta x^p, \Delta w^p, \Delta s^p)$:

$$\Delta w^p = M^{-1}rhs = \begin{bmatrix} 1.4724e - 06 \\ -5.7164e - 06 \end{bmatrix},$$

$$\Delta s^p = r_d - A^T\Delta w^p = \begin{bmatrix} 1.6045e - 05 \\ 7.5573e - 06 \\ 4.6125e - 06 \\ -1.9726e - 05 \end{bmatrix},$$

$$\Delta x^p = S^{-1}(r_c - X\Delta s^p) = \begin{bmatrix} 0.0005 \\ 9.6780e - 05 \\ -0.0010 \\ 0.0003 \end{bmatrix}$$

Next, we calculate the largest possible step lengths α_p^p, α_d^p:

$$\alpha_p^p = \min \left\{ 1, \min_{\Delta x_i^p > 0} \frac{x_i}{\Delta x_i^p} \right\} = 1.0000,$$

$$\alpha_d^p = \min \left\{ 1, \min_{\Delta s_i^p > 0} \frac{s_i}{\Delta s_i^p} \right\} = 0.9997$$

Then, we compute the centering parameter σ:

$$\sigma = \left(\frac{\left(x - \alpha_p^p \Delta x^p \right)^T \left(s - \alpha_d^p \Delta s^p \right)}{n\mu} \right)^3 = 3.5402e - 14$$

In the corrector step, we solve the system in (11.25) for $(\Delta x, \Delta w, \Delta s)$:

$$r_c = r_c - \sigma\mu + \Delta X^p \Delta s^p = \begin{bmatrix} 3.2116e-05 \\ 0.0002 \\ 1.3824-05 \\ 0.0004 \end{bmatrix},$$

$$rhs = r_p + AS^{-1}(-r_c + Xr_d)) = \begin{bmatrix} 19.4376 \\ -18.7045 \end{bmatrix},$$

$$\Delta w = M^{-1}rhs = \begin{bmatrix} 1.4703e-06 \\ -5.7193e-06 \end{bmatrix},$$

$$\Delta s = r_d - A^T \Delta w = \begin{bmatrix} 1.6050e-05 \\ 7.5515e-06 \\ 4.6110e-06 \\ -1.9736e-05 \end{bmatrix},$$

$$\Delta x = S^{-1}(r_c - X\Delta s) = \begin{bmatrix} 0.0005 \\ 9.6781e-05 \\ -0.0009 \\ 0.0002 \end{bmatrix}$$

Next, we calculate the primal and dual step lengths α_p, α_d:

$$\alpha_p = \min\left\{1, \eta \min_{\Delta x_i > 0} \frac{x_i}{\Delta x_i}\right\} = 0.9998,$$

$$\alpha_d = \min\left\{1, \eta \min_{\Delta s_i > 0} \frac{s_i}{\Delta s_i}\right\} = 0.9998$$

where $\eta = \max(0.995, 1 - \mu) = 0.9998$.
Finally, we update the solution (x, w, s):

$$x = x - \alpha_p \Delta x = \begin{bmatrix} 2.0000 \\ 0.0000 \\ 3.0000 \\ 0.0000 \end{bmatrix},$$

$$w = w - \alpha_d \Delta w = \begin{bmatrix} 0.5000 \\ 1.0000 \end{bmatrix},$$

$$s = s - \alpha_d \Delta s = \begin{bmatrix} 0.0000 \\ 2.0000 \\ 0.0000 \\ 1.5000 \end{bmatrix}$$

5th Iteration

Initially, we calculate the primal, dual, and complementarity residuals and the duality measure:

$$r_d = A^T w + s - c = \begin{bmatrix} 2 & 0 \\ 0 & -2 \\ 2 & -2 \\ 3 & -6 \end{bmatrix} \begin{bmatrix} 0.5000 \\ 1.0000 \end{bmatrix} + \begin{bmatrix} 0.0000 \\ 2.0000 \\ 0.0000 \\ 1.5000 \end{bmatrix} - \begin{bmatrix} 1 \\ 0 \\ -1 \\ -3 \end{bmatrix} = \begin{bmatrix} 3.1969e-09 \\ 3.1969e-09 \\ 3.1969e-09 \\ 3.1969e-09 \end{bmatrix},$$

$$r_p = Ax - b = \begin{bmatrix} 2 & 0 & 2 & 3 \\ 0 & -2 & -2 & -6 \end{bmatrix} \begin{bmatrix} 2.0000 \\ 0.0000 \\ 3.0000 \\ 0.0000 \end{bmatrix} - \begin{bmatrix} 10 \\ -6 \end{bmatrix} = \begin{bmatrix} 5.3291e-15 \\ -3.5527e-15 \end{bmatrix},$$

$$r_c = Xs = \begin{bmatrix} 2.0000 & 0 & 0 & 0 \\ 0 & 0.0000 & 0 & 0 \\ 0 & 0 & 3.0000 & 0 \\ 0 & 0 & 0 & 0.0000 \end{bmatrix} \begin{bmatrix} 0.0000 \\ 2.0000 \\ 0.0000 \\ 1.5000 \end{bmatrix} = \begin{bmatrix} 5.4059e-09 \\ 3.2599e-08 \\ 2.3302e-09 \\ 7.3118e-08 \end{bmatrix},$$

$$\mu = x^T s/n = \begin{bmatrix} 2.0000 & 0.0000 & 3.0000 & 0.0000 \end{bmatrix} \begin{bmatrix} 0.0000 \\ 2.0000 \\ 0.0000 \\ 1.5000 \end{bmatrix} /4 = 2.8363e-08$$

Then, we perform the test of optimality:

$$\max(\mu, ||r_p||, ||r_d||) = \max(2.8363e-08, 6.4047e-15, 6.3938e-09)$$
$$= 2.8363e-08 > 10^{-8}$$

The LP is not optimal. Next, we compute the coefficient matrix M of the systems (11.21) and (11.25):

$$M = AXS^{-1}A^T = \begin{bmatrix} 18,408,951,851.68 & -15,449,236,545.203 \\ -15,449,236,545.203 & 15,449,236,545.203 \end{bmatrix}$$

Then, we compute the right-hand side *rhs* of the system (11.21):

$$rhs = r_p + AS^{-1}(-r_c + Xr_d)) = \begin{bmatrix} 19.4258 \\ -18.6948 \end{bmatrix}$$

In the predictor step, we solve the system in (11.21) for $(\Delta x^p, \Delta w^p, \Delta s^p)$:

$$\Delta w^p = M^{-1} rhs = \begin{bmatrix} 2.4697e - 10 \\ -9.6311e - 10 \end{bmatrix},$$

$$\Delta s^p = r_d - A^T \Delta w^p = \begin{bmatrix} 2.7030e - 09 \\ 1.2707e - 09 \\ 7.7674e - 10 \\ -3.3227e - 09 \end{bmatrix},$$

$$\Delta x^p = S^{-1}(r_c - X\Delta s^p) = \begin{bmatrix} 8.9418e - 08 \\ 1.6300e - 08 \\ -1.6254e - 07 \\ 4.8746e - 08 \end{bmatrix}.$$

Next, we calculate the largest possible step lengths α_p^p, α_d^p:

$$\alpha_p^p = \min \left\{ 1, \min_{\Delta x_i^p > 0} \frac{x_i}{\Delta x_i^p} \right\} = 1.0000,$$

$$\alpha_d^p = \min \left\{ 1, \min_{\Delta s_i^p > 0} \frac{s_i}{\Delta s_i^p} \right\} = 1.0000$$

Then, we compute the centering parameter σ:

$$\sigma = \left(\frac{\left(x - \alpha_p^p \Delta x^p\right)^T \left(s - \alpha_d^p \Delta s^p\right)}{n\mu} \right)^3 = 1.6920e - 25$$

In the corrector step, we solve the system in (11.25) for $(\Delta x, \Delta w, \Delta s)$:

$$r_c = r_c - \sigma\mu + \Delta X^p \Delta s^p = \begin{bmatrix} 5.4059e - 09 \\ 3.2600e - 08 \\ 2.3302e - 09 \\ 7.3118e - 08 \end{bmatrix},$$

$$rhs = r_p + AS^{-1}(-r_c + Xr_d)) = \begin{bmatrix} 19.4258 \\ -18.6948 \end{bmatrix},$$

$$\Delta w = M^{-1} rhs = \begin{bmatrix} 2.4697e - 10 \\ -9.6311e - 10 \end{bmatrix},$$

$$\Delta s = r_d - A^T \Delta w = \begin{bmatrix} 2.7030e - 09 \\ 1.2707e - 09 \\ 7.7674e - 10 \\ -3.3227e - 09 \end{bmatrix},$$

$$\Delta x = S^{-1}(r_c - X\Delta s) = \begin{bmatrix} 8.9418e - 08 \\ 1.6300e - 08 \\ -1.6254e - 07 \\ 4.8746e - 08 \end{bmatrix}$$

Next, we calculate the primal and dual step lengths α_p, α_d:

$$\alpha_p = \min\left\{1, \eta \min_{\Delta x_i > 0} \frac{x_i}{\Delta x_i}\right\} = 1.0000,$$

$$\alpha_d = \min\left\{1, \eta \min_{\Delta s_i > 0} \frac{s_i}{\Delta s_i}\right\} = 1.0000$$

where $\eta = \max(0.995, 1 - \mu) = 1.0000$.

Finally, we update the solution (x, w, s):

$$x = x - \alpha_p \Delta x = \begin{bmatrix} 2.0000 \\ 0.0000 \\ 3.0000 \\ 0.0000 \end{bmatrix},$$

$$w = w - \alpha_d \Delta w = \begin{bmatrix} 0.5000 \\ 1.0000 \end{bmatrix},$$

$$s = s - \alpha_d \Delta s = \begin{bmatrix} 0.0000 \\ 2.0000 \\ 0.0000 \\ 1.5000 \end{bmatrix}$$

6th Iteration

Initially, we calculate the primal, dual, and complementarity residuals and the duality measure:

$$r_d = A^T w + s - c = \begin{bmatrix} 2 & 0 \\ 0 & -2 \\ 2 & -2 \\ 3 & -6 \end{bmatrix} \begin{bmatrix} 0.5000 \\ 1.0000 \end{bmatrix} + \begin{bmatrix} 0.0000 \\ 2.0000 \\ 0.0000 \\ 1.5000 \end{bmatrix} - \begin{bmatrix} 1 \\ 0 \\ -1 \\ -3 \end{bmatrix} = \begin{bmatrix} 0 \\ 0 \\ 0 \\ 0 \end{bmatrix},$$

$$r_p = Ax - b = \begin{bmatrix} 2 & 0 & 2 & 3 \\ 0 & -2 & -2 & -6 \end{bmatrix} \begin{bmatrix} 2.0000 \\ 0.0000 \\ 3.0000 \\ 0.0000 \end{bmatrix} - \begin{bmatrix} 10 \\ -6 \end{bmatrix} = \begin{bmatrix} 0 \\ -6.2172e - 15 \end{bmatrix},$$

$$r_c = Xs = \begin{bmatrix} 2.0000 & 0 & 0 & 0 \\ 0 & 0.0000 & 0 & 0 \\ 0 & 0 & 3.0000 & 0 \\ 0 & 0 & 0 & 0.0000 \end{bmatrix} \begin{bmatrix} 0.0000 \\ 2.0000 \\ 0.0000 \\ 1.5000 \end{bmatrix} = \begin{bmatrix} 1.5333e - 16 \\ 9.2462e - 16 \\ 6.6093e - 17 \\ 2.0739e - 15 \end{bmatrix},$$

$$\mu = x^T s/n = [2.0000 \; 0.0000 \; 3.0000 \; 0.0000] \begin{bmatrix} 0.0000 \\ 2.0000 \\ 0.0000 \\ 1.5000 \end{bmatrix} /4 = 8.0448e - 16$$

Then, we perform the test of optimality:

$$\max(\mu, \|r_p\|, \|r_d\|) = \max(8.0448e - 16, 6.2172e - 15, 0) = 6.2172e - 15 < 10^{-8}$$

The LP is optimal. The solution of the LP problem is:

$$x = \begin{bmatrix} 2 \\ 0 \\ 3 \\ 0 \end{bmatrix}, w = \begin{bmatrix} 0.5 \\ 1 \end{bmatrix}, s = \begin{bmatrix} 0 \\ 2 \\ 0 \\ 1.5 \end{bmatrix}$$

$$z = c^T x = -1$$

Example 2 The LP problem that will be solved is the following:

$$\begin{aligned} \min z = & \; 5x_1 + 2x_2 - 4x_3 \\ \text{s.t.} \quad & 6x_1 + x_2 - 2x_3 \geq 5 \\ & x_1 + x_2 + x_3 \leq 4 \\ & 6x_1 + 4x_2 - 2x_3 \geq 10 \\ x_j \geq 0, \quad & (j = 1, 2, 3) \end{aligned}$$

In matrix notation the above LP problem is written as follows:

$$A = \begin{bmatrix} 6 & 1 & -2 \\ 1 & 1 & 1 \\ 6 & 4 & -2 \end{bmatrix}, c = \begin{bmatrix} 5 \\ 2 \\ -4 \end{bmatrix}, b = \begin{bmatrix} 5 \\ 4 \\ 10 \end{bmatrix}, Eqin = \begin{bmatrix} 1 \\ -1 \\ 1 \end{bmatrix}$$

Initially, we need to convert the LP problem in its standard form:

$$\begin{aligned} \min z = & \; 5x_1 + 2x_2 - 4x_3 \\ \text{s.t.} \quad & 6x_1 + x_2 - 2x_3 - x_4 \qquad\qquad = 5 \\ & x_1 + x_2 + x_3 \qquad + x_5 \quad = 4 \\ & 6x_1 + 4x_2 - 2x_3 \qquad\quad - x_6 = 10 \\ x_j \geq 0, \quad & (j = 1, 2, 3, 4, 5, 6) \end{aligned}$$

In matrix notation the above LP problem is written as follows:

$$A = \begin{bmatrix} 6 & 1 & -2 & -1 & 0 & 0 \\ 1 & 1 & 1 & 0 & 1 & 0 \\ 6 & 4 & -2 & 0 & 0 & -1 \end{bmatrix}, c = \begin{bmatrix} 5 \\ 2 \\ -4 \\ 0 \\ 0 \\ 0 \end{bmatrix}, b = \begin{bmatrix} 5 \\ 4 \\ 10 \end{bmatrix}, Eqin = \begin{bmatrix} 0 \\ 0 \\ 0 \end{bmatrix}$$

Using Equation (11.19), we compute a starting point:

$$\bar{x} = A^T(AA^T)^{-1}b = \begin{bmatrix} 6 & 1 & 6 \\ 1 & 1 & 4 \\ -2 & 1 & -2 \\ -1 & 0 & 0 \\ 0 & 1 & 0 \\ 0 & 0 & -1 \end{bmatrix} \left(\begin{bmatrix} 6 & 1 & -2 & -1 & 0 & 0 \\ 1 & 1 & 1 & 0 & 1 & 0 \\ 6 & 4 & -2 & 0 & 0 & -1 \end{bmatrix} \begin{bmatrix} 6 & 1 & 6 \\ 1 & 1 & 4 \\ -2 & 1 & -2 \\ -1 & 0 & 0 \\ 0 & 1 & 0 \\ 0 & 0 & -1 \end{bmatrix} \right)^{-1} \begin{bmatrix} 5 \\ 4 \\ 10 \end{bmatrix}$$

$$= \begin{bmatrix} 0.8805 \\ 1.5165 \\ 0.7902 \\ 0.2195 \\ 0.8128 \\ -0.2308 \end{bmatrix},$$

$$\bar{w} = (AA^T)^{-1}Ac =$$

$$\left(\begin{bmatrix} 6 & 1 & -2 & -1 & 0 & 0 \\ 1 & 1 & 1 & 0 & 1 & 0 \\ 6 & 4 & -2 & 0 & 0 & -1 \end{bmatrix} \begin{bmatrix} 6 & 1 & 6 \\ 1 & 1 & 4 \\ -2 & 1 & -2 \\ -1 & 0 & 0 \\ 0 & 1 & 0 \\ 0 & 0 & -1 \end{bmatrix} \right)^{-1} \begin{bmatrix} 6 & 1 & -2 & -1 & 0 & 0 \\ 1 & 1 & 1 & 0 & 1 & 0 \\ 6 & 4 & -2 & 0 & 0 & -1 \end{bmatrix} \begin{bmatrix} 5 \\ 2 \\ -4 \\ 0 \\ 0 \\ 0 \end{bmatrix} =$$

$$\begin{bmatrix} 0.4076 \\ -1.03470.6376 \end{bmatrix},$$

$$\bar{s} = c - A^T\bar{w} = \begin{bmatrix} 5 \\ 2 \\ -4 \\ 0 \\ 0 \\ 0 \end{bmatrix} - \begin{bmatrix} 6 & 1 & 6 \\ 1 & 1 & 4 \\ -2 & 1 & -2 \\ -1 & 0 & 0 \\ 0 & 1 & 0 \\ 0 & 0 & -1 \end{bmatrix} \begin{bmatrix} 0.4076 \\ -1.03470.6376 \end{bmatrix} = \begin{bmatrix} -0.2365 \\ 0.0767 \\ -0.8749 \\ 0.4076 \\ 1.0347 \\ 0.6376 \end{bmatrix},$$

$$\delta x = \max(-1.5\min(\bar{x}), 0) = \max(-1.5 \times -0.2308), 0) = 0.3462,$$

$$\delta s = \max(-1.5\min(\bar{s}), 0) = \max(-1.5 \times -0.8749, 0) = 1.3123,$$

$$\bar{\delta}_x = \delta_x + 0.5\frac{(\bar{x} + \delta_x e)^T(\bar{s} + \delta se)}{\sum_{i=1}^{n}\bar{s}_i + \delta_s} = 0.8128,$$

$$\bar{\delta}_s = \delta_s + 0.5\frac{(\bar{x} + \delta_x e)^T(\bar{s} + \delta se)}{\sum_{i=1}^{n}\bar{x}_i + \delta_x} = 1.9984,$$

$$x = \bar{x} + \bar{\delta}_x e = \begin{bmatrix} 1.6934 \\ 2.3294 \\ 1.6030 \\ 1.0323 \\ 1.6256 \\ 0.5820 \end{bmatrix},$$

$$w = \bar{w} = \begin{bmatrix} 0.4076 \\ -1.0347 \\ 0.6376 \end{bmatrix},$$

$$s = \bar{s} + \bar{\delta}_s e = \begin{bmatrix} 1.7619 \\ 2.0750 \\ 1.1235 \\ 2.4059 \\ 3.0331 \\ 2.6360 \end{bmatrix}$$

1st Iteration

Initially, we calculate the primal, dual, and complementarity residuals and the duality measure:

$$r_d = A^T w + s - c = \begin{bmatrix} 6 & 1 & 6 \\ 1 & 1 & 4 \\ -2 & 1 & -2 \\ -1 & 0 & 0 \\ 0 & 1 & 0 \\ 0 & 0 & -1 \end{bmatrix}\begin{bmatrix} 0.4076 \\ -1.0347 \\ 0.6376 \end{bmatrix} + \begin{bmatrix} 1.7619 \\ 2.0750 \\ 1.1235 \\ 2.4059 \\ 3.0331 \\ 2.6360 \end{bmatrix} - \begin{bmatrix} 5 \\ 2 \\ -4 \\ 0 \\ 0 \\ 0 \end{bmatrix} = \begin{bmatrix} 1.9984 \\ 1.9984 \\ 1.9984 \\ 1.9984 \\ 1.9984 \\ 1.9984 \end{bmatrix},$$

$$r_p = Ax - b = \begin{bmatrix} 6 & 1 & -2 & -1 & 0 & 0 \\ 1 & 1 & 1 & 0 & 1 & 0 \\ 6 & 4 & -2 & 0 & 0 & -1 \end{bmatrix} \begin{bmatrix} 1.6934 \\ 2.3294 \\ 1.6030 \\ 1.0323 \\ 1.6256 \\ 0.5820 \end{bmatrix} - \begin{bmatrix} 5 \\ 4 \\ 10 \end{bmatrix} = \begin{bmatrix} 3.2512 \\ 3.2512 \\ 5.6897 \end{bmatrix},$$

$$r_c = Xs = \begin{bmatrix} 1.6934 & 0 & 0 & 0 & 0 & 0 \\ 0 & 2.3294 & 0 & 0 & 0 & 0 \\ 0 & 0 & 1.6030 & 0 & 0 & 0 \\ 0 & 0 & 0 & 1.0323 & 0 & 0 \\ 0 & 0 & 0 & 0 & 1.6256 & 0 \\ 0 & 0 & 0 & 0 & 0 & 0.5820 \end{bmatrix} \begin{bmatrix} 1.7619 \\ 2.0750 \\ 1.1235 \\ 2.4059 \\ 3.0331 \\ 2.6360 \end{bmatrix} = \begin{bmatrix} 2.9835 \\ 4.8335 \\ 1.8009 \\ 2.4838 \\ 4.9304 \\ 1.5341 \end{bmatrix},$$

$$\mu = x^T s/n = \begin{bmatrix} 1.6934 & 2.3294 & 1.6030 & 1.0323 & 1.6256 & 0.5820 \end{bmatrix} \begin{bmatrix} 1.7619 \\ 2.0750 \\ 1.1235 \\ 2.4059 \\ 3.0331 \\ 2.6360 \end{bmatrix} /4 = 3.0943$$

Then, we perform the test of optimality:

$$\max(\mu, ||r_p||, ||r_d||) = \max(3.0943, 7.3153, 4.8949) = 7.3153 > 10^{-8}$$

The LP is not optimal. Next, we compute the coefficient matrix M of the systems (11.21) and (11.25):

$$M = AXS^{-1}A^T = \begin{bmatrix} 41.8591 & 4.0356 & 44.7977 \\ 4.0356 & 4.0465 & 7.4033 \\ 44.7977 & 7.4033 & 58.4893 \end{bmatrix}$$

Then, we compute the right-hand side rhs of the system (11.21):

$$rhs = r_p + AS^{-1}(-r_c + Xr_d)) = \begin{bmatrix} 2.2071 \\ 4.0862 \\ 4.3532 \end{bmatrix}$$

In the predictor step, we solve the system in (11.21) for $(\Delta x^p, \Delta w^p, \Delta s^p)$:

$$\Delta w^p = M^{-1} rhs = \begin{bmatrix} 0.1094 \\ 1.1945 \\ -0.1605 \end{bmatrix},$$

$$\Delta s^p = r_d - A^T \Delta w^p = \begin{bmatrix} 1.1109 \\ 1.3366 \\ 0.7016 \\ 2.1077 \\ 0.8039 \\ 1.8378 \end{bmatrix},$$

$$\Delta x^p = S^{-1}(r_c - X\Delta s^p) = \begin{bmatrix} 0.6257 \\ 0.8289 \\ 0.6019 \\ 0.1280 \\ 1.1947 \\ 0.1762 \end{bmatrix}$$

Next, we calculate the largest possible step lengths α_p^p, α_d^p:

$$\alpha_p^p = \min\left\{ 1, \min_{\Delta x_i^p > 0} \frac{x_i}{\Delta x_i^p} \right\} = 1,$$

$$\alpha_d^p = \min\left\{ 1, \min_{\Delta s_i^p > 0} \frac{s_i}{\Delta s_i^p} \right\} = 1$$

Then, we compute the centering parameter σ:

$$\sigma = \left(\frac{\left(x - \alpha_p^p \Delta x^p\right)^T \left(s - \alpha_d^p \Delta s^p\right)}{n\mu} \right)^3 = 0.0084$$

In the corrector step, we solve the system in (11.25) for $(\Delta x, \Delta w, \Delta s)$:

$$r_c = r_c - \sigma\mu + \Delta X^p \Delta s^p = \begin{bmatrix} 3.6524 \\ 5.9153 \\ 2.1971 \\ 2.7274 \\ 5.8648 \\ 1.8318 \end{bmatrix},$$

$$rhs = r_p + AS^{-1}(-r_c + Xr_d)) = \begin{bmatrix} 0.2142 \\ 2.5245 \\ 0.8080 \end{bmatrix},$$

$$\Delta w = M^{-1}rhs = \begin{bmatrix} 0.1299 \\ 0.8473 \\ -0.1929 \end{bmatrix},$$

$$\Delta s = r_d - A^T \Delta w = \begin{bmatrix} 1.5292 \\ 1.7929 \\ 1.0250 \\ 2.1282 \\ 1.1511 \\ 1.8054 \end{bmatrix},$$

$$\Delta x = S^{-1}(r_c - X\Delta s) = \begin{bmatrix} 0.6033 \\ 0.8381 \\ 0.4931 \\ 0.2204 \\ 1.3167 \\ 0.2963 \end{bmatrix}$$

Next, we calculate the primal and dual step lengths α_p, α_d:

$$\alpha_p = \min \left\{ 1, \eta \min_{\Delta x_i > 0} \frac{x_i}{\Delta x_i} \right\} = 1,$$

$$\alpha_d = \min \left\{ 1, \eta \min_{\Delta s_i > 0} \frac{s_i}{\Delta s_i} \right\} = 1$$

where $\eta = \max(0.995, 1 - \mu) = 0.995$.

Finally, we update the solution (x, w, s):

$$x = x - \alpha_p \Delta x = \begin{bmatrix} 1.0901 \\ 1.4912 \\ 1.1098 \\ 0.8119 \\ 0.3089 \\ 0.2857 \end{bmatrix},$$

$$w = w - \alpha_d \Delta w = \begin{bmatrix} 0.2777 \\ -1.8820 \\ 0.8305 \end{bmatrix},$$

$$s = s - \alpha_d \Delta s = \begin{bmatrix} 0.2327 \\ 0.2822 \\ 0.0984 \\ 0.2777 \\ 1.8820 \\ 0.8305 \end{bmatrix}$$

2nd Iteration

Initially, we calculate the primal, dual, and complementarity residuals and the duality measure:

$$
r_d = A^T w + s - c = \begin{bmatrix} 6 & 1 & 6 \\ 1 & 1 & 4 \\ -2 & 1 & -2 \\ -1 & 0 & 0 \\ 0 & 1 & 0 \\ 0 & 0 & -1 \end{bmatrix} \begin{bmatrix} 0.2777 \\ -1.8820 \\ 0.8305 \end{bmatrix} + \begin{bmatrix} 0.2327 \\ 0.2822 \\ 0.0984 \\ 0.2777 \\ 1.8820 \\ 0.8305 \end{bmatrix} - \begin{bmatrix} 5 \\ 2 \\ -4 \\ 0 \\ 0 \\ 0 \end{bmatrix} = \begin{bmatrix} 0 \\ 0 \\ 0.4441 \\ -0.1110 \\ 0 \\ 0.1110 \end{bmatrix},
$$

$$
r_p = Ax - b = \begin{bmatrix} 6 & 1 & -2 & -1 & 0 & 0 \\ 1 & 1 & 1 & 0 & 1 & 0 \\ 6 & 4 & -2 & 0 & 0 & -1 \end{bmatrix} \begin{bmatrix} 1.0901 \\ 1.4912 \\ 1.1098 \\ 0.8119 \\ 0.3089 \\ 0.2857 \end{bmatrix} - \begin{bmatrix} 5 \\ 4 \\ 10 \end{bmatrix} = \begin{bmatrix} -2.6645e - 15 \\ 0 \\ 0 \end{bmatrix},
$$

$$
r_c = Xs = \begin{bmatrix} 1.0901 & 0 & 0 & 0 & 0 & 0 \\ 0 & 1.4912 & 0 & 0 & 0 & 0 \\ 0 & 0 & 1.1098 & 0 & 0 & 0 \\ 0 & 0 & 0 & 0.8119 & 0 & 0 \\ 0 & 0 & 0 & 0 & 0.3089 & 0 \\ 0 & 0 & 0 & 0 & 0 & 0.2857 \end{bmatrix} \begin{bmatrix} 0.2327 \\ 0.2822 \\ 0.0984 \\ 0.2777 \\ 1.8820 \\ 0.8305 \end{bmatrix} = \begin{bmatrix} 0.2536 \\ 0.4208 \\ 0.1093 \\ 0.2255 \\ 0.5813 \\ 0.2372 \end{bmatrix},
$$

$$
\mu = x^T s/n = \begin{bmatrix} 1.0901 & 1.4912 & 1.1098 & 0.8119 & 0.3089 & 0.2857 \end{bmatrix} \begin{bmatrix} 0.2327 \\ 0.2822 \\ 0.0984 \\ 0.2777 \\ 1.8820 \\ 0.8305 \end{bmatrix} /4 = 0.3046
$$

Then, we perform the test of optimality:

$$
\max(\mu, ||r_p||, ||r_d||) = \max(0.3046, 2.6645e - 15, 4.7103e - 16) = 0.3046 > 10^{-8}
$$

The LP is not optimal. Next, we compute the coefficient matrix M of the systems (11.21) and (11.25):

$$
M = AXS^{-1}A^T = \begin{bmatrix} 221.9758 & 10.8483 & 234.9062 \\ 10.8483 & 21.4083 & 26.7027 \\ 234.9062 & 26.7027 & 298.6675 \end{bmatrix}
$$

Then, we compute the right-hand side *rhs* of the system (11.21):

$$rhs = r_p + AS^{-1}(-r_c + Xr_d)) = \begin{bmatrix} -5.0000 \\ -4.0000 \\ -10.0000 \end{bmatrix}$$

In the predictor step, we solve the system in (11.21) for $(\Delta x^p, \Delta w^p, \Delta s^p)$:

$$\Delta w^p = M^{-1}rhs = \begin{bmatrix} 0.0380 \\ -0.1430 \\ -0.0506 \end{bmatrix},$$

$$\Delta s^p = r_d - A^T \Delta w^p = \begin{bmatrix} 0.2186 \\ 0.3073 \\ 0.1178 \\ 0.0380 \\ 0.1430 \\ -0.0506 \end{bmatrix},$$

$$\Delta x^p = S^{-1}(r_c - X\Delta s^p) = \begin{bmatrix} 0.0660 \\ -0.1326 \\ -0.2188 \\ 0.7009 \\ 0.2854 \\ 0.3030 \end{bmatrix}$$

Next, we calculate the largest possible step lengths α_p^p, α_d^p:

$$\alpha_p^p = \min \left\{ 1, \min_{\Delta x_i^p > 0} \frac{x_i}{\Delta x_i^p} \right\} = 0.9426,$$

$$\alpha_d^p = \min \left\{ 1, \min_{\Delta s_i^p > 0} \frac{s_i}{\Delta s_i^p} \right\} = 0.8353$$

Then, we compute the centering parameter σ:

$$\sigma = \left(\frac{\left(x - \alpha_p^p \Delta x^p\right)^T \left(s - \alpha_d^p \Delta s^p\right)}{n\mu} \right)^3 = 0.0013$$

In the corrector step, we solve the system in (11.25) for $(\Delta x, \Delta w, \Delta s)$:

$$r_c = r_c - \sigma\mu + \Delta X^p \Delta s^p = \begin{bmatrix} 0.2676 \\ 0.3796 \\ 0.0831 \\ 0.2517 \\ 0.6217 \\ 0.2215 \end{bmatrix},$$

$$rhs = r_p + AS^{-1}(-r_c + Xr_d)) = \begin{bmatrix} -5.6534 \\ -3.6700 \\ -10.3292 \end{bmatrix},$$

$$\Delta w = M^{-1} rhs = \begin{bmatrix} 0.0316 \\ -0.1275 \\ -0.0480 \end{bmatrix},$$

$$\Delta s = r_d - A^T \Delta w = \begin{bmatrix} 0.2262 \\ 0.2881 \\ 0.0947 \\ 0.0316 \\ 0.1275 \\ -0.0480 \end{bmatrix},$$

$$\Delta x = S^{-1}(r_c - X\Delta s) = \begin{bmatrix} 0.0907 \\ -0.1769 \\ -0.2232 \\ 0.8140 \\ 0.3094 \\ 0.2832 \end{bmatrix}.$$

Next, we calculate the primal and dual step lengths α_p, α_d:

$$\alpha_p = \min\left\{1, \eta \min_{\Delta x_i > 0} \frac{x_i}{\Delta x_i}\right\} = 0.9925,$$

$$\alpha_d = \min\left\{1, \eta \min_{\Delta s_i > 0} \frac{s_i}{\Delta s_i}\right\} = 0.9747$$

where $\eta = \max(0.995, 1 - \mu) = 0.995$.
Finally, we update the solution (x, w, s):

$$x = x - \alpha_p \Delta x = \begin{bmatrix} 1.0000 \\ 1.6668 \\ 1.3314 \\ 0.0041 \\ 0.0018 \\ 0.0045 \end{bmatrix},$$

$$w = w - \alpha_d \Delta w = \begin{bmatrix} 0.2469 \\ -1.7577 \\ 0.8773 \end{bmatrix},$$

$$s = s - \alpha_d \Delta s = \begin{bmatrix} 0.0122 \\ 0.0014 \\ 0.0062 \\ 0.2469 \\ 1.7577 \\ 0.8773 \end{bmatrix}$$

3rd Iteration

Initially, we calculate the primal, dual, and complementarity residuals and the duality measure:

$$r_d = A^T w + s - c = \begin{bmatrix} 6 & 1 & 6 \\ 1 & 1 & 4 \\ -2 & 1 & -2 \\ -1 & 0 & 0 \\ 0 & 1 & 0 \\ 0 & 0 & -1 \end{bmatrix} \begin{bmatrix} 0.2469 \\ -1.7577 \\ 0.8773 \end{bmatrix} + \begin{bmatrix} 0.0122 \\ 0.0014 \\ 0.0062 \\ 0.2469 \\ 1.7577 \\ 0.8773 \end{bmatrix} - \begin{bmatrix} 5 \\ 2 \\ -4 \\ 0 \\ 0 \\ 0 \end{bmatrix} = \begin{bmatrix} 0.8882 \\ 0 \\ 0 \\ -0.0278 \\ 0 \\ 0 \end{bmatrix},$$

$$r_p = Ax - b = \begin{bmatrix} 6 & 1 & -2 & -1 & 0 & 0 \\ 1 & 1 & 1 & 0 & 1 & 0 \\ 6 & 4 & -2 & 0 & 0 & -1 \end{bmatrix} \begin{bmatrix} 1.0000 \\ 1.6668 \\ 1.3314 \\ 0.0041 \\ 0.0018 \\ 0.0045 \end{bmatrix} - \begin{bmatrix} 5 \\ 4 \\ 10 \end{bmatrix} = \begin{bmatrix} -2.6645e - 15 \\ 0 \\ 0 \end{bmatrix},$$

$$r_c = Xs = \begin{bmatrix} 1.0000 & 0 & 0 & 0 & 0 & 0 \\ 0 & 1.6668 & 0 & 0 & 0 & 0 \\ 0 & 0 & 1.3314 & 0 & 0 & 0 \\ 0 & 0 & 0 & 0.0041 & 0 & 0 \\ 0 & 0 & 0 & 0 & 0.0018 & 0 \\ 0 & 0 & 0 & 0 & 0 & 0.0045 \end{bmatrix} \begin{bmatrix} 0.0122 \\ 0.0014 \\ 0.0062 \\ 0.2469 \\ 1.7577 \\ 0.8773 \end{bmatrix} = \begin{bmatrix} 0.0122 \\ 0.0024 \\ 0.0082 \\ 0.0010 \\ 0.0031 \\ 0.0040 \end{bmatrix},$$

$$\mu = x^T s / n = \begin{bmatrix} 1.0000 & 1.6668 & 1.3314 & 0.0041 & 0.0018 & 0.0045 \end{bmatrix} \begin{bmatrix} 0.0122 \\ 0.0014 \\ 0.0062 \\ 0.2469 \\ 1.7577 \\ 0.8773 \end{bmatrix} / 4 = 0.0051$$

Then, we perform the test of optimality:

$$\max(\mu, ||r_p||, ||r_d||) = \max(0.0051, 3.9721e - 15, 8.8861e - 16) = 0.0051 > 10^{-8}$$

The LP is not optimal. Next, we compute the coefficient matrix M of the systems (11.21) and (11.25):

$$M = AXS^{-1}A^T = \begin{bmatrix} 4,989.2032 & 1,241.8051 & 8,533.4015 \\ 1,241.8051 & 1,478.5877 & 4,786.0199 \\ 8,533.4014 & 4,786.0199 & 22,710.2656 \end{bmatrix}$$

Then, we compute the right-hand side *rhs* of the system (11.21):

$$rhs = r_p + AS^{-1}(-r_c + Xr_d)) = \begin{bmatrix} -5.0000 \\ -4.0000 \\ -10.0000 \end{bmatrix}$$

In the predictor step, we solve the system in (11.21) for $(\Delta x^p, \Delta w^p, \Delta s^p)$:

$$\Delta w^p = M^{-1} rhs = \begin{bmatrix} -0.0031 \\ -0.0077 \\ 0.0023 \end{bmatrix},$$

$$\Delta s^p = r_d - A^T \Delta w^p = \begin{bmatrix} 0.0122 \\ 0.0014 \\ 0.0062 \\ -0.0031 \\ 0.0077 \\ 0.0023 \end{bmatrix},$$

$$\Delta x^p = S^{-1}(r_c - X\Delta s^p) = \begin{bmatrix} 0.0000 \\ 0.0001 \\ -0.0019 \\ 0.0041 \\ 0.0018 \\ 0.0045 \end{bmatrix}$$

Next, we calculate the largest possible step lengths α_p^p, α_d^p:

$$\alpha_p^p = \min\left\{ 1, \min_{\Delta x_i^p > 0} \frac{x_i}{\Delta x_i^p} \right\} = 0.9876,$$

$$\alpha_d^p = \min\left\{ 1, \min_{\Delta s_i^p > 0} \frac{s_i}{\Delta s_i^p} \right\} = 0.9986$$

Then, we compute the centering parameter σ:

$$\sigma = \left(\frac{\left(x - \alpha_p^P \Delta x^p\right)^T \left(s - \alpha_d^P \Delta s^p\right)}{n\mu} \right)^3 = 7.9151e - 08$$

In the corrector step, we solve the system in (11.25) for $(\Delta x, \Delta w, \Delta s)$:

$$r_c = r_c - \sigma\mu + \Delta X^P \Delta s^P = \begin{bmatrix} 0.0122 \\ 0.0024 \\ 0.0082 \\ 0.0010 \\ 0.0031 \\ 0.0040 \end{bmatrix},$$

$$rhs = r_p + AS^{-1}(-r_c + Xr_d)) = \begin{bmatrix} -5.0042 \\ -3.9982 \\ -10.0045 \end{bmatrix},$$

$$\Delta w = M^{-1}rhs = \begin{bmatrix} -0.0031 \\ -0.0077 \\ 0.0023 \end{bmatrix},$$

$$\Delta s = r_d - A^T \Delta w = \begin{bmatrix} 0.0122 \\ 0.0014 \\ 0.0062 \\ -0.0031 \\ 0.0077 \\ 0.0023 \end{bmatrix},$$

$$\Delta x = S^{-1}(r_c - X\Delta s) = \begin{bmatrix} 0.0000 \\ 0.0002 \\ -0.0019 \\ 0.0041 \\ 0.0018 \\ 0.0045 \end{bmatrix}$$

Next, we calculate the primal and dual step lengths α_p, α_d:

$$\alpha_p = \min\left\{1, \eta \min_{\Delta x_i > 0} \frac{x_i}{\Delta x_i}\right\} = 0.9950,$$

$$\alpha_d = \min\left\{1, \eta \min_{\Delta s_i > 0} \frac{s_i}{\Delta s_i}\right\} = 0.9950$$

where $\eta = \max(0.995, 1 - \mu) = 0.995$.

Finally, we update the solution (x, w, s):

$$x = x - \alpha_p \Delta x = \begin{bmatrix} 1.0000 \\ 1.6667 \\ 1.3333 \\ 0.0000 \\ 0.0000 \\ 0.0000 \end{bmatrix},$$

$$w = w - \alpha_d \Delta w = \begin{bmatrix} 0.2500 \\ -1.7500 \\ 0.8750 \end{bmatrix},$$

$$s = s - \alpha_d \Delta s = \begin{bmatrix} 0.0001 \\ 0.0000 \\ 0.0000 \\ 0.2500 \\ 1.7500 \\ 0.8750 \end{bmatrix}$$

4th Iteration

Initially, we calculate the primal, dual, and complementarity residuals and the duality measure:

$$r_d = A^T w + s - c = \begin{bmatrix} 6 & 1 & 6 \\ 1 & 1 & 4 \\ -2 & 1 & -2 \\ -1 & 0 & 0 \\ 0 & 1 & 0 \\ 0 & 0 & -1 \end{bmatrix} \begin{bmatrix} 0.2500 \\ -1.7500 \\ 0.8750 \end{bmatrix} + \begin{bmatrix} 0.0001 \\ 0.0000 \\ 0.0000 \\ 0.2500 \\ 1.7500 \\ 0.8750 \end{bmatrix} - \begin{bmatrix} 5 \\ 2 \\ -4 \\ 0 \\ 0 \\ 0 \end{bmatrix} = \begin{bmatrix} 0 \\ 0 \\ 0 \\ 0 \\ 0 \\ 0 \end{bmatrix},$$

$$r_p = Ax - b = \begin{bmatrix} 6 & 1 & -2 & -1 & 0 & 0 \\ 1 & 1 & 1 & 0 & 1 & 0 \\ 6 & 4 & -2 & 0 & 0 & -1 \end{bmatrix} \begin{bmatrix} 1.0000 \\ 1.6667 \\ 1.3333 \\ 0.0000 \\ 0.0000 \\ 0.0000 \end{bmatrix} - \begin{bmatrix} 5 \\ 4 \\ 10 \end{bmatrix} = \begin{bmatrix} -5.3290e - 15 \\ -1.7763e - 15 \\ -3.5527e - 15 \end{bmatrix},$$

$$r_c = Xs = \begin{bmatrix} 1.0000 & 0 & 0 & 0 & 0 & 0 \\ 0 & 1.6667 & 0 & 0 & 0 & 0 \\ 0 & 0 & 1.3333 & 0 & 0 & 0 \\ 0 & 0 & 0 & 0.0000 & 0 & 0 \\ 0 & 0 & 0 & 0 & 0.0000 & 0 \\ 0 & 0 & 0 & 0 & 0 & 0.0000 \end{bmatrix} \begin{bmatrix} 0.0001 \\ 0.0000 \\ 0.0000 \\ 0.2500 \\ 1.7500 \\ 0.8750 \end{bmatrix} = \begin{bmatrix} 0.6109 \\ 0.1183 \\ 0.4129 \\ 0.0523 \\ 0.1546 \\ 0.1979 \end{bmatrix},$$

$$\mu = x^T s/n = \begin{bmatrix} 1.0000 & 1.6667 & 1.3333 & 0.0000 & 0.0000 & 0.0000 \end{bmatrix} \begin{bmatrix} 0.0001 \\ 0.0000 \\ 0.0000 \\ 0.2500 \\ 1.7500 \\ 0.8750 \end{bmatrix} /4 = 2.5782e - 05$$

Then, we perform the test of optimality:

$$\max(\mu, ||r_p||, ||r_d||) = \max(2.5782e-05, 6.6465e-15, 0) = 2.5782e-05 > 10^{-8}$$

The LP is not optimal. Next, we compute the coefficient matrix M of the systems (11.21) and (11.25):

$$M = AXS^{-1}A^T = \begin{bmatrix} 996,385.1948 & 247,001.9395 & 1,701,064.5302 \\ 247,001.9395 & 294,314.5095 & 951,681.2750 \\ 1,701,064.5302 & 951,681.2750 & 4,519,781.8723 \end{bmatrix}$$

Then, we compute the right-hand side rhs of the system (11.21):

$$rhs = r_p + AS^{-1}(-r_c + Xr_d)) = \begin{bmatrix} -5 \\ -4 \\ -10 \end{bmatrix}$$

In the predictor step, we solve the system in (11.21) for $(\Delta x^p, \Delta w^p, \Delta s^p)$:

$$\Delta w^p = M^{-1}rhs = \begin{bmatrix} -1.5489e - 05 \\ -3.8500e - 05 \\ 1.1723e - 05 \end{bmatrix},$$

$$\Delta s^p = r_d - A^T \Delta w^p = \begin{bmatrix} 6.1091e - 05 \\ 7.0954e - 06 \\ 3.0970e - 05 \\ -1.5489e - 05 \\ 3.8500e - 05 \\ 1.1723e - 05 \end{bmatrix},$$

$$
\Delta x^p = S^{-1}(r_c - X\Delta s^p) = \begin{bmatrix} 1.9658e - 07 \\ 5.6184e - 07 \\ -9.5932e - 06 \\ 2.0928e - 05 \\ 8.8348e - 06 \\ 2.2613e - 05 \end{bmatrix}
$$

Next, we calculate the largest possible step lengths α_p^p, α_d^p:

$$
\alpha_p^p = \min \left\{ 1, \min_{\Delta x_i^p > 0} \frac{x_i}{\Delta x_i^p} \right\} = 0.9999,
$$

$$
\alpha_d^p = \min \left\{ 1, \min_{\Delta s_i^p > 0} \frac{s_i}{\Delta s_i^p} \right\} = 1.0000
$$

Then, we compute the centering parameter σ:

$$
\sigma = \left(\frac{\left(x - \alpha_p^p \Delta x^p \right)^T \left(s - \alpha_d^P \Delta s^p \right)}{n\mu} \right)^3 = 9.9722e - 15
$$

In the corrector step, we solve the system in (11.25) for $(\Delta x, \Delta w, \Delta s)$:

$$
r_c = r_c - \sigma\mu + \Delta X^p \Delta s^p = \begin{bmatrix} 0.6109 \\ 0.1183 \\ 0.4129 \\ 0.0523 \\ 0.1546 \\ 0.1979 \end{bmatrix},
$$

$$
rhs = r_p + AS^{-1}(-r_c + Xr_d)) = \begin{bmatrix} -5.0000 \\ -4.0000 \\ -10.0000 \end{bmatrix},
$$

$$
\Delta w = M^{-1}rhs = \begin{bmatrix} -1.5488e - 05 \\ -3.8500e - 05 \\ 1.1723e - 05 \end{bmatrix},
$$

$$\Delta s = r_d - A^T \Delta w = \begin{bmatrix} 6.1091e-05 \\ 7.0954e-06 \\ 3.0970e-05 \\ -1.5488e-05 \\ 3.8500e-05 \\ 1.1723e-05 \end{bmatrix},$$

$$\Delta x = S^{-1}(r_c - X\Delta s) = \begin{bmatrix} 1.9617e-07 \\ 5.6238e-07 \\ -9.5935e-06 \\ 2.0927e-05 \\ 8.8350e-06 \\ 2.2614e-05 \end{bmatrix}$$

Next, we calculate the primal and dual step lengths α_p, α_d:

$$\alpha_p = \min\left\{1, \eta \min_{\Delta x_i > 0} \frac{x_i}{\Delta x_i}\right\} = 1.0000,$$

$$\alpha_d = \min\left\{1, \eta \min_{\Delta s_i > 0} \frac{s_i}{\Delta s_i}\right\} = 1.0000$$

where $\eta = \max(0.995, 1 - \mu) = 1.0000$.

Finally, we update the solution (x, w, s):

$$x = x - \alpha_p \Delta x = \begin{bmatrix} 1.0000 \\ 1.6667 \\ 1.3333 \\ 0.0000 \\ 0.0000 \\ 0.0000 \end{bmatrix},$$

$$w = w - \alpha_d \Delta w = \begin{bmatrix} 0.2500 \\ -1.7500 \\ 0.8750 \end{bmatrix},$$

$$s = s - \alpha_d \Delta s = \begin{bmatrix} 0.0000 \\ 0.0000 \\ 0.0000 \\ 0.2500 \\ 1.7500 \\ 0.8750 \end{bmatrix}$$

5th Iteration

Initially, we calculate the primal, dual, and complementarity residuals and the duality measure:

$$
r_d = A^T w + s - c = \begin{bmatrix} 6 & 1 & 6 \\ 1 & 1 & 4 \\ -2 & 1 & -2 \\ -1 & 0 & 0 \\ 0 & 1 & 0 \\ 0 & 0 & -1 \end{bmatrix} \begin{bmatrix} 0.2500 \\ -1.7500 \\ 0.8750 \end{bmatrix} + \begin{bmatrix} 0.0000 \\ 0.0000 \\ 0.0000 \\ 0.2500 \\ 1.7500 \\ 0.8750 \end{bmatrix} - \begin{bmatrix} 5 \\ 2 \\ -4 \\ 0 \\ 0 \\ 0 \end{bmatrix} = \begin{bmatrix} 0 \\ 0.4441 \\ 0 \\ 0 \\ 0 \\ 0 \end{bmatrix},
$$

$$
r_p = Ax - b = \begin{bmatrix} 6 & 1 & -2 & -1 & 0 & 0 \\ 1 & 1 & 1 & 0 & 1 & 0 \\ 6 & 4 & -2 & 0 & 0 & -1 \end{bmatrix} \begin{bmatrix} 1.0000 \\ 1.6667 \\ 1.3333 \\ 0.0000 \\ 0.0000 \\ 0.0000 \end{bmatrix} - \begin{bmatrix} 5 \\ 4 \\ 10 \end{bmatrix} = \begin{bmatrix} 2.6645e - 15 \\ -2.2204e - 15 \\ -1.7764e - 15 \end{bmatrix},
$$

$$
r_c = Xs = \begin{bmatrix} 1.0000 & 0 & 0 & 0 & 0 & 0 \\ 0 & 1.6667 & 0 & 0 & 0 & 0 \\ 0 & 0 & 1.3333 & 0 & 0 & 0 \\ 0 & 0 & 0 & 0.0000 & 0 & 0 \\ 0 & 0 & 0 & 0 & 0.0000 & 0 \\ 0 & 0 & 0 & 0 & 0 & 0.0000 \end{bmatrix} \begin{bmatrix} 0.0000 \\ 0.0000 \\ 0.0000 \\ 0.2500 \\ 1.7500 \\ 0.8750 \end{bmatrix} = \begin{bmatrix} 0.1575 \\ 0.0305 \\ 0.1065 \\ 0.0135 \\ 0.0399 \\ 0.0510 \end{bmatrix},
$$

$$
\mu = x^T s/n = \begin{bmatrix} 1.0000 & 1.6667 & 1.3333 & 0.0000 & 0.0000 & 0.0000 \end{bmatrix} \begin{bmatrix} 0.0000 \\ 0.0000 \\ 0.0000 \\ 0.2500 \\ 1.7500 \\ 0.8750 \end{bmatrix} /4 = 6.6470e - 10
$$

Then, we perform the test of optimality:

$$
\max(\mu, ||r_p||, ||r_d||) = \max(6.6470e - 10, 3.8969e - 15, 4.4409e - 16)
$$
$$
= 6.6470e - 10 < 10^{-8}
$$

The LP is optimal. The solution of the LP problem is:

$$x = \begin{bmatrix} 1 \\ 5/3 \\ 4/3 \\ 0 \\ 0 \\ 0 \end{bmatrix}, w = \begin{bmatrix} 1/4 \\ -7/4 \\ 7/8 \end{bmatrix}, s = \begin{bmatrix} 0 \\ 0 \\ 0 \\ 1/4 \\ 7/4 \\ 7/8 \end{bmatrix}$$

$$z = c^T x = 3$$

11.6 Implementation in MATLAB

This subsection presents the implementation in MATLAB of Mehrotra's Predictor-Corrector method (filename: `ipdipm.m`). Some necessary notations should be introduced before the presentation of the implementation of Mehrotra's Predictor-Corrector method. Let A be a $m \times n$ matrix with the coefficients of the constraints, c be a $n \times 1$ vector with the coefficients of the objective function, b be a $m \times 1$ vector with the right-hand side of the constraints, $Eqin$ be a $m \times 1$ vector with the type of the constraints, $MinMaxLP$ be the type of optimization, $c0$ be the constant term of the objective function, $maxIterations$ be the number of the maximum number of iterations to perform if an optimal solution is not found, tol be the tolerance for the termination criterion, $etaMin$ be the parameter η, $scalingTechnique$ be the scaling method (0: no scaling, 1: arithmetic mean, 2: de Buchet for the case p = 1, 3: de Buchet for the case p = 2, 4: de Buchet for the case p = Inf, 5: entropy, 6: equilibration, 7: geometric mean, 8: IBM MPSX, 9: LP-norm for the case p = 1, 10: LP-norm for the case p = 2, 11: LP-norm for the case p = Inf), $xsol$ be a $n \times 1$ vector with the solution found, $fval$ be the value of the objective function at the solution $xsol$, $exitflag$ be the reason that the algorithm terminated (0: optimal solution found, 1: the LP problem is infeasible, -1: the input data is not logically or numerically correct), and $iterations$ be the number of iterations.

The function for the implementation of Mehrotra's Predictor-Corrector method takes as input the matrix with the coefficients of the constraints (matrix A), the vector with the coefficients of the objective function (vector c), the vector with the right-hand side of the constraints (vector b), the vector with the type of the constraints (vector $Eqin$), the type of optimization (variable $MinMaxLP$), the constant term of the objective function (variable $c0$), the number of the maximum number of iterations to perform if an optimal solution is not found (variable $maxIterations$), the tolerance for the termination criterion (variable tol), the parameter η (variable $etaMin$), and the scaling method (variable $scalingTechnique$), and returns as output the solution found (vector $xsol$), the value of the objective function at the solution $xsol$ (variable $fval$), the reason that the algorithm terminated (variable $exitflag$), and the number of iterations (variable $iterations$).

In lines 54–57, the output variables are initialized. In lines 59–70, we set default values to the missing inputs (if any). Next, we check if the input data is logically correct (lines 76–107). If the type of optimization is maximization, then we multiply vector c and constant $c0$ by -1 (lines 110–113). Then, the presolve analysis is performed (lines 116–127). Next, we scale the LP problem using the selected scaling technique (lines 134–162). Then, we add slack variables if necessary (lines 163–187). If the density of matrix A is less than 20%, then we use sparse algebra for faster computations (lines 193–197). Then, we calculate the tolerance for detecting infeasible LPs (lines 199–200). Next, we calculate the starting point using Mehrotra's heuristic (lines 203–216).

At each iteration of the algorithm, we perform the following steps. Initially, we calculate the residuals and the duality measure (lines 222–225). Then, we use a heuristic to detect infeasible LPs (lines 228–233). Next, we use Mehrotra's termination criterion to check if the LP problem is optimal (lines 235–246). Then, we formulate the coefficient matrix of the linear systems to solve and factorize matrix M using the Cholesky decomposition (lines 249–250). Note that we do not deal with numerical instabilities on this code, so if the coefficient matrix M is not positive definite, we assume that the LP problem is optimal (lines 251–263). Next, we perform the predictor step (lines 265–270). Then, we find the centering parameter σ (lines 272–273) and perform the corrector step (lines 275–282). Next, we update the solution (x, w, s) (lines 284–286). Finally, we print intermediate results every 5 iterations (lines 288–298).

```
1.    function [xsol, fval, exitflag, iterations] = ...
2.        ipdipm(A, c, b, Eqin, MinMaxLP, c0, ...
3.        maxIterations, tol, etaMin, scalingTechnique)
4.    % Filename: ipdipm.m
5.    % Description: the function is an implementation of
6.    % Mehrotra's Predictor-Corrector infeasible
7.    % primal-dual interior point method
8.    % Authors: Ploskas, N., & Samaras, N.
9.    %
10.   % Syntax: [xsol, fval, exitflag, iterations] = ...
11.   %    ipdipm(A, c, b, Eqin, MinMaxLP, c0, ...
12.   %    maxIterations, tol, etaMin, scalingTechnique)
13.   %
14.   % Input:
15.   % -- A: matrix of coefficients of the constraints
16.   %    (size m x n)
17.   % -- c: vector of coefficients of the objective function
18.   %    (size n x 1)
19.   % -- b: vector of the right-hand side of the constraints
20.   %    (size m x 1)
21.   % -- Eqin: vector of the type of the constraints
22.   %    (size m x 1)
23.   % -- MinMaxLP: the type of optimization (optional:
24.   %    default value -1 - minimization)
25.   % -- c0: constant term of the objective function
26.   %    (optional: default value 0)
27.   % -- maxIterations: maximum number of iterations to
28.   %    perform if an optimal solution is not found
29.   %    (optional: default value 100)
30.   % -- tole: tolerance for the termination criterion
```

```
31.  %      (optional: default value 1e-08)
32.  % -- etaMin: parameter eta (optional: default
33.  %      value 0.995)
34.  % -- scalingTechnique: the scaling method to be used
35.  %      (0: no scaling, 1: arithmetic mean, 2: de Buchet for
36.  %      the case p = 1, 3: de Buchet for the case p = 2,
37.  %      4: de Buchet for the case p = Inf, 5: entropy,
38.  %      6: equilibration, 7: geometric mean, 8: IBM MPSX,
39.  %      9: LP-norm for the case p = 1, 10: LP-norm for
40.  %      the case p = 2, 11: LP-norm for the case p = Inf)
41.  %      (optional: default value 6)
42.  %
43.  % Output:
44.  % -- xsol: the solution found (size n x 1)
45.  % -- fval: the value of the objective function at the
46.  %      solution x
47.  % -- exitflag: the reason that the algorithm terminated
48.  %      (0: optimal solution found, 1: the LP problem is
49.  %      infeasible, -1: the input data is not logically
50.  %      or numerically correct)
51.  % -- iterations: the number of iterations
52.
53.  % initialize output variables
54.  xsol = [];
55.  fval = 0;
56.  exitflag = 0;
57.  iterations = 0;
58.  % set default values to missing inputs
59.  if ~exist('maxIterations')
60.      maxIterations = 100;
61.  end
62.  if ~exist('tol')
63.      tol = 1e-8;
64.  end
65.  if ~exist('etaMin')
66.      etaMin = 0.995;
67.  end
68.  if ~exist('scalingTechnique')
69.      scalingTechnique = 6;
70.  end
71.  [m, n] = size(A); % find the size of matrix A
72.  [m2, n2] = size(c); % find the size of vector c
73.  [m3, n3] = size(Eqin); % find the size of vector Eqin
74.  [m4, n4] = size(b); % find the size of vector b
75.  % check if input data is logically correct
76.  if n2 ~= 1
77.      disp('Vector c is not a column vector.')
78.      exitflag = -1;
79.      return
80.  end
81.  if n ~= m2
82.      disp(['The number of columns in matrix A and ' ...
83.            'the number of rows in vector c do not match.'])
84.      exitflag = -1;
```

```
85.      return
86.  end
87.  if m4 ~= m
88.      disp(['The number of the right-hand side values ' ...
89.          'is not equal to the number of constraints.'])
90.      exitflag = -1;
91.      return
92.  end
93.  if n3 ~= 1
94.      disp('Vector Eqin is not a column vector')
95.      exitflag = -1;
96.      return
97.  end
98.  if n4 ~= 1
99.      disp('Vector b is not a column vector')
100.     exitflag = -1;
101.     return
102. end
103. if m4 ~= m3
104.     disp('The size of vectors Eqin and b does not match')
105.     exitflag = -1;
106.     return
107. end
108. % if the type of optimization is maximization, then multiply
109. % vector c and constant c0 by -1
110. if MinMaxLP == 1
111.     c = -c;
112.     c0 = -c0;
113. end
114. % perform the presolve analysis
115. disp('---- P R E S O L V E    A N A L Y S I S ----')
116. [A, c, b, Eqin, c0, infeasible, unbounded] = ...
117.     presolve(A, c, b, Eqin, c0);
118. if infeasible == 1 % the LP problem is infeasible
119.      disp('The LP problem is infeasible')
120.     exitflag = 1;
121.     return
122. end
123. if unbounded == 1 % the LP problem is unbounded
124.      disp('The LP problem is unbounded')
125.     exitflag = 2;
126.     return
127. end
128. [m, n] = size(A); % find the new size of matrix A
129. [m2, ~] = size(c); % find the new size of vector c
130. [m3, ~] = size(Eqin); % find the size of vector Eqin
131. % scale the LP problem using the selected scaling
132. % technique
133. disp('---- S C A L I N G ----')
134. if scalingTechnique == 1 % arithmetic mean
135.     [A, c, b, ~, ~] = arithmeticMean(A, c, b);
136. % de buchet for the case p = 1
137. elseif scalingTechnique == 2
138.     [A, c, b, ~, ~] = debuchet1(A, c, b);
```

```
139. % de buchet for the case p = 2
140. elseif scalingTechnique == 3
141.    [A, c, b, ~, ~] = debuchet2(A, c, b);
142. % de buchet for the case p = Inf
143. elseif scalingTechnique == 4
144.    [A, c, b, ~, ~] = debuchetinf(A, c, b);
145. elseif scalingTechnique == 5 % entropy
146.    [A, c, b, ~, ~] = entropy(A, c, b);
147. elseif scalingTechnique == 6 % equilibration
148.    [A, c, b, ~, ~] = equilibration(A, c, b);
149. elseif scalingTechnique == 7 % geometric mean
150.    [A, c, b, ~, ~] = geometricMean(A, c, b);
151. elseif scalingTechnique == 8 % IBM MPSX
152.    [A, c, b, ~, ~, ~, ~] = ibmmpsx(A, c, b);
153. % LP-norm for the case p = 1
154. elseif scalingTechnique == 9
155.    [A, c, b, ~, ~] = lpnorm1(A, c, b);
156. % LP-norm for the case p = 2
157. elseif scalingTechnique == 10
158.    [A, c, b, ~, ~] = lpnorm2(A, c, b);
159. % LP-norm for the case p = Inf
160. elseif scalingTechnique == 11
161.    [A, c, b, ~, ~] = lpnorminf(A, c, b);
162. end
163. flag = isequal(Eqin, zeros(m3, 1));
164. if flag ~= 1
165.    % some or all constraints are inequalities
166.    % add slack variables
167.    axm = nnz(Eqin);
168.    c(m2 + 1:m2 + axm, :) = 0;
169.    A(:, n + 1:n + axm) = zeros(m, axm);
170.    curcol = 1;
171.    for i = 1:m3
172.        % 'greater than or equal to' inequality constraint
173.        if Eqin(i, 1) == 1
174.            A(i, n + curcol) = -1;
175.            curcol = curcol + 1;
176.        % 'less than or equal to' inequality constraint
177.        elseif Eqin(i, 1) == -1
178.            A(i, n + curcol) = 1;
179.            curcol = curcol + 1;
180.        % unrecognized type of constraint
181.        elseif Eqin(i,1) ~= 0
182.            disp('Vector Eqin is not numerically correct.')
183.            exitflag = -1;
184.            return
185.        end
186.    end
187. end
188. [m, n] = size(A); % new size of matrix A
189. % calculate the density of matrix A
190. density = (nnz(A) / (m * n)) * 100;
191. % if the density of matrix A is less than 20%, then
192. % use sparse algebra for faster computations
```

```
193. if density < 20
194.     A = sparse(A);
195.     c = sparse(c);
196.     b = sparse(b);
197. end
198. % tolerance for detecting infeasible LPs
199. infTole = 1.e15 * max([normest(A), normest(b), ...
200.     normest(c)]);
201. % calculate the starting point using Mehrotra's
202. % heuristic
203. x = A' * ((A * A') \ b);
204. w = (A * A') \ (A * c);
205. s = c - A' * w;
206. delta_x = max(-1.5 * min(x), 0);
207. delta_s = max(-1.5 * min(s), 0);
208. e = ones(n, 1);
209. delta_x_s = 0.5 * (x + delta_x * e)' * ...
210.     (s + delta_s * e);
211. delta_x_c = delta_x + delta_x_s / ...
212.     (sum(s) + n * delta_s);
213. delta_s_c = delta_s + delta_x_s / ...
214.     (sum(x) + n * delta_x);
215. x = x + delta_x_c * e;
216. s = s + delta_s_c * e;
217. % iterate until maxIterations or the LP problem is
218. % optimal or infeasible
219. for i = 1:maxIterations
220.     iterations = iterations + 1;
221.     % calculate the residuals and the duality measure
222.     rd = A' * w + s - c; % dual residual
223.     rp = A * x - b; % primal residual
224.     rc = x .* s; % complementarity residual
225.     mu = x' * s / n; % duality measure
226.     residual = normest(rc, 1) / (1 + abs(b' * w));
227.     % heuristic to detect infeasible LPs
228.     if isnan(residual) || normest(x) + normest(s) ...
229.             >= infTole
230.         disp('The LP problem is infeasible!');
231.         exitflag = 1;
232.         break;
233.     end
234.     % the termination criterion
235.     if max(mu, max(normest(rp), normest(rd))) <= tol
236.         xsol = x;
237.         fval = c' * x; % calculate the objective function value
238.         if MinMaxLP == 1 % maximization
239.             fval = full(-(fval + c0));
240.         else % minimization
241.             fval = full(fval + c0);
242.         end
243.         exitflag = 0;
244.         disp('The LP problem is optimal')
245.         break;
246.     end
```

```
247.        % formulate the coefficient matrix of the
248.        % linear systems
249.        M = A * diag(x ./ s) * A';
250.        [R, p] = chol(M); % factorize M
251.        if p > 0 % if matrix M is positive definite, we
252.            % assume that the LP problem is optimal
253.            xsol = x;
254.            fval = c' * x; % calculate the objective function value
255.            if MinMaxLP == 1 % maximization
256.                fval = full(-(fval + c0));
257.            else % minimization
258.                fval = full(fval + c0);
259.            end
260.            exitflag = 0;
261.            disp('The LP problem is optimal')
262.            break;
263.        end
264.        % predictor step
265.        rhs = rp - A * ((rc - x .* rd) ./ s);
266.        dw = R \ (R' \ rhs);
267.        ds = rd - A' * dw;
268.        dx = (rc - x .* ds) ./ s;
269.        alpha_p = 1 / max([1; dx ./ x]);
270.        alpha_d = 1 / max([1; ds ./ s]);
271.        % centering parameter step
272.        mun = ((x - alpha_p * dx)' * (s - alpha_d * ds)) / n;
273.        sigma = (mun / mu) ^ 3;
274.        % corrector step
275.        rc = rc - sigma * mu + dx .* ds;
276.        rhs = rp - A * ((rc - x .* rd) ./ s);
277.        dw = R \ (R' \ rhs);
278.        ds = rd - A' * dw;
279.        dx = (rc - x .* ds) ./ s;
280.        eta = max(etaMin, 1 - mu);
281.        alpha_p = eta / max([eta; dx ./ x]);
282.        alpha_d = eta / max([eta; ds ./ s]);
283.        % update step
284.        x = x - alpha_p * dx;
285.        w = w - alpha_d * dw;
286.        s = s - alpha_d * ds;
287.        % print intermediate results every 5 iterations
288.        if mod(iterations, 5) == 0
289.            % calculate the value of the objective
290.            % function
291.            if MinMaxLP == 1 % maximization
292.                fval = full(-(c' * x + c0));
293.            else % minimization
294.                fval = full(c' *x + c0);
295.            end
296.            fprintf(['Iteration %i - objective value: ' ...
297.                '%f\n'], iterations, fval);
298.        end
299. end
300. end
```

11.7   Computational Study

In this section, we present a computational study. The aim of the computational study is to compare the efficiency of the proposed implementation with MATLAB's IPM. The test set used in the computational study is 19 LPs from the Netlib (optimal, Kennington, and infeasible LPs) and Mészáros problem sets that were presented in Section 3.6 and a set of randomly generated sparse LPs. Tables 11.2 and 11.3 present the total execution time and the number of iterations of Mehrotra's Predictor-Corrector method that was presented in this chapter and MATLAB's interior point method that will be presented in Appendix A over a set of benchmark LPs and over a set of randomly generated sparse LPs, respectively. The following abbreviations are used: (i) MPC—the proposed implementation of Mehrotra's Predictor-Corrector method, and (ii) MATLAB's IPM—MATLAB's interior point method. For each instance, we averaged times over 10 runs. All times are measured in seconds. MPC method is executed without presolve and scaling.

The two algorithms have similar number of iterations, while MATLAB's IPM performs almost 1.4 times faster than MPC method over the benchmark set. On the

Table 11.2 Total execution time and number of iterations of the proposed implementation of Mehrotra's predictor-corrector method and MATLAB's interior point method over the benchmark set

Problem				MPC		MATLAB's IPM	
Name	Constraints	Variables	Nonzeros	Time	Iterations	Time	Iterations
beaconfd	173	262	3,375	0.02	9	0.03	12
cari	400	1,200	152,800	0.96	13	1.27	15
farm	7	12	36	0.01	16	0.01	7
itest6	11	8	20	0.02	33	0.01	7
klein2	477	54	4,585	0.02	1	1.00	38
nsic1	451	463	2,853	0.06	12	0.03	11
nsic2	465	463	3,015	0.45	87	0.05	15
osa-07	1,118	23,949	143,694	1.01	33	0.85	30
osa-14	2,337	52,460	314,760	2.68	37	2.46	40
osa-30	4,350	100,024	600,138	5.45	36	4.05	30
rosen1	520	1,024	23,274	0.18	17	0.06	13
rosen10	2,056	4,096	62,136	3.99	19	0.16	15
rosen2	1,032	2,048	46,504	1.00	21	0.12	15
rosen7	264	512	7,770	0.05	15	0.03	12
rosen8	520	1,024	15,538	0.19	19	0.05	15
sc205	205	203	551	0.01	11	0.02	11
scfxm1	330	457	2,589	0.04	21	0.04	18
sctap2	1,090	1,880	6,714	0.10	13	0.12	18
sctap3	1,480	2,480	8,874	0.19	14	0.17	18
Geometric mean				0.17	16.97	0.12	15.90

Table 11.3 Total execution time and number of iterations of the proposed implementation of Mehrotra's predictor-corrector method and MATLAB's interior point method over a set of randomly generated sparse LPs (5% density)

Problem	MPC		MATLAB's IPM	
	Time	Iterations	Time	Iterations
100 × 100	0.02	11	0.03	19
200 × 200	0.04	17	0.08	19
300 × 300	0.12	25	0.23	18
400 × 400	0.24	25	0.61	20
500 × 500	0.44	25	1.13	21
600 × 600	0.81	28	2.03	22
700 × 700	1.35	30	2.75	19
800 × 800	2.09	33	4.47	21
900 × 900	2.59	29	6.11	21
1,000 × 1,000	3.51	30	8.79	23
1,100 × 1,100	4.34	32	12.48	25
1,200 × 1,200	6.23	35	13.50	21
1,300 × 1,300	7.95	39	18.58	23
1,400 × 1,400	7.59	32	23.19	23
1,500 × 1,500	9.01	31	28.49	23
1,600 × 1,600	12.09	35	33.04	22
1,700 × 1,700	12.09	30	36.52	20
1,800 × 1,800	12.54	29	50.54	24
1,900 × 1,900	16.90	35	56.94	23
2,000 × 2,000	20.30	35	65.76	23
Geometric mean	2.05	28.37	5.22	21.42

other hand, the proposed implementation is almost 2.5 times faster than MATLAB's IPM over the set of the randomly generated sparse LPs, while performing more iterations than MATLAB's IPM.

11.8 Chapter Review

Primal-dual IPMs have shown great computational performance. Most primal-dual IPMs are based on Mehrotra's Predictor-Corrector method. In this chapter, a presentation of the basic concepts of primal-dual IPMs was performed. Next, we presented the MPC method. The various steps of the algorithm were presented. Numerical examples were also presented in order for the reader to understand better the algorithm. Furthermore, an implementation of the algorithm in MATLAB was presented. Finally, a computational study was performed over benchmark LPs and randomly generated sparse LPs in order to compare the efficiency of the proposed implementation with MATLAB's IPM solver. Computational results showed that the two algorithms have similar number of iterations, while MATLAB's IPM performs almost 1.4 times faster than MPC method over the benchmark set. On the other hand, the proposed implementation is almost 2.5 times faster than MATLAB's IPM over the set of the randomly generated sparse LPs, while performing more iterations than MATLAB's IPM.

References

1. Bixby, E. R. (1992). Implementing the simplex method: The initial basis. *ORSA Journal on Computing, 4*, 267–284.
2. Dantzig, G. B. (1949). Programming in linear structure. *Econometrica, 17*, 73–74.
3. Frisch, K. F. (1955). *The logarithmic potential method of convex programming.* Technical report, University Institute of Economics, Oslo, Norway.
4. Gondzio, J. (1992). Splitting dense columns of constraint matrix in interior point methods for large-scale linear programming. *Optimization, 24*, 285–297.
5. Gondzio, J. (1996). Multiple centrality corrections in a primal-dual method for linear programming. *Computational Optimization and Applications, 6*, 137–156.
6. Gondzio, J. (2012). Interior point methods 25 years later. *European Journal of Operational Research, 218*(3), 587–601.
7. Gonzaga, C. C. (1992). Path-following methods for linear programming. *SIAM Review, 34*(2), 167–224.
8. Hoffman, A. J., Mannos, M., Sokolowsky, D., & Wiegman, N. (1953). Computational experience in solving linear programs. *Journal of the Society for Industrial and Applied Mathematics, 1*, 17–33.
9. Karmarkar, N. K. (1984). A new polynomial-time algorithm for linear programming. *Combinatorica, 4*, 373–395.
10. Khachiyan, L. G. (1979). A polynomial algorithm in linear programming. *Soviet Mathematics Doklady, 20*, 191–194.
11. Lustig, I. J., Marsten, R. E., & Shanno, D. F. (1992). On implementing Mehrotra's predictor corrector interior point method for linear programming. *SIAM Journal on Optimization, 2*, 435–449.
12. Lustig, I. J., Marsten, R. E., & Shanno, D. F. (1994). Interior point methods for linear programming: Computational state of the art. *ORSA Journal on Computing, 6*(1), 1–14.
13. Mehrotra, S. (1992). On the implementation of a primal-dual interior point method. *SIAM Journal on Optimization, 2*, 575–601.
14. Potra, F. A., & Wright, S. J. (2000). Interior-point methods. *Journal of Computational and Applied Mathematics, 124*(1), 281–302.
15. Von Neumann, J. (1947). *On a maximization problem.* Technical report, Institute for Advanced Study, Princeton, NJ, USA.

Chapter 12
Sensitivity Analysis

Abstract In many cases, after solving an LP problem with the simplex method, there is a change in the data of the LP problem. With sensitivity analysis, we can find if the input data of the LP problem can change without affecting the optimal solution. This chapter discusses how to deal with such changes efficiently. This topic is called sensitivity analysis. Sensitivity analysis is very useful in two situations: (i) when we wish to know how the solution will be affected if we perform a small change in the LP problem, and (ii) when we have already solved an LP problem and we also want to solve a second LP problem in which the data is only slightly different. Rather than restarting the simplex method from scratch for the modified LP problem, we want to solve the modified LP problem starting with the optimal basis of the original LP problem and perform only a few iterations to solve the modified LP problem (if necessary). We examine how the solution of an LP problem is affected when changes are made to the input data of the LP problem. Moreover, we examine changes in: (i) the cost vector, (ii) the right-hand side vector, and (iii) the coefficient of the constraints.

12.1 Chapter Objectives

- Present the classical sensitivity analysis.
- Understand when an LP problem should not be solved from scratch when small changes have been made to the initial data.
- Implement codes in MATLAB to identify if an LP problem should not be solved from scratch when small changes have been made to the initial data.

12.2 Introduction

In many cases, after solving an LP problem with the simplex method, there is a change in the data of the LP problem. This chapter discusses how to deal with such changes efficiently. This topic is called sensitivity analysis. Sensitivity analysis is very useful in two situations:

© Springer International Publishing AG 2017   541
N. Ploskas, N. Samaras, *Linear Programming Using MATLAB®*,
Springer Optimization and Its Applications 127, DOI 10.1007/978-3-319-65919-0_12

- when we wish to know how the optimal solution will be affected if we perform a small change in the original data of the LP problem, and
- when we have already solved an LP problem and we also want to solve a second LP problem in which the original data is only slightly different.

Rather than restarting the simplex method from scratch for the modified LP problem, we want to solve the modified LP problem starting with the optimal basis of the original LP problem and perform only a few iterations in order to solve the modified LP problem (if necessary).

This chapter presents topics about sensitivity analysis. We examine how the solution of an LP problem is affected when changes are made to the input data of the LP problem. Moreover, we examine changes in: (i) the cost vector, (ii) the right-hand side vector, and (iii) the coefficient of the constraints. Interested readers are referred to [1–5] for more details.

The structure of this chapter is as follows. In Section 12.3, we compute the range of the cost and right-hand side vectors where the optimal basis of the original LP problem remains optimal for the modified LP problem. Moreover, we present a method to detect if the optimal basis of the original LP problem remains optimal if changes are made to the cost vector and the right-hand side vector, respectively. Sections 12.4 and 12.5 present a method to detect if the optimal basis of the original LP problem remains optimal after adding a new variable and a new constraint, respectively. Finally, conclusions and further discussion are presented in Section 12.6.

12.3   Change in the Cost and Right-Hand Side Vectors

12.3.1   Range of the Cost and Right-Hand Side Vectors

There are cases where we have already solved an LP problem and a coefficient cost c_j or a right-hand side value b_i has changed. We can solve it again and get the solution of the modified LP problem. However, if it is a large LP problem, then we can save time by using the optimal solution of the original LP problem. At some other cases, we want to know if a change to a coefficient cost c_j or a right-hand side value b_i can affect the optimal solution of an LP problem. In this section, we compute the range of the coefficient cost vector and the right-hand side vector in which the optimal basis of an LP problem remains optimal. Using this approach, we compute these ranges along with the shadow prices for each right-hand side coefficient (w) and the reduced costs for each variable x_j (s).

12.3.1.1   Range of the Cost Vector

We assume that a change occurs in only one element of the cost coefficients. Moreover, the LP problem is in its standard form. In addition, the following two relations hold:

$$\max\{\emptyset\} = -\infty \text{ and } \min\{\emptyset\} = \infty$$

Let $\left[c_j + \alpha_j, c_j + \beta_j\right]$ be the range of the cost coefficient c_j in which the optimal basis of an LP problem remains optimal. Let $cl_j = c_j + \alpha_j$ and $cu_j = c_j + \beta_j$. There are two cases that we should consider:

- Variable j is basic ($j \in B$) in the optimal solution: Variable j is basic, so the change Δc_j results in a change to the simplex multiplier w and the reduced costs of each nonbasic variable. Hence:

$$cl_j = c_j + \max\left\{\frac{s_k}{h_{ik}} : k \in N \wedge h_{ik} < 0\right\}$$

$$cu_j = c_j + \min\left\{\frac{s_k}{h_{ik}} : k \in N \wedge h_{ik} > 0\right\}$$

where h_{ik} are elements of the pivoting column $H_{rN} = \left(A_B^{-1}\right)_{r.} A_N$.

- Variable j is not basic ($j \in N$) in the optimal solution: Variable j is not basic, so the change Δc_j results in a change to the reduced cost of c_j. Hence:

$$cl_j = c_j - s_j$$

$$cu_j = \infty$$

12.3.1.2   Range of the Right-Hand Side Vector

We assume that a change occurs in only one element of the right-hand side vector. Moreover, the LP problem is in its standard form. Let $[b_i + \gamma_i, b_i + \delta_i]$ be the range of the right-hand side coefficient b_i in which the optimal basis of an LP problem remains optimal. Let $bl_i = b_i + \gamma_i$ and $bu_i = b_i + \delta_i$. Hence:

$$bl_i = b_i + \max\left\{\frac{-x_{B[k]}}{\left(A_B^{-1}\right)_{ki}} : \left(A_B^{-1}\right)_{ki} > 0\right\}$$

$$bu_i = b_i + \min\left\{\frac{-x_{B[k]}}{\left(A_B^{-1}\right)_{ki}} : \left(A_B^{-1}\right)_{ki} < 0\right\}$$

where $1 \le k \le n$.

12.3.1.3   Example

Consider the following LP problem:

$$\min z = -2x_1 + 18x_2 + 3x_3 + 4x_4$$
$$\text{s.t.} \quad -x_1 + 3x_2 + 2x_3 \qquad - x_5 \qquad = 0$$
$$-x_1 + 6x_2 \qquad + 2x_4 \qquad - x_6 = 1$$
$$x_j \geq 0, \quad (j = 1, 2, 3, 4, 5, 6)$$

In matrix notation the above LP problem is written as follows:

$$A = \begin{bmatrix} -1 & 3 & 2 & 0 & -1 & 0 \\ -1 & 6 & 0 & 2 & 0 & -1 \end{bmatrix}, c = \begin{bmatrix} -2 \\ 18 \\ 3 \\ 4 \\ 0 \\ 0 \end{bmatrix}, b = \begin{bmatrix} 0 \\ 1 \end{bmatrix}, Eqin = \begin{bmatrix} 0 \\ 0 \end{bmatrix}$$

The LP problem is in its standard form. The basic list is $B = [5, 4]$ and the nonbasic list is $N = [1, 2, 3, 6]$. The basic list $B = [5, 4]$ is optimal:

$$A_B = \begin{bmatrix} -1 & 0 \\ 0 & 2 \end{bmatrix},$$

$$A_B^{-1} = \begin{bmatrix} -1 & 0 \\ 0 & 1/2 \end{bmatrix},$$

$$x_B = A_B^{-1}b = \begin{bmatrix} -1 & 0 \\ 0 & 1/2 \end{bmatrix} \begin{bmatrix} 0 \\ 1 \end{bmatrix} = \begin{bmatrix} 0 \\ 1/2 \end{bmatrix},$$

$$w^T = c_B^T A_B^{-1} = \begin{bmatrix} 0 & 4 \end{bmatrix} \begin{bmatrix} -1 & 0 \\ 0 & 1/2 \end{bmatrix} = \begin{bmatrix} 0 & 2 \end{bmatrix},$$

$$s^T = c^T - w^T A = \begin{bmatrix} -2 & 18 & 3 & 4 & 0 & 0 \end{bmatrix} - \begin{bmatrix} 0 & 2 \end{bmatrix} \begin{bmatrix} -1 & 3 & 2 & 0 & -1 & 0 \\ -1 & 6 & 0 & 2 & 0 & -1 \end{bmatrix} = \begin{bmatrix} 0 & 6 & 3 & 0 & 0 & 2 \end{bmatrix} \geq 0$$

Let's calculate the range for the cost coefficient vector. For the basic variable x_4, we find that $B[2] = 4, r = 2$:

$$H_{2N} = \left(A_B^{-1}\right)_{2.} A_N = \begin{bmatrix} 0 & 1/2 \end{bmatrix} \begin{bmatrix} -1 & 3 & 2 & 0 \\ -1 & 6 & 0 & -1 \end{bmatrix} = \begin{bmatrix} -1/2 & 3 & 0 & -1/2 \end{bmatrix}$$

Hence:

$$[cl_4, cu_4] = [c_4 + \alpha_4, c_4 + \beta_4] =$$

$$\left[4 + \max\left\{ \frac{s_k}{h_{ik}} : k \in N \wedge h_{ik} < 0 \right\}, 4 + \min\left\{ \frac{s_k}{h_{ik}} : k \in N \wedge h_{ik} > 0 \right\} \right] =$$

$$\left[4 + \max\left\{\frac{s_1}{h_{21}}, \frac{s_6}{h_{26}}\right\}, 4 + \min\left\{\frac{s_2}{h_{22}}\right\}\right] = \left[4 + \max\left\{\frac{0}{-1/2}, \frac{2}{-1/2}\right\}, 4 + \min\left\{\frac{6}{3}\right\}\right] =$$

$$[4 + 0, 4 + 2] = [4, 6]$$

For the basic variable x_5, we find that $B[1] = 5, r = 1$:

$$H_{1N} = \left(A_B^{-1}\right)_{1.} A_N = [-1\ 0]\begin{bmatrix} -1 & 3 & 2 & 0 \\ -1 & 6 & 0 & -1 \end{bmatrix} = [1\ -3\ -2\ 0]$$

Hence:

$$[cl_5, cu_5] = [c_5 + \alpha_5, c_5 + \beta_5] =$$

$$\left[0 + \max\left\{\frac{s_k}{h_{ik}} : k \in N \wedge h_{ik} < 0\right\}, 0 + \min\left\{\frac{s_k}{h_{ik}} : k \in N \wedge h_{ik} > 0\right\}\right] =$$

$$\left[0 + \max\left\{\frac{s_2}{h_{12}}, \frac{s_3}{h_{13}}\right\}, 0 + \min\left\{\frac{s_1}{h_{11}}\right\}\right] = \left[0 + \max\left\{\frac{6}{-3}, \frac{3}{-2}\right\}, 0 + \min\left\{\frac{0}{1}\right\}\right]$$

$$= [0 + (-3/2), 0 + 0] = [-3/2, 0]$$

For the nonbasic variables x_1, x_2, x_3, and x_6, we calculate the range as $[cl_j, cu_j] = [c_j - s_j, \infty), j = 1, 2, 3, 6$:

$$[cl_1, cu_1] = [c_1 - s_1, \infty) = [-2 - 0, \infty) = [-2, \infty)$$

$$[cl_2, cu_2] = [c_2 - s_2, \infty) = [18 - 6, \infty) = [12, \infty)$$

$$[cl_3, cu_3] = [c_3 - s_3, \infty) = [3 - 3, \infty) = [0, \infty)$$

$$[cl_6, cu_6] = [c_6 - s_6, \infty) = [0 - 2, \infty) = [-2, \infty)$$

Let's calculate the range for the right-hand side vector. For the coefficient b_1 ($i = 1$), we find that:

$$\left(A_B^{-1}\right)_{.i} = \left(A_B^{-1}\right)_{.1} = \begin{bmatrix} -1 \\ 0 \end{bmatrix}$$

Hence:

$$[bl_1 + \gamma_1, bu_1 + \delta_1] =$$

$$\left[0 + \max\left\{\frac{-x_{B[k]}}{\left(A_B^{-1}\right)_{ki}} : \left(A_B^{-1}\right)_{ki} > 0\right\}, 0 + \min\left\{\frac{-x_{B[k]}}{\left(A_B^{-1}\right)_{ki}} : \left(A_B^{-1}\right)_{ki} < 0\right\}\right] =$$

$$\left[0 + \max\{\emptyset\}, 0 + \min\left\{\frac{-x_{B[1]}}{\left(A_B^{-1}\right)_{11}}\right\}\right] = \left(-\infty, 0 + \min\left\{\frac{0}{-1}\right\}\right] = (-\infty, 0]$$

Table 12.1 Ranges of the cost and right-hand side vectors, shadow prices, and reduced costs

Coefficient	c_1	c_2	c_3	c_4	c_5	c_6	b_1	b_2
Lower value of the range	-2	12	0	4	$-3/2$	-2	$-\infty$	0
Upper value of the range	∞	∞	∞	6	0	∞	0	∞
Shadow price	–	–	–	–	–	–	0	2
Reduced cost	0	6	3	0	0	2	–	–

For the coefficient b_2 ($i = 2$), we find that:

$$\left(A_B^{-1}\right)_{.i} = \left(A_B^{-1}\right)_{.2} = \begin{bmatrix} 0 \\ 1/2 \end{bmatrix}$$

Hence:

$$[bl_2 + \gamma_2, bu_2 + \delta_2] =$$

$$\left[1 + \max\left\{\frac{-x_{B[k]}}{\left(A_B^{-1}\right)_{ki}} : \left(A_B^{-1}\right)_{ki} > 0\right\}, 1 + \min\left\{\frac{-x_{B[k]}}{\left(A_B^{-1}\right)_{ki}} : \left(A_B^{-1}\right)_{ki} < 0\right\}\right] =$$

$$\left[1 + \max\left\{\frac{-x_{B[2]}}{\left(A_B^{-1}\right)_{22}}\right\}, 1 + \min\{\emptyset\}\right] = \left[1 + \max\left\{\frac{-1/2}{1/2}\right\}, \infty\right) = [1 - 1, \infty) = [0, \infty)$$

The results from the sensitivity analysis can be summarized in Table 12.1:

12.3.1.4 Implementation in MATLAB

Below, we present the implementation in MATLAB of the sensitivity analysis. Some necessary notations should be introduced before the presentation of the implementation of the sensitivity analysis. Let A be a $m \times n$ matrix with the coefficients of the constraints, c be a $n \times 1$ vector with the coefficients of the objective function, b be a $m \times 1$ vector with the right-hand side of the constraints, *Eqin* be a $m \times 1$ vector with the type of the constraints, *MinMaxLP* be a variable denoting the type of optimization (-1 for minimization and 1 for maximization), *BasicList* be a $1 \times m$ vector of the indices of the basic variables of the original LP problem, *cBounds* be a $2 \times n$ matrix of the range of the coefficients of the objective function, *bBounds* be a $2 \times m$ matrix of the range of the right-hand side coefficients, *sp* be a $1 \times m$ vector of the shadow prices, and *rc* be a $1 \times n$ vector of the shadow prices.

The sensitivity analysis is implemented with three functions: (i) one function that implements the sensitivity analysis (filename: `sensitivityAnalysis.m`), (ii) one function that implements the sensitivity analysis for the coefficients of the objective function (filename: `sensitivityAnalysisc.m`), and (iii) one function that implements the sensitivity analysis for the right-hand side coefficients (filename: `sensitivityAnalysisb.m`).

The function for the implementation of the sensitivity analysis (filename: *sensitivityAnalysis.m*) takes as input the matrix with the coefficients of the constraints (matrix *A*), the vector with the coefficients of the objective function (vector *c*), the vector with the right-hand side of the constraints (vector *b*), the vector with the type of the constraints (vector *Eqin*), a variable denoting the type of optimization (variable *MinMaxLP*), and the vector of the indices of the basic variables (vector *BasicList*), and returns as output the matrix of the range of the coefficients of the objective function (matrix *cBounds*), the matrix of the range of the right-hand side coefficients (matrix *bBounds*), the vector of the shadow prices (vector *sp*), and the vector of the reduced costs (vector *rc*).

In lines 34–35, we transform the LP problem in its standard form (for more details, see Chapter 2). Then, we calculate the ranges of the coefficients of the objective function (lines 38–39) and the ranges of the right-hand side coefficients (lines 42–43). Finally, we calculate the shadow prices and the reduced costs (lines 45–49).

```
1.    function [cBounds, bBounds, sp, rc] = ...
2.        sensitivityAnalysis(A, c, b, Eqin, MinMaxLP, BasicList)
3.    % Filename: sensitivityAnalysis.m
4.    % Description: the function is an implementation of the
5.    % sensitivity analysis
6.    % Authors: Ploskas, N., & Samaras, N.
7.    %
8.    % Syntax: [cBounds, bBounds, sp, rc] = ...
9.    %    sensitivityAnalysis(A, c, b, Eqin, MinMaxLP, BasicList)
10.   %
11.   % Input:
12.   % -- A: matrix of coefficients of the constraints
13.   %     (size m x n)
14.   % -- c: vector of coefficients of the objective function
15.   %     (size n x 1)
16.   % -- b: vector of the right-hand side of the constraints
17.   %     (size m x 1)
18.   % -- Eqin: vector of the type of the constraints
19.   %     (size m x 1)
20.   % -- MinMaxLP: the type of optimization (optional:
21.   %     default value -1 - minimization)
22.   % -- BasicList: vector of the indices of the basic
23.   %     variables (size 1 x m)
24.   %
25.   % Output:
26.   % -- cBounds: matrix of the range of the coefficients
27.   %     of the objective function (size 2 x n)
28.   % -- bBounds: matrix of the range of the right-hand
29.   %     side coefficients (size 2 x m)
30.   % -- sp: vector of the shadow prices (size 1 x m)
31.   % -- rc: vector of the reduced costs (size 1 x n)
32.
33.   % transform the LP problem in its standard form
34.   [A, c, b, Eqin, MinMaxLP] = ...
35.       general2standard(A, c, b, Eqin, MinMaxLP);
```

```
36.  % find the ranges of the coefficients of the
37.  % objective function
38.  [cl, cu] = sensitivityAnalysisc(A, c, BasicList);
39.  cBounds = [cl; cu];
40.  % find the ranges of the right-hand side
41.  % coefficients
42.  [bl, bu] = sensitivityAnalysisb(A, b, BasicList);
43.  bBounds = [bl; bu];
44.  % calculate the shadow prices and reduced costs
45.  BasisInv = inv(A(:, BasicList)); % invert the basis
46.  % calculate the reduced costs
47.  sp = c(BasicList)' * BasisInv;
48.  % calculate the shadow prices
49.  rc = c' - sp * A;
50.  end
```

The function (filename: sensitivityAnalysisc.m) for the implementation of the sensitivity analysis for the coefficients of the objective function takes as input the matrix with the coefficients of the constraints (matrix A), the vector with the coefficients of the objective function (vector c), and the vector of the indices of the basic variables (vector *BasicList*), and returns as output the vector of the lower values of the range of the coefficients of the objective function (vector *cl*), and the vector of the upper values of the range of the coefficients of the objective function (vector *cu*).

In lines 27–28, we initialize the output variables. Then, we find the nonbasic list and calculate the simplex multiplier and the reduced costs (lines 30–34). Finally, we calculate the lower and upper values of the range of the coefficients of the objective function (lines 37–61).

```
1.   function [cl, cu] = sensitivityAnalysisc(A, c, BasicList)
2.   % Filename: sensitivityAnalysisc.m
3.   % Description: the function is an implementation of the
4.   % sensitivity analysis for the coefficients of the
5.   % objective function
6.   % Authors: Ploskas, N., & Samaras, N.
7.   %
8.   % Syntax: [cl, cu, sp] = sensitivityAnalysisc(A, c, ...
9.   %    MinMaxLP, BasicList)
10.  %
11.  % Input:
12.  % -- A: matrix of coefficients of the constraints
13.  %    (size m x n)
14.  % -- c: vector of coefficients of the objective function
15.  %    (size n x 1)
16.  % -- BasicList: vector of the indices of the basic
17.  %    variables (size 1 x m)
18.  %
19.  % Output:
20.  % -- cl: vector of the lower values of the range of the
21.  %    coefficients of the objective function (size 1 x n)
22.  % -- cu: vector of the upper values of the range of the
23.  %    coefficients of the objective function (size 1 x n)
```

```
24.
25.   n = length(c); % find the number of variables
26.   % initialize output
27.   cl = zeros(1, n);
28.   cu = zeros(1, n);
29.   % find the nonbasic list
30.   NonBasicList = setdiff([1:n], BasicList);
31.   BasusInv = inv(A(:, BasicList)); % invert the basis
32.   w = c(BasicList)' * BasisInv; % calculate the simplex multiplier
33.   % calculate the reduced costs
34.   Sn = c(NonBasicList)' - w * A(:, NonBasicList);
35.   % calculate the lower and upper values of the range
36.   % of the coefficients of the objective function
37.   for j = 1:n
38.        % if it is a nonbasic variable
39.        if ismember(j, NonBasicList)
40.              t = find(NonBasicList == j);
41.              cu(j) = Inf;
42.              cl(j) = c(j) - Sn(t(1));
43.        else % if it is a basic variable
44.              r = find(BasicList == j);
45.              HrN = BasisInv(r, :) * A(:, NonBasicList);
46.              HrNplus = find(HrN > 0);
47.              HrNminus = find(HrN < 0);
48.              if isempty(HrNplus)
49.                   cu(j) = Inf;
50.              else
51.                   cu(j) = c(j) + min(Sn(HrNplus) ...
52.                        ./ HrN(HrNplus));
53.              end
54.              if isempty(HrNminus)
55.                   cl(j) = -Inf;
56.              else
57.                   cl(j) = c(j) + max(Sn(HrNminus) ...
58.                        ./ HrN(HrNminus));
59.              end
60.        end
61.   end
62.   end
```

The function (filename: sensitivityAnalysisb.m) for the implementation of the sensitivity analysis for the right-hand side coefficients takes as input the matrix with the coefficients of the constraints (matrix A), the vector with the right-hand side of the constraints (vector b), and the vector of the indices of the basic variables (vector *BasicList*), and returns as output the vector of the lower values of the range of the right-hand side coefficients (vector *bl*) and the vector of the upper values of the range of the right-hand side coefficients (vector *bu*).

In lines 26–27, we initialize the output variables. Then, we calculate the lower and upper values of the range of the right-hand side coefficients (lines 32–49).

```
1.   function [bl, bu] = sensitivityAnalysisb(A, b, BasicList)
2.   % Filename: sensitivityAnalysisb.m
3.   % Description: the function is an implementation of the
4.   % sensitivity analysis for the right-hand side
5.   % coefficients
6.   % Authors: Ploskas, N., & Samaras, N.
7.   %
8.   % Syntax: [bl, bu] = sensitivityAnalysisb(A, b, BasicList)
9.   %
10.  % Input:
11.  % -- A: matrix of coefficients of the constraints
12.  %      (size m x n)
13.  % -- b: vector of the right-hand side of the constraints
14.  %      (size m x 1)
15.  % -- BasicList: vector of the indices of the basic
16.  %      variables (size 1 x m)
17.  %
18.  % Output:
19.  % -- bl: vector of the lower values of the range of the
20.  %      right-hand side coefficients (size 1 x m)
21.  % -- bu: vector of the upper values of the range of the
22.  %      right-hand side coefficients (size 1 x m)
23.
24.  m = length(BasicList); % find the number of constraints
25.  % initialize output
26.  bl = zeros(1, m);
27.  bu = zeros(1, m);
28.  BasisInv = inv(A(:, BasicList)); % invert the basis
29.  Xb = BasisInv * b; % compute vector Xb
30.  % calculate the lower and upper values of the range
31.  % of the right-hand side coefficients
32.  for i = 1:m
33.     % find the positive values in the basis inverse
34.     % of the specific constraint
35.     t = find(BasisInv(:, i) > 0);
36.     if isempty(t) % if no positive value exists
37.         bl(i) = -Inf;
38.     else % calculate the lower value
39.         bl(i) = b(i) + max(-Xb(t) ./ BasisInv(t, i));
40.     end
41.     % find the negative values in the basis inverse
42.     % of the specific constraint
43.     t = find(BasisInv(:, i) < 0);
44.     if isempty(t) % if no negative value exists
45.         bu(i) = Inf;
46.     else % calculate the upper value
47.         bu(i) = b(i) + min(-Xb(t) ./ BasisInv(t, i));
48.     end
49.  end
50.  end
```

12.3.2 Change in the Cost Vector

After solving an LP problem, we find out that the cost coefficient of variable x_j is not c_j but $c_j + \Delta c_j$. We can apply the methodology presented in Section 12.3.1 in order to find out the range of the cost coefficient of the specific variable. If the amount $c_j + \Delta c_j$ is in that range, then the optimal basis for the original LP problem remains optimal for the modified LP problem. Otherwise, we solve again the modified LP problem. However, we do not need to solve the LP problem from scratch but we can use the optimal basis for the original LP problem as the initial basis for the revised primal simplex algorithm. That way, we will probably need fewer iterations to solve the modified LP problem.

12.3.2.1 Example

Let's assume that the cost coefficient of the variable x_2 is not 18 but 2 in the example presented in Section 12.3.1. Does the optimal basis $B = [5, 4]$, $N = [1, 2, 3, 6]$ for the original LP problem remain optimal for the modified LP problem?

We already know that the range of cost coefficient c_2 is $[12, \infty)$. The new value of $c_2 + \Delta c_2$ is 2, so it does not belong to the range $[12, \infty)$. That means that the optimal basis for the original LP problem does not remain optimal for the modified LP problem. Hence, we can apply the revised primal simplex algorithm starting from the previous optimal basis $B = [5, 4]$ in order to solve the modified LP problem.

12.3.2.2 Implementation in MATLAB

Below, we present the implementation in MATLAB of the sensitivity analysis when a change in the cost coefficients occurs. Some necessary notations should be introduced before the presentation of the implementation of the sensitivity analysis when a change in the cost coefficients occurs. We assume that the LP problem is in its standard form. Let A be a $m \times n$ matrix with the coefficients of the constraints, c be a $n \times 1$ vector with the coefficients of the objective function, b be a $m \times 1$ vector with the right-hand side of the constraints, *MinMaxLP* be a variable denoting the type of optimization (-1 for minimization and 1 for maximization), *BasicList* be a $1 \times m$ vector of the indices of the basic variables of the original LP problem, *index* be the index of the variable where the change in the cost coefficients occurs, *DeltaC* be the amount of change in the cost vector, and *optimal* be a flag variable showing if the LP problem remains optimal or not (0—the current basis is not optimal, 1—the current basis remains optimal).

The function for the implementation of the sensitivity analysis (filename: sensitivityAnalysis Changec.m) when a change in the cost coefficients occurs takes as input the matrix with the coefficients of the constraints (matrix A), the vector with the coefficients of the objective function (vector c), the vector with the right-hand side of the constraints (vector *b*), a variable denoting the type of optimization (variable *MinMaxLP*), the vector of the indices of the basic variables (vector *BasicList*), the index of the variable where the change in the cost coefficient occurs (variable *index*), and the amount of change in the cost vector (*DeltaC*), and returns as output a flag variable showing if the LP problem remains optimal or not (variable *optimal*).

In lines 35–41, we calculate the solution vector *x* and vectors *w* and *s*. Then, we check if the current basis is not optimal for the original LP problem (lines 43–52). Next, we find the ranges of the cost coefficients (lines 54–55) and change the cost coefficient of the variable index (line 58). Finally, we check if the current basis remains optimal for the modified LP problem (lines 60–68).

```
1.    function [optimal] = sensitivityAnalysisChangec(A, ...
2.        c, b, MinMaxLP, BasicList, index, DeltaC)
3.    % Filename: sensitivityAnalysisChangec.m
4.    % Description: the function is an implementation of the
5.    % sensitivity analysis when a change in the cost
6.    % coefficients occurs
7.    % Authors: Ploskas, N., & Samaras, N.
8.    %
9.    % Syntax: [optimal] = sensitivityAnalysisChangec(A, ...
10.   %    c, b, MinMaxLP, BasicList, index, DeltaC)
11.   %
12.   % Input:
13.   % -- A: matrix of coefficients of the constraints
14.   %    (size m x n)
15.   % -- c: vector of coefficients of the objective function
16.   %    (size n x 1)
17.   % -- b: vector of the right-hand side of the constraints
18.   %    (size m x 1)
19.   % -- MinMaxLP: the type of optimization (optional:
20.   %    default value -1 - minimization)
21.   % -- BasicList: vector of the indices of the basic
22.   %    variables (size 1 x m)
23.   % -- index: the index of the variable where the change
24.   %    in the cost coefficients occurs
25.   % -- DeltaC: the amount of change in the cost vector
26.   %
27.   % Output:
28.   % -- optimal: a flag variable showing if the LP problem
29.   %    remains optimal or not (0 - the current basis is
30.   %    not optimal, 1 - the current basis remains
31.   %    optimal)
32.
33.   [~, n] = size(A); % the size of matrix A
34.   % calculate the basis inverse
35.   BasisInv = inv(A(:, BasicList));
36.   % compute vector x
37.   x = zeros(1, n);
38.   x(BasicList) = BasisInv * b;
39.   % calculate vectors w and s
40.   w = c(BasicList)' * BasisInv;
```

```
41.    s = c' - w * A;
42.    % check if the current basis is not optimal
43.    if any(x(BasicList) < 0)
44.        disp('The current basis is not optimal!')
45.        optimal = 0;
46.        return
47.    end
48.    if any(s < 0)
49.        disp('The current basis is not optimal!')
50.        optimal = 0;
51.        return
52.    end
53.    % find the ranges of the cost coefficients
54.    [cl, cu] = sensitivityAnalysisc(A, c, ...
55.          MinMaxLP, BasicList);
56.    % change the cost coefficient of the variable
57.    % index
58.    c(index) = c(index) + DeltaC;
59.    % check if the current basis remains optimal
60.    if c(index) >= cl(index) && c(index) <= cu(index)
61.        disp('The current basis remains optimal!')
62.        optimal = 1;
63.        return
64.    else
65.        disp('The current basis is not optimal!')
66.        optimal = 0;
67.        return
68.    end
69.    end
```

12.3.3 Change in the Right-Hide Side Vector

After solving an LP problem, we find out that the right-hand side of constraint b_i has changed to $b_i + \Delta b_i$. We can apply the methodology presented in Section 12.3.1 in order to find out the range of the right-hand side of the specific constraint. If the amount $b_i + \Delta b_i$ is in that range, then the optimal basis for the original LP problem remains optimal for the modified LP problem. Otherwise, we solve again the modified LP problem. However, we do not need to solve the LP problem from scratch but we can use the optimal basis for the original LP problem as the initial basis for the revised dual simplex algorithm. That way, we will probably need fewer iterations to solve the modified LP problem.

12.3.3.1 Example

Let's assume that the right-hand side of the first constraint b_1 is not 0 but 3 in the example presented in Section 12.3.1. Does the optimal basis $B = [5, 4]$, $N = [1, 2, 3, 6]$ for the original LP problem remain optimal for the modified LP problem?

We already know that the range of the right-hand side of the first constraint b_1 is $(-\infty, 0]$. The new value of $b_1 + \Delta b_1$ is 3, so it does not belong to the range $(-\infty, 0]$. That means that the optimal basis for the original LP problem does not remain optimal for the modified LP problem. Moreover, the previous optimal basis $B = [5, 4]$ is not feasible for the modified LP problem ($x_B \not\geq 0$):

$$x_B = A_B^{-1} b = \begin{bmatrix} -1 & 0 \\ 0 & 1/2 \end{bmatrix} \begin{bmatrix} 3 \\ 1 \end{bmatrix} = \begin{bmatrix} -3 \\ 1/2 \end{bmatrix}$$

Hence, we can apply the revised dual simplex algorithm starting from the previous optimal basis $B = [5, 4]$ in order to solve the modified LP problem.

12.3.3.2   Implementation in MATLAB

Below, we present the implementation in MATLAB of the sensitivity analysis when a change in the right-hand side occurs (filename: `sensitivityAnalysis Changeb.m`). Some necessary notations should be introduced before the presentation of the implementation of the sensitivity analysis when a change in the right-hand side occurs. We assume that the LP problem is in its standard form. Let A be a $m \times n$ matrix with the coefficients of the constraints, c be a $n \times 1$ vector with the coefficients of the objective function, b be a $m \times 1$ vector with the right-hand side of the constraints, *BasicList* be a $1 \times m$ vector of the indices of the basic variables of the original LP problem, *index* be the index of the constraint where the change in the right-hand side occurs, *DeltaB* be the amount of change in the right-hand side, and *optimal* be a flag variable showing if the LP problem remains optimal or not (0—the current basis is not optimal, 1—the current basis remains optimal).

The function for the implementation of the sensitivity analysis when a change in the right-hand side occurs takes as input the matrix with the coefficients of the constraints (matrix A), the vector with the coefficients of the objective function (vector c), the vector with the right-hand side of the constraints (vector b), the vector of the indices of the basic variables (vector *BasicList*), the index of the constraint where the change in the right-hand side occurs (variable *index*), and the amount of change in the right-hand side (*DeltaB*), and returns as output a flag variable showing if the LP problem remains optimal or not (variable *optimal*).

In lines 33–39, we calculate the solution vector x and vectors w and s. Then, we check if the current basis is not optimal for the original LP problem (lines 41–50). Next, we find the ranges of the right-hand side coefficients (line 52) and change the right-hand side of the specific constraint (line 54). Finally, we check if the current basis remains optimal for the modified LP problem (lines 56–64).

```
1.    function [optimal] = sensitivityAnalysisChangeb(A, ...
2.        c, b, BasicList, index, DeltaB)
3.    % Filename: sensitivityAnalysisChangeb.m
4.    % Description: the function is an implementation of the
5.    % sensitivity analysis when a change in the right-hand
```

```
6.    % side occurs
7.    % Authors: Ploskas, N., & Samaras, N.
8.    %
9.    % Syntax: [optimal] = sensitivityAnalysisChangeb(A, ...
10.   %    c, b, BasicList, index, DeltaB)
11.   %
12.   % Input:
13.   % -- A: matrix of coefficients of the constraints
14.   %    (size m x n)
15.   % -- c: vector of coefficients of the objective function
16.   %    (size n x 1)
17.   % -- b: vector of the right-hand side of the constraints
18.   %    (size m x 1)
19.   % -- BasicList: vector of the indices of the basic
20.   %    variables (size 1 x m)
21.   % -- index: the index of the constraint where the change
22.   %    in the right-hand side occurs
23.   % -- DeltaB: the amount of change in the right-hand side
24.   %
25.   % Output:
26.   % -- optimal: a flag variable showing if the LP problem
27.   %    remains optimal or not (0 - the current basis is
28.   %    not optimal, 1 - the current basis remains
29.   %    optimal)
30.
31.   [~, n] = size(A); % the size of matrix A
32.   % calculate the basis inverse
33.   BasisInv = inv(A(:, BasicList));
34.   % compute vector x
35.   x = zeros(1, n);
36.   x(BasicList) = BasisInv * b;
37.   % calculate vectors w and s
38.   w = c(BasicList)' * BasisInv;
39.   s = c' - w * A;
40.   % check if the current basis is not optimal
41.   if any(x(BasicList) < 0)
42.       disp('The current basis is not optimal!')
43.       optimal = 0;
44.       return
45.   end
46.   if any(s < 0)
47.       disp('The current basis is not optimal!')
48.       optimal = 0;
49.       return
50.   end
51.   % find the ranges of the right-hand side coefficients
52.   [bl, bu] = sensitivityAnalysisb(A, b, BasicList);
53.   % change the right-hand side of the specific constraint
54.   b(index) = b(index) + DeltaB;
55.   % check if the current basis remains optimal
56.   if b(index) >= bl(index) && b(index) <= bu(index)
57.       disp('The current basis remains optimal!')
58.       optimal = 1;
59.       return
```

```
60.   else
61.        disp('The current basis is not optimal!')
62.        optimal = 0;
63.        return
64.   end
65.   end
```

12.4   Adding New Variables

After solving an LP problem, we find out that there is a new variable x_{n+1} that must be added. Hence, the modified LP problem is the following:

$$
\begin{array}{rlll}
\min z = & c^T x & + & c_{n+1} x_{n+1} \\
\text{s.t.} & Ax & + & A_{.,n+1} x_{n+1} = b \\
& x, x_{n+1} \geq 0
\end{array}
$$

The optimal basis for the original LP problem remains optimal if $s_{n+1} = c_{n+1} - w^T A_{.,n+1} \geq 0$. Otherwise, we solve again the modified LP problem. However, we do not need to solve the LP problem from scratch but we can use the optimal basis for the original LP problem as the initial basis for the revised primal simplex algorithm and select the new variable x_{n+1} as the entering variable. That way, we will probably need fewer iterations in order to solve the modified LP problem.

12.4.1   Example

Let's assume that a new variable x_7 must be added in the example presented in Section 12.3.1. The new cost coefficient $c_7 = -4$ and $A_{.7} = \begin{bmatrix} -1 & 1 \end{bmatrix}^T$. Does the optimal basis $B = [5, 4], N = [1, 2, 3, 6]$ for the original LP problem remain optimal for the modified LP problem?

We calculate the reduced cost s_7:

$$
s_7 = c_7 - wA_{.,n+1} = -4 - \begin{bmatrix} 0 & 2 \end{bmatrix} \begin{bmatrix} -1 & 1 \end{bmatrix}^T = -4 - 2 = -6
$$

$s_7 \ngeq 0$, so the basis $B = [5, 4], N = [1, 2, 3, 6]$ is not optimal for the modified LP problem. Hence, we can apply the revised primal simplex algorithm starting from the previous optimal basis $B = [5, 4]$ and select the new variable x_{n+1} as the entering variable in order to solve the modified LP problem.

12.4.2 Implementation in MATLAB

Below, we present the implementation in MATLAB of the sensitivity analysis when adding a new variable (filename: `sensitivityAnalysisAddVariable.m`). Some necessary notations should be introduced before the presentation of the implementation of the sensitivity analysis when adding a new variable. We assume that the LP problem is in its standard form. Let A be a $m \times n$ matrix with the coefficients of the constraints, c be a $n \times 1$ vector with the coefficients of the objective function, b be a $m \times 1$ vector with the right-hand side of the constraints, $BasicList$ be a $1 \times m$ vector of the indices of the basic variables of the original LP problem, $Anplus1$ be a $m \times 1$ vector of the new column in the matrix of coefficients of the constraints, $cnplus1$ be a variable with the cost coefficient of the new variable, and $optimal$ be a flag variable showing if the LP problem remains optimal or not (0—the current basis is not optimal, 1—the current basis remains optimal).

The function for the implementation of the sensitivity analysis when adding a new variable occurs takes as input the matrix with the coefficients of the constraints (matrix A), the vector with the coefficients of the objective function (vector c), the vector with the right-hand side of the constraints (vector b), the vector of the indices of the basic variables (vector $BasicList$), the vector of the new column in the matrix of coefficients of the constraints (vector $Anplus1$), and the cost coefficient of the new variable (variable $cnplus1$), and returns as output a flag variable showing if the LP problem remains optimal or not (variable $optimal$).

In lines 34–40, we calculate the solution vector x and vectors w and s. Then, we check if the current basis is not optimal for the original LP problem (lines 42–51). Next, we calculate the new reduced cost (line 53). Finally, we check if the current basis remains optimal for the modified LP problem (lines 55–63).

```
1.    function [optimal] = ...
2.        sensitivityAnalysisAddVariable(A, c, b, ...
3.        BasicList, Anplus1, cnplus1)
4.    % Filename: sensitivityAnalysisAddVariable.m
5.    % Description: the function is an implementation of the
6.    % sensitivity analysis when adding a new variable
7.    % Authors: Ploskas, N., & Samaras, N.
8.    %
9.    % Syntax: [optimal] = ...
10.   %    sensitivityAnalysisAddVariable(A, c, b, ...
11.   %    BasicList, Anplus1, cnplus1)
12.   %
13.   % Input:
14.   % -- A: matrix of coefficients of the constraints
15.   %    (size m x n)
16.   % -- c: vector of coefficients of the objective function
17.   %    (size n x 1)
18.   % -- b: vector of the right-hand side of the constraints
19.   %    (size m x 1)
20.   % -- BasicList: vector of the indices of the basic
21.   %    variables (size 1 x m)
```

```
22.   % -- Anplus1: the new column in the matrix of coefficients
23.   %      of the constraints (size m x 1)
24.   % -- cnplus1: the cost coefficient of the new variable
25.   %
26.   % Output:
27.   % -- optimal: a flag variable showing if the LP problem
28.   %      remains optimal or not (0 - the current basis is
29.   %      not optimal, 1 - the current basis remains
30.   %      optimal)
31.
32.   [~, n] = size(A); % the size of matrix A
33.   % calculate the basis inverse
34.   BasisInv = inv(A(:, BasicList));
35.   % compute vector x
36.   x = zeros(1, n);
37.   x(BasicList) = BasisInv * b;
38.   % calculate vectors w and s
39.   w = c(BasicList)' * BasisInv;
40.   s = c' - w * A;
41.   % check if the current basis is not optimal
42.   if any(x(BasicList) < 0)
43.       disp('The current basis is not optimal!')
44.       optimal = 0;
45.       return
46.   end
47.   if any(s < 0)
48.       disp('The current basis is not optimal!')
49.       optimal = 0;
50.       return
51.   end
52.   % find the value of the new reduced cost
53.   Snplus1 = cnplus1 - w * Anplus1;
54.   % check if the current basis remains optimal
55.   if Snplus1 >= 0
56.       disp('The current basis remains optimal!')
57.       optimal = 1;
58.       return
59.   else
60.       disp('The current basis is not optimal!')
61.       optimal = 0;
62.       return
63.   end
64.   end
```

12.5 Adding New Constraints

After solving an LP problem, we find out that there is a new constraint that must be added:

$$A_{m+1,1}x_1 + A_{m+1,2}x_2 + \cdots + A_{m+1,n}x_n + x_{n+1} = b_{m+1}$$

where x_{n+1} is a slack variable. We add one more variable to the basis. We add to the basis the variable x_{n+1}. Hence, $B = [B, n + 1]$ and the new basic matrix will be:

$$A_{\overline{B}} = \begin{bmatrix} A_B & 0 \\ A_{m+1,B} & 1 \end{bmatrix}$$

The new basis inverse will be:

$$(A_{\overline{B}})^{-1} = \begin{bmatrix} A_B^{-1} & 0 \\ -A_{m+1,B}A_B^{-1} & 1 \end{bmatrix}$$

The new basic cost coefficient vector will be:

$$\overline{c}_B = \begin{bmatrix} c_B \\ 0 \end{bmatrix}$$

Hence:

$$\overline{w} = \begin{bmatrix} \overline{c}_B^T (A_{\overline{B}})^{-1} & 0 \end{bmatrix}$$

$$\overline{x}_B = (A_{\overline{B}})^{-1} \begin{bmatrix} b \\ b_{m+1} \end{bmatrix} = \begin{bmatrix} A_B^{-1} & 0 \\ -A_{m+1,B}A_B^{-1} & 1 \end{bmatrix} \begin{bmatrix} b \\ b_{m+1} \end{bmatrix} = \begin{bmatrix} A_B^{-1}b \\ -A_{m+1,B}A_B^{-1}b + b_{m+1} \end{bmatrix}$$

If $x_{n+1} = x_{B[m+1]} = -A_{m+1,B}A_B^{-1}b + b_{m+1} \geq 0$, then the basic list $[B, n + 1]$ is optimal for the modified LP problem. Otherwise, we solve again the modified LP problem. However, we do not need to solve the LP problem from scratch but we can use the basis $[B, n + 1]$ as the initial basis for the revised dual simplex algorithm and select the new variable x_{n+1} as the leaving variable. That way, we will probably need fewer iterations to solve the modified LP problem.

12.5.1 Example

Let's assume that a new constraint $-2x_1 + 3x_3 + 5x_5 - 3x_6 \leq -8$ must be added in the example presented in Section 12.3.1. Does the optimal basis $B = [5, 4]$, $N = [1, 2, 3, 6]$ for the original LP problem remain optimal for the modified LP problem?

We add the slack variable x_7, so the constraint is now $-2x_1 + 3x_3 + 5x_5 - 3x_6 + x_7 = -8$. So, $A_{3.} = \begin{bmatrix} -2 & 0 & 3 & 0 & 5 & -3 & 1 \end{bmatrix}$. The new basic list is $B = [B, n + 1] = [5, 4, 7]$ and the nonbasic list is $N = [1, 2, 3, 6]$. Hence, the new basis inverse is:

$$(A_{\overline{B}})^{-1} = \begin{bmatrix} A_B^{-1} & 0 \\ -A_{3,B}A_B^{-1} & 1 \end{bmatrix} = \begin{bmatrix} -1 & 0 & 0 \\ 0 & 1/2 & 0 \\ 5 & 0 & 1 \end{bmatrix}$$

$x_7 = -A_{3,B}A_B^{-1}b + b_3 = -8 \ngeq 0$. So, the basic list $B = [5, 4, 7]$ is not optimal for the modified LP problem. Hence, we can apply the revised dual simplex algorithm starting from the optimal basis $B = [5, 4, 7]$ and select the new variable x_7 as the leaving variable in order to solve the modified LP problem.

12.5.2 Implementation in MATLAB

Below, we present the implementation in MATLAB of the sensitivity analysis when adding a new constraint (filename: `sensitivityAnalysisAddConstr -aint.m`). Some necessary notations should be introduced before the presentation of the implementation of the sensitivity analysis when adding a new constraint. We assume that the LP problem is in its standard form. Let A be a $m \times n$ matrix with the coefficients of the constraints, c be a $n \times 1$ vector with the coefficients of the objective function, b be a $m \times 1$ vector with the right-hand side of the constraints, *BasicList* be a $1 \times m$ vector of the indices of the basic variables of the original LP problem, *Amplus*1 be a $1 \times n$ vector of the new row in the matrix of coefficients of the constraints, *bmplus*1 be the right-hand side value of the new constraint, *constrEqin* be the type of the new constraint (-1: constraint type is 'less than or equal to', 0: equality constraint, 1: constraint type is 'greater than or equal to'), and *optimal* be a flag variable showing if the LP problem remains optimal or not (0—the current basis is not optimal, 1—the current basis remains optimal).

The function for the implementation of the sensitivity analysis when adding a new constraint takes as input the matrix with the coefficients of the constraints (matrix A), the vector with the coefficients of the objective function (vector c), the vector with the right-hand side of the constraints (vector b), the vector of the indices of the basic variables (vector *BasicList*), the vector of the new row in the matrix of coefficients of the constraints (vector *Amplus*1), the right-hand side value of the new constraint (variable *bmplus*1), and a flag variable showing the type of the new constraint (variable *constrEqin*), and returns as output a flag variable showing if the LP problem remains optimal or not (variable *optimal*).

In lines 39–45, we calculate the solution vector x and vectors w and s. Then, we check if the current basis is not optimal for the original LP problem (lines 47–56). Next, we add the new constraint to the LP problem (lines 58–69) and add the new constraint to the basic list (line 71). Then, we compute the new solution vector x (lines 73–76). Finally, we check if the current basis remains optimal for the modified LP problem (lines 78–85).

```
1.    function [optimal] = ...
2.        sensitivityAnalysisAddConstraint(A, c, b, ...
3.        BasicList, Amplus1, bmplus1, constrEqin)
4.    % Filename: sensitivityAnalysisAddConstraint.m
5.    % Description: the function is an implementation of the
6.    % sensitivity analysis when adding a new constraint
7.    % Authors: Ploskas, N., & Samaras, N.
```

```
8.   %
9.   % Syntax: [optimal] = ...
10.  %    sensitivityAnalysisAddConstraint(A, c, b, ...
11.  %    BasicList, Amplus1, bmplus1, constrEqin)
12.  %
13.  % Input:
14.  % -- A: matrix of coefficients of the constraints
15.  %    (size m x n)
16.  % -- c: vector of coefficients of the objective function
17.  %    (size n x 1)
18.  % -- b: vector of the right-hand side of the constraints
19.  %    (size m x 1)
20.  % -- BasicList: vector of the indices of the basic
21.  %    variables (size 1 x m)
22.  % -- Amplus1: the new row in the matrix of coefficients
23.  %    of the constraints (size 1 x n)
24.  % -- bmplus1: the right-hand side value of the new
25.  %    constraint
26.  % -- constrEqin: the type of the new constraint (-1:
27.  %    constraint type is 'less than or equal to', 0:
28.  %    equality constraint, 1: constraint type is
29.  %    'greater than or equal to')
30.  %
31.  % Output:
32.  % -- optimal: a flag variable showing if the LP problem
33.  %    remains optimal or not (0 - the current basis is
34.  %    not optimal, 1 - the current basis remains
35.  %    optimal)
36.
37.  [m, n] = size(A); % the size of matrix A
38.  % calculate the basis inverse
39.  BasisInv = inv(A(:, BasicList));
40.  % compute vector x
41.  x = zeros(1, n);
42.  x(BasicList) = BasisInv * b;
43.  % calculate vectors w and s
44.  w = c(BasicList)' * BasisInv;
45.  s = c' - w * A;
46.  % check if the current basis is not optimal
47.  if any(x(BasicList) < 0)
48.      disp('The current basis is not optimal!')
49.      optimal = 0;
50.      return
51.  end
52.  if any(s < 0)
53.      disp('The current basis is not optimal!')
54.      optimal = 0;
55.      return
56.  end
57.  % add the new constraint
58.  b = [b; bmplus1];
59.  c = [c; 0];
60.  if constrEqin == -1 % 'less than or equal to' constraint
61.      A = [A zeros(m, 1); Amplus1 1];
```

```
62.  elseif constrEqin == 0 % equality constraint
63.      % we do not need to add a new variable, but
64.      % we add a new variable for simplicity's sake
65.      A = [A zeros(m, 1); Amplus1 0];
66.  % 'greater than or equal to' constraint
67.  elseif constrEqin == 1
68.      A = [A zeros(m, 1); Amplus1 -1];
69.  end
70.  % add the new constraint to the basic list
71.  BasicList = [BasicList n + 1];
72.  % calculate the new basis inverse
73.  BasisInv = inv(A(:, BasicList));
74.  % compute the new vector x
75.  x = zeros(1, n + 1);
76.  x(BasicList) = BasisInv * b;
77.  % check if the current basis is not optimal
78.  if x(n + 1) >= 0
79.      disp('The current basis remains optimal!')
80.      optimal = 1;
81.      return
82.  else
83.      disp('The current basis is not optimal!')
84.      optimal = 0;
85.  end
86.  end
```

12.6   Chapter Review

In many cases, after solving an LP problem with the simplex method, there is a change in the data of the LP problem. This chapter discusses how to deal with such changes efficiently. This topic is called sensitivity analysis. Sensitivity analysis is very useful in two situations: (i) when we wish to know how the solution will be affected if we perform a small change in the LP problem, and (ii) when we have already solved an LP problem and we want also to solve a second LP problem in which the data is only slightly different. Rather than restarting the simplex method from scratch for the modified LP problem, we want to solve the modified LP problem starting with the optimal basis of the original LP problem and perform only a few iterations to solve the modified LP problem (if necessary). In this chapter, we presented topics about sensitivity analysis. We examined how the solution of an LP problem is affected when changes are made to the input data of the LP problem. Moreover, we examined changes in: (i) the cost vector, (ii) the right-hand side vector, and (iii) the coefficient of the constraints.

References

1. Bazaraa, M. S., Jarvis, J. J., & Sherali, H. D. (2011). *Linear programming and network flows.* Hoboken: John Wiley & Sons.
2. Bertsimas, D., & Tsitsiklis, J. N. (1997). *Introduction to linear optimization* (Vol. 6). Belmont, MA: Athena Scientific.
3. Borgonovo, E., & Plischke, E. (2016). Sensitivity analysis: A review of recent advances. *European Journal of Operational Research, 248*(3), 869–887.
4. Vanderbei, R. J. (2015). *Linear programming.* Heidelberg: Springer.
5. Winston, W. L. (2004). *Operations research: Applications and algorithms* (Vol. 3). Belmont: Thomson Brooks/Cole.

Errata to: Linear Programming Using MATLAB®

Nikolaos Ploskas and Nikolaos Samaras

Errata to:
N. Ploskas, N. Samaras, *Linear Programming Using MATLAB®*,
Springer Optimization and Its Applications 127,
https://doi.org/10.1007/978-3-319-65919-0

The original version of the book was inadvertently published without updating the following corrections:

Preface:
On page ix, the last line reads:

April 2017

It should read:

November 2017

The updated online version of this book can be found at
https://doi.org/10.1007/10.1007/978-3-319-65919-0_1
https://doi.org/10.1007/10.1007/978-3-319-65919-0_2
https://doi.org/10.1007/10.1007/978-3-319-65919-0_4
https://doi.org/10.1007/10.1007/978-3-319-65919-0_8
https://doi.org/10.1007/10.1007/978-3-319-65919-0_10
https://doi.org/10.1007/978-3-319-65919-0

© Springer International Publishing AG 2017
N. Ploskas, N. Samaras, *Linear Programming Using MATLAB®*,
Springer Optimization and Its Applications 127, DOI 10.1007/978-3-319-65919-0_13

Chapter 1:
On page 5, 11th line from top reads:

A more efficient approach is the Primal-Dual Exterior Point Simplex Algorithm (PDEPSA) proposed by Samaras [23] and Paparrizos [22].

It should read:

A more efficient approach is the Primal-Dual Exterior Point Simplex Algorithm (PDEPSA) proposed by Samaras [23] and Paparrizos et al. [22].

Chapter 2:
On page 68, alignment of the following equations in Table 2.8 were as follows:

$s_0 = \sum_{j \in P} \lambda_j s_j$
and the direction
$d_B = -\sum_{j \in P} \lambda_j h_j$, where $h_j = A_B^{-1} A_{.j}$.

It should be as follows:

$s_0 = \sum_{j \in P} \lambda_j s_j$
and the direction
$d_B = -\sum_{j \in P} \lambda_j h_j$, where $h_j = A_B^{-1} A_{.j}$.

And

if $d_B \geq 0$ then
    if $s_0 = 0$ then STOP. The LP problem is optimal.
    else
    choose the leaving variable $x_{B[r]} = x_k$ using the following relation:
    $a = \frac{x_{B[r]}}{-d_{B[r]}} = \min\left\{\frac{x_{B[i]}}{-d_{B[i]}} : d_{B[i]} < 0\right\}, i = 1, 2, \cdots, m$
    if $a = \infty$, the LP problem is unbounded.

It should be as follows:

if $d_B \geq 0$ then
    if $s_0 = 0$ then STOP. The LP problem is optimal.
    else
        choose the leaving variable $x_{B[r]} = x_k$ using the following relation:
    $a = \frac{x_{B[r]}}{-d_{B[r]}} = \min\left\{\frac{x_{B[i]}}{-d_{B[i]}} : d_{B[i]} < 0\right\}, i = 1, 2, \cdots, m$
    if $a = \infty$, the LP problem is unbounded.

Chapter 4:
On page 211, 15th line from top, the sentence reads:

The column "Total size reduction" in Table 4.2 is calculated as follows: $-(m_{new} + n_{new} - m - n)/((m + n))$.

It should read:

The column "Total size reduction" in Table 4.2 is calculated as follows: $-(m_{new} + n_{new} - m - n)/(m + n)$.

Chapter 8:
On page 345, 7th line from bottom, the sentence reads:

There are elements in vector h_1 that are greater than 0, so we perform the minimum ratio test (where the letter x is used below to represent that $h_i l \leq 0$, therefore $\frac{x_{B[i]}}{h_{il}}$ is not defined):

It should read:

There are elements in vector h_1 that are greater than 0, so we perform the minimum ratio test (where the letter x is used below to represent that $h_{il} \leq 0$, therefore $\frac{x_{B[i]}}{h_{il}}$ is not defined):

Chapter 10:
On page 439, alignment of the following equation in Table 10.1 was as follows:

$s_0 = \sum_{j \in P} \lambda_j s_j$
and the direction
$d_B = -\sum_{j \in P} \lambda_j h_j$, where $h_j = A_B^{-1} A_j$.
Step 2.1. *(Test of Optimality)*.
if $P = \emptyset$ then STOP. (LP.1) is optimal.
else
  if $d_B \geq 0$ then
    if $s_0 = 0$ then STOP. (LP.1) is optimal.

It should be as follows:

$s_0 = \sum_{j \in P} \lambda_j s_j$
and the direction
$d_B = -\sum_{j \in P} \lambda_j h_j$, where $h_j = A_B^{-1} A_j$.
Step 2.1. *(Test of Optimality)*.
if $P = \emptyset$ then STOP. (LP.1) is optimal.
else
  if $d_B \geq 0$ then
    if $s_0 = 0$ then STOP. (LP.1) is optimal.

Appendix A
MATLAB's Optimization Toolbox Algorithms

Abstract MATLAB's Optimization Toolbox (version 7.2) includes a family of algorithms for solving optimization problems. The toolbox provides functions for solving linear programming, mixed-integer linear programming, quadratic programming, nonlinear programming, and nonlinear least squares problems. This chapter presents the available functions of this toolbox for solving LPs. A computational study is performed. The aim of the computational study is to compare the efficiency of MATLAB's Optimization Toolbox solvers.

A.1 Chapter Objectives

- Present the linear programming algorithms of MATLAB's Optimization Toolbox.
- Understand how to use the linear programming solver of MATLAB's Optimization Toolbox.
- Present all different ways to solve linear programming problems using MATLAB.
- Compare the computational performance of the linear programming algorithms of MATLAB's Optimization Toolbox.

A.2 Introduction

MATLAB is a matrix language intended primarily for numerical computing. MATLAB is especially designed for matrix computations like solving systems of linear equations or factoring matrices. Users that are not familiar with MATLAB are referred to [3, 7]. Experienced MATLAB users that want to further optimize their codes are referred to [1].

MATLAB's Optimization Toolbox includes a family of algorithms for solving optimization problems. The toolbox provides functions for solving linear programming, mixed-integer linear programming, quadratic programming, nonlinear programming, and nonlinear least squares problems. MATLAB's Optimization Toolbox includes four categories of solvers [9]:

© Springer International Publishing AG 2017

N. Ploskas, N. Samaras, *Linear Programming Using MATLAB®*,
Springer Optimization and Its Applications 127, DOI 10.1007/978-3-319-65919-0

- **Minimizers**: solvers that attempt to find a local minimum of the objective function. These solvers handle unconstrained optimization, linear programming, quadratic programming, and nonlinear programming problems.
- **Multiobjective minimizers**: solvers that attempt to either minimize the value of a set of functions or find a location where a set of functions is below some certain values.
- **Equation solvers**: solvers that attempt to find a solution to a nonlinear equation $f(x) = 0$.
- **Least-squares solvers**: solvers that attempt to minimize a sum of squares.

A complete list of the type of optimization problems, which MATLAB's Optimization Toolbox solvers can handle, along with the name of the solver can be found in Table A.1 [9]:

Table A.1 Optimization problems handled by MATLAB's Optimization Toolbox Solver

Type of problem	Solver	Description
Scalar minimization	fminbnd	Find the minimum of a single-variable function on fixed interval
Unconstrained minimization	fminunc, fminsearch	Find the minimum of an unconstrained multivariable function (fminunc) using derivative-free optimization (fminsearch)
Linear programming	linprog	Solve linear programming problems
Mixed-integer linear programming	intlinprog	Solve mixed-integer linear programming problems
Quadratic programming	quadprog	Solve quadratic programming problems
Constrained minimization	fmincon	Find the minimum of a constrained nonlinear multivariable function
Semi-infinite minimization	fseminf	Find the minimum of a semi-infinitely constrained multivariable nonlinear function
Goal attainment	fgoalattain	Solve multiobjective goal attainment problems
Minimax	fminimax	Solve minimax constraint problems
Linear equations	\(matrix left division)	Solve systems of linear equations
Nonlinear equation of one variable	fzero	Find the root of a nonlinear function
Nonlinear equations	fsolve	Solve a system of nonlinear equations
Linear least-squares	\(matrix left division)	Find a least-squares solution to a system of linear equations
Nonnegative linear-least-squares	lsqnonneg	Solve nonnegative least-squares constraint problems
Constrained linear-least-squares	lsqlin	Solve constrained linear least-squares problems
Nonlinear least-squares	lsqnonlin	Solve nonlinear least-squares problems
Nonlinear curve fitting	lsqcurvefit	Solve nonlinear curve-fitting problems in least-squares sense

MATLAB's Optimization Toolbox offers four LP algorithms:

- an interior point method,
- a primal simplex algorithm,
- a dual simplex algorithm, and
- an active-set algorithm.

This chapter presents the LP algorithms implemented in MATLAB's Optimization Toolbox. Moreover, the Optimization App, a graphical user interface for solving optimization problems, is presented. In addition, LPs are solved using MATLAB in various ways. Finally, a computational study is performed. The aim of the computational study is to compare the execution time of MATLAB's Optimization Toolbox solvers for LPs.

The structure of this chapter is as follows. In Section A.3, the LP algorithms implemented in MATLAB's Optimization Toolbox are presented. Section A.4 presents the linprog solver, while Section A.5 presents the Optimization App, a graphical user interface for solving optimization problems. Section A.6 provides different ways to solve LPs using MATLAB. In Section A.7, a MATLAB code to solve LPs using the linprog solver is provided. Section A.8 presents a computational study that compares the execution time of MATLAB's Optimization Toolbox solvers. Finally, conclusions and further discussion are presented in Section A.9.

A.3   MATLAB's Optimization Toolbox Linear Programming Algorithms

As already mentioned, MATLAB's Optimization Toolbox offer four algorithms for the solution of LPs. These algorithms are presented in this section.

A.3.1   Interior Point Method

The interior point method used in MATLAB is based on LIPSOL [12], which is a variant of Mehrotra's predictor-corrector algorithm [10], a primal-dual interior point method presented in Chapter 11. The method starts by applying the following preprocessing techniques [9]:

- All variables are bounded below zero.
- All constraints are equalities.
- Fixed variables, these with equal lower and upper bounds, are removed.
- Empty rows are removed.
- The constraint matrix has full structural rank.
- Empty columns are removed.

- When a significant number of singleton rows exist, the associated variables are solved for and the associated rows are removed.

After preprocessing, the LP problem has the following form:

$$\begin{aligned} \min \quad & c^T x \\ \text{s.t.} \quad & Ax = b \\ & 0 \leq x \leq u \end{aligned} \qquad (LP.1)$$

The upper bounds are implicitly included in the constraint matrix A. After adding the slack variables s, problem (LP.1) becomes:

$$\begin{aligned} \min \quad & c^T x \\ \text{s.t.} \quad & Ax = b \\ & x + s = u \\ & x \geq 0, s \geq 0 \end{aligned} \qquad (LP.2)$$

Vector x consists of the primal variables and vector s consists of the primal slack variables. The dual problem can be written as:

$$\begin{aligned} \max \quad & b^T y - u^T w \\ \text{s.t.} \quad & A^T y - w + z = c \\ & z \geq 0, w \geq 0 \end{aligned} \qquad (DP.1)$$

Vectors y and w consist of the simplex multiplier and vector z consists of the reduced costs.

This method is a primal-dual algorithm, meaning that both the primal (LP.2) and the dual (DP.1) problems are solved simultaneously. Let us assume that $v = [x; y; z; s; w]$, where $[x; z; s; w] \geq 0$. This method first calculates the prediction or Newton direction:

$$\Delta v_p = - \left(F^T (v) \right)^{-1} F (v) \qquad (A.1)$$

Then, the corrector direction is computed as:

$$\Delta v_c = - \left(F^T (v) \right)^{-1} F \left(v + \Delta v_p \right) - \mu e \qquad (A.2)$$

where $\mu \geq 0$ is called the centering or barrier parameter and must be chosen carefully and e is a zero-one vector with the ones corresponding to the quadratic equation in $F (v)$. The two directions are combined with a step length parameter $\alpha \geq 0$ and update v to get a new iterate v^+:

$$v^+ = v + a\left(\Delta v_p + \Delta v_c\right) \tag{A.3}$$

where the step length parameter α is chosen so that $v^+ = \left[x^+;y^+;z^+;s^+;w^+\right]$ satisfies $\left[x^+;y^+;z^+;s^+;w^+\right] \geq 0$.

This method computes a sparse direct factorization on a modification of the Cholesky factors of AA^T. If A is dense, the Sherman-Morrison formula [11] is used. If the residual is too large, then an LDL factorization of an augmented system form of the step equations to find a solution is performed.

The algorithm then loops until the iterates converge. The main stopping criterion is the following:

$$\frac{\|r_b\|}{\max\left(1,\|b\|\right)} + \frac{\|r_c\|}{\max\left(1,\|c\|\right)} + \frac{\|r_u\|}{\max\left(1,\|u\|\right)} + \frac{\left|c^T x - b^T y + u^T w\right|}{\max\left(1,\left|c^T x\right|,\left|b^T y - u^T w\right|\right)} \leq tol \tag{A.4}$$

where $r_b = Ax - b$, $r_c = A^T y - w + z - c$, and $r_u = x + s - u$ are the primal residual, dual residual, and upper-bound feasibility, respectively, $c^T x - b^T y + u^T w$ is the difference between the primal and dual objective values (duality gap), and tol is the tolerance (a very small positive number).

A.3.2   Primal Simplex Algorithm

This algorithm is a primal simplex algorithm that solves the following LP problem:

$$\begin{aligned} &\min &&c^T x \\ &\text{s.t.} &&Ax \leq b \\ & &&Aeqx = beq \\ & &&lb \leq x \leq ub \end{aligned} \tag{LP.3}$$

The algorithm applies the same preprocessing steps as the ones presented in Section A.3.1. In addition, the algorithm performs two more preprocessing methods:

- Eliminates singleton columns and their corresponding rows.
- For each constraint equation $A_i x = b_i$, $i = 1, 2, \cdots, m$, the algorithm calculates the lower and upper bounds of the linear combination $A_i b_i$ as rlb and rub if the lower and upper bounds are finite. If either rlb or rub equals b_i, then the constraint is called a forcing constraint. The algorithm sets each variable corresponding to a nonzero coefficient of $A_i x$ equal to its lower or upper bound, depending on the forcing constraint. Then, the algorithm deletes the columns corresponding to these variables and deletes the rows corresponding to the forcing constraints.

This is a two-phase algorithm. In Phase I, the algorithm finds an initial basic feasible solution by solving an auxiliary piecewise LP problem. The objective function of the auxiliary problem is the linear penalty function $P = \sum_j P_j(x_j)$, where $P_j(x_j), j = 1, 2, \cdots, n$, is defined by:

$$P_j(x_j) = \begin{cases} x_j - u_j & , \text{if } x_j > u_j \\ 0 & , \text{if } l_j \leq x_j \leq u_j \\ l_j - x_j & , \text{if } l_j > x_j \end{cases} \tag{A.5}$$

The penalty function $P(x)$ measures how much a point x violates the lower and upper bound conditions. The auxiliary problem that is solved is the following:

$$\min \quad \sum_j P_j$$

$$\text{s.t.} \quad Ax \leq b \tag{LP.4}$$

$$Aeqx = beq$$

The original problem (LP.3) has a feasible basis point iff the auxiliary problem (LP.4) has a minimum optimal value of 0. The algorithm finds an initial point for the auxiliary problem (LP.4) using a heuristic method that adds slack and artificial variables. Then, the algorithm solves the auxiliary problem using the initial (feasible) point along with the simplex algorithm. The solution found is the initial (feasible) point for Phase II.

In Phase II, the algorithm applies the revised simplex algorithm using the initial point from Phase I to solve the original problem. As already presented in Chapter 8, the algorithm tests the optimality condition at each iteration and stops if the current solution is optimal. If the current solution is not optimal, the following steps are performed:

- The entering variable is chosen from the nonbasic variables and the corresponding column is added to the basis.
- The leaving variable is chosen from the basic variables and the corresponding column is removed from the basis.
- The current solution and objective value are updated.

The algorithm detects when there is no progress in Phase II process and it attempts to continue by performing bound perturbation [2].

A.3.3   Dual Simplex Algorithm

This algorithm is a dual simplex algorithm. It starts by preprocessing the LP problem using the preprocessing techniques that presented in Section A.3.1. This procedure reduces the original LP problem to the form of Equation (LP.1). This is a two-phase algorithm. In Phase I, the algorithm finds a dual feasible point. This is achieved by solving an auxiliary LP problem, similar to (LP.4). In Phase II, the algorithm chooses an entering and a leaving variable at each iteration. The entering variable is chosen using the variation of Harris ratio test proposed by Koberstein [8] and the leaving variable is chosen using the dual steepest edge pricing technique proposed by Forrest and Goldfarb [4]. The algorithm can apply additional perturbations during Phase II trying to alleviate degeneracy.

The algorithm iterates until the solution to the reduced problem is both primal and dual feasible. If the solution to the reduced problem is dual infeasible for the original problem, then the solver restores dual feasibility using primal simplex or Phase I algorithms. Finally, the algorithm unwinds the preprocessing steps to return the solution for the original problem.

A.3.4   Active-Set Algorithm

The active-set LP algorithm is a variant of the sequential quadratic programming (SQP) solver, called *fmincon*. The only difference is that the quadratic term is set to zero. At each major iteration of the SQP solver, a quadratic programming problem of the following form is solved:

$$\min_{d \in \mathbb{R}^n} \quad q(d) = c^T d$$

$$\text{s.t.} \quad A_{i.} d = b_i, i = 1, \ldots, m_e \qquad (QP.1)$$

$$A_{i.} d \leq b_i, i = m_e + 1, \ldots, m$$

where $A_{i.}$ refers to the ith row of constraint matrix A. This method is similar to that proposed by Gill et al. [5, 6]. This algorithm consists of two phases. In the first phase, a feasible point is constructed. The second phase generates an iterative sequence of feasible points that converge to the solution.

An active set, $\bar{A}_k$, which is an estimate of the active constraints at the solution point, is maintained. $\bar{A}_k$ is updated at each iteration and is used to form a basis for a search direction $\hat{d}_k$. The search direction $\hat{d}_k$ minimizes the objective function while remaining on any active constraint boundaries. The feasible subspace for $\hat{d}_k$ is formed from a basis Z_k whose columns are orthogonal to the estimate of the active set $\bar{A}_k$. Thus, a search direction, which is formed from a linear summation of any combination of the columns of Z_k, is guaranteed to remain on the boundaries of the active constraints.

Matrix Z_k is created from the last $m - l$ columns of the QR decomposition of the matrix $\bar{A}_k^T$, where l is the number of the active constraints ($l < m$). Z_k is given by

$$Z_k = Q[:, l + 1 : m]$$

where $Q^T \bar{A}_k^T = \begin{bmatrix} R \\ 0 \end{bmatrix}$.

Once Z_k is found, a new search direction $\hat{d}_k$ is sought that minimizes $q(d)$, where $\hat{d}_k$ is in the null space of the active constraints. Thus, $\hat{d}_k$ is a linear combination of the columns of Z_k and $\hat{d}_k = Z_k p$ for some vector p. Writing the quadratic term as a function of p by substituting $\hat{d}_k$, we have:

$$q(p) = \frac{1}{2} p^T Z_k^T H Z_k p + c^T Z_k p \tag{A.6}$$

Differentiating this with respect to p, we have:

$$\nabla q(p) = Z_k^T H Z_k p + Z_k^T c \tag{A.7}$$

where $\nabla q(p)$ is the projected gradient of the quadratic function and $Z_k^T H Z_k$ is the projected Hessian. Assuming that the Hessian matrix H is positive definite, then the minimum of the function $q(p)$ in the subspace defined by Z_k occurs when $\nabla q(p) = 0$, which is the solution of the linear equations $Z_k^T H Z_k p = -Z_k^T c$.

A step is then taken as $x_{k+1} = x_k + \alpha \hat{d}_k$, where $\hat{d}_k = Z_k^T p$. There are only two possible choices of step length α at each iteration. A step of unity along $\hat{d}_k$ is the exact step to the minimum of the function restricted to the null space of $\hat{A}_k$. If such a step can be taken without violation of the constraints, then this is the solution to (QP.1). Otherwise, the step along $\hat{d}_k$ to the nearest constraint is less than unity and a new constraint is included in the active set at the next iteration. The distance to the constraint boundaries in any direction $\hat{d}_k$ is given by:

$$\alpha = \min_{i \in \{1 \dots m\}} \left\{ \frac{-(A_i x_k - b_i)}{A_i d_k} \right\} \tag{A.8}$$

where $A_i \hat{d}_k > 0, i = 1 \dots m$.

When n independent constraints are included in the active set without locating the minimum, then Lagrange multipliers that satisfy the nonsingular set of linear equations $\bar{A}_k^T \lambda_k = c$ are calculated. If all elements of λ_k are positive, then x_k is the optimal solution of (QP.1). On the other hand, if any element of λ_k is negative and the element does not correspond to an equality constraint, then the corresponding element is deleted from the active set and a new iterate is sought.

A.4   MATLAB's Linear Programming Solver

MATLAB provides the linprog solver to optimize LPs. The input arguments to the linprog solver are the following:

- f: the linear objective function vector
- A: the matrix of the linear inequality constraints
- b: the right-hand side vector of the linear inequality constraints

- *Aeq*: the matrix of the linear equality constraints
- *beq*: the right-hand side vector of the linear equality constraints
- *lb*: the vector of the lower bounds
- *ub*: the vector of the upper bounds
- *x0*: the initial point for *x*, only applicable to the active-set algorithm
- *options*: options created with *optimoptions*

All `linprog` algorithms use the following options through the `optimoptions` input argument:

- *Algorithm*: the LP algorithm that will be used:

  1. 'interior-point' (default)
  2. 'dual-simplex'
  3. 'active-set'
  4. 'simplex'

- *Diagnostics*: the diagnostic information about the function to be minimized or solved:

  1. 'off' (default)
  2. 'on'

- *Display*: the level of display:

  1. 'off': no output
  2. 'none': no output
  3. 'iter': output at each iteration (not applicable for the active-set algorithm)
  4. 'final' (default): final output

- *LargeScale*: the 'interior-point' algorithm is used when this option is set to 'on' (default) and the primal simplex algorithm is used when set to 'off'
- *MaxIter*: the maximum number of iterations allowed. The default is:

  1. 85 for the interior-point algorithm
  2. $10 \times (numberOfEqualities + numberOfInequalities + numberOfVariables)$ for the dual simplex algorithm
  3. $10 \times numberOfVariables$ for the primal simplex algorithm
  4. $10 \times \max(numberOfVariables, numberOfInequalities + numberOfBounds)$ for the active-set algorithm

- *TolFun*: the termination tolerance on the function values:

  1. $1e - 8$ for the interior-point algorithm
  2. $1e - 7$ for the dual simplex algorithm
  3. $1e - 6$ for the primal simplex algorithm
  4. The option is not available for the active-set algorithm

- *MaxTime*: the maximum amount of time in seconds that the algorithm runs. The default value is ∞ (only available for the dual simplex algorithm)

- *Preprocess*: the level of preprocessing (only available for the dual simplex algorithm):

    1. 'none': no preprocessing is applied
    2. 'basic': preprocessing is applied

- *TolCon*: the feasibility tolerance for the constraints. The default value is $1e-4$. It can take a scalar value from $1e-10$ to $1e-3$ (only available for the dual simplex algorithm)

The output arguments of the `linprog` solver are the following:

- *fval*: the value of the objective function at the solution x
- *x*: the solution found by the solver. If *exitflag* > 0, then x is a solution; otherwise, x is the value of the variables when the solver terminated prematurely
- *exitflag*: the reason the algorithm terminates:

    - 1: the solver converged to a solution x
    - 0: the number of iterations exceeded the *MaxIter* option
    - −2: no feasible point was found
    - −3: the LP problem is unbounded
    - −4: *NaN* value was encountered during the execution of the algorithm
    - −5: both primal and dual problems are infeasible
    - −7: the search direction became too small and no further progress could be made

- *lambda*: a structure containing the Lagrange multipliers at the solution x:

    1. *lower*: the lower bounds
    2. *upper*: the upper bounds
    3. *ineqlin*: the linear inequalities
    4. *eqlin*: the linear equalities

- *output*: a structure containing information about the optimization:

    1. *iterations*: the number of iterations
    2. *algorithm*: the LP algorithm that was used
    3. *cgiterations*: 0 (included for backward compatibility and it is only used for the interior point algorithm)
    4. *message*: the exit message
    5. *constrviolation*: the maximum number of constraint functions
    6. *firstorderopt*: the first-order optimality measure

The following list provides the different syntax that can be used when calling the `linprog` solver:

- $x = linprog(f, A, b)$: solves $\min f^T x$ such that $Ax = b$ (it is used when there are only equality constraints)
- $x = linprog(f, A, b, Aeq, beq)$: solves the problem above while additionally satisfying the equality constraints $Aeqx = beq$. Set $A = []$ and $b = []$ if no inequalities exist

- $x = linprog(f, A, b, Aeq, beq, lb, ub)$: defines a set of lower and upper bounds on the design variables, x, so that the solution is always in the range $lb \leq x \leq ub$. Set $Aeq = []$ and $beq = []$ if no equalities exist
- $x = linprog(f, A, b, Aeq, beq, lb, ub, x0)$: sets the starting point to $x0$ (only used for the active-set algorithm)
- $x = linprog(f, A, b, Aeq, beq, lb, ub, x0, options)$: minimizes with the optimization options specified in options
- $x = linprog(problem)$: finds the minimum for the *problem*, where *problem* is a structure containing the following fields:

  - f: the linear objective function vector
  - *Aineq*: the matrix of the linear inequality constraints
  - *bineq*: the right-hand side vector of the linear inequality constraints
  - *Aeq*: the matrix of the linear equality constraints
  - *beq*: the right-hand side vector of the linear equality constraints
  - *lb*: the vector of the lower bounds
  - *ub*: the vector of the upper bounds
  - *x0*: the initial point for x, only applicable to the active-set algorithm
  - *options*: options created with *optimoptions*

- $[x, fval] = linprog(\dots)$: returns the value of the objective function at the solution x
- $[x, fval, exitflag] = linprog(\dots)$: returns a value *exitflag* that describes the exit condition
- $[x, fval, exitflag, output] = linprog(\dots)$: returns a structure *output* that contains information about the optimization process
- $[x, fval, exitflag, output, lambda] = linprog(\dots)$: returns a structure *lambda* whose fields contain the Lagrange multipliers at the solution x

Examples using the `linprog` solver are presented in the following sections of this chapter.

A.5 MATLAB's Optimization App

MATLAB's Optimization App (Figure A.1) is an application included in Optimization Toolbox that simplifies the solution of an optimization problem. It allows users to [9]:

- Define an optimization problem
- Select a solver and set optimization options for the selected solver
- Solve optimization problems and visualize intermediate and final results
- Import and export an optimization problem, solver options, and results
- Generate MATLAB code to automate tasks

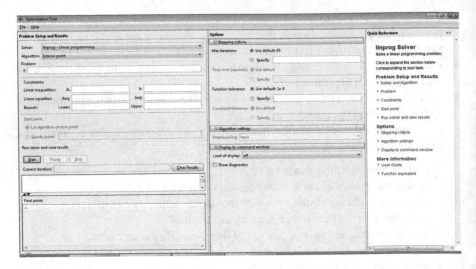

Fig. A.1 MATLAB's Optimization App

Fig. A.2 Optimization App
in MATLAB Apps tab

To open the Optimization App, type *optimtool* in the Command Window or start it from MATLAB Apps tab (Figure A.2).

Users can select the solver and the specific algorithm that they want to use in the 'Problem Setup and Results' tab. All algorithms presented in Table A.1 are available. Moreover, users can select one of the four LP algorithms presented in Section A.3. In addition, users can enter the objective function, the linear equalities, the linear inequalities, and the bounds of the LP problem using matrix format in the 'Problem' and 'Constraints' tabs. Only for the active-set algorithm, users can either let the algorithm choose an initial point or specify the initial point in the 'Start point' tab.

There are further options in the 'Options' tab that users can modify when solving LPs:

- **Stopping criteria**: users can set the stopping criteria for the LP solver. There are four criteria available for LP algorithms:

1. Max iterations (available for all LP algorithms)
2. Time limit (available only for the dual simplex algorithm)
3. Function tolerance (available for the interior point method and the primal and dual simplex algorithms)
4. Constraint tolerance (available only for the dual simplex algorithm)

- **Algorithm settings**: When the dual simplex algorithm is selected, users can choose if the specified LP problem will be preprocessed or not.
- **Display to command window**: Users can select the amount of the information displayed when they run the algorithm. Three options are available for the LP algorithms:

1. off: display no output
2. final: display the reason for stopping at the end of the run
3. iterative: display information at each iteration of the algorithm and the reason for stopping at the end of the run.

Selecting the 'Show diagnostics' checkbox, problem information and options that have changed are listed.

While a solver is running, users can click the 'Pause' button to temporarily suspend the algorithm or press the 'Stop' button to stop the algorithm from 'Run solver and view results' tab. When a solver terminates, the 'Run solver and view results' tab displays the results of the optimization process, including the objective function value, the solution of the LP problem, and the reason the algorithm terminated.

Users can import or export their work using the appropriate options from the 'File' menu. The 'Export to Workspace' dialog box enables users to send the problem information to MATLAB workspace as a structure or object. The 'Generate Code' in the 'File' menu generates MATLAB code as a function that users can run it and solve the specified LP problem.

Examples using the Optimization App are presented in the Section A.6.2.

A.6   Solving Linear Programming Problems with MATLAB's Optimization Toolbox

This section shows how to solve LPs with MATLAB's Optimization Toolbox using three different ways:

- By converting them to solver's form
- By using MATLAB's Optimization App
- By using custom code

A.6.1   Solving Linear Programming Problems by Converting Them to Solver's Form

1st example

Consider the following LP problem:

$$\min z = -2x_1 + 4x_2 - 2x_3 + 2x_4$$

$$\text{s.t.} \qquad 2x_2 + 3x_3 \qquad\qquad \geq\ 6$$
$$4x_1 - 3x_2 + 8x_3 - \ x_4 = 20$$
$$-3x_1 + 2x_2 \qquad\quad -\ 4x_4 \leq -8$$
$$4x_1 \qquad\quad -\ x_3 + 4x_4 = 18$$
$$x_1 \qquad\qquad\qquad\qquad \geq\ 1$$
$$x_3 \qquad\quad \geq\ 2$$
$$x_3 \qquad\quad \leq\ 10$$

$$x_j \geq 0, \quad (j = 1, 2, 3, 4)$$

Initially, we convert the previous LP problem from mathematical form into MATLAB's Optimization Toolbox solver syntax. There are 4 variables in the previous LP problem and we put these variables into one vector, named *variables*:

```
>> variables = {'x1', 'x2', 'x3', 'x4'}; % construct a cell array
>> % with the names of the variables
>> n = length(variables); % find the number of the variables
>> for i = 1:n % create x for indexing
>>    eval([variables{i}, ' = ', num2str(i), ';']);
>> end
```

By executing the above commands, the following named variables are created in MATLAB's workspace (Figure A.3):

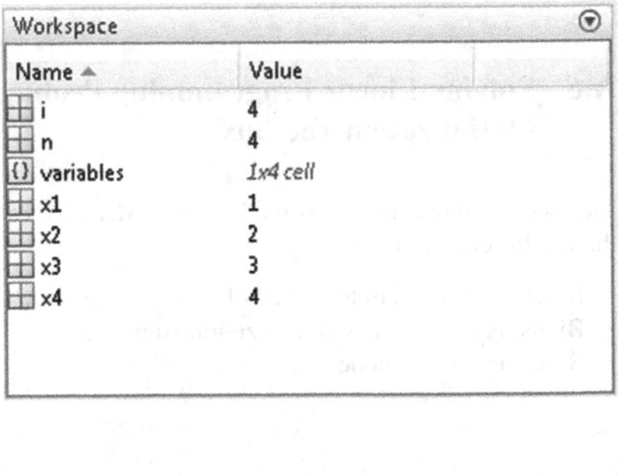

Fig. A.3 MATLAB's workspace

Workspace		⊙
Name ▲	Value	
⊞ i	4	
⊞ n	4	
{} variables	1x4 cell	
⊞ x1	1	
⊞ x2	2	
⊞ x3	3	
⊞ x4	4	

Then, we create two vectors, one for the lower bounds and one for the upper bounds. There are two variables with lower bounds:

$$x_1 \geq 1$$

$$x_3 \geq 2$$

Moreover, all variables are nonnegative, i.e., they have a lower bound of zero. We create the lower bound vector *lb* using the following code:

```
>> lb = zeros(1, n); % create a vector and add a lower bound of
>> % zero to all variables
>> lb([x1, x3]) = [1, 2]; % add lower bounds to variables x1, x3
```

There is one variable with an upper bound:

$$x_3 \leq 10$$

We create the upper bound vector *ub* using the following code:

```
>> ub = Inf(1, n); % create a vector and add an upper bound of
>> % Inf to all variables
>> ub(x3) = 10; % add an upper bound to variable x3
```

Then, we create two vectors, one for the linear inequality constraints and one for the linear equality constraints. There are two linear inequality constraints:

$$2x_1 + 3x_3 \geq 6$$

$$-3x_1 + 2x_2 - 4x_4 \leq -8$$

In order to have the equations in the form $Ax \leq b$, let's put all the variables on the left side of the inequality. All these equations already have this form. The next step is to ensure that each inequality is in 'less than or equal to' form by multiplying by -1 wherever appropriate:

$$-2x_1 - 3x_3 \leq -6$$

$$-3x_1 + 2x_2 - 4x_4 \leq -8$$

Then, we create matrix A as a 2x4 matrix, corresponding to 2 linear inequalities in 4 variables, and vector b with 2 elements using the following code:

```
>> A = zeros(2, n); % create the matrix A
>> b = zeros(2, 1); % create the vector b
>> % define the 1st inequality constraint
>> A(1, [x1, x3]) = [-2, -3]; b(1) = -6;
>> % define the 2nd inequality constraint
>> A(2, [x1, x2, x4]) = [-3, 2, -4]; b(2) = -8;
```

There are two linear equality constraints:

$$4x_1 - 3x_2 + 8x_3 - x_4 = 20$$

$$4x_1 - x_3 + 4x_4 = 18$$

In order to have the equations in the form $Aeqx = beq$, let's put all the variables on the left side of the equality. All these equations already have this form. Then, we create matrix Aeq as a 2x4 matrix, corresponding to 2 linear equalities in 4 variables, and vector beq with 2 elements using the following code:

```
>> Aeq = zeros(2, n); % create the matrix Aeq
>> beq = zeros(2, 1); % create the vector beq
>> % define the 1st equality constraint
>> Aeq(1, [x1, x2, x3, x4]) = [4, -3, 8, -1]; beq(1) = 20;
>> % define the 2nd equality constraint
>> Aeq(2, [x1, x3, x4]) = [4, -1, 4]; beq(2) = 18;
```

Then, we create the objective function $c^T x = -2x_1 + 4x_2 - 2x_3 + 2x_4$ using the following code:

```
>> c = zeros(n, 1); % create the objective function vector c
>> c([x1 x2 x3 x4]) = [-2; 4; -2; 2];
```

Finally, we call the linprog solver and print the output in formatted form using the following code:

```
>> % call the linprog solver
>> [x, objVal] = linprog(c, A, b, Aeq, beq, lb, ub);
>> for i = 1:n % print results in formatted form
>>     fprintf('%s \t %20.4f\n', variables{i}, x(i))
>> end
>> fprintf(['The value of the objective function ' ...
        'is %.4f\n'], objVal)
```

The complete code (filename: example1a.m) to solve the above LP problem is listed below:

```
1.    variables = {'x1', 'x2', 'x3', 'x4'}; % construct a cell
2.    % array with the names of the variables
3.    n = length(variables); % find the number of the variables
4.    for i = 1:n % create x for indexing
5.         eval([variables{i}, ' = ', num2str(i), ';']);
6.    end
7.    lb = zeros(1, n); % create a vector and add a lower bound
8.    % of zero to all variables
9.    lb([x1, x3]) = [1, 2]; % add lower bounds to variables x1,
10.   % x3
11.   ub = Inf(1, n); % create a vector and add an upper bound of
12.   % Inf to all variables
13.   ub(x3) = 10; % add an upper bound to variable x3
14.   A = zeros(2, n); % create the matrix A
15.   b = zeros(2, 1); % create the vector b
16.   % define the 1st constraint
```

```
Optimization terminated.
x1                    1.8000
x2                    0.0000
x3                    2.0000
x4                    3.2000
The value of the objective function is -1.2000
```

Fig. A.4 Solution of the LP problem

```
17.  A(1, [x1, x3]) = [-2, -3]; b(1) = -6;
18.  % define the 2nd constraint
19.  A(2, [x1, x2, x4]) = [-3, 2, -4]; b(2) = -8;
20.  Aeq = zeros(2, n); % create the matrix Aeq
21.  beq = zeros(2, 1); % create the vector beq
22.  % define the 1st equality constraint
23.  Aeq(1, [x1, x2, x3, x4]) = [4, -3, 8, -1]; beq(1) = 20;
24.  % define the 2nd equality constraint
25.  Aeq(2, [x1, x3, x4]) = [4, -1, 4]; beq(2) = 18;
26.  c = zeros(n, 1); % create the objective function vector c
27.  c([x1 x2 x3 x4]) = [-2; 4; -2; 2];
28.  % call the linprog solver
29.  [x, objVal] = linprog(c, A, b, Aeq, beq, lb, ub);
30.  for i = 1:n % print results in formatted form
31.      fprintf('%s \t %20.4f\n', variables{i}, x(i))
32.  end
33.  fprintf(['The value of the objective function ' ...
34.      'is %.4f\n'], objVal)
```

The result is presented in Figure A.4. The optimal value of the objective function is -1.2.

2nd example

Consider the following LP problem:

$$\min z = -2x_1 - 3x_2 + x_3 + 4x_4 + x_5$$

$$\begin{aligned}
\text{s.t.} \quad & x_1 + 4x_2 - 2x_3 - x_4 && \leq && 8 \\
& x_1 + 3x_2 - 4x_3 - x_4 + x_5 && = && 7 \\
& x_1 + 3x_2 + 2x_3 - x_4 && \leq && 10 \\
& -2x_1 - x_2 - 2x_3 - 3x_4 + x_5 && \geq && -20 \\
& x_1 && \geq && 5 \\
& x_2 && \geq && 1 \\
& x_j \geq 0, \quad (j = 1, 2, 3, 4, 5)
\end{aligned}$$

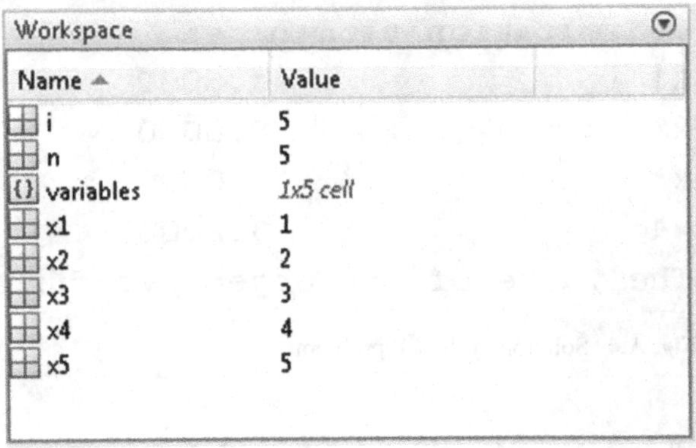

Fig. A.5 MATLAB's
workspace

Workspace	⊙
Name ▲	Value
⊞ i	5
⊞ n	5
{} variables	1x5 cell
⊞ x1	1
⊞ x2	2
⊞ x3	3
⊞ x4	4
⊞ x5	5

Initially, we convert the previous LP problem from mathematical form into MATLAB's Optimization Toolbox solver syntax. There are 5 variables in the previous LP problem and we put these variables into one vector, named *variables*:

```
>> variables = {'x1', 'x2', 'x3', 'x4', 'x5'}; % construct a cell
>> % array with the names of the variables
>> n = length(variables); % find the number of the variables
>> for i = 1:n % create x for indexing
>>    eval([variables{i}, ' = ', num2str(i), ';']);
>> end
```

By executing the above commands, the following named variables are created in MATLAB's workspace (Figure A.5):

Then, we create two vectors, one for the lower bounds and one for the upper bounds. There are two variables with lower bounds:

$$x_1 \geq 5$$

$$x_2 \geq 1$$

Moreover, all variables are nonnegative, i.e., they have a lower bound of zero. We create the lower bound vector *lb* using the following code:

```
>> lb = zeros(1, n); % create a vector and add a lower bound of
>> % zero to all variables
>> lb([x1, x2]) = [5, 1]; % add lower bounds to variables x1, x2
```

There is not any variable with an upper bound. We create the upper bound vector *ub* using the following code:

```
>> ub = Inf(1, n); % create a vector and add an upper bound of
>> % Inf to all variables
```

Then, we create two vectors, one for the linear inequality constraints and one for the linear equality constraints. There are three linear inequality constraints:

$$x_1 + 4x_2 - 2x_3 - x_4 \leq 8$$

$$x_1 + 3x_2 + 2x_3 - x_4 \leq 10$$

$$-2x_1 - x_2 - 2x_3 - 3x_4 + x_5 \geq -20$$

In order to have the equations in the form $Ax \leq b$, let's put all the variables on the left side of the inequality. All these equations already have this form. The next step is to ensure that each inequality is in 'less than or equal to' form by multiplying by -1 wherever appropriate:

$$x_1 + 4x_2 - 2x_3 - x_4 \leq 8$$

$$x_1 + 3x_2 + 2x_3 - x_4 \leq 10$$

$$2x_1 + x_2 + 2x_3 + 3x_4 - x_5 \leq 20$$

Then, we create matrix A as a 3x5 matrix, corresponding to 3 linear inequalities in 5 variables, and vector b with 3 elements using the following code:

```
>> A = zeros(3, n); % create the matrix A
>> b = zeros(3, 1); % create the vector b
>> % define the 1st inequality constraint
>> A(1, [x1, x2, x3, x4]) = [1, 4, -2, -1]; b(1) = 8;
>> % define the 2nd inequality constraint
>> A(2, [x1, x2, x3, x4]) = [1, 3, 2, -1]; b(2) = 10;
>> % define the 3rd inequality constraint
>> A(3, [x1, x2, x3, x4, x5]) = [2, 1, 2, 3, -1]; b(3) = 20;
```

There is one linear equality constraint:

$$x_1 + 3x_2 - 4x_3 - x_4 + x_5 = 7$$

In order to have the equations in the form $Aeqx = beq$, let's put all the variables on the left side of the equality. All these equations already have this form. Then, we create matrix Aeq as a 1x5 matrix, corresponding to 1 linear equality in 5 variables, and vector beq with 1 element using the following code:

```
>> Aeq = zeros(1, n); % create the matrix Aeq
>> beq = zeros(1, 1); % create the vector beq
>> % define the equality constraint
>> Aeq(1, [x1, x2, x3, x4, x5]) = [1, 3, -4, -1, 1]; beq(1) = 7;
```

Then, we create the objective function $c^T x = -2x_1 + -3x_2 + x_3 + 4x_4 + x_5$ using the following code:

```
>> c = zeros(n, 1); % create the objective function vector c
>> c([x1 x2 x3 x4 x5]) = [-2; -3; 1; 4; 1];
```

Finally, we call the `linprog` solver and print the output in formatted form using the following code:

```
>> % call the linprog solver
>> [x, objVal] = linprog(c, A, b, Aeq, beq, lb, ub);
>> for i = 1:n % print results in formatted form
>>     fprintf('%s \t %20.4f\n', variables{i}, x(i))
>> end
>> fprintf(['The value of the objective function ' ...
>>     'is %.4f\n'], objVal)
```

The complete code (filename: `example2a.m`) to solve the aforementioned LP problem is listed below:

```
1.    variables = {'x1', 'x2', 'x3', 'x4', 'x5'}; % construct a
2.    % cell array with the names of the variables
3.    n = length(variables); % find the number of the variables
4.    for i = 1:n % create x for indexing
5.        eval([variables{i}, ' = ', num2str(i), ';']);
6.    end
7.    lb = zeros(1, n); % create a vector and add a lower bound
8.    % of zero to all variables
9.    lb([x1, x2]) = [5, 1]; % add lower bounds to variables x1,
10.   % x2
11.   ub = Inf(1, n); % create a vector and add an upper bound
12.   % of Inf to all variables
13.   A = zeros(3, n); % create the matrix A
14.   b = zeros(3, 1); % create the vector b
15.   % define the 1st inequality constraint
16.   A(1, [x1, x2, x3, x4]) = [1, 4, -2, -1]; b(1) = 8;
17.   % define the 2nd inequality constraint
18.   A(2, [x1, x2, x3, x4]) = [1, 3, 2, -1]; b(2) = 10;
19.   % define the 3rd inequality constraint
20.   A(3, [x1, x2, x3, x4, x5]) = [2, 1, 2, 3, -1]; b(3) = 20;
21.   Aeq = zeros(1, n); % create the matrix Aeq
22.   beq = zeros(1, 1); % create the vector beq
23.   % define the equality constraint
24.   Aeq(1, [x1, x2, x3, x4, x5]) = [1, 3, -4, -1, 1]; beq(1) = 7;
25.   c = zeros(n, 1); % create the objective function vector c
26.   c([x1 x2 x3 x4 x5]) = [-2; -3; 1; 4; 1];
27.   % call the linprog solver
28.   [x, objVal] = linprog(c, A, b, Aeq, beq, lb, ub);
29.   for i = 1:n % print results in formatted form
30.       fprintf('%s \t %20.4f\n', variables{i}, x(i))
31.   end
32.   fprintf(['The value of the objective function ' ...
33.       'is %.4f\n'], objVal)
```

The result is presented in Figure A.6. The optimal value of the objective function is -11.75.

```
Optimization terminated.
x1                      5.5000
x2                      1.0000
x3                      0.7500
x4                      0.0000
x5                      1.5000
The value of the objective function is -11.7500
```

Fig. A.6 Solution of the LP problem

A.6.2 Solving Linear Programming Problems Using MATLAB's Optimization App

1st example

The LP problem that will be solved is the following:

$$
\begin{aligned}
\min z = -2x_1 + 4x_2 - 2x_3 + 2x_4 \\
\text{s.t.} \qquad\qquad 2x_2 + 3x_3 &\geq 6 \\
4x_1 - 3x_2 + 8x_3 - x_4 &= 20 \\
-3x_1 + 2x_2 \qquad\quad - 4x_4 &\leq -8 \\
4x_1 \qquad\quad - x_3 + 4x_4 &= 18 \\
x_1 \qquad\qquad\qquad\qquad &\geq 1 \\
x_3 \qquad\qquad &\geq 2 \\
x_3 \qquad\qquad &\leq 10 \\
x_j \geq 0, \quad (j = 1, 2, 3, 4)
\end{aligned}
$$

Initially, we convert the previous LP problem from mathematical form into MATLAB's Optimization App solver syntax. There are 4 variables in the previous LP problem. Initially, we create the objective function $f^T x = -2x_1 + 4x_2 - 2x_3 + 2x_4$, which in vector format can be defined as:

$$
f = \begin{bmatrix} -2 & 4 & -2 & 2 \end{bmatrix}
$$

Then, we create two vectors, one for the lower bounds and one for the upper bounds. There are two variables with lower bounds:

$$
x_1 \geq 1
$$

$$
x_3 \geq 2
$$

Moreover, all variables are nonnegative, i.e., they have a lower bound of zero. We can define the lower bound vector *lb* in vector format as:

$$lb = \begin{bmatrix} 1 & 0 & 2 & 0 \end{bmatrix}$$

There is one variable with an upper bound:

$$x_3 \leq 10$$

We can define the upper bound vector *ub* in vector format as:

$$ub = \begin{bmatrix} \infty & \infty & 10 & \infty \end{bmatrix}$$

Then, we create two vectors, one for the linear inequality constraints and one for the linear equality constraints. There are two linear inequality constraints:

$$2x_1 + 3x_3 \geq 6$$

$$-3x_1 + 2x_2 - 4x_4 \leq -8$$

In order to have the equations in the form $Ax \leq b$, let's put all the variables on the left side of the inequality. All these equations already have this form. The next step is to ensure that each inequality is in 'less than or equal to' form by multiplying by -1 wherever appropriate:

$$-2x_1 - 3x_3 \leq -6$$

$$-3x_1 + 2x_2 - 4x_4 \leq -8$$

Then, we define matrix A as a 2x4 matrix, corresponding to 2 linear inequalities in 4 variables, and vector b with 2 elements:

$$A = \begin{bmatrix} -2 & 0 & -3 & 0 \\ -3 & 2 & 0 & -4 \end{bmatrix}$$

$$b = \begin{bmatrix} -6 \\ -8 \end{bmatrix}$$

There are two linear equality constraints:

$$4x_1 - 3x_2 + 8x_3 - x_4 = 20$$

$$4x_1 - x_3 + 4x_4 = 18$$

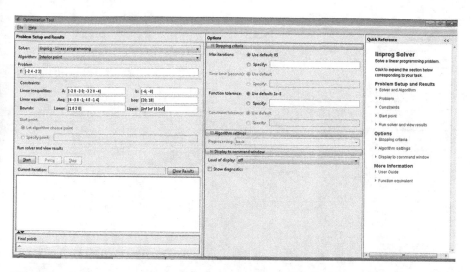

Fig. A.7 Write the LP problem using MATLAB's Optimization App

In order to have the equations in the form $Aeqx = beq$, let's put all the variables on the left side of the equality. All these equations already have this form. Then, we define matrix Aeq as a 2x4 matrix, corresponding to 2 linear equalities in 4 variables, and vector beq with 2 elements:

$$Aeq = \begin{bmatrix} 4 & -3 & 8 & -1 \\ 4 & 0 & -1 & 4 \end{bmatrix}$$

$$beq = \begin{bmatrix} 20 \\ 18 \end{bmatrix}$$

We can write these vectors to MATLAB's Optimization App and solve the LP problem. Open the Optimization App by typing *optimtool* in the Command Window and fill out the required fields, as shown in Figure A.7.

Finally, by pressing the 'Start' button, Optimization App solves the LP problem and outputs the results (Figure A.8). The optimal value of the objective function is -1.19.

2nd example

Consider the following LP problem:

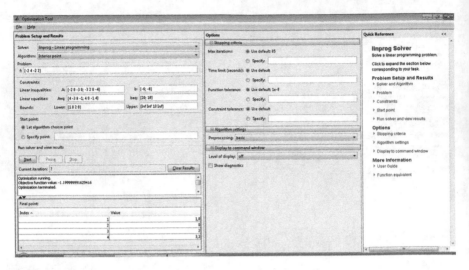

Fig. A.8 Solution of the LP problem using MATLAB's Optimization App

$$\min z = -2x_1 - 3x_2 + x_3 + 4x_4 + x_5$$

$$\text{s.t.} \quad x_1 + 4x_2 - 2x_3 - x_4 \leq 8$$

$$x_1 + 3x_2 - 4x_3 - x_4 + x_5 = 7$$

$$x_1 + 3x_2 + 2x_3 - x_4 \leq 10$$

$$-2x_1 - x_2 - 2x_3 - 3x_4 + x_5 \geq -20$$

$$x_1 \geq 5$$

$$x_2 \geq 1$$

$$x_j \geq 0, \quad (j = 1, 2, 3, 4, 5)$$

Initially, we convert the previous LP problem from mathematical form into MATLAB's Optimization App solver syntax. There are 5 variables in the previous LP problem. Initially, we create the objective function $f^T x = -2x_1 + -3x_2 + x_3 + 4x_4 + x_5$, which in vector format can be defined as:

$$f = \begin{bmatrix} -2 & -3 & 1 & 4 & 1 \end{bmatrix}$$

Then, we create two vectors, one for the lower bounds and one for the upper bounds. There are two variables with lower bounds:

$$x_1 \geq 5$$

$$x_2 \geq 1$$

Moreover, all variables are nonnegative, i.e., they have a lower bound of zero. We can define the lower bound vector lb in vector format as:

$$lb = \begin{bmatrix} 5 & 1 & 0 & 0 \end{bmatrix}$$

There is not any variable with an upper bound. We can define the upper bound vector ub in vector format as:

$$ub = \begin{bmatrix} \infty & \infty & \infty & \infty & \infty \end{bmatrix}$$

Then, we create two vectors, one for the linear inequality constraints and one for the linear equality constraints. There are three linear inequality constraints:

$$x_1 + 4x_2 - 2x_3 - x_4 \leq 8$$
$$x_1 + 3x_2 + 2x_3 - x_4 \leq 10$$
$$-2x_1 - x_2 - 2x_3 - 3x_4 + x_5 \geq -20$$

In order to have the equations in the form $Ax \leq b$, let's put all the variables on the left side of the inequality. All these equations already have this form. The next step is to ensure that each inequality is in 'less than or equal to' form by multiplying by -1 wherever appropriate:

$$x_1 + 4x_2 - 2x_3 - x_4 \leq 8$$
$$x_1 + 3x_2 + 2x_3 - x_4 \leq 10$$
$$2x_1 + x_2 + 2x_3 + 3x_4 - x_5 \leq 20$$

Then, we define matrix A as a 3x5 matrix, corresponding to 3 linear inequalities in 5 variables, and vector b with 3 elements:

$$A = \begin{bmatrix} 1 & 4 & -2 & -1 & 0 \\ 1 & 3 & 2 & -1 & 0 \\ 2 & 1 & 2 & 3 & -1 \end{bmatrix}$$

$$b = \begin{bmatrix} 8 \\ 10 \\ 20 \end{bmatrix}$$

There is one linear equality constraint:

$$x_1 + 3x_2 - 4x_3 - x_4 + x_5 = 7$$

In order to have the equations in the form $Aeqx = beq$, let's put all the variables on the left side of the equality. This equation has already this form. Then, we define matrix Aeq as a 2x4 matrix, corresponding to 1 linear equality in 5 variables, and vector beq with 1 element:

$$Aeq = \begin{bmatrix} 1 & 3 & -4 & -1 & 1 \end{bmatrix}$$

$$beq = \begin{bmatrix} 7 \end{bmatrix}$$

We can write these vectors to MATLAB's Optimization App and solve the LP problem. Open the Optimization App by typing *optimtool* in the Command Window and fill out the required fields, as shown in Figure A.9.

Finally, by pressing the 'Start' button, Optimization App solves the LP problem and outputs the results (Figure A.10). The optimal value of the objective function is -11.75.

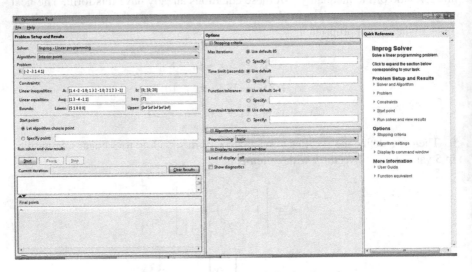

Fig. A.9 Write the LP problem using MATLAB's Optimization App

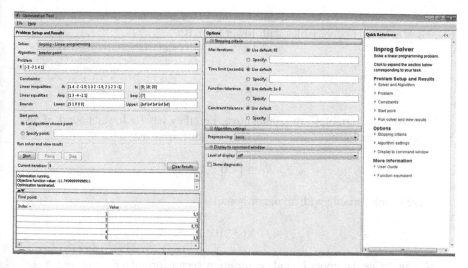

Fig. A.10 Solution of the LP problem using MATLAB's Optimization App

A.6.3 Solving Linear Programming Problems Using Custom Code

1st example

Consider the following LP problem:

$$
\begin{aligned}
\min z = -2x_1 + 4x_2 - 2x_3 + 2x_4 \\
\text{s.t.} \qquad\qquad 2x_2 + 3x_3 \quad &\geq 6 \\
4x_1 - 3x_2 + 8x_3 - x_4 &= 20 \\
-3x_1 + 2x_2 \qquad\quad - 4x_4 &\leq -8 \\
4x_1 \qquad\quad - x_3 + 4x_4 &= 18 \\
x_1 \qquad\qquad\qquad &\geq 1 \\
x_3 \qquad &\geq 2 \\
x_3 \qquad &\leq 10 \\
x_j \geq 0, \quad (j = 1, 2, 3, 4)
\end{aligned}
$$

Initially, we convert the previous LP problem from mathematical form into vector form. There are 4 variables in the previous LP problem. Initially, we create the objective function $c^T x = -2x_1 + 4x_2 - 2x_3 + 2x_4$, which in vector format can be defined as:

$$
c = \begin{bmatrix} -2 & 4 & -2 & 2 \end{bmatrix}
$$

Then, we create two vectors, one for the lower bounds and one for the upper bounds. There are two variables with lower bounds:

$$x_1 \geq 1$$

$$x_3 \geq 2$$

Moreover, all variables are nonnegative, i.e., they have a lower bound of zero. We can define the lower bound vector lb in vector format as:

$$lb = \begin{bmatrix} 1 & 0 & 2 & 0 \end{bmatrix}$$

There is one variable with an upper bound:

$$x_3 \leq 10$$

We can define the upper bound vector ub in vector format as:

$$ub = \begin{bmatrix} \infty & \infty & 10 & \infty \end{bmatrix}$$

Then, we create two vectors, one for the linear inequality constraints and one for the linear equality constraints. There are two linear inequality constraints:

$$2x_1 + 3x_3 \geq 6$$

$$-3x_1 + 2x_2 - 4x_4 \leq -8$$

In order to have the equations in the form $Ax \leq b$, let's put all the variables on the left side of the inequality. All these equations already have this form. The next step is to ensure that each inequality is in 'less than or equal to' form by multiplying by -1 wherever appropriate:

$$-2x_1 - 3x_3 \leq -6$$

$$-3x_1 + 2x_2 - 4x_4 \leq -8$$

Then, we define matrix A as a 2x4 matrix, corresponding to 2 linear inequalities in 4 variables, and vector b with 2 elements:

$$A = \begin{bmatrix} -2 & 0 & -3 & 0 \\ -3 & 2 & 0 & -4 \end{bmatrix}$$

$$b = \begin{bmatrix} -6 \\ -8 \end{bmatrix}$$

There are two linear equality constraints:

$$4x_1 - 3x_2 + 8x_3 - x_4 = 20$$

$$4x_1 - x_3 + 4x_4 = 18$$

In order to have the equations in the form $Aeqx = beq$, let's put all the variables on the left side of the equality. All these equations already have this form. Then, we define matrix Aeq as a 2x4 matrix, corresponding to 2 linear equalities in 4 variables, and vector beq with 2 elements:

$$Aeq = \begin{bmatrix} 4 & -3 & 8 & -1 \\ 4 & 0 & -1 & 4 \end{bmatrix}$$

$$beq = \begin{bmatrix} 20 \\ 18 \end{bmatrix}$$

Finally, we call the linprog solver and print the output in formatted form. The complete code (filename: example1b.m) to solve the aforementioned LP problem is listed below:

```
1.    A = [-2 0 -3 0; -3 2 0 -4]; % create the matrix A
2.    b = [-6; -8]; % create the vector b
3.    Aeq = [4 -3 8 -1; 4 0 -1 4]; %create the matrix Aeq
4.    beq = [20; 18]; % create the vector beq
5.    lb = [1 0 2 0]; % create the vector lb
6.    ub = [Inf Inf 10 Inf]; % create the vector ub
7.    c = [-2; 4; -2; 2]; %create the vector c
8.    % call the linprog solver
9.    [x, objVal] = linprog(c, A, b, Aeq, beq, lb, ub);
10.   for i = 1:4 % print results in formatted form
11.       fprintf('x%d \t %20.4f\n', i, x(i))
12.   end
13.   fprintf(['The value of the objective function ' ...
14.       'is %.4f\n'], objVal)
```

The result is presented in Figure A.11. The optimal value of the objective function is -1.2.

```
Optimization terminated.
x1                    1.8000
x2                    0.0000
x3                    2.0000
x4                    3.2000
The value of the objective function is -1.2000
```

Fig. A.11 Solution of the LP problem

2nd example

The LP problem that will be solved is the following:

$$
\begin{aligned}
\min z = -2x_1 - 3x_2 + & \ x_3 + 4x_4 + x_5 \\
\text{s.t.} \qquad x_1 + 4x_2 - 2x_3 - & \ x_4 && \leq && 8 \\
x_1 + 3x_2 - 4x_3 - & \ x_4 + x_5 && = && 7 \\
x_1 + 3x_2 + 2x_3 - & \ x_4 && \leq && 10 \\
-2x_1 - x_2 - 2x_3 - & \ 3x_4 + x_5 && \geq && -20 \\
x_1 \qquad\qquad\qquad & && \geq && 5 \\
x_2 \qquad\qquad & && \geq && 1 \\
x_j \geq 0, \quad (j = 1, 2, 3, 4, 5) &
\end{aligned}
$$

Initially, we convert the previous LP problem from mathematical form into vector form. There are 5 variables in the previous LP problem. Initially, we create the objective function $c^T x = -2x_1 + -3x_2 + x_3 + 4x_4 + x_5$, which in vector format can be defined as:

$$
c = \begin{bmatrix} -2 & -3 & 1 & 4 & 1 \end{bmatrix}
$$

Then, we create two vectors, one for the lower bounds and one for the upper bounds. There are two variables with lower bounds:

$$
x_1 \geq 5
$$

$$
x_2 \geq 1
$$

Moreover, all variables are nonnegative, i.e., they have a lower bound of zero. We can define the lower bound vector *lb* in vector format as:

$$
lb = \begin{bmatrix} 5 & 1 & 0 & 0 & 0 \end{bmatrix}
$$

There is not any variable with an upper bound. We can define the upper bound vector *ub* in vector format as:

$$
ub = \begin{bmatrix} \infty & \infty & \infty & \infty & \infty \end{bmatrix}
$$

Then, we create two vectors, one for the linear inequality constraints and one for the linear equality constraints. There are three linear inequality constraints:

$$
x_1 + 4x_2 - 2x_3 - x_4 \leq 8
$$

$$
x_1 + 3x_2 + 2x_3 - x_4 \leq 10
$$

$$
-2x_1 - x_2 - 2x_3 - 3x_4 + x_5 \geq -20
$$

In order to have the equations in the form $Ax \leq b$, let's put all the variables on the left side of the inequality. All these equations already have this form. The next step is to ensure that each inequality is in 'less than or equal to' form by multiplying by -1 wherever appropriate:

$$x_1 + 4x_2 - 2x_3 - x_4 \leq 8$$

$$x_1 + 3x_2 + 2x_3 - x_4 \leq 10$$

$$2x_1 + x_2 + 2x_3 + 3x_4 - x_5 \leq 20$$

Then, we define matrix A as a 3x5 matrix, corresponding to 3 linear inequalities in 5 variables, and vector b with 3 elements:

$$A = \begin{bmatrix} 1 & 4 & -2 & -1 & 0 \\ 1 & 3 & 2 & -1 & 0 \\ 2 & 1 & 2 & 3 & -1 \end{bmatrix}$$

$$b = \begin{bmatrix} 8 \\ 10 \\ 20 \end{bmatrix}$$

There is one linear equality constraint:

$$x_1 + 3x_2 - 4x_3 - x_4 + x_5 = 7$$

In order to have the equations in the form $Aeqx = beq$, let's put all the variables on the left side of the equality. This equation has already this form. Then, we define matrix Aeq as a 1x5 matrix, corresponding to 1 linear equality in 5 variables, and vector beq with 1 element:

$$Aeq = \begin{bmatrix} 1 & 3 & -4 & -1 & 1 \end{bmatrix}$$

$$beq = \begin{bmatrix} 7 \end{bmatrix}$$

Finally, we call the linprog solver and print the output in formatted form. The complete code (filename: example2b.m) to solve the aforementioned LP problem is listed below:

```
1.    A = [1 4 -2 -1 0; 1 3 2 -1 0; 2 1 2 3 -1]; % create the matrix A
2.    b = [8; 10; 20]; % create the vector b
3.    Aeq = [1 3 -4 -1 1]; %create the matrix Aeq
4.    beq = [7]; % create the vector beq
5.    lb = [5 1 0 0 0]; % create the vector lb
6.    ub = [Inf Inf Inf Inf Inf]; % create the vector ub
7.    c = [-2; -3; 1; 4; 1]; %create the vector c
8.    % call the linprog solver
9.    [x, objVal] = linprog(c, A, b, Aeq, beq, lb, ub);
```

```
Optimization terminated.
x1                    5.5000
x2                    1.0000
x3                    0.7500
x4                    0.0000
x5                    1.5000
The value of the objective function is -11.7500
```

Fig. A.12 Solution of the LP problem

```
10.  for i = 1:5 % print results in formatted form
11.      fprintf('x%d \t %20.4f\n', i, x(i))
12.  end
13.  fprintf(['The value of the objective function ' ...
14.      'is %.4f\n'], objVal)
```

The result is presented in Figure A.12. The optimal value of the objective function is −11.75.

A.7   MATLAB Code to Solve LPs Using the `linprog` Solver

This section presents the implementation in MATLAB of a code to solve LPs using `linprog` solver (filename: `linprogSolver.m`). Some necessary notations should be introduced before the presentation of the implementation of the code to solve LPs using linprog solver. Let A be a $m \times n$ matrix with the coefficients of the constraints, c be a $n \times 1$ vector with the coefficients of the objective function, b be a $m \times 1$ vector with the right-hand side of the constraints, $Eqin$ be a $m \times 1$ vector with the type of the constraints, $MinMaxLP$ be a variable showing the type of optimization, $c0$ be the constant term of the objective function, $algorithm$ be a variable showing the LP algorithm that will be used, $xsol$ be a $m \times 1$ vector with the solution found by the solver, $fval$ be the value of the objective function at the solution $xsol$, $exitflag$ be the reason that the algorithm terminated, and $iterations$ be the number of iterations.

The function for the solution of LPs using the linprog solver takes as input the matrix with the coefficients of the constraints (matrix A), the vector with the coefficients of the objective function (vector c), the vector with the right-hand side of the constraints (vector b), the vector with the type of the constraints (vector $Eqin$), the type of optimization (variable $MinMaxLP$), the constant term of the objective function (variable $c0$), and the LP algorithm that will be used (variable $algorithm$), and returns as output the solution found (vector $xsol$), the value of the objective function at the solution $xsol$ (variable $fval$), the reason that the algorithm terminated (variable $exitflag$), and the number of iterations (variable $iterations$).

In lines 47–52, we set default values to the missing inputs (if any). Next, we check if the input data is logically correct (lines 58–89). If the type of optimization is maximization, then we multiply vector c and constant $c0$ by -1 (lines 92–94). Next, we formulate matrices A and Aeq and vectors b, beq, and lb (lines 95–124). Then, we choose the LP algorithm and solve the LP problem using the linprog solver (lines 126–129).

```
1.    function [xsol, fval, exitflag, iterations] = ...
2.        linprogSolver(A, c, b, Eqin, MinMaxLP, c0, ...
3.        algorithm)
4.    % Filename: linprogSolver.m
5.    % Description: the function is a MATLAB code to solve LPs
6.    % using the linprog solver
7.    % Authors: Ploskas, N., & Samaras, N.
8.    %
9.    % Syntax: [x, fval, exitflag, iterations] = ...
10.   %    linprogSolver(A, c, b, Eqin, MinMaxLP, c0, ...
11.   %    algorithm)
12.   %
13.   % Input:
14.   % -- A: matrix of coefficients of the constraints
15.   %    (size m x n)
16.   % -- c: vector of coefficients of the objective function
17.   %    (size n x 1)
18.   % -- b: vector of the right-hand side of the constraints
19.   %    (size m x 1)
20.   % -- Eqin: vector of the type of the constraints
21.   %    (size m x 1)
22.   % -- MinMaxLP: the type of optimization (optional:
23.   %    default value -1 - minimization)
24.   % -- c0: constant term of the objective function
25.   %    (optional: default value 0)
26.   % -- algorithm: the LP algorithm that will be used
27.   %    (possible values: 'interior-point', 'dual-simplex',
28.   %    'simplex', 'active-set')
29.   %
30.   % Output:
31.   % -- xsol: the solution found by the solver (size m x 1)
32.   % -- fval: the value of the objective function at the
33.   %    solution xsol
34.   % -- exitflag: the reason that the algorithm terminated
35.   %    (1: the solver converged to a solution x, 0: the
36.   %    number of iterations exceeded the MaxIter option,
37.   %    -1: the input data is not logically or numerically
38.   %    correct, -2: no feasible point was found, -3: the LP
39.   %    problem is unbounded, -4: NaN value was encountered
40.   %    during the execution of the algorithm, -5: both primal
41.   %    and dual problems are infeasible, -7: the search
42.   %    direction became too small and no further progress
43.   %    could be made)
44.   % -- iterations: the number of iterations
45.
46.   % set default values to missing inputs
```

```
47.   if ~exist('MinMaxLP')
48.       MinMaxLP = -1;
49.   end
50.   if ~exist('c0')
51.       c0 = 0;
52.   end
53.   [m, n] = size(A); % find the size of matrix A
54.   [m2, n2] = size(c); % find the size of vector c
55.   [m3, n3] = size(Eqin); % find the size of vector Eqin
56.   [m4, n4] = size(b); % find the size of vector b
57.   % check if input data is logically correct
58.   if n2 ~= 1
59.       disp('Vector c is not a column vector.')
60.       exitflag = -1;
61.       return
62.   end
63.   if n ~= m2
64.       disp(['The number of columns in matrix A and ' ...
65.             'the number of rows in vector c do not match.'])
66.       exitflag = -1;
67.       return
68.   end
69.   if m4 ~= m
70.       disp(['The number of the right-hand side values ' ...
71.             'is not equal to the number of constraints.'])
72.       exitflag = -1;
73.       return
74.   end
75.   if n3 ~= 1
76.       disp('Vector Eqin is not a column vector')
77.       exitflag = -1;
78.       return
79.   end
80.   if n4 ~= 1
81.       disp('Vector b is not a column vector')
82.       exitflag = -1;
83.       return
84.   end
85.   if m4 ~= m3
86.       disp('The size of vectors Eqin and b does not match')
87.       exitflag = -1;
88.       return
89.   end
90.   % if the problem is a maximization problem, transform it to
91.   % a minimization problem
92.   if MinMaxLP == 1
93.       c = -c;
94.   end
95.   Aeq = []; % matrix constraint of equality constraints
96.   beq = []; % right-hand side of equality constraints
97.   % check if all constraints are equalities
98.   flag = isequal(zeros(m3, 1), Eqin);
99.   if flag == 0 % some or all constraints are inequalities
100.      indices = []; % indices of the equality constraints
```

```
101.    for i = 1:m3
102.        if Eqin(i, 1) == 0 % find the equality constraints
103.            indices = [indices i];
104.        % convert 'greater than or equal to' type of
105.        % constraints to 'less than or equal to' type
106.        % of constraints
107.        elseif Eqin(i, 1) == 1
108.            A(i, :) = -A(i, :);
109.            b(i) = -b(i);
110.        end
111.    end
112.    % create the matrix constraint of equality constraints
113.    Aeq = A(indices, :);
114.    A(indices, :) = []; % delete equality constraints
115.    % create the right-hand side of equality constraints
116.    beq = b(indices);
117.    b(indices) = []; % delete equality constraints
118. else % all constraints are equalities
119.    Aeq = A;
120.    beq = b;
121.    A = [];
122.    b = [];
123. end
124. lb = zeros(n, 1); % create zero lower bounds
125. % choose LP algorithm
126. options = optimoptions(@linprog, 'Algorithm', algorithm);
127. % call the linprog solver
128. [xsol, fval, exitflag, output] = linprog(c, A, b, ...
129.    Aeq, beq, lb, [], [], options);
130. iterations = output.iterations;
131. % calculate the value of the objective function
132. if MinMaxLP == 1 % maximization
133.    fval = -(fval + c0);
134. else % minimization
135.    fval = fval + c0;
136. end
137. end
```

A.8   Computational Study

In this section, we present a computational study. The aim of the computational study is to compare the execution time of MATLAB's Optimization Toolbox solvers. The test set used in the computational study were 19 LPs from the Netlib (optimal, Kennington, and infeasible LPs) and Mészáros problem sets that were presented in Section 3.6. Table A.2 presents the total execution time of four LP algorithms of the linprog solver that were presented in Section A.3. The following abbreviations are used: (i) IPM—Interior-Point Method, (ii) DSA—Dual-Simplex Algorithm,

Table A.2 Total execution time of the four LP algorithms of the `linprog` solver over the benchmark set

Name	Problem Constraints	Variables	Nonzeros	IPM	DSA	ASA	PSA
beaconfd	173	262	3,375	0.02	0.02	–	0.05
cari	400	1,200	152,800	1.26	0.05	–	0.74
farm	7	12	36	0.01	0.02	0.01	0.01
itest6	11	8	20	0.01	0.02	0.01	0.01
klein2	477	54	4,585	0.98	0.05	0.69	0.31
nsic1	451	463	2,853	0.03	0.03	8.09	0.22
nsic2	465	463	3,015	0.04	0.03	–	0.22
osa-07	1,118	23,949	143,694	0.85	0.37	–	3.71
osa-14	2,337	52,460	314,760	2.46	1.13	–	13.42
osa-30	4,350	100,024	600,138	4.02	3.87	–	50.68
rosen1	520	1,024	23,274	0.05	0.05	39.56	2.46
rosen10	2,056	4,096	62,136	0.15	0.26	–	10.62
rosen2	1,032	2,048	46,504	0.11	0.11	–	6.24
rosen7	264	512	7,770	0.02	0.03	–	0.60
rosen8	520	1,024	15,538	0.05	0.05	–	1.49
sc205	205	203	551	0.01	0.02	0.47	0.09
scfxm1	330	457	2,589	0.04	0.03	–	0.34
sctap2	1,090	1,880	6,714	0.11	0.09	–	1.74
sctap3	1,480	2,480	8,874	0.17	0.14	–	2.29
Geometric mean				0.10	0.08	0.47	0.69

(iii) ASA—Active-Set Algorithm, and (iv) PSA—Primal-Simplex Algorithm. For each instance, we averaged times over 10 runs. All times are measured in seconds using MATLAB's functions `tic` and `toc`.

The dual simplex algorithm is the fastest algorithm, while the active-set algorithm is the worst since it fails to solve 13 LPs.

A.9   Chapter Review

MATLAB's Optimization Toolbox includes a family of algorithms for solving optimization problems. The toolbox provides functions for solving linear programming, mixed-integer linear programming, quadratic programming, nonlinear programming, and nonlinear least squares problems. In this chapter, we presented the available functions of this toolbox for solving LPs. Finally, we performed a computational study in order to compare the execution time of MATLAB's Optimization Toolbox solvers. The dual simplex algorithm is the fastest algorithm, while the active-set algorithm is the worst. Finally, the active-set algorithm fails to solve 13 LPs.

References

1. Altman, Y. M. (2014). Accelerating MATLAB performance: 1001 tips to speed up MATLAB programs. CRC Press.
2. Applegate, D. L., Bixby, R. E., Chvatal, V., & Cook, W. J. (2007). *The traveling salesman problem: a computational study*. Princeton: Princeton University Press.
3. Davis, T. A. (2011). *MATLAB primer* (8th edition). Chapman & Hall/CRC.
4. Forrest, J. J., & Goldfarb, D. (1992). Steepest-edge simplex algorithms for linear programming. *Mathematical Programming, 57*(1–3), 341–374.
5. Gill, P. E., Murray, W., Saunders, M. A., & Wright, M. H. (1984). Procedures for optimization problems with a mixture of bounds and general linear constraints. *ACM Transactions on Mathematical Software (TOMS), 10*(3), 282–298.
6. Gill, P. E., Murray, W., & Wright, M. H. (1991). *Numerical linear algebra and optimization (Vol. 5)*. Redwood City: Addison-Wesley.
7. Hahn, B., & Valentine, D. T. (2016). *Essential MATLAB for engineers and scientists* (5th ed.). Academic.
8. Koberstein, A. (2008). Progress in the dual simplex algorithm for solving large scale LP problems: techniques for a fast and stable implementation. *Computational Optimization and Applications, 41*(2), 185–204.
9. MathWorks. (2017). *Optimization Toolbox: user's guide*. Available online at: http://www.mathworks.com/help/releases/R2015a/pdf_doc/optim/optim_tb.pdf (Last access on March 31, 2017).
10. Mehrotra, S. (1992). On the implementation of a primal-dual interior point method. *SIAM Journal on Optimization, 2*(4), 575–601.
11. Sherman, J., & Morrison, W. J. (1950). Adjustment of an inverse matrix corresponding to a change in one element of a given matrix. *The Annals of Mathematical Statistics, 21*(1), 124–127.
12. Zhang, Y. (1998). Solving large-scale linear programs by interior-point methods under the Matlab environment. *Optimization Methods and Software, 10*(1), 1–31.

Appendix B
State-of-the-Art Linear Programming Solvers: CLP and CPLEX

Abstract CLP and CPLEX are two of the most powerful linear programming solvers. CLP is an open-source linear programming solver, while CPLEX is a commercial one. Both solvers include a plethora of algorithms and methods. This chapter presents these solvers and provides codes to access them through MATLAB. A computational study is performed. The aim of the computational study is to compare the efficiency of CLP and CPLEX.

B.1 Chapter Objectives

- Present CLP and CPLEX.
- Understand how to use CLP and CPLEX.
- Provide codes to access CLP and CPLEX through MATLAB.
- Compare the computational performance of CLP and CPLEX.

B.2 Introduction

Many open-source and commercial linear programming solvers are available. The most well-known LP solvers are listed in Table B.1.

CLP and CPLEX are two of the most powerful LP solvers. CLP is an open-source LP solver, while CPLEX is a commercial one. Both solvers include a plethora of algorithms and methods. This chapter presents these solvers and provides codes to access them through MATLAB. A computational study is performed. The aim of the computational study is to compare the efficiency of CLP and CPLEX.

The structure of this chapter is as follows. We present CLP in Section B.3 and CPLEX in Section B.4. For each solver, we present the implemented algorithms and methods, and provide codes to call them through MATLAB. Section B.5 presents a computational study that compares the execution time of CLP and CPLEX. Finally, conclusions and further discussion are presented in Section B.6.

© Springer International Publishing AG 2017
N. Ploskas, N. Samaras, *Linear Programming Using MATLAB®*,
Springer Optimization and Its Applications 127, DOI 10.1007/978-3-319-65919-0

Table B.1 Open-source and commercial linear programming solvers

Name	Open-source/Commercial	Website
CLP	Open-source	https://projects.coin-or.org/Clp
CPLEX	Commercial	http://www-01.ibm.com/software/commerce/ optimization/cplex-optimizer/index.html
GLPK	Open-source	https://www.gnu.org/software/glpk/glpk.html
GLOP	Open-source	https://developers.google.com/optimization/lp/glop
GUROBI	Commercial	http://www.gurobi.com/
LINPROG	Commercial	http://www.mathworks.com/
LP_SOLVE	Open-source	http://lpsolve.sourceforge.net/
MOSEK	Commercial	https://mosek.com/
SOPLEX	Open-source	http://soplex.zib.de/
XPRESS	Commercial	http://www.fico.com/en/products/fico-xpress- optimization-suite

B.3 CLP

The Computational Infrastructure for Operations Research (COIN-OR) project [1] is an initiative to develop open-source software for the operations research community. CLP (Coin-or Linear Programming) [2] is part of this project. CLP is a powerful LP solver. CLP includes both primal and dual simplex solvers, and an interior point method. Various options are available to tune the performance of CLP solver for specific problems.

CLP is written in C++ and is primarily intended to be used as a callable library. However, a stand-alone executable is also available. Users can download and build the source code of CLP or download and install a binary executable (https://projects. coin-or.org/Clp). CLP is available on Microsoft Windows, Linux, and MAC OS.

By typing "?" in CLP command prompt, the following list of options is shown (Figure B.1).

In order to solve an LP problem with CLP, type "*import filename*" and the algorithm that will solve the problem ("*primalS*" for the primal simplex algorithm, "*dualS*" for the dual simplex algorithm, and "*barr*" for the interior point method—Figure B.2).

Figure B.2 displays problem statistics (after using the command "*import*") and the solution process. CLP solved this instance, afiro from the Netlib benchmark library, in six iterations and 0.012 s; the optimal objective value is −464.75.

CLP is a callable library that can be utilized in other applications. Next, we call CLP through MATLAB. We use OPTI Toolbox [5], a free MATLAB toolbox for Optimization. OPTI provides a MATLAB interface over a suite of open source and free academic solvers, including CLP and CPLEX. OPTI Toolbox provides a MEX interface for CLP. The syntax is the following:

```
[xsol, fval, exitflag, info] = opti_clp(H, c, A, bl...
    bu, lb, ub, opts);
```

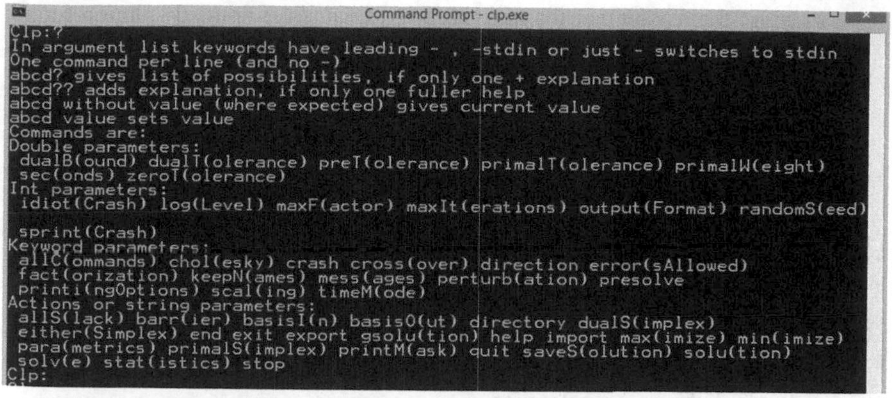

Fig. B.1 CLP options

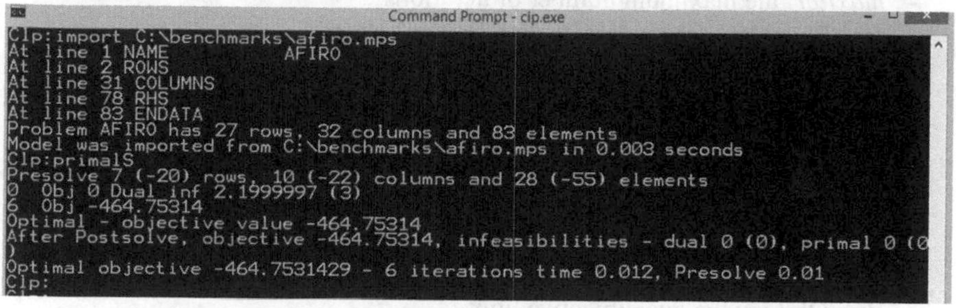

Fig. B.2 Solution of an LP Problem with CLP

where:

- Input arguments:

  - H: the quadratic objective matrix when solving quadratic LP problems (optional)
  - c: the vector of coefficients of the objective function (size $n \times 1$)
  - A: the matrix of coefficients of the constraints (size $m \times n$)
  - bl: the vector of the right-hand side of the constraints (size $m \times 1$)
  - bu: the vector of the left-hand side of the constraints (size $m \times 1$)
  - lb: the lower bounds of the variables (optional, (size $n \times 1$)
  - ub: the upper bounds of the variables (optional, (size $n \times 1$)
  - $opts$: solver options (see below)

- Output arguments:

  - $xsol$: the solution found (size $m \times 1$)
  - $fval$: the value of the objective function at the solution x

- *exitflag*: the reason that the algorithm terminated (1: optimal solution found, 0: maximum iterations or time reached, and −1: the primal LP problem is infeasible or unbounded)
- *info*: an information structure containing the number of iterations (*Iterations*), the execution time (*Time*), the algorithm used to solve the LP problem (*Algorithm*), the status of the problem that was solved (*Status*), and a structure containing the Lagrange multipliers at the solution (*lambda*)

• Option fields (all optional):

- *algorithm*: the linear programming algorithm that will be used (possible values: *DualSimplex, PrimalSimplex, PrimalSimplexOrSprint, Barrier, BarrierNoCross, Automatic*)
- *primalTol*: the primal tolerance
- *dualTol*: the dual tolerance
- *maxIter*: the maximum number of iterations
- *maxTime*: the maximum execution time in seconds
- *display*: the display level (possible values: 0, 1 → 100 increasing)
- *objBias*: the objective bias term
- *numPresolvePasses*: the number of presolver passes
- *factorFreq*: every *factorFreq* number of iterations, the basis inverse is recomputed from scratch
- *numberRefinements*: the number of iterative simplex refinements
- *primalObjLim*: the primal objective limit
- *dualObjLim*: the dual objective limit
- *numThreads*: the number of Cilk worker threads (only with Aboca CLP build)
- *abcState*: Aboca's partition size

Next, we present an implementation in MATLAB of a code to solve LPs using CLP solver via OPTI Toolbox (filename: `clpOptimizer.m`). The function for the solution of LPs using the CLP solver takes as input the full path to the MPS file that will be solved (variable *mpsFullPath*), the LP algorithm that will be used (variable *algorithm*), the primal tolerance (variable *primalTol*), the dual tolerance (variable *dualTol*), the maximum number of iterations (variable *maxIter*), the maximum execution time in seconds (variable *maxTime*), the display level (variable *displayLevel*), the objective bias term (variable *objBias*), the number of presolver passes (variable *numPresolvePasses*), the number of iterations that the basis inverse is recomputed from scratch (variable *factorFreq*), the number of iterative simplex refinements (variable *numberRefinements*), the primal objective limit (variable *primalObjLim*), the dual objective limit (variable *dualObjLim*), the number of Cilk worker threads (variable *numThreads*), and Aboca's partition size (variable *abcState*), and returns as output the solution found (vector *xsol*), the value of the objective function at the solution *xsol* (variable *fval*), the reason that the algorithm terminated (variable *exitflag*), and the number of iterations (variable *iterations*).

In lines 57–104, we set user-defined values to the options (if any). Next, we read the MPS file (line 106) and solve the LP problem using the CLP solver (lines 108–109).

```
1.    function [xsol, fval, exitflag, iterations] = ...
2.        clpOptimizer(mpsFullPath, algorithm, primalTol, ...
3.        dualTol, maxIter, maxTime, displayLevel, objBias, ...
4.        numPresolvePasses, factorFreq, numberRefinements, ...
5.        primalObjLim, dualObjLim, numThreads, abcState)
6.    % Filename: clpOptimizer.m
7.    % Description: the function is a MATLAB code to solve LPs
8.    % using CLP (via OPTI Toolbox)
9.    % Authors: Ploskas, N., & Samaras, N.
10.   %
11.   % Syntax: [xsol, fval, exitflag, iterations] = ...
12.   %     clpOptimizer(mpsFullPath, algorithm, primalTol, ...
13.   %     dualTol, maxIter, maxTime, display, objBias, ...
14.   %     numPresolvePasses, factorFreq, numberRefinements, ...
15.   %     primalObjLim, dualObjLim, numThreads, abcState)
16.   %
17.   % Input:
18.   % -- mpsFullPath: the full path of the MPS file
19.   % -- algorithm: the CLP algorithm that will be used
20.   %     (optional, possible values: 'DualSimplex',
21.   %     'PrimalSimplex', 'PrimalSimplexOrSprint',
22.   %     'Barrier', 'BarrierNoCross', 'Automatic')
23.   % -- primalTol: the primal tolerance (optional)
24.   % -- dualTol: the dual tolerance (optional)
25.   % -- maxIter: the maximum number of iterations
26.   %     (optional)
27.   % -- maxTime: the maximum execution time in seconds
28.   %     (optional)
29.   % -- displayLevel: the display level (optional,
30.   %     possible values: 0, 1 -> 100 increasing)
31.   % -- objBias: the objective bias term (optional)
32.   % -- numPresolvePasses: the number of presolver passes
33.   %     (optional)
34.   % -- factorFreq: every factorFreq number of iterations,
35.   %     the basis inverse is re-computed from scratch
36.   %     (optional)
37.   % -- numberRefinements: the number of iterative simplex
38.   %     refinements (optional)
39.   % -- primalObjLim: the primal objective limit (optional)
40.   % -- dualObjLim: the dual objective limit (optional)
41.   % -- numThreads: the number of Cilk worker threads
42.   %     (optional, only with Aboca CLP build)
43.   % -- abcState: Aboca's partition size (optional)
44.   %
45.   % Output:
46.   % -- xsol: the solution found by the solver (size m x 1)
47.   % -- fval: the value of the objective function at the
48.   %     solution xsol
49.   % -- exitflag: the reason that the algorithm terminated
50.   %     (1: the solver converged to a solution x, 0: the
```

```
51.  %      number of iterations exceeded the maxIter option or
52.  %      time reached, -1: the LP problem is infeasible or
53.  %      unbounded)
54.  % -- iterations: the number of iterations
55.
56.  % set user defined values to options
57.  opts = optiset('solver', 'clp');
58.  if exist('algorithm')
59.      opts.solverOpts.algorithm = algorithm;
60.  end
61.  if exist('primalTol')
62.      opts.solverOpts.primalTol = primalTol;
63.      opts.tolrfun = primalTol;
64.  end
65.  if exist('dualTol')
66.      opts.solverOpts.dualTol = dualTol;
67.  end
68.  if exist('maxIter')
69.      opts.maxiter = maxIter;
70.  else
71.       opts.maxiter = 1000000;
72.  end
73.  if exist('maxTime')
74.      opts.maxtime = maxTime;
75.  else
76.      opts.maxtime = 1000000;
77.  end
78.  if exist('displayLevel')
79.      opts.display = displayLevel;
80.  end
81.  if exist('objBias')
82.      opts.solverOpts.objbias = objBias;
83.  end
84.  if exist('numPresolvePasses')
85.      opts.solverOpts.numPresolvePasses = numPresolvePasses;
86.  end
87.  if exist('factorFreq')
88.      opts.solverOpts.factorFreq = factorFreq;
89.  end
90.  if exist('numberRefinements')
91.      opts.solverOpts.numberRefinements = numberRefinements;
92.  end
93.  if exist('primalObjLim')
94.      opts.solverOpts.primalObjLim = primalObjLim;
95.  end
96.  if exist('dualObjLim')
97.      opts.solverOpts.dualObjLim = dualObjLim;
98.  end
99.  if exist('numThreads')
100.     opts.solverOpts.numThreads = numThreads;
101. end
102. if exist('abcState')
103.     opts.solverOpts.abcState = abcState;
104. end
```

```
105. % read the MPS file
106. prob = coinRead(mpsFullPath);
107. % call CLP solver
108. [xsol, fval, exitflag, info] = opti_clp([], prob.f, ...
109.     prob.A, prob.rl, prob.ru, prob.lb, prob.ub, opts);
110. iterations = info.Iterations;
111. end
```

B.4   CPLEX

IBM ILOG CPLEX Optimizer [3] is a commercial solver for solving linear programming and mixed integer linear programming problems. IBM ILOG CPLEX Optimizer automatically determines smart settings for a wide range of algorithm parameters. However, many parameters can be manually tuned, including several LP algorithms and pricing rules.

CPLEX can be used both as a stand-alone executable and as a callable library. CPLEX can be called through several programming languages, like MATLAB, C, C++, Java, and Python. Users can download a fully featured free academic license (https://www-304.ibm.com/ibm/university/academic/pub/jsps/assetredirector.jsp? asset_id=1070), download a problem size-limited community edition (https://www-01.ibm.com/software/websphere/products/optimization/cplex-studio-community-edition/) or buy CPLEX. CPLEX is available on Microsoft Windows, Linux, and MAC OS.

By typing "*help*" in CPLEX interactive optimizer, the following list of options is shown (Figure B.3).

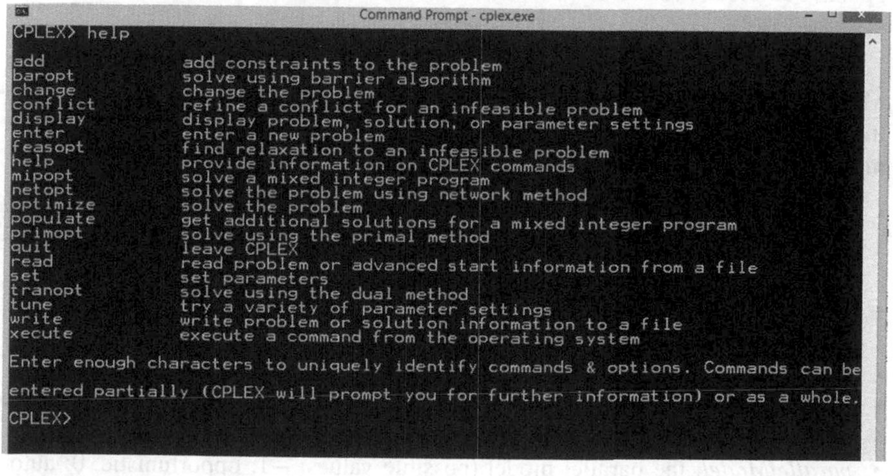

Fig. B.3 CPLEX options

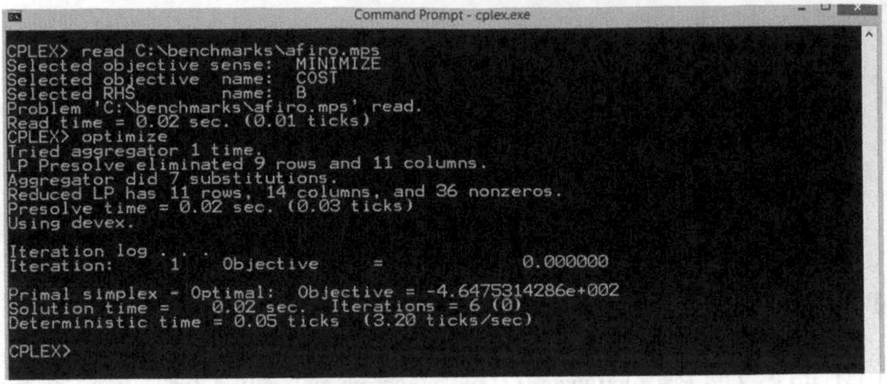

```
Command Prompt - cplex.exe

CPLEX> read C:\benchmarks\afiro.mps
Selected objective sense:  MINIMIZE
Selected objective  name:  COST
Selected RHS        name:  B
Problem 'C:\benchmarks\afiro.mps' read.
Read time = 0.02 sec. (0.01 ticks)
CPLEX> optimize
Tried aggregator 1 time.
LP Presolve eliminated 9 rows and 11 columns.
Aggregator did 7 substitutions.
Reduced LP has 11 rows, 14 columns, and 36 nonzeros.
Presolve time = 0.02 sec. (0.03 ticks)
Using devex.

Iteration log . .
Iteration:      1    Objective      =          0.000000

Primal simplex - Optimal:  Objective = -4.6475314286e+002
Solution time =    0.02 sec.  Iterations = 6 (0)
Deterministic time = 0.05 ticks  (3.20 ticks/sec)

CPLEX>
```

Fig. B.4   Solution of an LP Problem with CPLEX

In order to solve an LP problem with CPLEX, type *"read filename"* and (*"optimize"* (Figure B.4).

Figure B.4 displays the solution process. CPLEX solved this instance, afiro from the Netlib benchmark library, in six iterations and 0.02 s; the optimal objective value is −464.75.

CPLEX is a callable library that can be utilized in other applications. Next, we call CPLEX through MATLAB. We use its own interface, instead of OPTI Toolbox. In order to use that interface, we only have to add to MATLAB's path the folder that stores CPLEX interface. This folder is named *matlab* and is located under the CPLEX installation folder. The code that we need to call CPLEX in order to solve LPs is the following:

```
cplex = Cplex;
cplex.readModel(filename);
cplex.solve;
```

Initially, we create an object of type *Cplex*. Then we read the problem from an MPS file through the *readModel* command. Finally, we solve the problem using the command *solve*. As already mentioned, we can adjust several options. Next, we mention the most important options for solving LPs (for a full list, see [4]):

- *algorithm*: the linear programming algorithm that will be used (possible values: 0—automatic, 1—primal simplex, 2—dual simplex, 3—network simplex, 4—barrier, 5—sifting, 6—concurrent optimizers)
- *simplexTol*: the tolerance for simplex-type algorithms
- *barrierTol*: the tolerance for the barrier algorithm
- *maxTime*: the maximum execution time in seconds
- *simplexMaxIter*: the maximum number of iterations for simplex-type algorithms
- *barrierMaxIter*: the maximum number of iterations for the barrier algorithm
- *parallelModel*: the parallel model (possible values: −1: opportunistic, 0: automatic, 1: deterministic)

- *presolve*: decides whether CPLEX applies presolve or not (possible values: 0: no, 1: yes)
- *numPresolvePasses*: the number of presolver passes
- *factorFreq*: every factorFreq number of iterations, the basis inverse is recomputed from scratch

Next, we present an implementation in MATLAB of a code to solve LPs using CPLEX solver (filename: `cplexOptimizer.m`). The function for the solution of LPs using the CPLEX solver takes as input the full path to the MPS file that will be solved (variable *mpsFullPath*), the LP algorithm that will be used (variable *algorithm*), the tolerance for simplex-type algorithms (variable *simplexTol*), the tolerance for the barrier algorithm (variable *barrierTol*), the maximum execution time in seconds (variable *maxTime*), the maximum number of iterations for simplex-type algorithms (variable *simplexMaxIter*), the maximum number of iterations for the barrier algorithm (variable *barrierMaxIter*), the parallel model (variable *parallelMode*), whether to perform the presolve analysis or not (variable *presolve*), the number of presolver passes (variable *numPresolvePasses*), and the number of iterations that the basis inverse is recomputed from scratch (variable *factorFreq*), and returns as output the solution found (vector *xsol*), the value of the objective function at the solution *xsol* (variable *fval*), the reason that the algorithm terminated (variable *exitflag*), and the number of iterations (variable *iterations*).

In lines 57–93, we set user-defined values to the options (if any). Next, we read the MPS file (line 95) and solve the LP problem using the CPLEX solver (line 97).

```
1.    function [xsol, fval, exitflag, iterations] = ...
2.          cplexOptimizer(mpsFullPath, algorithm, ...
3.          simplexTol, barrierTol, maxTime, simplexMaxIter, ...
4.          barrierMaxIter, parallelMode, presolve, ...
5.          numPresolvePasses, factorFreq)
6.    % Filename: cplexOptimizer.m
7.    % Description: the function is a MATLAB code to solve LPs
8.    % using CPLEX
9.    % Authors: Ploskas, N., & Samaras, N.
10.   %
11.   % Syntax: [xsol, fval, exitflag, iterations] = ...
12.   %     cplexOptimizer(mpsFullPath, algorithm, ...
13.   %     simplexTol, barrierTol, maxTime, simplexMaxIter, ...
14.   %     barrierMaxIter, parallelMode, presolve, ...
15.   %     numPresolvePasses, factorFreq)
16.   %
17.   % Input:
18.   % -- mpsFullPath: the full path of the MPS file
19.   % -- algorithm: the CPLEX algorithm that will be used
20.   %     (optional, possible values: 0 - automatic, 1 -
21.   %     primal simplex, 2 - dual simplex, 3 - network
22.   %     simplex, 4 - barrier, 5 - sifting, 6 - concurrent
23.   %     optimizers)
24.   % -- simplexTol: the tolerance for simplex type
25.   %     algorithms (optional)
26.   % -- barrierTol: the tolerance for the barrier
```

```
27.  %      algorithm (optional)
28.  % -- maxTime: the maximum execution time in seconds
29.  % -- (optional)
30.  % -- simplexMaxIter: the maximum number of iterations
31.  % -- for simplex type algorithms (optional)
32.  % -- barrierMaxIter: the maximum number of iterations
33.  % -- for the barrier algorithm (optional)
34.  % -- parallelModel: the parallel model (optional,
35.  %      possible values: -1: opportunistic, 0: automatic,
36.  %      1: deterministic)
37.  % -- presolve: decides whether CPLEX applies presolve
38.  %      (optional, possible values: 0: no, 1: yes)
39.  % -- numPresolvePasses: the number of presolver passes
40.  %      (optional)
41.  % -- factorFreq: every factorFreq number of iterations,
42.  %      the basis inverse is re-computed from scratch
43.  %      (optional)
44.  %
45.  % Output:
46.  % -- xsol: the solution found by the solver (size m x 1)
47.  % -- fval: the value of the objective function at the
48.  %      solution xsol
49.  % -- exitflag: the reason that the algorithm terminated
50.  %      (1: the solver converged to a solution x, 2: the
51.  %      LP problem is unbounded, 3: the LP problem is
52.  %      infeasible, 10: the number of iterations exceeded
53.  %      the maxIter option, 11: time reached)
54.  % -- iterations: the number of iterations
55.
56.  % set user defined values to options
57.  cplex = Cplex;
58.  if exist('algorithm')
59.      cplex.Param.lpmethod.Cur = algorithm;
60.  end
61.  if exist('simplexTol')
62.      cplex.Param.simplex.tolerances.optimality.Cur ...
63.          = simplexTol;
64.  end
65.  if exist('barrierTol')
66.      cplex.Param.barrier.convergetol.Cur = barrierTol;
67.  end
68.  if exist('maxTime')
69.      cplex.Param.timelimit.Cur = maxTime;
70.  end
71.  if exist('simplexMaxIter')
72.      cplex.Param.simplex.limits.iterations.Cur = ...
73.          simplexMaxIter;
74.  end
75.  if exist('barrierMaxIter')
76.      cplex.Param.barrier.limits.iteration.Cur = ...
77.          barrierMaxIter;
78.  end
79.  if exist('parallelMode')
80.      cplex.Param.parallel.Cur = parallelMode;
```

```
81.  end
82.  if exist('presolve')
83.      cplex.Param.preprocessing.presolve.Cur = ...
84.          presolve;
85.  end
86.  if exist('numPresolvePasses')
87.      cplex.Param.preprocessing.numpass.Cur = ...
88.          numPresolvePasses;
89.  end
90.  if exist('factorFreq')
91.      cplex.Param.simplex.refactor.Cur = factorFreq;
92.  end
93.  cplex.DisplayFunc = [];
94.  % read the MPS file
95.  cplex.readModel(mpsFullPath);
96.  % call CPLEX solver
97.  cplex.solve;
98.  % export solution
99.  xsol = cplex.Solution.x;
100. fval = cplex.Solution.objval;
101. exitflag = cplex.Solution.status;
102. if isfield(cplex.Solution, 'itcnt')
103.     iterations = cplex.Solution.itcnt;
104. else
105.     iterations = cplex.Solution.baritcnt;
106. end
107. end
```

B.5   Computational Study

In this section, we present a computational study. The aim of the computational study is to compare the execution time of CLP and CPLEX solvers. The test set used in the computational study is all instances from the Netlib (optimal, Kennington, and infeasible LPs), Mészáros, and Mittelmann problem sets that were presented in Section 3.6. Tables B.2, B.3, B.4, B.5, B.6, B.7, B.8, and B.9 present the total execution time of the primal simplex algorithm, the dual simplex algorithm, and the interior point method of each solver on the problems from the Netlib library; the Kennington library; the infeasible problems from the Netlib library; the misc section of the Mészáros library; the problematic section of the Mészáros library; the new section of the Mészáros library; and the Mittelmann library, respectively. For each instance, we averaged times over 10 runs. All times are measured in seconds using MATLAB's functions tic and toc. A time limit of 1, 000 s was set. A dash "-" indicates that a problem was not solved because the solver reached the time limit.

Table B.2 Total execution time of CLP and CPLEX solvers over the Netlib LP test set

Problem				CLP			CPLEX		
Name	Constraints	Variables	Nonzeros	Primal	Dual	Barrier	Primal	Dual	Barrier
25fv47	821	1,571	10,400	0.17	0.19	0.09	0.11	0.10	0.06
80bau3b	2,262	9,799	21,002	0.27	0.24	0.24	0.13	0.11	0.14
adlittle	56	97	383	0.01	0.01	0.01	0.01	0.01	0.01
afiro	27	32	83	0.01	0.01	0.01	0.01	0.01	0.01
agg	488	163	2,410	0.02	0.01	0.02	0.01	0.01	0.02
agg2	516	302	4,284	0.02	0.01	0.03	0.01	0.01	0.02
agg3	516	302	4,300	0.02	0.01	0.03	0.01	0.01	0.02
bandm	305	472	2,494	0.02	0.01	0.02	0.01	0.01	0.02
beaconfd	173	262	3,375	0.01	0.01	0.01	0.01	0.01	0.01
blend	74	83	491	0.01	0.01	0.01	0.01	0.01	0.01
bnl1	643	1,175	5,121	0.06	0.04	0.05	0.04	0.03	0.05
bnl2	2,324	3,489	13,999	0.15	0.06	0.20	0.15	0.05	0.15
boeing1	351	384	3,485	0.02	0.01	0.02	0.01	0.01	0.02
boeing2	166	143	1,196	0.01	0.01	0.01	0.01	0.01	0.02
bore3d	233	315	1,429	0.01	0.01	0.01	0.01	0.01	0.01
brandy	220	249	2,148	0.01	0.01	0.01	0.01	0.01	0.02
capri	271	353	1,767	0.02	0.01	0.01	0.01	0.01	0.02
cycle	1,903	2,857	20,720	0.08	0.03	0.11	0.03	0.02	0.06
czprob	929	3,523	10,669	0.07	0.02	0.04	0.02	0.03	0.03
d2q06c	2,171	5,167	32,417	1.24	1.30	0.41	0.79	0.51	0.19
d6cube	415	6,184	37,704	1.20	0.08	0.24	1.00	0.06	0.10
degen2	444	534	3,978	0.04	0.02	0.04	0.03	0.02	0.03
degen3	1,503	1,818	24,646	0.35	0.18	0.32	0.34	0.12	0.29
dfl001	6,071	12,230	35,632	11.45	6.04	20.75	5.10	3.55	1.67
e226	223	282	2,578	0.02	0.01	0.02	0.01	0.01	0.02
etamacro	400	688	2,409	0.02	0.02	0.03	0.01	0.02	0.03
fffff800	524	854	6,227	0.03	0.02	0.04	0.01	0.01	0.03
finnis	497	614	2,310	0.02	0.01	0.02	0.01	0.01	0.02
fit1d	24	1,026	13,404	0.06	0.02	0.03	0.02	0.02	0.03
fit1p	627	1,677	9,868	0.03	0.04	0.50	0.03	0.03	0.04
fit2d	25	10,500	129,018	2.17	0.15	0.28	0.21	0.16	0.15
fit2p	3,000	13,525	50,284	0.78	1.47	45.19	0.87	0.88	0.16
forplan	161	421	4,563	0.02	0.01	0.02	0.01	0.01	0.02
ganges	1,309	1,681	6,912	0.03	0.02	0.02	0.02	0.02	0.03
gfrd-pnc	616	1,092	2,377	0.02	0.01	0.02	0.01	0.01	0.02
greenbea	2,392	5,405	30,877	0.74	0.77	0.25	0.26	0.22	0.12
greenbeb	2,392	5,405	30,877	0.48	0.40	0.20	0.21	0.29	0.10
grow15	300	645	5,620	0.03	0.02	0.03	0.03	0.03	0.03
grow22	440	946	8,252	0.03	0.03	0.05	0.05	0.09	0.04
grow7	140	301	2,612	0.01	0.01	0.02	0.01	0.01	0.02

(continued)

Table B.2 (continued)

Name	Problem Constraints	Variables	Nonzeros	CLP Primal	Dual	Barrier	CPLEX Primal	Dual	Barrier
kb2	43	41	286	0.01	0.01	0.01	0.01	0.01	0.01
israel	174	142	2,269	0.01	0.01	0.02	0.01	0.01	0.02
lotfi	153	308	1,078	0.02	0.01	0.01	0.01	0.01	0.01
maros	846	1,443	9,614	1.12	1.13	1.15	0.84	0.58	0.42
maros-r7	3,136	9,408	144,848	0.07	0.04	0.04	0.03	0.03	0.03
modszk1	687	1,620	3,168	0.04	0.02	0.04	0.02	0.01	0.03
nesm	662	2,923	13,288	0.11	0.11	0.09	0.06	0.11	0.09
perold	625	1,376	6,018	0.11	0.11	0.16	0.10	0.05	0.06
pilot	410	1,000	5,141	0.23	0.24	0.17	0.24	0.09	0.11
pilot.ja	2,030	4,883	73,152	1.60	1.20	1.19	0.90	0.56	0.40
pilot.we	940	1,988	14,698	0.19	0.16	0.26	0.12	0.12	0.06
pilot4	1,441	3,652	43,167	0.07	0.05	0.07	0.04	0.03	0.05
pilot87	975	2,172	13,057	4.62	6.21	3.46	2.52	6.16	1.42
pilotnov	722	2,789	9,126	0.13	0.10	0.11	0.10	0.07	0.08
qap8	912	1,632	7,296	0.51	0.67	0.60	0.34	0.35	0.21
qap12	3,192	8,856	38,304	35.07	218.22	25.23	20.51	38.86	1.80
qap15	6,330	22,275	94,950	531.54	–	169.93	256.25	543.92	11.26
recipe	91	180	663	0.01	0.01	0.01	0.01	0.01	0.01
sc105	105	103	280	0.01	0.01	0.01	0.01	0.01	0.01
sc205	205	203	551	0.01	0.01	0.01	0.01	0.01	0.01
sc50a	50	48	130	0.01	0.01	0.01	0.01	0.01	0.01
sc50b	50	48	118	0.01	0.01	0.01	0.01	0.01	0.01
scagr25	471	500	1,554	0.02	0.01	0.01	0.01	0.01	0.02
scagr7	129	140	420	0.01	0.01	0.01	0.01	0.01	0.01
scfxm1	330	457	2,589	0.02	0.01	0.02	0.01	0.01	0.02
scfxm2	660	914	5,183	0.03	0.03	0.03	0.02	0.02	0.03
scfxm3	990	1,371	7,777	0.04	0.04	0.04	0.03	0.03	0.04
scorpion	388	358	1,426	0.01	0.01	0.01	0.01	0.01	0.01
scrs8	490	1,169	3,182	0.02	0.02	0.02	0.02	0.01	0.02
scsd1	77	760	2,388	0.02	0.01	0.01	0.01	0.01	0.02
scsd6	147	1,350	4,316	0.02	0.01	0.02	0.01	0.01	0.02
scsd8	397	2,750	8,584	0.07	0.04	0.03	0.03	0.04	0.03
sctap1	300	480	1,692	0.01	0.01	0.01	0.01	0.01	0.02
sctap2	1,090	1,880	6,714	0.02	0.02	0.03	0.02	0.02	0.03
sctap3	1,480	2,480	8,874	0.04	0.02	0.04	0.02	0.02	0.04
seba	515	1,028	4,352	0.02	0.01	0.01	0.01	0.01	0.01
share1b	117	225	1,151	0.01	0.01	0.01	0.01	0.01	0.02
share2b	96	79	694	0.01	0.01	0.01	0.01	0.01	0.01
shell	536	1,775	3,556	0.02	0.01	0.02	0.01	0.01	0.02
ship04l	402	2,118	6,332	0.07	0.01	0.02	0.01	0.01	0.02
ship04s	402	1,458	4,352	0.02	0.01	0.02	0.01	0.01	0.02

(continued)

Table B.2 (continued)

| Problem | | | CLP | | | CPLEX | | |
Name	Constraints	Variables	Nonzeros	Primal	Dual	Barrier	Primal	Dual	Barrier
ship08l	778	4,283	12,802	0.03	0.02	0.04	0.02	0.02	0.03
ship08s	778	2,387	7,114	0.03	0.02	0.02	0.02	0.02	0.02
ship12l	1,151	5,427	16,170	0.03	0.03	0.05	0.03	0.03	0.04
ship12s	1,151	2,763	8,178	0.03	0.02	0.03	0.02	0.02	0.03
sierra	1,227	2,036	7,302	0.04	0.02	0.04	0.02	0.02	0.03
stair	356	467	3,856	0.03	0.02	0.04	0.02	0.01	0.03
standata	359	1,075	3,031	0.02	0.01	0.01	0.01	0.01	0.02
standgub	361	1,184	3,139	0.01	0.01	0.01	0.01	0.01	0.02
standmps	467	1,075	3,679	0.02	0.01	0.02	0.01	0.01	0.02
stocfor1	117	111	447	0.01	0.01	0.01	0.01	0.01	0.01
stocfor2	2,157	2,031	8,343	0.06	0.03	0.06	0.03	0.03	0.04
stocfor3	16,675	15,695	64,875	1.08	0.53	0.58	0.51	0.36	0.32
truss	1,000	8,806	27,836	0.81	3.03	0.16	0.50	2.10	0.09
tuff	333	587	4,520	0.04	0.01	0.02	0.01	0.01	0.02
vtp.base	198	203	908	0.01	0.01	0.01	0.01	0.01	0.01
wood1p	244	2,594	70,215	0.09	0.06	0.11	0.04	0.04	0.07
woodw	1,098	8,405	37,474	0.08	0.07	0.11	0.04	0.05	0.07
Geometric mean				0.05	0.03	0.05	0.04	0.03	0.04

Table B.3 Total execution time of CLP and CPLEX solvers over the Kennington LP test set

| Problem | | | CLP | | | CPLEX | | |
Name	Constraints	Variables	Nonzeros	Primal	Dual	Barrier	Primal	Dual	Barrier
cre-a	3,516	4,067	14,987	0.17	0.07	0.14	0.16	0.06	0.11
cre-b	9,648	72,447	256,095	4.22	2.07	2.40	1.03	0.88	1.20
cre-c	3,068	3,678	13,244	0.15	0.07	0.12	0.13	0.05	0.10
cre-d	8,926	69,980	242,646	2.57	0.94	1.86	0.76	0.46	0.53
ken-07	2,426	3,602	8,404	0.04	0.03	0.04	0.03	0.02	0.04
ken-11	14,694	21,349	49,058	0.36	0.21	0.29	0.19	0.13	0.19
ken-13	28,632	42,659	97,246	2.34	0.51	0.74	1.14	0.34	0.43
ken-18	105,127	154,699	358,171	15.23	2.61	5.48	10.77	2.13	2.48
osa-07	1,118	23,949	143,694	0.21	0.14	0.31	0.11	0.12	0.15
osa-14	2,337	52,460	314,760	0.51	0.43	0.76	0.26	0.25	0.35
osa-30	4,350	100,024	600,138	1.57	1.27	1.61	0.61	0.52	0.65
osa-60	10,280	232,966	1,397,793	6.86	4.79	5.74	1.86	1.38	1.64
pds-02	2,953	7,535	16,390	0.05	0.05	0.11	0.04	0.03	0.06
pds-06	9,881	28,655	62,524	0.27	0.32	2.04	0.19	0.14	0.49
pds-10	16,558	48,763	106,436	0.57	1.04	11.23	0.48	0.30	1.01
pds-20	33,874	105,728	230,200	3.30	8.64	42.96	3.60	1.20	5.55
Geometric mean				0.71	0.45	0.98	0.40	0.24	0.40

Table B.4 Total execution time of CLP and CPLEX solvers over the infeas section of the Netlib LP test set

Name	Problem			CLP			CPLEX		
	Constraints	Variables	Nonzeros	Primal	Dual	Barrier	Primal	Dual	Barrier
bgdbg1	348	407	1,440	0.01	0.01	0.07	0.01	0.01	0.01
bgetam	400	688	2,409	0.02	0.06	0.24	0.01	0.01	0.02
bgindy	2,671	10,116	65,502	0.11	0.17	9.84	0.04	0.04	0.04
bgprtr	20	34	64	0.01	0.01	0.01	0.01	0.01	0.01
box1	231	261	651	0.01	0.01	0.01	0.01	0.01	0.01
ceria3d	3,576	824	17,602	0.07	0.71	140.26	0.01	0.01	0.01
chemcom	288	720	1,566	0.01	0.02	0.08	0.01	0.01	0.01
cplex1	3,005	3,221	8,944	0.14	0.14	87.14	0.01	0.01	0.01
cplex2	224	221	1,058	0.01	0.02	0.02	0.02	0.01	0.03
ex72a	197	215	467	0.01	0.01	0.01	0.01	0.01	0.01
ex73a	193	211	457	0.01	0.01	0.01	0.01	0.01	0.01
forest6	66	95	210	0.01	0.01	0.01	0.01	0.01	0.01
galenet	8	8	16	0.01	0.01	0.01	0.01	0.01	0.01
gosh	3,792	10,733	97,231	1.62	1.72	–	0.15	0.06	0.41
gran	2,658	2,520	20,106	0.10	0.09	4.97	0.02	0.02	0.02
greenbea	2,393	5,405	30,883	0.73	5.66	1.78	0.02	0.02	0.02
itest2	9	4	17	0.01	0.01	0.01	0.01	0.01	0.01
itest6	11	8	20	0.01	0.01	0.01	0.01	0.01	0.01
klein1	54	54	696	0.02	0.01	0.02	0.01	0.01	0.01
klein2	477	54	4,585	0.03	0.03	2.36	0.04	0.02	0.02
klein3	994	88	12,107	0.06	0.09	15.99	0.11	0.05	0.03
mondou2	312	604	1,208	0.01	0.01	0.03	0.01	0.01	0.01
pang	361	460	2,652	0.02	0.01	0.03	0.01	0.01	0.02
pilot4i	410	1,000	5,141	0.03	0.02	0.13	0.01	0.01	0.01
qual	323	464	1,646	0.02	0.02	0.07	0.01	0.01	0.02
reactor	318	637	2,420	0.02	0.01	0.12	0.01	0.01	0.01
refinery	323	464	1,626	0.02	0.03	0.07	0.01	0.01	0.01
vol1	323	464	1,646	0.02	0.03	0.07	0.01	0.01	0.02
woodinfe	35	89	140	0.01	0.01	0.01	0.01	0.01	0.01
Geometric mean				0.02	0.03	0.12	0.01	0.01	0.01

Table B.5 Total execution time of CLP and CPLEX solvers over the misc section of the Mészáros LP test set

Name	Problem			CLP			CPLEX		
	Constraints	Variables	Nonzeros	Primal	Dual	Barrier	Primal	Dual	Barrier
aa01	823	8,904	72,965	2.04	0.65	1.13	1.52	0.48	0.70
aa03	825	8,627	70,806	1.71	0.32	1.23	1.84	0.26	0.36
aa3	825	8,627	70,806	1.40	0.31	1.11	1.65	0.26	0.30
aa4	426	7,195	52,121	0.36	0.21	0.37	0.30	0.15	0.32
aa5	801	8,308	65,953	1.05	0.33	0.94	1.32	0.30	0.32
aa6	646	7,292	51,728	0.83	0.18	0.91	1.14	0.18	0.32
air02	50	6,774	61,555	0.07	0.06	0.14	0.04	0.04	0.09
air03	124	10,757	91,028	0.16	0.12	0.37	0.08	0.08	0.13
air04	823	8,904	72,965	2.03	0.65	1.12	1.52	0.49	0.41
air05	426	7,195	52,121	0.41	0.22	0.37	0.34	0.15	0.28
air06	825	8,627	70,806	1.70	0.32	1.24	1.84	0.27	1.09
aircraft	3,754	7,517	20,267	0.14	0.13	4.23	0.08	0.11	0.17
bas1lp	5,411	4,461	582,411	1.19	0.81	21.02	0.51	4.06	2.92
baxter	27,441	15,128	95,971	4.04	0.48	282.93	0.93	0.41	1.65
car4	16,384	33,052	63,724	0.30	0.24	0.43	0.25	0.19	0.19
cari	400	1,200	152,800	0.11	0.10	0.24	0.06	0.06	0.06
ch	3,700	5,062	20,873	0.47	0.67	0.35	0.45	0.19	0.61
co5	5,774	7,993	53,661	1.42	1.61	0.77	1.23	0.60	0.35
co9	10,789	14,851	101,578	8.20	6.74	2.59	4.48	2.92	1.46
complex	1,023	1,408	46,463	0.66	0.61	1.41	0.50	0.59	0.13
cq5	5,048	7,530	47,353	1.05	1.03	0.75	0.72	0.48	0.26
cq9	9,278	13,778	88,897	4.01	3.97	2.64	2.63	1.73	0.71
cr42	905	1,513	6,614	0.03	0.03	0.31	0.02	0.02	0.03
crew1	135	6,469	46,950	0.09	0.09	0.12	0.07	0.08	0.07
dano3mip	3,202	13,873	79,655	10.61	37.60	40.38	4.51	6.49	1.97
dbic1	43,200	183,235	1,038,761	28.80	283.38	44.07	16.07	25.07	14.64
dbir1	18,804	27,355	1,058,605	1.23	0.83	31.98	0.73	0.53	6.39
dbir2	18,906	27,355	1,139,637	1.84	1.42	66.76	1.24	0.77	6.29
delf000	3,128	5,464	12,606	0.12	0.14	0.61	0.07	0.04	0.09
delf001	3,098	5,462	13,214	0.14	0.15	0.37	0.08	0.04	0.09
delf002	3,135	5,460	13,287	0.14	0.14	0.38	0.08	0.05	0.09
delf003	3,065	5,460	13,269	0.13	0.15	0.39	0.07	0.05	0.10
delf004	3,142	5,464	13,546	0.15	0.16	0.43	0.10	0.06	0.11
delf005	3,103	5,464	13,494	0.12	0.15	0.41	0.09	0.06	0.11
delf006	3,147	5,469	13,604	0.19	0.16	0.57	0.09	0.06	0.10
delf007	3,137	5,471	13,758	0.16	0.21	0.55	0.12	0.08	0.10
delf008	3,148	5,472	13,821	0.18	0.22	0.62	0.12	0.08	0.11
delf009	3,135	5,472	13,750	0.16	0.20	0.53	0.09	0.12	0.11
delf010	3,147	5,472	13,802	0.19	0.25	0.70	0.10	0.06	0.11

(continued)

Table B.5 (continued)

Name	Problem Constraints	Variables	Nonzeros	CLP Primal	Dual	Barrier	CPLEX Primal	Dual	Barrier
delf011	3,134	5,471	13,777	0.16	0.20	0.47	0.09	0.06	0.10
delf012	3,151	5,471	13,793	0.19	0.23	0.62	0.10	0.07	0.11
delf013	3,116	5,472	13,809	0.17	0.22	0.63	0.09	0.07	0.11
delf014	3,170	5,472	13,866	0.17	0.18	0.49	0.10	0.07	0.11
delf015	3,161	5,471	13,793	0.18	0.23	0.41	0.10	0.06	0.11
delf017	3,176	5,471	13,732	0.18	0.17	0.57	0.10	0.06	0.10
delf018	3,196	5,471	13,774	0.17	0.16	0.61	0.10	0.05	0.10
delf019	3,185	5,471	13,762	0.15	0.15	0.60	0.09	0.05	0.10
delf020	3,213	5,472	14,070	0.17	0.19	0.47	0.11	0.06	0.11
delf021	3,208	5,471	14,068	0.19	0.24	0.50	0.14	0.07	0.12
delf022	3,214	5,472	14,060	0.17	0.24	0.61	0.10	0.07	0.11
delf023	3,214	5,472	14,098	0.17	0.24	0.68	0.17	0.08	0.12
delf024	3,207	5,466	14,456	0.20	0.29	0.59	0.13	0.08	0.12
delf025	3,197	5,464	14,447	0.16	0.24	0.57	0.14	0.07	0.13
delf026	3,190	5,462	14,220	0.18	0.27	0.49	0.13	0.07	0.12
delf027	3,187	5,457	14,200	0.16	0.17	0.43	0.11	0.07	0.12
delf028	3,177	5,452	14,402	0.19	0.23	0.58	0.11	0.07	0.12
delf029	3,179	5,454	14,402	0.17	0.21	0.54	0.14	0.07	0.13
delf030	3,199	5,469	14,262	0.16	0.26	0.69	0.15	0.06	0.13
delf031	3,176	5,455	14,205	0.17	0.27	0.56	0.12	0.06	0.12
delf032	3,196	5,467	14,251	0.17	0.27	0.84	0.16	0.06	0.13
delf033	3,173	5,456	14,205	0.16	0.19	0.53	0.16	0.06	0.12
delf034	3,175	5,455	14,208	0.17	0.28	0.64	0.15	0.06	0.13
delf035	3,193	5,468	14,284	0.15	0.30	0.64	0.16	0.06	0.13
delf036	3,170	5,459	14,202	0.16	0.30	0.75	0.14	0.06	0.12
df2177	630	9,728	21,706	0.13	139.65	0.35	0.09	0.18	0.26
disp3	2,182	1,856	6,407	0.03	0.02	0.05	0.05	0.02	0.05
dsbmip	1,182	1,886	7,366	0.04	0.09	0.07	0.03	0.06	0.07
e18	24,617	14,231	132,095	1.92	0.92	–	2.29	0.96	81.41
ex3sta1	17,443	8,156	59,419	3.48	3.49	337.95	0.99	7.79	0.74
farm	7	12	36	0.01	0.01	0.01	0.01	0.01	0.01
gams10a	114	61	297	0.01	0.01	0.01	0.01	0.01	0.01
gams30a	354	181	937	0.01	0.01	0.01	0.01	0.01	0.01
ge	10,099	11,098	39,554	1.30	1.35	0.82	0.75	0.31	0.42
iiasa	669	2,970	6,648	0.03	0.02	0.03	0.02	0.03	0.03
jendrec1	2,109	4,228	89,608	1.82	3.09	28.80	1.02	0.43	0.43
kent	31,300	16,620	184,710	0.16	0.15	0.17	0.16	0.14	0.17
kl02	71	36,699	212,536	0.35	0.31	0.51	0.15	0.15	0.21
kleemin3	3	3	6	0.01	0.01	0.01	0.01	0.01	0.01
kleemin4	4	4	10	0.01	0.01	0.01	0.01	0.01	0.01
kleemin5	5	5	15	0.01	0.01	0.01	0.01	0.01	0.01

(continued)

Table B.5 (continued)

Name	Problem			CLP			CPLEX		
	Constraints	Variables	Nonzeros	Primal	Dual	Barrier	Primal	Dual	Barrier
kleemin6	6	6	21	0.01	0.01	0.01	0.01	0.01	0.01
kleemin7	7	7	28	0.01	0.01	0.01	0.01	0.01	0.01
kleemin8	8	8	36	0.01	0.01	0.01	0.01	0.01	0.01
l9	244	1,401	4,577	0.03	0.03	0.02	0.04	0.03	0.03
large000	4,239	6,833	16,573	0.20	0.24	0.53	0.15	0.06	0.12
large001	4,162	6,834	17,225	0.23	0.21	0.59	0.11	0.58	0.11
large002	4,249	6,835	18,330	0.41	0.44	0.73	0.16	0.10	0.17
large003	4,200	6,835	18,016	0.33	0.30	0.67	0.16	0.13	0.14
large004	4,250	6,836	17,739	0.28	0.39	0.67	0.17	0.09	0.15
large005	4,237	6,837	17,575	0.29	0.30	0.64	0.14	0.08	0.13
large006	4,249	6,837	17,887	0.28	0.34	0.77	0.19	0.10	0.14
large007	4,236	6,836	17,856	0.26	0.35	0.78	0.18	0.11	0.14
large008	4,248	6,837	17,898	0.28	0.41	0.80	0.16	0.11	0.15
large009	4,237	6,837	17,878	0.28	0.33	0.70	0.19	0.09	0.13
large010	4,247	6,837	17,887	0.28	0.36	0.75	0.18	0.13	0.14
large011	4,236	6,837	17,878	0.30	0.36	0.65	0.18	0.09	0.14
large012	4,253	6,838	17,919	0.29	0.46	0.85	0.19	0.09	0.14
large013	4,248	6,838	17,941	0.31	0.39	0.73	0.20	0.10	0.14
large014	4,271	6,838	17,979	0.27	0.33	0.62	0.18	0.08	0.13
large015	4,265	6,838	17,957	0.24	0.29	0.56	0.17	0.08	0.13
large016	4,287	6,838	18,029	0.27	0.67	0.46	0.17	0.08	0.14
large017	4,277	6,837	17,983	0.27	0.49	0.53	0.13	0.07	0.14
large018	4,297	6,837	17,791	0.22	0.24	0.51	0.16	0.07	0.15
large019	4,300	6,836	17,786	0.23	0.18	0.48	0.18	0.07	0.14
large020	4,315	6,837	18,136	0.26	0.56	0.52	0.21	0.08	0.15
large021	4,311	6,838	18,157	0.28	0.56	0.59	0.20	0.10	0.17
large022	4,312	6,834	18,104	0.35	0.76	0.58	0.24	0.10	0.14
large023	4,302	6,835	18,123	0.31	1.16	0.89	0.23	0.10	0.17
large024	4,292	6,831	18,599	0.34	0.82	0.77	0.19	0.15	0.18
large025	4,297	6,832	18,743	0.32	0.41	0.61	0.23	0.13	0.19
large026	4,284	6,824	18,631	0.32	0.42	0.80	0.22	0.13	0.18
large027	4,275	6,821	18,562	0.29	0.46	0.40	0.18	0.14	0.18
large028	4,302	6,833	18,886	0.31	0.76	0.66	0.20	0.22	0.18
large029	4,301	6,832	18,952	0.27	0.37	0.54	0.16	0.18	0.17
large030	4,285	6,823	18,843	0.33	0.49	0.53	0.17	0.13	0.18
large031	4,294	6,826	18,867	0.36	0.51	0.64	0.19	0.20	0.18
large032	4,292	6,827	18,850	0.27	0.47	0.73	0.30	0.18	0.18
large033	4,273	6,817	18,791	0.27	0.36	0.67	0.18	0.18	0.17
large034	4,294	6,831	18,855	0.27	0.34	0.68	0.18	0.15	0.18
large035	4,293	6,829	18,881	0.27	0.40	0.76	0.21	0.14	0.17
large036	4,282	6,822	18,840	0.29	0.41	0.70	0.18	0.19	0.17

(continued)

Table B.5 (continued)

Name	Problem			CLP			CPLEX		
	Constraints	Variables	Nonzeros	Primal	Dual	Barrier	Primal	Dual	Barrier
lp22	2,958	13,434	65,560	13.43	11.93	9.45	15.14	7.54	1.25
lpl1	39,951	125,000	381,259	177.34	14.10	21.50	3.41	6.16	3.52
lpl2	3,294	10,755	32,106	0.08	0.06	0.16	0.05	0.07	0.08
lpl3	10,828	33,538	100,377	0.41	0.34	0.82	0.19	0.51	0.62
mod2	34,774	31,728	165,129	71.07	24.08	116.34	31.87	12.10	1.75
model1	4,400	15,447	149,000	0.01	0.01	0.02	0.02	0.01	0.02
model10	2,896	6,464	25,277	7.38	10.64	2.83	2.99	2.95	0.59
model11	2,879	10,257	55,274	1.01	1.08	0.74	0.43	0.47	0.47
model2	7,056	18,288	55,859	0.04	0.07	0.05	0.04	0.03	0.03
model3	362	798	3,028	0.20	0.33	0.18	0.13	0.22	0.09
model4	379	1,212	7,498	0.60	1.49	0.44	0.25	0.62	0.24
model5	1,609	3,840	23,236	0.77	1.65	0.81	0.49	0.57	0.61
model6	1,337	4,549	45,340	0.51	0.91	0.29	0.35	0.64	0.19
model7	1,888	11,360	89,483	0.84	1.62	0.51	0.61	0.85	0.79
model8	2,096	5,001	27,340	0.10	0.24	0.43	0.12	0.14	0.67
model9	3,358	8,007	49,452	0.57	0.57	0.36	0.34	0.29	0.20
multi	61	102	961	0.01	0.01	0.01	0.01	0.01	0.01
nemsafm	334	2,252	2,730	0.01	0.01	0.02	0.02	0.01	0.02
nemscem	651	1,570	3,698	0.02	0.02	0.02	0.06	0.02	0.03
nemsemm1	3,945	71,413	1,050,047	1.20	0.96	4.35	0.80	0.63	1.40
nemsemm2	6,943	42,133	175,267	0.51	0.32	0.61	0.29	0.23	0.37
nemspmm1	2,372	8,622	55,586	1.31	1.24	0.51	0.85	0.99	0.21
nemspmm2	2,301	8,413	67,904	1.76	1.72	0.71	1.03	1.03	0.62
nemswrld	7,138	27,174	190,907	24.45	18.55	9.82	10.89	8.67	0.93
nl	7,039	9,718	41,428	5.48	4.86	1.22	2.19	0.65	0.56
nsct1	22,901	14,981	656,259	0.71	0.51	32.03	0.42	0.31	7.33
nsct2	23,003	14,981	675,156	0.93	0.80	155.42	0.77	0.49	7.40
nsic1	451	463	2,853	0.01	0.01	0.01	0.03	0.01	0.02
nsic2	465	463	3,015	0.01	0.01	0.03	0.02	0.01	0.02
nsir1	4,407	5,717	138,955	0.16	0.11	2.87	0.14	0.07	0.40
nsir2	4,453	5,717	150,599	0.34	0.19	7.39	0.34	0.14	0.48
nug05	210	225	1,050	0.01	0.01	0.02	0.02	0.01	0.02
nug06	372	486	2,232	0.03	0.02	0.08	0.04	0.02	0.02
nug07	602	931	4,214	0.13	0.09	0.28	0.13	0.10	0.18
nug08	912	1,632	7,296	0.46	0.50	0.65	0.35	0.43	0.19
nug12	3,192	8,856	38,304	34.80	156.13	30.08	19.49	32.89	1.93
nug15	6,330	22,275	94,950	642.37	–	250.91	239.14	531.05	8.03
nw14	73	123,409	904,910	2.34	1.58	4.31	0.82	0.59	1.19
orna1	882	882	3,108	0.05	0.03	0.04	0.05	0.03	0.04
orna2	882	882	3,108	0.05	0.04	0.04	0.05	0.03	0.04
orna3	882	882	3,108	0.06	0.03	0.03	0.05	0.03	0.05

(continued)

Table B.5 (continued)

Name	Problem			CLP			CPLEX		
	Constraints	Variables	Nonzeros	Primal	Dual	Barrier	Primal	Dual	Barrier
orna4	882	882	3,108	0.06	0.05	0.04	0.04	0.03	0.05
orna7	882	882	3,108	0.08	0.04	0.03	0.05	0.02	0.04
orswq2	80	80	264	0.01	0.01	0.01	0.01	0.01	0.01
p0033	15	33	98	0.01	0.01	0.01	0.01	0.01	0.01
p0040	23	40	110	0.01	0.01	0.01	0.01	0.01	0.01
p010	10,090	19,000	117,910	1.40	0.33	0.34	0.50	0.26	0.34
p0201	133	201	1,923	0.01	0.01	0.01	0.02	0.01	0.01
p0282	241	282	1,966	0.01	0.01	0.01	0.01	0.01	0.02
p0291	252	291	2,031	0.01	0.01	0.01	0.04	0.01	0.01
p05	176	548	1,711	0.54	0.14	0.16	0.20	0.12	0.12
p0548	5,090	9,500	58,955	0.01	0.01	0.01	0.02	0.01	0.02
p19	284	586	5,305	0.01	0.01	0.02	0.02	0.01	0.02
p2756	755	2,756	8,937	0.02	0.02	0.04	0.05	0.02	0.04
p6000	2,095	5,872	17,731	0.05	0.03	0.20	0.04	0.03	0.12
pcb1000	1,565	2,428	20,071	0.10	0.09	0.15	0.09	0.07	0.09
pcb3000	3,960	6,810	56,557	0.36	0.31	0.57	0.36	0.27	0.30
pf2177	9,728	900	21,706	0.27	1.63	88.52	0.16	0.16	0.13
pldd000b	3,069	3,267	8,980	0.08	0.06	0.07	0.05	0.04	0.07
pldd001b	3,069	3,267	8,981	0.08	0.06	0.07	0.06	0.04	0.07
pldd002b	3,069	3,267	8,982	0.09	0.06	0.06	0.05	0.04	0.06
pldd003b	3,069	3,267	8,983	0.10	0.06	0.06	0.05	0.04	0.06
pldd004b	3,069	3,267	8,984	0.09	0.06	0.06	0.06	0.04	0.07
pldd005b	3,069	3,267	8,985	0.09	0.06	0.07	0.06	0.04	0.07
pldd006b	3,069	3,267	8,986	0.09	0.06	0.07	0.04	0.04	0.06
pldd007b	3,069	3,267	8,987	0.10	0.06	0.06	0.06	0.04	0.07
pldd008b	3,069	3,267	9,047	0.09	0.06	0.07	0.04	0.04	0.07
pldd009b	3,069	3,267	9,050	0.08	0.06	0.06	0.04	0.04	0.07
pldd010b	3,069	3,267	9,053	0.10	0.06	0.06	0.04	0.04	0.07
pldd011b	3,069	3,267	9,055	0.08	0.06	0.07	0.05	0.04	0.07
pldd012b	3,069	3,267	9,057	0.09	0.06	0.07	0.05	0.04	0.07
primagaz	1,554	10,836	21,665	0.07	0.07	0.08	0.05	0.05	0.06
problem	12	46	86	0.01	0.01	0.01	0.01	0.01	0.01
progas	1,650	1,425	8,422	0.12	0.09	0.06	0.10	0.07	0.05
qiulp	1,192	840	3,432	0.04	0.04	0.09	0.05	0.03	0.02
r05	5,190	9,500	103,955	0.56	0.15	0.18	0.22	0.13	0.13
rat1	3,136	9,408	88,267	0.38	0.19	0.25	0.16	0.14	0.14
rat5	3,136	9,408	137,413	0.90	0.57	0.72	0.39	0.40	0.63
rat7a	3,136	9,408	268,908	7.04	5.69	5.97	3.70	3.27	1.29
refine	29	33	124	0.01	0.01	0.01	0.01	0.01	0.01
rlfddd	4,050	57,471	260,577	0.25	0.26	3.53	0.29	0.20	0.60

(continued)

Table B.5 (continued)

Name	Problem			CLP			CPLEX		
	Constraints	Variables	Nonzeros	Primal	Dual	Barrier	Primal	Dual	Barrier
rlfdual	8,052	66,918	273,979	0.62	370.29	3.88	0.40	1.11	0.82
rlfprim	58,866	8,052	265,927	2.96	0.55	–	0.74	0.44	0.88
rosen1	2,056	4,096	62,136	0.06	0.05	0.06	0.06	0.04	0.05
rosen10	2,056	4,096	62,136	0.36	0.10	0.15	0.32	0.12	0.11
rosen2	1,032	2,048	46,504	0.15	0.15	0.11	0.21	0.11	0.08
rosen7	264	512	7,770	0.02	0.02	0.02	0.03	0.02	0.02
rosen8	520	1,024	15,538	0.04	0.04	0.04	0.03	0.02	0.03
route	20,894	23,923	187,686	1.92	0.38	6.17	4.51	0.17	1.03
seymourl	4,944	1,372	33,549	0.33	0.72	79.32	0.53	0.47	0.35
slptsk	2,861	3,347	72,465	1.36	1.41	60.95	0.55	0.32	0.91
small000	709	1,140	2,749	0.02	0.01	0.03	0.02	0.01	0.03
small001	687	1,140	2,871	0.02	0.02	0.03	0.02	0.01	0.03
small002	713	1,140	2,946	0.02	0.03	0.04	0.02	0.01	0.03
small003	711	1,140	2,945	0.02	0.02	0.03	0.02	0.01	0.03
small004	717	1,140	2,983	0.02	0.02	0.03	0.02	0.01	0.03
small005	717	1,140	3,017	0.02	0.04	0.04	0.02	0.01	0.03
small006	710	1,138	3,024	0.02	0.02	0.04	0.02	0.01	0.03
small007	711	1,137	3,079	0.02	0.02	0.03	0.02	0.01	0.03
small008	712	1,134	3,042	0.02	0.02	0.03	0.02	0.01	0.03
small009	710	1,135	3,030	0.02	0.02	0.03	0.02	0.01	0.03
small010	711	1,138	3,027	0.02	0.02	0.03	0.03	0.01	0.03
small011	705	1,133	3,005	0.02	0.02	0.02	0.02	0.01	0.03
small012	706	1,134	3,014	0.02	0.02	0.03	0.02	0.01	0.03
small013	701	1,131	2,989	0.01	0.02	0.02	0.02	0.01	0.03
small014	687	1,130	2,927	0.01	0.02	0.02	0.02	0.01	0.03
small015	683	1,130	2,967	0.02	0.02	0.03	0.02	0.01	0.03
small016	677	1,130	2,937	0.01	0.02	0.02	0.02	0.01	0.03
south31	18,425	35,421	111,498	5.93	8.04	–	4.00	11.48	1.37
stat96v1	5,995	197,472	588,798	69.98	–	74.88	36.90	29.91	41.25
stat96v2	29,089	957,432	2,852,184	–	–	–	–	–	–
stat96v3	33,841	1,113,780	3,317,736	–	–	–	–	–	–
stat96v4	3,173	62,212	490,472	222.72	274.51	4.80	120.48	52.35	2.35
stat96v5	2,307	75,779	233,921	6.51	16.47	11.32	3.06	4.91	1.09
sws	14,310	12,465	93,015	0.09	0.08	0.09	0.08	0.07	0.08
t0331-4l	664	46,915	430,982	8.77	5.35	3.68	6.06	3.90	1.03
testbig	17,613	31,223	61,639	0.30	0.26	14.54	0.12	0.11	0.12
ulevimin	6,590	44,605	162,206	18.37	12.69	82.41	19.99	6.62	2.36
us04	163	28,016	297,538	0.30	0.23	0.37	0.22	0.17	0.30
world	34,506	32,734	164,470	80.10	79.22	52.72	44.60	11.98	2.77
zed	116	43	567	0.01	0.01	0.01	0.01	0.01	0.01
Geometric mean				0.20	0.20	0.35	0.16	0.11	0.14

Table B.6 Total execution time of CLP and CPLEX solvers over the problematic section of the Mészáros LP test set

Name	Problem Constraints	Variables	Nonzeros	CLP Primal	Dual	Barrier	CPLEX Primal	Dual	Barrier
degme	185,501	659,415	8,127,528	–	–	–	–	–	181.14
karted	46,502	133,115	1,770,349	–	–	–	–	–	227.62
tp-6	142,752	1,014,301	11,537,419	–	–	–	–	–	168.22
ts-palko	22,002	47,235	1,076,903	–	–	553.61	–	–	113.52
Geometric mean				–	–	553.61	–	–	167.51

Table B.7 Total execution time of CLP and CPLEX solvers over the new section of the Mészáros LP test set

Name	Problem Constraints	Variables	Nonzeros	CLP Primal	Dual	Barrier	CPLEX Primal	Dual	Barrier
de063155	852	1,487	4,553	0.05	0.05	0.13	0.02	0.03	0.05
de063157	936	1,487	4,699	0.05	0.05	0.23	0.03	0.03	0.05
de080285	936	1,487	4,662	0.04	0.05	0.08	0.02	0.02	0.05
gen	769	2,560	63,085	7.42	1.40	2.06	1.09	0.07	0.45
gen1	769	2,560	63,085	7.41	1.40	2.06	1.09	0.07	0.34
gen2	1,121	3,264	81,855	14.95	10.13	2.04	11.46	3.38	1.51
gen4	1,537	4,297	107,102	62.74	30.32	15.97	13.52	0.39	3.32
iprob	3,001	3,001	9,000	0.09	0.22	0.09	0.03	0.02	0.03
l30	2,701	15,380	51,169	13.44	27.50	1.52	10.05	4.87	2.89
Geometric mean				1.23	0.91	0.70	0.45	0.13	0.27

Both solvers have very efficient algorithms. All CPLEX algorithms perform better on these benchmark libraries. In addition, we cannot find an LP algorithm that is better on all instances. There exist LPs for which one of these three algorithms outperforms the others.

B.6   Chapter Review

CLP and CPLEX are two of the most powerful LP solvers. CLP is an open-source LP solver, while CPLEX is a commercial one. Both solvers include a plethora of algorithms and methods. This chapter presented these solvers and provided codes to access them through MATLAB. A computational study was performed. The aim of the computational study was to compare the efficiency of CLP and CPLEX. Both solvers have very efficient algorithms. All CPLEX algorithms perform better on these benchmark libraries. In addition, we cannot find an LP algorithm that is better on all instances. There exist LPs for which one of the three algorithms, i.e., the primal simplex algorithm, the dual simplex algorithm, and the interior point method, outperforms the others.

Table B.8 Total execution time of CLP and CPLEX solvers over the stochlp section of the Mészáros LP test set

Problem			CLP			CPLEX		
Name	Constraints	Variables	Nonzeros	Primal	Dual	Barrier	Primal	Dual	Barrier
aircraft	3,754	7,517	20,267	0.14	0.13	4.23	0.21	0.12	0.19
cep1	1,521	3,248	6,712	0.05	0.04	1.05	0.03	0.03	0.04
deter0	1,923	5,468	11,173	0.11	0.07	0.11	0.03	0.04	0.07
deter1	5,527	15,737	32,187	0.61	0.29	1.03	0.09	0.17	0.19
deter2	6,095	17,313	35,731	1.12	0.35	2.40	0.09	0.14	0.20
deter3	7,647	21,777	44,547	1.23	0.21	2.08	0.12	0.18	0.28
deter4	3,235	9,133	19,231	0.28	0.09	1.54	0.05	0.06	0.11
deter5	5,103	14,529	29,715	0.59	0.39	0.78	0.08	0.12	0.18
deter6	4,255	12,113	24,771	0.38	0.19	0.50	0.07	0.10	0.15
deter7	6,375	18,153	37,131	0.83	0.31	1.51	0.10	0.16	0.22
deter8	3,831	10,905	22,299	0.31	0.18	0.41	0.06	0.09	0.13
fxm2-6	1,520	2,172	12,139	0.22	0.22	0.16	0.07	0.04	0.04
fxm2-16	3,900	5,602	32,239	0.06	0.07	0.06	0.10	0.14	0.10
fxm3_6	6,200	9,492	54,589	8.75	8.52	2.83	0.17	0.27	0.18
fxm3_16	41,340	64,162	370,839	0.32	0.34	0.26	3.01	2.45	1.46
fxm4_6	22,400	30,732	248,989	1.97	1.62	1.27	0.95	0.69	0.98
pgp2	4,034	9,220	18,440	0.28	0.10	22.93	0.11	0.07	0.09
pltexpa2-6	686	1,820	3,708	0.01	0.01	0.02	0.02	0.01	0.02
pltexpa2-16	1,726	4,540	9,223	0.02	0.03	0.11	0.03	0.02	1.54
pltexpa3-6	4,430	11,612	23,611	0.05	0.08	0.22	0.04	0.06	0.13
pltexpa3-16	28,350	74,172	150,801	0.37	0.72	3.68	0.22	0.43	3.12
pltexpa4-6	26,894	70,364	143,059	0.37	0.81	2.57	0.21	0.57	0.65
sc205-2r-4	101	102	270	0.01	0.01	0.01	0.01	0.01	0.01
sc205-2r-8	189	190	510	0.01	0.01	0.01	0.01	0.01	0.01
sc205-2r-16	365	366	990	0.01	0.01	0.01	0.01	0.01	0.01
sc205-2r-27	607	608	1,650	0.01	0.01	0.01	0.02	0.01	0.01
sc205-2r-32	717	718	1,950	0.01	0.01	0.01	0.01	0.01	0.01
sc205-2r-50	1,113	1,114	3,030	0.01	0.01	0.02	0.02	0.01	0.02
sc205-2r-64	1,421	1,422	3,870	0.01	0.01	0.02	0.02	0.01	0.02
sc205-2r-100	2,213	2,214	6,030	0.02	0.02	0.05	0.02	0.02	0.02
sc205-2r-200	4,413	4,414	12,030	0.03	0.03	0.34	0.03	0.03	0.03
sc205-2r-400	8,813	8,814	24,030	0.07	0.07	2.25	0.05	0.05	0.05
sc205-2r-800	17,613	17,614	48,030	0.20	0.21	17.08	0.09	0.09	0.11
sc205-2r-1600	35,213	35,214	96,030	0.70	0.71	126.70	0.20	0.19	0.27
scagr7-2b-4	167	180	546	0.01	0.01	0.01	0.01	0.01	0.01
scagr7-2b-16	623	660	2,058	0.01	0.01	0.02	0.02	0.01	0.02
scagr7-2b-64	9,743	10,260	32,298	0.33	0.12	47.98	0.10	0.08	0.08
scagr7-2c-4	167	180	546	0.01	0.01	0.01	0.01	0.01	0.01
scagr7-2c-16	623	660	2,058	0.01	0.01	0.02	0.02	0.01	0.02

(continued)

Table B.8 (continued)

Name	Problem			CLP			CPLEX		
	Constraints	Variables	Nonzeros	Primal	Dual	Barrier	Primal	Dual	Barrier
scagr7-2c-64	2,447	2,580	8,106	0.04	0.02	0.64	0.04	0.03	0.03
scagr7-2r-4	167	180	546	0.01	0.01	0.01	0.01	0.01	0.01
scagr7-2r-8	319	340	1,050	0.01	0.01	0.01	0.01	0.01	0.01
scagr7-2r-16	623	660	2,058	0.01	0.01	0.03	0.07	0.01	0.02
scagr7-2r-27	1,041	1,100	3,444	0.02	0.01	0.07	0.02	0.02	0.02
scagr7-2r-32	1,231	1,300	4,074	0.02	0.01	0.11	0.02	0.02	0.02
scagr7-2r-54	2,067	2,180	6,846	0.03	0.02	0.45	0.05	0.03	0.03
scagr7-2r-64	2,447	2,580	8,106	0.05	0.03	0.72	0.03	0.03	0.03
scagr7-2r-108	4,119	4,340	13,542	0.07	0.04	2.55	0.05	0.05	0.05
scagr7-2r-216	8,223	8,660	27,042	0.23	0.09	17.78	0.09	0.12	0.08
scagr7-2r-432	16,431	17,300	54,042	0.86	0.23	182.72	0.29	0.14	0.15
scagr7-2r-864	32,847	34,580	108,042	3.17	0.72	–	1.03	0.40	0.31
scfxm1-2b-4	684	1,014	3,999	0.03	0.02	0.03	0.02	0.02	0.02
scfxm1-2b-16	2,460	3,714	13,959	0.08	0.05	0.09	0.07	0.05	0.07
scfxm1-2b-64	19,036	28,914	106,919	2.74	1.09	4.91	0.99	0.74	0.51
scfxm1-2c-4	684	1,014	3,999	0.03	0.02	0.03	0.02	0.02	0.03
scfxm1-2r-4	684	1,014	3,999	0.03	0.02	0.02	0.02	0.02	0.03
scfxm1-2r-8	1,276	1,914	7,319	0.04	0.03	0.04	0.03	0.03	0.04
scfxm1-2r-16	2,460	3,714	13,959	0.08	0.06	0.09	0.07	0.05	0.07
scfxm1-2r-27	4,088	6,189	23,089	0.17	0.12	0.19	0.09	0.09	0.11
scfxm1-2r-32	4,828	7,314	27,239	0.27	0.15	0.23	0.11	0.13	0.13
scfxm1-2r-64	9,564	14,514	53,799	0.76	0.33	0.88	0.38	0.28	0.34
scfxm1-2r-96	14,300	21,714	80,359	1.41	0.70	2.31	0.58	0.49	0.36
scfxm1-2r-128	19,036	28,914	106,919	2.83	1.07	4.69	1.32	0.65	0.50
scfxm1-2r-256	37,980	57,714	213,159	8.99	3.73	39.67	2.99	2.27	1.21
scrs8-2b-4	140	189	457	0.01	0.01	0.01	0.01	0.01	0.01
scrs8-2b-16	476	645	1,633	0.02	0.01	0.01	0.01	0.01	0.02
scrs8-2b-64	1,820	2,469	6,337	0.02	0.01	0.02	0.02	0.01	0.02
scrs8-2c-4	140	189	457	0.01	0.01	0.01	0.01	0.01	0.01
scrs8-2c-8	252	341	849	0.01	0.01	0.01	0.01	0.01	0.01
scrs8-2c-16	476	645	1,633	0.07	0.01	0.01	0.01	0.01	0.01
scrs8-2c-32	924	1,253	3,201	0.03	0.01	0.01	0.01	0.01	0.01
scrs8-2c-64	1,820	2,469	6,337	0.03	0.01	0.02	0.01	0.01	0.02
scrs8-2r-4	140	189	457	0.01	0.01	0.01	0.01	0.01	0.01
scrs8-2r-8	252	341	849	0.01	0.01	0.01	0.01	0.01	0.01
scrs8-2r-16	476	645	1,633	0.02	0.01	0.01	0.01	0.01	0.01
scrs8-2r-27	784	1,063	2,711	0.04	0.01	0.01	0.01	0.01	0.01
scrs8-2r-32	924	1,253	3,201	0.02	0.01	0.01	0.01	0.01	0.01
scrs8-2r-64	1,820	2,469	6,337	0.03	0.01	0.02	0.02	0.01	0.02
scrs8-2r-64b	1,820	2,469	6,337	0.03	0.02	0.02	0.02	0.01	0.02
scrs8-2r-128	3,612	4,901	12,609	0.03	0.02	0.05	0.02	0.02	0.03

(continued)

Table B.8 (continued)

Name	Problem			CLP			CPLEX		
	Constraints	Variables	Nonzeros	Primal	Dual	Barrier	Primal	Dual	Barrier
scrs8-2r-256	7,196	9,765	25,153	0.05	0.05	0.28	0.04	0.04	0.04
scrs8-2r-512	14,364	19,493	50,241	0.14	0.14	1.87	0.08	0.07	0.08
scsd8-2b-4	90	630	1,890	0.02	0.01	0.01	0.01	0.01	0.01
scsd8-2b-16	330	2,310	7,170	0.04	0.01	0.02	0.02	0.01	0.02
scsd8-2b-64	5,130	35,910	112,770	4.93	0.17	14.37	0.33	0.14	0.29
scsd8-2c-4	90	630	1,890	0.02	0.01	0.01	0.01	0.01	0.01
scsd8-2c-16	330	2,310	7,170	0.04	0.01	0.02	0.02	0.01	0.02
scsd8-2c-64	5,130	35,910	112,770	4.29	0.17	13.22	0.34	0.15	0.31
scsd8-2r-4	90	630	1,890	0.02	0.01	0.01	0.01	0.01	0.01
scsd8-2r-8	170	1,190	3,650	0.08	0.01	0.01	0.01	0.01	0.02
scsd8-2r-8b	170	1,190	3,650	0.07	0.01	0.01	0.01	0.01	0.02
scsd8-2r-16	330	2,310	7,170	0.04	0.01	0.02	0.02	0.01	0.03
scsd8-2r-27	550	3,850	12,010	0.06	0.02	0.06	0.03	0.02	0.05
scsd8-2r-32	650	4,550	14,210	0.09	0.02	0.07	0.03	0.02	0.08
scsd8-2r-54	1,090	7,630	23,890	0.17	0.03	0.25	0.08	0.03	1.27
scsd8-2r-64	1,290	9,030	28,290	0.23	0.04	0.30	0.09	0.03	0.11
scsd8-2r-108	2,170	15,190	47,650	0.51	0.06	1.45	0.11	0.05	1.35
scsd8-2r-216	4,330	30,310	95,170	2.51	0.15	10.32	0.29	0.11	0.61
scsd8-2r-432	8,650	60,550	190,210	8.83	0.37	78.26	0.91	0.25	0.64
sctap1-2b-4	270	432	1,516	0.01	0.01	0.01	0.01	0.01	0.02
sctap1-2b-16	990	1,584	5,740	0.05	0.01	0.04	0.01	0.01	0.03
sctap1-2b-64	15,390	24,624	90,220	0.23	0.19	43.44	0.20	0.13	0.30
sctap1-2c-4	270	432	1,516	0.01	0.01	0.01	0.01	0.01	0.02
sctap1-2c-16	990	1,584	5,740	0.05	0.01	0.04	0.01	0.01	0.03
sctap1-2c-64	3,390	5,424	19,820	0.04	0.03	0.52	0.03	0.03	0.07
sctap1-2r-4	270	432	1,516	0.01	0.01	0.01	0.01	0.01	0.01
sctap1-2r-8	510	816	2,924	0.02	0.01	0.02	0.01	0.01	0.02
sctap1-2r-8b	510	816	2,924	0.02	0.01	0.02	0.02	0.01	0.02
sctap1-2r-16	990	1,584	5,740	0.05	0.01	0.03	0.01	0.01	0.03
sctap1-2r-27	1,650	2,640	9,612	0.03	0.02	0.09	0.02	0.02	0.04
sctap1-2r-32	1,950	3,120	11,372	0.03	0.02	0.12	0.02	0.02	0.04
sctap1-2r-54	3,270	5,232	19,116	0.04	0.04	0.42	0.03	0.03	0.06
sctap1-2r-64	3,870	6,192	22,636	0.05	0.04	0.60	0.03	0.03	0.07
sctap1-2r-108	6,510	10,416	38,124	0.08	0.07	3.25	0.06	0.05	0.12
sctap1-2r-216	12,990	20,784	76,140	0.18	0.16	26.33	0.15	0.14	0.24
sctap1-2r-480	28,830	46,128	169,068	0.51	0.40	279.32	0.69	0.40	0.65
stormg2_1000	528,185	1,259,121	3,341,696	–	154.61	–	387.89	92.40	46.46
stormg2-8	4,409	10,193	27,424	0.32	0.10	0.67	0.07	0.07	0.16
stormg2-27	14,441	34,114	90,903	2.81	0.46	22.47	0.28	0.34	0.98
stormg2-125	66,185	157,496	418,321	81.09	3.89	–	2.93	2.50	3.18
Geometric mean				0.09	0.05	0.19	0.05	0.04	0.07

Table B.9 Total execution time of CLP and CPLEX solvers over the Mittelmann LP test set

Name	Problem Constraints	Variables	Nonzeros	CLP Primal	Dual	Barrier	CPLEX Primal	Dual	Barrier
16_n14	3,754	7,517	20,267	639.58	29.68	67.03	255.42	22.76	43.19
cont1	1,521	3,248	6,712	–	833.17	864.86	129.31	362.06	250.37
cont1_1	1,923	5,468	11,173	–	–	–	–	–	–
cont4	5,527	15,737	32,187	958.03	753.82	–	132.99	246.61	158.12
cont11	6,095	17,313	35,731	–	–	–	–	–	902.37
cont11_1	7,647	21,777	44,547	–	–	–	–	–	–
fome11	3,235	9,133	19,231	27.45	15.64	44.22	16.45	9.81	3.53
fome12	5,103	14,529	29,715	70.66	34.32	85.51	45.02	27.96	2.25
fome13	4,255	12,113	24,771	163.45	73.73	190.53	149.81	81.13	3.11
fome20	6,375	18,153	37,131	3.20	8.46	42.84	3.59	1.20	7.90
fome21	3,831	10,905	22,299	9.05	16.91	111.61	10.60	3.24	7.10
i_n13	1,520	2,172	12,139	–	72.63	–	312.07	26.98	77.61
L1_sixm 250obs	3,900	5,602	32,239	–	–	–	–	–	146.44
Linf_520c	6,200	9,492	54,589	–	–	–	–	–	143.54
lo10	41,340	64,162	370,839	–	25.14	–	234.00	4.08	459.82
long15	22,400	30,732	248,989	–	22.83	–	194.16	7.79	–
neos	4,034	9,220	18,440	314.32	108.34	–	29.99	5.62	12.16
neos1	686	1,820	3,708	8.01	7.96	–	3.63	113.41	4.65
neos2	1,726	4,540	9,223	23.91	18.69	64.65	5.61	86.40	3.48
neos3	4,430	11,612	23,611	–	–	–	–	–	35.50
netlarge1	28,350	74,172	150,801	–	860.23	–	–	57.84	–
netlarge2	26,894	70,364	143,059	–	78.78	–	325.44	8.43	–
netlarge3	101	102	270	–	536.25	–	472.98	52.32	–
netlarge6	189	190	510	–	–	–	161.09	138.77	–
ns1687037	365	366	990	–	920.78	–	–	–	153.08
ns1688926	607	608	1,650	245.96	16.65	–	501.31	8.96	32.17
nug08-3rd	717	718	1,950	–	477.21	–	–	453.54	218.37
nug20	1,113	1,114	3,030	–	–	–	–	–	102.15
nug30	1,421	1,422	3,870	–	–	–	–	–	–
pds-20	2,213	2,214	6,030	3.22	8.54	42.70	3.59	1.21	5.07
pds-30	4,413	4,414	12,030	10.39	30.55	122.32	9.00	3.72	7.47
pds-40	8,813	8,814	24,030	32.70	66.58	333.19	27.36	5.39	13.90
pds-50	17,613	17,614	48,030	90.31	111.55	–	55.46	6.86	19.64
pds-60	35,213	35,214	96,030	162.20	109.70	–	123.09	8.89	33.14
pds-70	167	180	546	302.32	143.64	–	170.29	11.45	35.26
pds-80	623	660	2,058	428.24	157.31	–	231.36	13.80	49.57

(continued)

Table B.9 (continued)

Name	Problem			CLP			CPLEX		
	Constraints	Variables	Nonzeros	Primal	Dual	Barrier	Primal	Dual	Barrier
pds-90	9,743	10,260	32,298	502.75	197.72	–	282.10	20.45	52.30
pds-100	167	180	546	564.43	162.91	–	269.90	18.86	72.09
rail507	623	660	2,058	3.89	3.41	3.39	1.33	2.00	0.54
rail516	2,447	2,580	8,106	1.36	1.49	2.15	0.76	0.81	2.52
rail582	167	180	546	3.96	2.91	3.70	2.56	0.99	1.94
rail2586	319	340	1,050	859.19	510.86	419.16	279.94	427.01	16.06
rail4284	623	660	2,058	–	–	–	–	937.28	32.43
sgpf5y6	1,041	1,100	3,444	4.00	3.84	375.93	1.45	1.54	2.78
spal_004	1,231	1,300	4,074	–	–	–	–	–	–
square15	2,067	2,180	6,846	–	24.26	–	210.78	7.39	–
stormG2_1000	2,447	2,580	8,106	–	154.62	–	395.36	92.63	49.62
watson_1	4,119	4,340	13,542	141.88	23.47	26.61	25.18	4.27	13.59
watson_2	8,223	8,660	27,042	397.76	469.91	39.51	89.59	19.50	9.70
wide15	16,431	17,300	54,042	–	22.95	–	193.33	7.90	–
Geometric mean				56.47	55.10	59.96	50.87	16.48	22.64

References

1. COIN-OR. (2017). Available online at: http://www.coin-or.org/ (Last access on March 31, 2017).
2. COIN-OR. (2017). *CLP.* Available online at: https://projects.coin-or.org/Clp (Last access on March 31, 2017).
3. IBM. (2017). *IBM ILOG CPLEX Optimizer.* Available online at: http://www-01.ibm.com/software/commerce/optimization/cplex-optimizer/ (Last access on March 31, 2017).
4. IBM. (2017). *IBM ILOG CPLEX parameters reference.* Available online at: http://www.ibm.com/support/knowledgecenter/SSSA5P_12.6.3/ilog.odms.studio.help/pdf/paramcplex.pdf (Last access on March 31, 2017).
5. Industrial Information & Control Centre. (2017). *OPTI Toolbox.* Available online at: http://www.i2c2.aut.ac.nz/Wiki/OPTI/index.php (Last access on March 31, 2017).

List of Codes

© Springer International Publishing AG 2017
N. Ploskas, N. Samaras, *Linear Programming Using MATLAB®*,
Springer Optimization and Its Applications 127, DOI 10.1007/978-3-319-65919-0

Index

© Springer International Publishing AG 2017
N. Ploskas, N. Samaras, *Linear Programming Using MATLAB®*,
Springer Optimization and Its Applications 127, DOI 10.1007/978-3-319-65919-0

Printed in the United States
By Bookmasters